Comprehensive Forestry

(For RFO and GPSC Examinations)

વ્યાપક વાનિકી

(આર.એફ.ઓ. અને જી.પી.એસ.સી. પરીક્ષાઓ માટે)

Shri R. L. Meena
PCCF (Retd.), Gujarat Forest Department
Gandhinagar, Gujarat

Dr. Aadil A. Kazi
Asst Professor, College of Forestry
Navsari, Gujarat

Published by
SCIENTIFIC PUBLISHERS (INDIA)
5 A, New Pali Road, P.O. Box 91
Jodhpur 342 001 (India)
E-mail: info@scientificpub.com
Website: www.scientificpubonline.com

Cover Page Photos: Mr. Jivan B. Gadhvi, Range Forest Officer, Porbandar and Ms. Rashila P. Vadher, Range Forest Officer, Gir Somnath.

Asiatic Lion and lioness photo: Ms. Minal Patel.

ISBN : 978-93-90495-91-7

Printed in India

Preface

આમુખ

Success doesn't come overnight. It's a long journey with patience and perseverance. Cracking competitive exam demands consistent efforts, diligence and indomitable will. Failure in one or two exams may dishearten the candidate but cannot overpower when he is determined and dedicated. As a knowledge seeker, you may have gone through number of books learning about and Environment and Forestry. However, this book offers a one stop solution to imbibe the fundamental understanding of Forests, Forestry, Wildlife and Environment. In the current scenario, when we are facing numerous challenges of environmental degradation, forest and wildlife depletion, and high global need for conservation, the basic understanding of all these areas is expected in all competitive examinations and interviews. Moreover, it turns out to be the prime requirement when a candidate aspires for forest services. This book efforts to provide necessary information of forests and environment in lucid and simpler manner.

The book covers India's unique affinity with Forests and Environment, Forest Soils, Forest and Forest resources with special reference to Gujarat and India, Environment and Anthropogenic activities, Pollution and Environment, Wildlife Management, National Park and Sanctuaries of India, Forest policy and law, Global conventions related to Forest and Environment, Environmental science and current challenges of global warming and climate change. We aim to include the chapters that target for competitive exams and interviews, and hence we suggest students to treat this book as a competitive exam book and not a textbook for various degree courses. The book also contains Forest types of India, details of National Parks and Wildlife Sanctuaries of Gujarat, Organizational structure of MoEF & CC and Gujarat Forest Department, important Abbreviations related forests and environment; State animal and State Birds; Status of forests in India and Gujarat; Biogeographic Zones of India; Important environmental days; Significant historical environment conservation events and Important projects being executed by MoEF & CC.

In the world of google, the availability and access of information is plenty, but sometimes authenticity is questionable. If you are oblivious to the source, the possibility of getting either incorrect or obsolete information is high. Furthermore, internet cannot replace genuine books. We took utmost care and compiled all the information from reliable sources and with due acknowledgements. To cover every aspect of Forests, Wildlife and Environment in one book is a durable task. To err is human and readers' suggestions are cordially welcomed at rlmeena58@gmail.com and aadilforest@gmail.com for further improvement of this book. We express our sincere respect towards forest officers, researchers and foot soldiers who have been contributing to protect our Forests, Wildlife and Environment.

— **R. L. Meena**

— **Aadil Kazi**

વન અને પર્યાવરણ સંલગ્ન આર.એફ.ઓ. નો અભ્યાસક્રમ

પ્રાથમિક પરીક્ષા

1. પર્યાવરણ વિજ્ઞાન: રાષ્ટ્રીય અને આંતરરાષ્ટ્રીય કક્ષાએ પર્યાવરણની જાળવણી માટે નીતિઓ અને સંધિઓ; વન અને વન્યજીવન સંરક્ષણ; ક્લાઇમેટ ચેન્જ અને ડિઝાસ્ટર મેનેજમેન્ટ પર નેશનલ એક્શન પ્લાન.

2. વન અને પર્યાવરણ સાથે ભારતના અનન્ય સંબંધો: ઐતિહાસિક, સામાજિક અને આર્થિક દ્રષ્ટિએ.

3. વન્ય ભૂમિ: વર્ગીકરણ, ભૂમિ નિર્માણને અસર કરતા પરિબળો; શારીરિક, રાસાયણિક અને જૈવિક ગુણધર્મો.

4. ગુજરાત અને ભારતના સંદર્ભમાં વન અને વન્ય સ્ત્રોતો.

5. પર્યાવરણ; ઘટકો અને મહત્વ, સંરક્ષણના સિદ્ધાંતો, વનનાબૂદી; દાવાનળ અને વિવિધ માનવ પ્રવૃત્તિઓ જેવી કે ખાણકામ, બાંધકામ અને યોજનાઓ, વસ્તી વૃદ્ધિ વગેરેની પર્યાવરણ પાર અસરો.

6. પ્રદૂષણ - પ્રકારો, ગ્લોબલ વોર્મિંગ, ગ્રીન હાઉસ ઇફેક્ટ્સ, ઓઝોન લેયર ઘટાડા, એસિડ વરસાદ, અસર અને નિયંત્રણ પગલાં, પર્યાવરણીય દેખરેખ; ટકાઉ વિકાસ નો ખ્યાલ. પર્યાવરણીય સંરક્ષણમાં વૃક્ષો અને જંગલોની ભૂમિકા; હવા, પાણી અને અવાજ પ્રદૂષણનું નિયંત્રણ અને નિવારણ. ભારતમાં પર્યાવરણીય નીતિ અને કાયદો.

7. વન્યજીવ સંચાલન: ભારતમાં વન અને વન્યપ્રાણી સંચાલનનો આંતરસંબંધ. ભારતમાં વન્યજીવનના સંચાલન માટે કાનૂની સિદ્ધાંતો અને પદ્ધતિઓ. વન્યપ્રાણી સંચાલન ના અભિગમો. ખાસ નિવાસ્થાનોનું સંચાલન; તટવર્તી વિસ્તારો, ઘાસના મેદાનો વગેરે. વાવેતર અને શોષણ કરેલ જંગલોમાં

Syllabus of RFO that contains Forest and Environment part

Preliminary Examination

1. Environmental Science: Policies and Treaties for the protection of environment at the national and the international level; Forest and Wildlife Conservation; National Action plans on Climate Change and Disaster management.

2. India's unique affinity with Forests and Environment: from Historical, Social and Economic point of view.

3. Forests Soils: classification, factors affecting soil formation; physical, chemical and biological properties.

4. Forest and Forest resources with special reference to Gujarat and India.

5. Environment; components and importance, principles of conservation, impact of deforestation; forest fires and various human activities like mining, construction and developmental projects, population growth on environment.

6. Pollution - types, global warming, green house effects, ozone layer depletion, acid rain, impact and control measures, environmental monitoring; concept of sustainable development. Role of trees and forests in environmental conservation; control and prevention of air, water and noise pollution. Environmental policy and legislation in India.

7. Wildlife Management: Interface between forest and wildlife management in India. Legal instruments for managing wildlife in India. Principles and practices of wildlife management. Course and fine filter approaches for wildlife management. Management of special habitats; riparian zones. Grasslands etc. Analysis of wildlife

વન્યપ્રાણી વ્યવસ્થાપનની સમસ્યાઓનું વિશ્લેષણ; ભારતીય અને વૈશ્વિક દૃશ્ય વિશ્લેષણ, પ્રજાતિઓ સંરક્ષણ પ્રોજેક્ટ્સ; વાઘ, સિંહ, ગેંડો, મગર વગેરે.

8. રાષ્ટ્રીય ઉધાનો / અભયારણ્યો.

9. જંગલનું મહત્વ અને તેના પડકારો.

10. ભારતમાં વન અને પર્યાવરણ માટે કાનૂની, નૈતિક અને સંસ્થાકીય માળખું. રાષ્ટ્રીય ગ્રીન ટ્રિબ્યુનલ અને સંશોધન સંસ્થાઓ સહિત.

11. વન અને પર્યાવરણને લગતા વૈશ્વિક સંમેલનો જેમકે UNCED (યુનાઇટેડ નેશનન્સ કોન્ફરન્સ ઓન એન્વાયરન્મેન્ટ એન્ડ ડેવલોપમેન્ટ), CBD (કન્વેશન ઓન બાયોલોજીકલ ડાયવર્સીટી), UNFCCC (યુનાઇટેડ નેશન્સ ફ્રેમવર્ક કન્વેશન ઓન ક્લાઇમેટ ચેન્જ), UNCCD (યુનાઇટેડ નેશન્સ કન્વેશન ટૂ કોમ્બેટ ડેઝર્ટિફિકેશન).

12. ગુજરાતની અર્થવ્યવસ્થામાં જંગલોની ભૂમિકા.

13. આર્થિક વનીકરણ-વિશ્વ, ભારત અને ગુજરાત.

management problems in plantations and exploited forests; Indian and global scenario. Species conservation projects; tiger, lion, rhino, crocodile etc.

8. National Parks/ Sanctuaries.

9. Importance of forest and Challenges against it.

10. Legal, Policy and Institutional Framework in India for Forest and Environment including National Green Tribunal and Research institutes.

11. Global conventions related to Forest and Environment including UNCED (United Nations Conference on Environment and Development), CBD (Convention on Biological Diversity), UNFCCC (United Nations Framework Convention on Climate Change), UNCCD (United Nations Conventions to Combat Desertification).

12. Role of forests in Gujarat's economy.

13. Economic forestry- World, India and Gujarat.

આર.એફ.ઓ. ની મુખ્ય પરીક્ષા માટેનો અભ્યાસક્રમ

1. વન અને પર્યાવરણ સાથે ભારતનું અનોખું જોડાણ: ઐતિહાસિક, સામાજિક અને આર્થિક દૃષ્ટિકોણથી.

2. વનવૃક્ષવિજ્ઞાન- વ્યાપક ખ્યાલ:

વનવૃક્ષવિજ્ઞાનના વ્યાપક સિદ્ધાંતો: વનસ્પતિને અસર કરતા ઇકોલોજીકલ અને ફિઝિયોલોજીકલ પરિબળો, જંગલોના કુદરતી અને કૃત્રિમ પુનર્જીવન; પ્રચારની પદ્ધતિઓ, કલમ બનાવવાની તકનીકો; સાઇટ પરિબળો; નર્સરી અને રોપણી તકનીકો રોપ ઉછરે કચારા.

3. ભારત અને ગુજરાતમાં જંગલના પ્રકાર, ભારત અને ગુજરાતમાં મેંગ્રોવ અને એગ્રો-ક્લાઇમેટ ઝોન.

4. કૃષિ વનીકરણ, સામાજિક વનીકરણ અને સહભાગી વન વ્યવસ્થાપન:

કૃષિ વનીકરણ - કાર્ય ક્ષેત્ર અને આવશ્યકતાં; લોકો અને પાલતું પ્રાણીઓના જીવનમાં તથા સંકલિત જમીનના ઉપયોગમાં ભૂમિકા, ખાસ કરીને (i) જમીન અને જળસંચય (ii) પાણીનું રિચાર્જ; (iii) પાક માટે પોષક તત્વોની ઉપલબ્ધતા; (iv) જીવાત-ભક્ષક સંબંધો દ્વારા પરિસ્થિતિ વિજ્ઞાન સંતુલન સહિતની પ્રકૃતિ અને નિવસનતંત્રની જાળવણી અને (v) જૈવ વિવિધતા, ઔષધીય અને અન્ય વનસ્પતિ અને પ્રાણીસૃષ્ટિમાં વૃદ્ધિ માટે તકો પૂરી પાડતી બાબતોનું આયોજન, વિવિધ કૃષિ નિવસનતંત્ર ક્ષેત્રો હેઠળ કૃષિ વનીકરણ પ્રણાલીઓ; પ્રજાતિઓની પસંદગી અને વિવિધલક્ષી વૃક્ષો તથા NTFP (બિન લાકડાવાળા વન ઉત્પાદનો) ની ભૂમિકા, તકનીકો, ખોરાક, ઘાસચારો અને ઇંધણ સુરક્ષા.સંશોધન અને વિસ્તરણની જરૂરિયાતો.

સામાજિક/શહેરી વનીકરણ: ઉદ્દેશો, કાર્ય ક્ષેત્ર અને આવશ્યકતા; જન સહભાગિતા

સહભાગી વન વ્યવસ્થાપન - સિદ્ધાંતો, ઉદ્દેશો, પદ્ધતિ, કાર્ય ક્ષેત્ર, ફાયદાઓ અને NGOની ભૂમિકા.

ગુજરાતના અનામત વૃક્ષો અને તેમનું વ્યવસ્થાપન.

Syllabus for RFO Main Examination

1. India's unique affinity with Forests and Environment: from Historical, Social and Economic point of view.

2. Silviculture – General : General Silvicultural Principles:

Ecological and physiological factors influencing vegetation, natural and artificial regeneration of forests; methods of propagation, grafting techniques; site factors; nursery and planting techniques nursery beds.

3. Types of Forest in India and Gujarat, Mangrove and Agro-climatic zones in India and Gujarat.

4. Agroforestry, Social Forestry and Joint Forest Management :

Agroforestry - scope and necessity; role in the life of people and domestic animals and in integrated land use, planning especially related to (i) soil and water conservation; (ii) water recharge; (iii) nutrient availability to crops; (iv) nature and ecosystem preservation including ecological balances through pest-predator relationships and (v) providing opportunities for enhancing bio- diversity, medicinal and other flora and fauna. Agro forestry systems under different agro-ecological zones; selection of species and role of multipurpose trees and NTFPs (Non-timber forest products), techniques, food, fodder and fuel security. Research and Extension needs.

Social/Urban Forestry: objectives, scope and necessity; people's participation. JFM (Joint Forest Management)- principles, objectives, methodology, scope, benefits and role of NGOs.

Reserved Trees of Gujarat and their management.

5. વન જમીન, જમીન સંરક્ષણ અને જળસ્ત્રાવ વ્યવસ્થાન:

વન જમીન: વર્ગીકરણ, જમીનની રચનાને અસર કરતા પરિબળો; જમીનના ભૌતિક, રાસાયણિક અને જૈવિક ગુણધર્મો, જમીનની ફળદ્રુપતાને અસર કરતા પરિબળો.

જમીન સંરક્ષણ - વ્યાખ્યા, ધોવાણ માટેનાં કારણો; પ્રકારો - પવન અને જળ ધોવાણ; ઘસારાની જમીન/વિસ્તારો, પવન અવરોધક, આશ્રય પટ્ટાઓનું સંરક્ષણ અને વ્યવસ્થાપન, રેતીના ઢુવાઓ; ખારાશ અને ક્ષારયુક્ત જમીનનો સુધાર, પાણીનો ભરાવો અને અન્ય પડતર જમીન (ખરાબા). જમીન સંરક્ષણમાં વનોની ભૂમિકા જમીનના સેન્દ્રીય દ્રવ્યોની જાળવણી અને નિર્માણ, લીલા પાંદડાના ખાતર માટે કાપવાની જોગવાઇ, વન પર્ણ કચરો, ખાતર; જમીન સુધારણા માં સુક્ષ્મસજીવોની ભૂમિકા; N અને C ચક્રો, VAM.

વોટરશેડ મેનેજમેન્ટ - વોટરશેડની વિભાવનાઓ; એકંદર સંસાધન સંચાલનમાં મિનિ-વનો અને વન વૃક્ષોની ભૂમિકા, વન જળવિજ્ઞાન, પ્રવાહ નિયંત્રણના સંદર્ભમાં જળસંગ્રહ, નદીના નળાકાર સ્થિરતા, વેસ્ટલેન્ડ મેનેજમેન્ટ: કલ્પના, પ્રકારો અને કચરાના જમીનને સુધારવામાં ઝાડની ભૂમિકા.

6. પર્યાવરણીય સંરક્ષણ અને જૈવવિવિધતા:

પર્યાવરણ; ઘટકો અને મહત્વ, સંરક્ષણના સિદ્ધાંતો, વનોની અસર; વન અગ્નિ અને વિવિધ માનવ પ્રવૃત્તિઓ જેવી કે ખાણકામ, બાંધકામ અને વિકાસલક્ષી પ્રોજેક્ટ્સ, પર્યાવરણ પર વસ્તી વૃદ્ધિ. પ્રદૂષણ - પ્રકારો, ગ્લોબલ વોર્મિંગ, ગ્રીન હાઉસ ઇફેક્ટ્સ, ઓઝોન લેયર ઘટાડા, એસિડ વરસાદ, અસર અને નિયંત્રણ પગલાં, પર્યાવરણીય દેખરેખ; ટકાઉ વિકાસની કલ્પના. પર્યાવરણીય સંરક્ષણમાં ઝાડ અને જંગલોની ભૂમિકા; હવા, પાણી અને અવાજ પ્રદૂષણનું નિયંત્રણ અને નિવારણ. ભારતમાં

5. Forest Soils, Soil Conservation and Watershed management:

Forests Soils: classification, factors affecting soil formation; physical, chemical and biological properties of soil. Factors affecting Soil fertility.

Soil conservation - definition, causes for erosion; types - wind and water erosion; conservation and management of eroded soils/areas, wind breaks, shelter belts; sand dunes; reclamation of saline and alkaline soils, water logged and other waste lands. Role of forests in conserving soils. Maintenance and build up of soil organic matter, provision of loppings for green leaf manuring; forest leaf litter and composting; Role of microorganisms in ameliorating soils; N and C cycles, VAM.

Watershed Management - concepts of watershed; role of mini-forests and forest trees in overall resource management, forest hydrology, watershed development in respect of torrent control, river channel stabilization, Wasteland Management: Concept, types and role of trees in reclamation of wasteland.

6. Environmental Conservation and Bio-diversity:

Environment; components and importance, principles of conservation, impact of deforestation; forest fires and various human activities like mining, construction and developmental projects, population growth on environment. Pollution - types, global warming, green house effects, ozone layer depletion, acid rain, impact and control measures, environmental monitoring; concept of sustainable development. Role of trees and forests in environmental conservation; control and prevention of air, water and noise pollution. Environmental

પર્યાવરણીય નીતિ અને કાયદો. પર્યાવરણીય અસર મૂલ્યાંકન. ઇકોલોજીકલ અને પર્યાવરણીય સંરક્ષણની જેમ જળવિશેષ વિકાસનું અર્થશાસ્ત્ર મૂલ્યાંકન.

7. **વન કાર્યકારી યોજના:**

વન આયોજન, સંકલિત આયોજન માટે મૂલ્યાંકન અને નિરીક્ષણના સાધનો અને અભિગમો; વન સંસાધનો નો બહુહેતુક વિકાસ અને વન ઉધાગોનો વિકાસ; કાર્યકારી આયોજનો અને કાર્યકારી યોજનાઓ, પ્રકૃતિ સંરક્ષણમાં તેમની ભૂમિકા, જૈવ વિવિધતા અને અન્ય પરિમાણો; નિર્માણ અને નિયંત્રણ. વિભાગીય કાર્યકારી યોજનાઓ, કામગીરીની વાર્ષિક યોજના. વન આયોજનમાં રીમોટ સેન્સિંગની ભૂમિકા.

8. **વન પરિસ્થિતિ વિજ્ઞાન અને વંશીય વનસ્પતિશાસ્ત્ર:**

વન પરિસ્થિતિ વિજ્ઞાન- જૈવિક અને અજૈવિક ઘટકો, વન નિવસન પ્રણાલી; વન સમુદાય વિભાવના,વનસ્પતિ વિભાવના, પરિસ્થિતિ વિજ્ઞાન ઉત્તરાધિકાર અને ઉત્તરક્રમ, પ્રાથમિક ઉત્પાદકતા, પોષક ક્રિયાચક્ર અને જળ સંબંધો; તણાવ વાતાવરણમાં દેહધર્મવિધા (દુષ્કાળ, પાણીનો ભરાવો, ખારાશ અને ક્ષારીયતા). ભારતમાં વન પ્રકાર, પ્રજાતિઓની ઓળખ, બંધારણ અને જોડાણો; વૃક્ષવિજ્ઞાન,વિભાજન સંબંધી વર્ગીકરણ ઔષધિ સંગ્રહાલયો અને વૃક્ષોધાન ના સિદ્ધાંતો અને સ્થાપના, વન નિવસનતંત્રનું સંરક્ષણ. પ્રતિકૃતિ ઉધાનો, ભારતીય ઔષધિ પ્રણાલીમાં વંશીય વનસ્પતિશાસ્ત્રની ભુમિકા; આયુર્વેદ અને યુનાની - ઔષધીય છોડના પરિચય, નામકરણ, વસાહત, વિતરણ અને વનસ્પતિ વિશેષતાઓ.

મહત્વપૂર્ણ ઔષધીય છોડ અને તેમના ઉપયોગ.

9. **વન્યજીવ વ્યવસ્થાપન:**

ભારતમાં વન અને વન્યજીવ વ્યવસ્થાપન વચ્ચેનું આંતરગુંથન, ભારતમાં વન્યજીવોના સંરક્ષણ અને વ્યવસ્થાપન માટેનું કાયદાકીય માળખું: વિસ્તાર અને જોગવાઇઓ, વન્યજીવ વ્યવસ્થાપન માટેનો

policy and legislation in India. Environmental Impact Assessment. Economics assessment of watershed development vis-a-vis ecological and environmental protection.

7. **Forest Working Plan :**

Forest planning, evaluation and monitoring tools and approaches for integrated planning; multipurpose development of forest resources and forest industries development; working plans and working schemes, their role in nature conservation, bio-diversity and other dimensions; preparation and control. Divisional Working Plans, Annual Plan of Operations. Role of Remote Sensing in Forest Planning.

8. **Forest Ecology and Ethnobotany :**

Forest ecology - Biotic and aboitic components, forest eco-systems; forest community concepts; vegetation concepts, ecological succession and climax, primary productivity, nutrient cycling and water relations; physiology in stress environments (drought, water logging salinity and alkalinity). Forest types in India, identification of species, composition and associations; dendrology, taxonomic classification, principles and establishment of herbaria and arboreta. Conservation of forest ecosystems. Clonal parks, Role of Ethnobotany in Indian Systems of Medicine; Ayurveda and Unani - Introduction, nomenclature, habitat, distribution and botanical features of medicinal plants.

Important Medicinal Plants and their usage.

9. **Wildlife Management:**

Interface between forest and wildlife management in India. Legal framework for Conservation and Management of wildlife in India: Scope and Provision. Principles and practices of wildlife management.

સ્થૂળ અને સુક્ષ્મ શુધ્ધિ અભિગમ. વિશિષ્ટ વસાહતોનું સંચાલન; તટવર્તી વિસ્તારો, ઘાસના મેદાનો વગેરે. બગીચા-વાડીમાં અને શોષિત જંગલોમાં વન્યપ્રાણી વ્યવસ્થાપનની સમસ્યાઓનું વિશ્લેષણ, ભારતીય અને વૈશ્વિક પરિસ્થિતી, પ્રજાતિઓ સંરક્ષણ પ્રોજેક્ટ્સ, વાઘ, સિંહ, ગેંડો, મગર વગેરે વ્યવસ્થાપનમાં જીવવિજ્ઞાનની ભુમિકા, રાષ્ટ્રીય ઉધાનો અને અભયારણ્યો. સંરક્ષિત વિસ્તારો માટેની વ્યવસ્થાપન યોજના: વન કાર્યકારી યોજનાઓ અને વન્યપ્રાણી વ્યવસ્થાપન યોજનાઓ. વન્યજીવ વ્યવસ્થાપન આયોજનની જરૂરીયાત.

10. વન અર્થશાસ્ત્ર:

વન અર્થશાસ્ત્ર: મૂળભૂત સિદ્ધાંતો, ખર્ચ-લાભ વિશ્લેષણ; માંગ અને સપ્લાયનો અંદાજ; રાષ્ટ્રીય અને આંતરરાષ્ટ્રીય બજારમાં વલણોનું વિશ્લેષણ અને ઉત્પાદન અને વપરાશના દાખલામાં ફેરફાર; ગુજરાતના નાના વન ઉત્પાદનો (એમએફપી) અને સામાજિક આર્થિક વિકાસમાં તેની ભૂમિકા.

11. વન અને પર્યાવરણનું કાયદાકીય અને નીતિકીય માળખું:

ભારતમાં વન અને પર્યાવરણનું કાયદાકીય અને સંસ્થાકીય માળખું: ઇતિહાસ, મુખ્ય કાયદાઓ, સંસ્થાઓ વગેરે, ભારતની વન નીતિઓ.

પર્યાવરણ અને વન વ્યવસ્થાપન પ્રણાલીના સુદઢ સંચાલન માટેની રાજ્ય સરકારની અને કેન્દ્ર સરકારની પહેલ. વન અને પર્યાવરણના સંરક્ષણ માટેના વિવિધ સરકારી તંત્રો અને સંસ્થાઓ. વન અને પર્યાવરણના સંરક્ષણ માટે જન ભાગીદારી

12. વન અને પર્યાવરણને લગતા વૈશ્વિક સંમેલનો: યુએનસીઇડી (પર્યાવરણ અને વિકાસ પર યુનાઇટેડ નેશન્સ કોન્ફરન્સ), સીબીડી (જૈવિક વિવિધતા પરના સંમેલન), યુએનએફસીસીસી (યુનાઇટેડ નેશન્સ ફ્રેમવર્ક કન્વેશન ઓન ક્લાઇમેટ ચેન્જ), યુએનસીસીડી (યુનાઇટેડ નેશન્સ કન્વેશન્સ ટુ

Course and fine filter approaches for wildlife management. Management of special habitats; riparian zones. Grasslands etc. Analysis of wildlife management problems in plantations and exploited forests; Indian and global scenario. Species conservation projects; tiger, lion, rhino, crocodile etc. Role of Biology in management. National Parks and Sanctuaries. Management plan for Protected Areas: Forest working plans and wildlife management plans. Need for wildlife management planning.

10. Forest Economics :

Forest economics: fundamental principles, cost-benefit analyses; estimation of demand and supply; analysis of trends in the national and international market and changes in production and consumption patterns; Minor Forest Products (MFP) of Gujarat and its role in Socio Economic Development.

11. Legal and Policy Framework for Forest and Environment:

Legal and Institutional framework for Conservation and Management of forest and environment in India: History, Important Acts, Institutions etc., Forest policies in India.

State Government's and Central Government's initiatives for sound environment and forest management. Various Government bodies and institutes for conservation of forests and environment. People's participation for conservation of Forest and Environment.

12. Global conventions related to Forest and Environment including UNCED (United Nations Conference on Environment and Development), CBD (Convention on Biological Diversity), UNFCCC (United Nations Framework Convention on Climate Change), UNCCD (United Nations Conventions to Combat Desertification).

કોમ્બેટ ડિઝર્ટિફિકેશન) સહિત. સસ્ટેનેબલ ડેવેલોપમેન્ટ લક્ષ્યો. વૈશ્વિક સંમેલનો અને સંધિઓ અંગે ભારતનો પ્રતિસાદ.

Sustainable Development Goals. India's response to the Global Conventions and treaties.

અંગત મુલાકાત

1. વન સેવાઓ માટે ઉમેદવારની યોગ્યતા
2. સામાન્ય જાગૃતિ અને દૈનિક બાબતો
3. ભારતનું ઇકોલોજીકલ, સામાજિક-આર્થિક અને રાજકીય દૃશ્ય
4. વન સંરક્ષણ, સંચાલન અને વહીવટની સમજ
5. પર્યાવરણીય સંરક્ષણ અને સામુદાયિક વિકાસ

Personal Interview

1. Candidate's aptitude for Forest Services
2. General awareness and day to day affairs
3. India's ecological, socio-economic and political scenario
4. Understanding of forest conservation, management and administration
5. Environmental protection and community development

Contents

અનુક્રમણિકા

Chapter 1

પ્રકરણ ૧

Gujarat Forest Profile and State Forest Resources
ગુજરાતના વનોની રૂપરેખા અને રાજ્યની વન સંપદા

GUJARAT FACTS

Constituted as a Gujarat State	1 May 1961
Total Geographical area	1,96,244 sq km
Out of Geographical area of the country	5.97 %
Recorded Forest Area (RFA)	21,647 sq km
% of RFA to geographical area	11.03 %
Forest cover	14,857.33 sq km
% of forest cover to geographical area	7.57 %
Tree Cover Outside Forest Area	6,912 sq km
% of Tree Cover Outside Forest Area to geographical area	3.52 %
Reserved forest of Gujarat	66.39 %
Protected forest of Gujarat	13.33 %
Unclassed forest of Gujarat	20.28 %
Carbon stock in forests of Gujarat	107.25 million tons
% of total forest carbon of the country	1.51 %
District with highest Forest Area	Kachchh
District with lowest Forest Area	Anand
District with highest Forest Cover to its geographic area	Dangs - 77.16 %
District with lowest Forest Cover to its geographic area	Rajkot - 1.38 %
Total Protected Areas	28
National Parks	4
Wildlife Sanctuaries	23
Conservation Reserve	1 (Chhari Dhand CR)
Biosphere Reserve	1 (Kutch BR) Declared under UNESCO's MAB programme
Last population in forests	Asiatic Lion in Gir Wild Ass in Kachchh

State animal	Asiatic Lion
State bird	Greater Flamingo
State flower	Marigold
State tree	Mango
Top five tree species in Trees Outside Forests (TOF) of Rural Gujarat	Neem - 25.17 % Mango -12.06 % Babul - 6.1 % Gando Bawal - 5.79 % Teak - 6.62 %
Top five tree species in Trees Outside Forests (TOF) of Urban Gujarat	Neem - 26.91 % Pittosporum - 8.16 % Asopalav - 5.13 % Mango - 4.12 % Coconut - 3.73 %

ISFR 2019

ગુજરાત હકીકતો

ગુજરાત રાજ્ય સ્થાપના દિવસ	1 મે 1961
કુલ ભૌગોલિક ક્ષેત્ર	1,96,244 ચોરસ કિ.મી.
દેશના ભૌગોલિક ક્ષેત્રમાં ગુજરાતનો વિસ્તાર	5.97%
રેકોર્ડ કરેલ વન વિસ્તાર (RFA)	21,647 ચોરસ કિ.મી.
ભૌગોલિક ક્ષેત્રમાં RFA %	11.03%
વન આવરણ	14,857.33 ચો.કિ.મી.
ભૌગોલિક ક્ષેત્રમાં જંગલ આવરણ %	7.57%
વન વિસ્તારની બહાર વૃક્ષ આવરણ	6,912 ચોરસ કિ.મી.
ભૌગોલિક ક્ષેત્રમાં વન વિસ્તારની બહાર વૃક્ષ આવરણ %	3.52%
ગુજરાતનું અનામત વન (Reserve Forest)	66.39%
ગુજરાતનું રક્ષિત વન (Protected Forest)	13.33%
ગુજરાતનું અવર્ગીકૃત વન (Unclassed Forest)	20.28%
ગુજરાતના જંગલોમાં કાર્બન સ્ટોક	107.25 મિલિયન ટન
દેશના કુલ વન કાર્બનનો %	1.51%
સૌથી વધુ વન વિસ્તાર ધરાવતો જિલ્લો	કચ્છ
સૌથી ઓછો વન વિસ્તાર ધરાવતો જિલ્લો	આણંદ
ભૌગોલિક ક્ષેત્રમાં સૌથી વધુ વન આવરણ ધરાવતો જિલ્લો	ડાંગ- 77.16%
ભૌગોલિક ક્ષેત્રમાં સૌથી ઓછો વન આવરણ ધરાવતો જિલ્લો	રાજકોટ - 1.38%
કુલ રક્ષિત વિસ્તારો	28

રાષ્ટ્રીય ઉદ્યાનો	4
વન્યપ્રાણી અભયારણ્ય	23
કન્ઝર્વેશન રિઝર્વ	1 (છારી ધંડ)
બાયોસ્ફિયર રિઝર્વ	1 (કચ્છ) યુનેસ્કોના એમએબી પ્રોગ્રામ હેઠળ જાહેર
જંગલોમાં છેલ્લી વસ્તી	ગીરમાં એશિયાટીક સિંહ કચ્છમાં ઘુડખર
રાજ્ય પ્રાણી	એશિયાટીક સિંહ
રાજ્ય પક્ષી	સુરખાબ (ગ્રેટર ફ્લેમિંગો)
રાજ્ય ફૂલ	ગલગોટા
રાજ્ય વૃક્ષ	કેરી
ગ્રામીણ ગુજરાતના જંગલોની બહાર વૃક્ષોની (TOF) ટોચની પાંચ પ્રજાતિઓ	લીમડો - 25.17% આંબો -12.06% બાવળ - 6.11% ગાંડો બાવળ - 5.79% સાગ - 6.62%
શહેરી ગુજરાતના જંગલોની બહાર વૃક્ષોની (TOF) ટોચની પાંચ પ્રજાતિઓ	લીમડો - 26.91% પિટોસ્પોરમ - 8.16% આસોપાલવ - 5.13% કેરી - 4.12% નાળિયેર - 3.73%

ISFR 2019

Who's who

Minister, Forest and Environment, Gujarat	Ganpatbhai Vasava
State Minister, Forest, Gujarat	Ramanlal Patkar
State Minister, Environment, Gujarat	Jaydrathsinhji Parmar
Additional Chief Secretary, Forest and Environment Department, Gujarat	Rajiv Kumar Gupta
Principal Chief Conservator of Forests & Head of Forest Force, Gujarat State	Dinesh Kumar Sharma

GFD 2020

કોણ શું છે

મંત્રી, વન અને પર્યાવરણ, ગુજરાત	ગણપતભાઇ વસાવા
રાજ્યમંત્રી, વન, ગુજરાત	રમણલાલ પાટકર
રાજયમંત્રી, પર્યાવરણ, ગુજરાત	જયદ્રથસિંહજી પરમાર
અધિક મુખ્ય સચિવ, વન અને પર્યાવરણ વિભાગ, ગુજરાત	રાજીવકુમાર ગુપ્તા
અગ્ર મુખ્ય વન સંરક્ષક અને ગુજરાત રાજ્યના વન દળના વડા	દિનેશકુમાર શર્મા

GFD 2020

District wise Forest Cover of Gujarat | ગુજરાતનો જિલ્લાવાર વન આવરણ

District	Geographical Area (sq km)	Forest Cover (sq km)	% of Geographical Area
Ahmedabad	8,107	131.21	1.62
Amreli	7,397	260.40	3.52
Anand	3,204	50.48	1.58
Banas Kantha	10,743	822.50	7.66
Bharuch	6,509	277.52	4.26
Bhavnagar	10,034	293.21	2.92
Dohod	3,642	529.82	14.55
Gandhinagar	2,140	92.05	4.30
Jamnagar	14,184	471.11	3.32
Junagadh	8,831	1,693.31	19.17
Kachchh	45,674	2,345.29	5.13
Kheda	3,953	94.44	2.39
Mahesana	4,401	158.62	3.60
Narmada	2,817	948.00	33.65
Navsari	2,246	362.43	16.14
Panch Mahals	5,231	710.06	13.57
Patan	5,792	102.51	1.77
Porbandar	2,316	126.68	5.47
Rajkot	11,198	154.32	1.38
Sabarkantha	7,394	815.27	11.03
Surat	4,549	500.06	10.99
Surendranagar	10,423	173.99	1.67
Tapi	3,139	792.41	25.24
The Dangs	1,766	1,362.62	77.16
Vadodara	7,546	603.21	7.99
Valsad	3,008	985.81	32.77
Total	**1,96,244**	**14,857.33**	**7.57**

ISFR 2019

Forest Types of Gujarat | ગુજરાતના વન પ્રકાર

No.	Forest Type (as per Champion & Seth)	% of Forest cover	
1.	3B/C1b Moist Teak Forest	4.50	
2.	3B/C1c Slightly Moist Teak Forest	3.83	
3.	3B/C2 Southern Moist Mixed Deciduous Forest	6.79	
4.	**4B/TS1 Mangrove Scrub**	**3.86**	
5.	**4B/TS2 Mangrove Forest**	**3.37**	→ 7.23 % Mangroves
6.	5A/C1a Very Dry Teak Forest	4.60	
7.	5A/C1b Dry Teak Forest	11.77	
8.	**5A/C3 Southern Dry Mixed Deciduous Forest**	**12.75**	→ Highest
9.	5B/DS1 Dry Deciduous Scrub	5.00	
10.	5B/DS4 (Dry Grassland)	1.93	
11.	5/E1 Anogeissus Pendula Forest	0.35	
12.	**5/E2 Boswellia Forest**	**0.01**	→ Lowest
13.	5/E3 Babul Forest	0.81	
14.	5/E5 Butea Forest	0.52	
15.	5/E8c Salvadora - Tamarix Scrub	0.99	
16.	5/E9 Dry Bamboo Brake	0.66	
17.	**5/1S1 Dry Tropical Riverain Forest**	**0.01**	→ Lowest
18.	5/2S1 Secondary Dry Deciduous Forest	0.28	
19.	5B/C2 Northern Dry Mixed Deciduous Forest	5.28	
20.	6A/C1 Southern Thorn Forest	0.38	
21.	6B/C1 Desert Thorn Forest	9.22	
22.	6/DS2 Tropical Euphorbia Scrub	4.82	
23.	6/E2 Acacia Senegal Forest	5.20	
24.	6/E3 Rann Saline Thorn Forest	4.76	
25.	6/E4 Salvadora Scrub	0.10	
26.	Plantation/TOF	8.21	
	Total	**100.00**	

As per Champion & Seth 1968; ISFR 2019

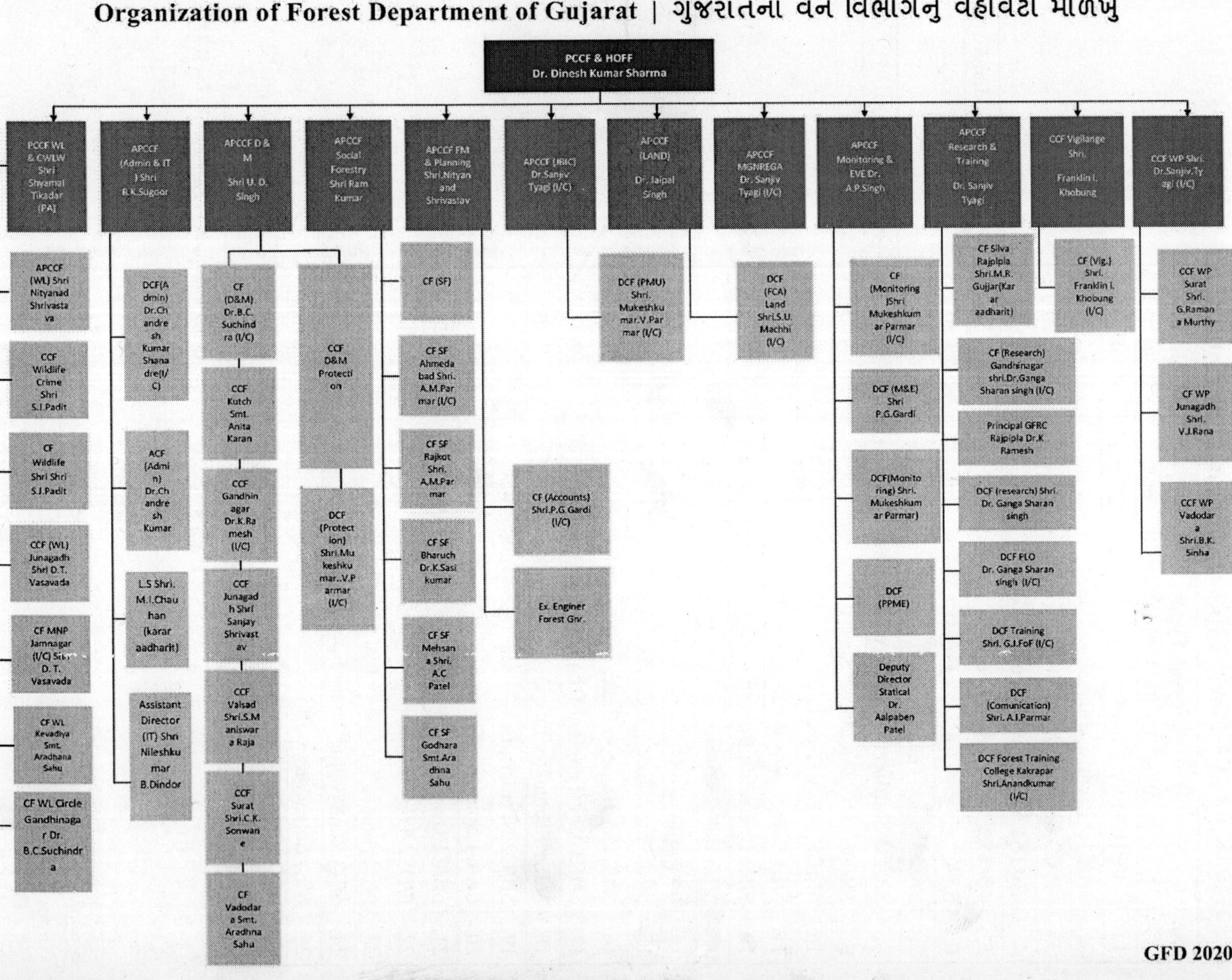
Organization of Forest Department of Gujarat | ગુજરાતના વન વિભાગનું વહીવટી માળખું
PCCF & HOFF Dr. Dinesh Kumar Sharma
PCCF WL & CWLW Shri Shyamal Tikadar (PA)
APCCF (WL) Shri Nityanad Shrivasta va
CCF Wildlife Crime Shri S.J.Padit
CF Wildlife Shri Shri S.J.Padit
CCF (WL) Junagadh Shri D.T. Vasavada
CF MNP Jamnagar (I/C) Shri D.T. Vasavada
CF WL Kevadiya Smt. Aradhana Sahu
CF WL Circle Gandhinagar Dr. B.C.Suchindra
APCCF (Admin & IT) Shri B.K.Sugoor
DCF(Admin) Dr.Chandresh Kumar Shanadre(I/C)
ACF (Admin) Dr.Chandresh Kumar
L.S Shri. M.I.Chauhan (karar aadharit)
Assistant Director (IT) Shri Nileshkumar B.Dindor
APCCF D & M Shri U. D. Singh
CF (D&M) Dr.B.C. Suchindra (I/C)
CCF Kutch Smt. Anita Karan
CCF Gandhinagar Dr.K.Ramesh (I/C)
CCF Junagadh Shri Sanjay Shrivastav
CCF Valsad Shri.S.Maniswara Raja
CCF Surat Shri.C.K. Sonwane
CF Vadodara Smt. Aradhna Sahu
CCF D&M Protection
DCF (Protection) Shri.Mukeshkumar..V.Parmar (I/C)
APCCF Social Forestry Shri Ram Kumar
CF (SF)
CF SF Ahmedabad Shri. A.M.Parmar (I/C)
CF SF Rajkot Shri. A.M.Parmar
CF SF Bharuch Dr.K.Sasi kumar
CF SF Mehsana Shri. A.C Patel
CF SF Godhara Smt.Aradhna Sahu
APCCF FM & Planning Shri.Nityan and Shrivastav
CF (Accounts) Shri.P.G Gardi (I/C)
Ex. Enginer Forest Gnr.
APCCF (JBIC) Dr.Sanjiv Tyagi (I/C)
DCF (PMU) Shri. Mukeshkumar.V.Parmar (I/C)
APCCF (LAND) Dr. Jaipal Singh
DCF (FCA) Land Shri.S.U. Machhi (I/C)
APCCF MGNREGA Dr. Sanjiv Tyagi (I/C)
APCCF Monitoring & EVE Dr. A.P.Singh
CF (Monitoring) Shri Mukeshkumar Parmar (I/C)
DCF (M&E) Shri P.G.Gardi
DCF(Monitoring) Shri. Mukeshkumar Parmar)
DCF (PPME)
Deputy Director Statical Dr. Aalpaben Patel
APCCF Research & Training Dr. Sanjiv Tyagi
CF Silva Rajpipla Shri.M.R. Gujjar(Karar aadharit)
CF (Research) Gandhinagar shri.Dr.Ganga Sharan singh (I/C)
Principal GFRC Rajpipla Dr.K Ramesh
DCF (research) Shri. Dr. Ganga Sharan singh
DCF FLO Dr. Ganga Sharan singh (I/C)
DCF Training Shri. G.J.FoF (I/C)
DCF (Comunication) Shri. A.I.Parmar
DCF Forest Training College Kakrapar Shri.Anandkumar (I/C)
CCF Vigilange Shri. Franklin I. Khobung
CF (Vig.) Shri. Franklin I. Khobung (I/C)
CCF WP Shri. Dr.Sanjiv.Tyagi (I/C)
CCF WP Surat Shri. G.Ramana Murthy
CF WP Junagadh Shri. V.J.Rana
CCF WP Vadodara Shri.B.K. Sinha

Authorities and Boards of Forest and Environment Department, Govt. of Gujarat
ગુજરાત સરકારના વન અને પર્યાવરણ વિભાગના ઓથોરિટી / બોર્ડ

Authority/ Board	HQ	Year of establishment	Head	Function
Gujarat Coastal Zone Management Authority	Gandhinagar	2016	Rajiv Kumar Gupta	Protect the coastal environment of Gujarat
Gujarat State Environment Impact Assessment Authority	Gandhinagar	2017	C. L. Meena	Impose legal prohibitions on new projects or activities that harm the environment of Gujarat.
Gujarat Biodiversity Board (GBB)	Gandhinagar	2006	M. M. Sharma	Conserve the biodiversity of Gujarat
Gujarat Pollution Control Board (GPCB)	Gandhinagar	1974	Sanjiv Kumar	Control the pollution and protect the environment in Gujarat

FED, GoG 2020

Organizations of Forest and Environment Department, Govt. of Gujarat
ગુજરાત સરકારના વન અને પર્યાવરણ વિભાગની સંસ્થાઓ

Organization	HQ	Year of establishment	Head	Function
Gujarat Ecological Education and Research (GEER) Foundation	Gandhinagar	1982	R. D. Kamboj	Ecological education and research
Gujarat Ecology Commission (GEC)	Gandhinagar	1992	Shyamal Tikadar	Restoration of degraded ecosystems in Gujarat
Gujarat Environmental Management Institute (GEMI)	Gandhinagar	1999	A. K. Jha	Management of ecology and environment of Gujarat
Gujarat Forestry Research Foundation (GFRF)	Gandhinagar	2008	Sanjiv Tyagi	Conducting research in forest and wildlife of Gujarat
Gujarat State Forest Development Corporation (GSFDC)	Vadodara	1976	S. K. Chaturvedi	Facilitate trade of minor forest products

FED, GoG 2020

Societies for Forest Management in Gujarat
ગુજરાત સરકારના વન અને પર્યાવરણ વિભાગની સોસાયટીઓ

Organization	HQ	Year of establishment	Chair	Function
Gujarat State Lion Conservation Society	Junagadh	2007	PCCF & CWLW	Conserve Asiatic Lion and its habitat
Gujarat Marine National Park and Marine Sanctuary Conservation Society	Jamnagar	2010	PCCF & CWLW	Conserve Marine biodiversity and its habitat
Sakkarbaug Zoo Management Advisory Society	Junagadh	2007	CCF, WL, Junagadh	Spread empathy for wild animals
Little Rann of Kutch Land Scape-LRK Society	Gandhinagar	2008	Principal Secretary, FED	Biodiversity conservation and rural livelihoods improvement project at the Little Rann of Kutch landscape
Kevadia Jungle Safari (Sardar Patel Zoological Park) Society	Kevadia	2019	Principal Secretary, FED	Manage the educative establishment of Kevadia
Kevadia Valley of Flowers Society	Kevadia	2019	Principal Secretary, FED	Conserve and improve floral diversity of Vally of Flowers, Kevadia

FED, GoG 2020

Gujarat State Board for Wildlife | ગુજરાત રાજ્યનું વન્યજીવ બોર્ડ

Chief Minister	Chairperson
Minister, Department of Forests	Vice Chairperson
Minister of State, Department of Forests	Co-Vice Chairperson
Three members of State Legislature	Members
Representatives of NGOs a. Director, GEER Foundation, Gandhinagar b. Member Secretary, Gujarat Ecology Commission, Gandhinagar c. Representative of Corbett Foundation, Kutch	Members
Eminent Conservationist, ecologist and environmentalist	Members
Permanent Invitee Chief Secretary, Gujarat State	Permanent Invitee
Additional Chief Secretary/ Principal Secretary, Forest & Environment Department, Gandhinagar	Member
Principal Chief Conservator of Forest, Gandhinagar	Member

Secretary, Tribal Development Department, Gandhinagar	Member
Managing Director, Gujarat Tourism Development Corporation, Gandhinagar	Member
Director General of Police, Gujarat State, Gandhinagar	Member
A representative of the Armed Force (to be nominated by Central Government)	Member
Director, Animal Husbandry, Gandhinagar	Member
Commissioner of Fisheries, Gandhinagar	Member
An officer to be nominated by Director, Wildlife Preservation, MoEFCC, New Delhi	Member
A representative of Wildlife Institute of India, Dehradun	Member
A representative of Botanical Survey of India	Member
A representative of Zoological Survey of India	Member
A representative of Wildlife Crime Control Bureau, MoEF, New Delhi	Member
Chief Wildlife Warden, Gujarat State, Gandhinagar	Member Secretary

FED 2020

Zoos of Gujarat | ગુજરાતના પ્રાણી સંગ્રહાલય

Name	Location	Established in	Head	Administered by
Ambardi Safari Park	Dhari	2017	Anshuman Sharma	Forest Department
Devalia Safari Park	Sasan Gir	1965	Mohan Ram	Forest Department
Dr. Shyamaprasad Mukherjee Zoological Garden	Surat	2003	P.C. Mehta	Surat Municipal Corporation
Indroda Nature Park	Gandhinagar	1978	R. D. Kamboj	Forest Department
Kamla Nehru Zoological Garden	Ahmedabad	1951	R. K. Sahu	Ahmedabad Municipal Corporation
Rajkot Municipal Zoo	Rajkot	1987	R. K. Hirpara	Rajkot Municipal Corporation
Sardar Patel Zoological Park (Kevadia Jungle Safari)	Kevadia	2019	R. R. Nala	Forest Department
Sayaji Baug Zoo	Vadodara	1875	Pratyush Patankar	Vadodara Municipal Corporation
Sakkar Baug Zoological Park	Junagadh	1863	Abhishek Kumar	Forest Department
Sundervan Nature Discovery Centre	Ahmedabad	1978	Kartikeya Sarabhai	Centre for Environment Education

CZA 2020

Agroclimatic Zones of Gujarat | ગુજરાતના કૃષિ વિસ્તારો

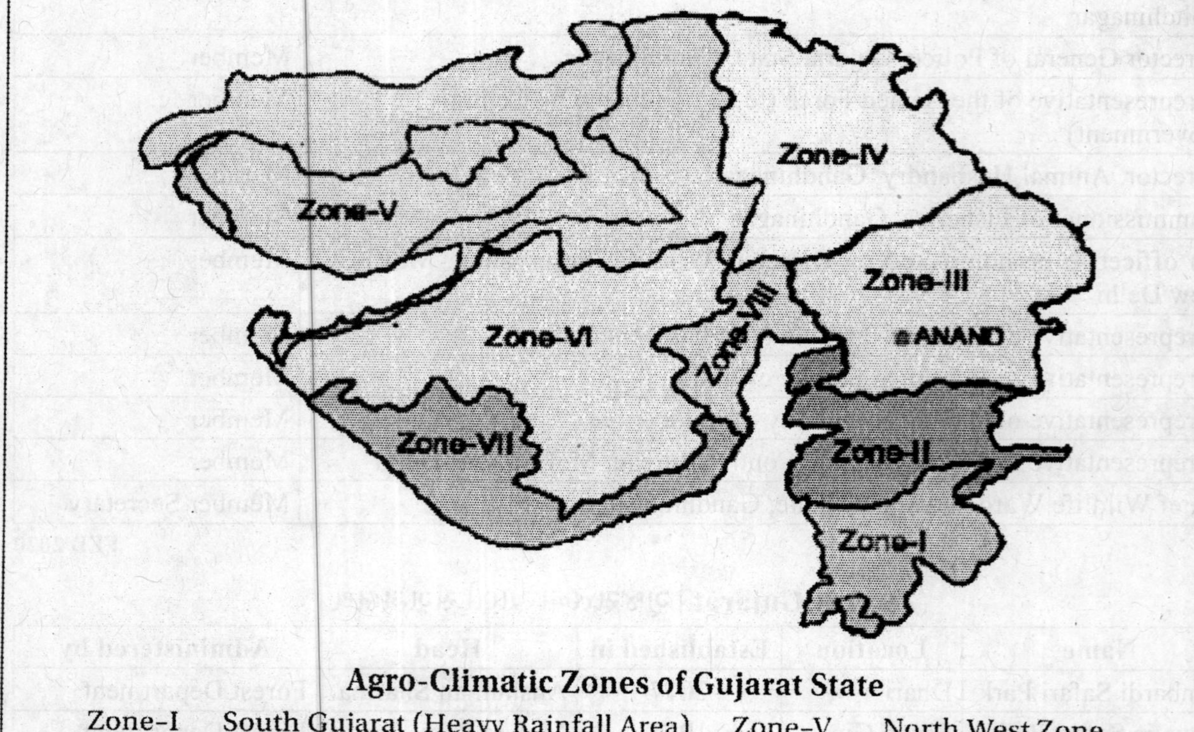

IMD 2020

Zone I	Southern Heavy Rainfall Area & Hilly Area	Semi-arid to dry sub-humid climate
Zone II	South Gujarat	Semi-arid to dry sub-humid climate
Zone III	Middle Gujarat	Semi-arid climate
Zone IV	North Gujarat	Arid to semi-arid climate
Zone V	North-West Arid	Arid to semi-arid climate
Zone VI	North Saurashtra	Dry sub-humid climate
Zone VII	South Saurashtra	Dry sub-humid climate
Zone VIII	Bhal & Coastal Area	–

Agriculture, Farmers Welfare & Co-operation Department, GoG

Wetlands of Gujarat | ગુજરાતના જલપ્લાવિત વિસ્તારો

No.	Name of Wetland	District
1.	Amipur Wetland	Porbandar
2.	Bajana-Tundi Wetland	Surendranagar
3.	Bardasagar Wetland	Porbandar
4.	Bhaskarpura Wetland	Surendranagar
5.	Charakla Salt- Works Wetland	Devbhoomi Dwarka
6.	Chhari Dhandh Wetland	Kachchh
7.	Gosabara-Mokarsagar Wetland	Porbandar
8.	Hanj bet (Flamingo City) Wetland	Kachchh
9.	Kaj Wetland	Gir Somnath
10.	Kanewal Wetland	Anand
11.	Khijadiya Wetland	Jamnagar
12.	Kuchhadi-Javar Wetland	Porbandar
13.	Kumbharwada Wetland	Bhavnagar
14.	Nal Sarovar Wetland	Ahemdabad
15.	Nanda bet Wetland	Kachchh
16.	Narda Wetland	Kheda
17.	Pariej Wetland	Kheda
18.	Porbandar Bird Sanctuary Wetland	Porbandar
19.	Thol Bird Sanctuary Wetland	Mehsana
20.	Wadhwana Wetland	Vadodara

GEER 2018

Important Ecotourism destinations of Gujarat
ગુજરાતના મહત્વપૂર્ણ પરિસરીય પ્રવાસન સ્થળો

Batsewar Ecotourism site	Jam Wada, Gir
Botanical Garden	Waghai, Dang
Devghatt Ecotourism site	Divtan, Surat
Dhanpari Eco Campsite	Jambughoda, Panchmahal
Gira falls	Waghai, Dang
Jungle Safari	Kevadia
Kaj-nanavada Wetland Site	Kodinar, Amreli
Kevdi Eco Campsite	Mandvi, Surat
Kilad Ecotourism Site	Waghai, Dang
Madhupur Turtle Hatching Site	Porbandar
Mahal Ecotourism Site	Mahal, Dang
Padamdungari Ecotourism Site	Unai, Surat
Pirotan Island	Jamnagar
Polo Forest Site	Vijaynagar, Sabarkantha
Sagai Eco Campsite	Dediapada, Narmada

GFD 2020

Tribal of Gujarat | ગુજરાતના આદિવાસી પ્રજાજનો

Bhil	Dang, Panchmahal, Bharuch, Sabarkantha
Chaudhri	Surat, Tapi, Bharuch
Dhanka	Vadodara, Bharuch, Panchmahal
Dhodia	Valsad, Surat
Gamit	Surat
Halpati	Surat, Valsad, Navsari, Bharuch
Kathodi	Surat, Navsari, Dang, Sabarkantha, Narmada
Koigha	Surat, Valsad, Navsari, Dang
Kokna	Navsari, Valsad, Dang
Kotwalia	Surat, Valsad, Navsari, Dang, Bharuch
Naikda	Panchmahal, Valsad
Padhar	Ahmedabad, Surendranagar
Patelia	Panchmahal
Rathwa	Chotaudepur
Siddi	Junagadh, Bhavnagar, Amreli Porbandar, Rajkot
Varli	Valsad, Navsari

Important Projects of Forests, Environment and Wildlife of Gujarat

Project	Key Points
Gujarat Forest Development Project (GFDP)	• To restore degraded forests and improve the livelihoods of people living in and around forest areas • The project aims to encourage the participation of village communities in Joint Forest Management Committees (JFMC) and, with them, to pursue the regeneration, conservation, development and maintenance of degraded forests. • The afforestation project funded by the Japan International Cooperation Agency (JICA) in Gujarat is looking at expanding a green canopy • Results achieved are formation of Peoples Organization: JFMC 1408, EDC 207 and SFDC 812 formed • SHGs 3484 have been established, which now have access to banks and saving facilities. • 6 eco-tourism sites have been created and transferred to eco-development committees near the Protected areas. • Thousands of micro-plans have been prepared and are in the process of implementation. • The population densities of key wild animal species have been maintained, as have as the density and richness of botanical species. • Training and capacity building activities are now available for local non-governmental organizations.
Bamboo Mission	• Under the centrally sponsored scheme- National Bamboo Mission • The mission aims to promote the growth of the Bamboo sector through an area based regionally differentiated state increase the coverage of area under Bamboo in potential areas, with suitable species to enhance yields; promote marketing of Bamboo and Bamboo based handicrafts; promote, develop and disseminate technologies through a seamless blend of traditional wisdom and modern scientific knowledge; and generate employment opportunities for skilled and unskilled persons, especially unemployed youth. • State and District level Bamboo Development Agency were established.
Mangrove Conservation	• The mangrove cover of the state is distributed over in 14 districts and in 4 regions, Kachchh, Gulf of Kachchh & Saurashtra and South Gujarat (including Gulf of Khambhat- Dumas Ubharat areas). • Kachchh has the highest mangrove cover (71.5%) of the State. Gulf of Kachchh (15.6%) Saurashtra (0.3 %) and South Gujarat, including areas of Gulf of Khambhat and Dumas Ubharat has 12.6% mangroves. • 15 mangrove species present in Gujarat. Among these 15 species, *Avicennia marina* is the most dominant species which represent 97% of the total mangroves of Gujarat.

Project	Key Points
	• 3 models of mangrove plantation in Gujarat: (i) Direct seed sowing, (ii) Raised bed plantation and (iii) Fishbone channel plantations
Leogene project	• Establishing Institute of Wildlife Genomics and DNA banking is being implemented in joint collaboration with GSBTM and Wildlife Wing of Gujarat Forest Department. • Collecting germplasm of important wildlife species of Gujarat.
Biodiversity Conservation and Rural Livelihoods Improvement Project (BCRLIP)	• To mainstreaming the traditional knowledge in conservation and livelihood generation. • To establish information technology and communication network to support conservation and livelihood planning. • Landscape level research and studies to fill the knowledge gaps.

GFD 2020

ગુજરાતના વન, પર્યાવરણ અને વન્યજીવનના મહત્વપૂર્ણ પ્રોજેક્ટ્સ

પ્રોજેક્ટ	મહત્વના મુદ્દાઓ
ગુજરાત વન વિકાસ યોજના (GFDP)	• ક્ષતિગ્રસ્ત જંગલોને પુનર્સ્થાપિત કરવા અને વન વિસ્તારોમાં અને આસપાસના લોકોની આજીવિકા સુધારવી • આ પ્રોજેક્ટનો હેતુ સંયુક્ત વન વ્યવસ્થાપન સમિતિઓ (JFMC) માં ગ્રામીણ સમુદાયોની ભાગીદારીને પ્રોત્સાહન આપવાનો છે અને તેમની સાથે જંગલોના પુનર્જીવન, સંરક્ષણ, વિકાસ અને જાળવણીને આગળ વધારવાનો છે. • ગુજરાતમાં જાપાન ઇન્ટરનેશનલ કોઓપરેશન એજન્સી (JICA) દ્વારા ભંડોળ પૂરું પાડવામાં આવેલ એફોરેસ્ટેશન પ્રોજેક્ટને વિસ્તૃત કરવો • પ્રાપ્ત થયેલા પરિણામો પીપલ્સ ઓર્ગેનાઇઝેશનના છે: JFMC 1408, EDC 207 અને SFDC 812 ની રચના કરવામાં આવી • HSGs 3484 ની સ્થાપના કરવામાં આવી છે, જેની પાસે હવે બેંકોમાં બચત સુવિધાઓ છે. • સંરક્ષિત વિસ્તારો નજીક ઇકો-ડેવલપમેન્ટ સમિતિઓમાં 6 ઇકો-ટુરિઝમ સાઇટ્સ બનાવવામાં આવી છે અને સ્થાળાંતરિત કરવામાં આવી છે. • હજારો માઇક્રો-યોજનાઓ તૈયાર કરવામાં આવી છે અને અમલીકરણની પ્રક્રિયામાં છે. • વનસ્પતિ પ્રજાતિઓની ગીચતા અને સમૃદ્ધિની જેમ મહત્વના જંગલી પ્રાણીઓની પ્રજાતિઓ જાળવવામાં આવી છે. • સ્થાનિક બિન-સરકારી સંસ્થાઓ માટે હવે તાલીમ માટેની પ્રવૃત્તિઓ ઉપલબ્ધ છે.

પ્રોજેક્ટ	મહત્વના મુદ્દાઓ
વાંસ મિશન	• કેન્દ્રિય પ્રાયોજિત યોજના હેઠળ- રાષ્ટ્રીય વાંસ મિશન • આ મિશનનો હેતુ વાંસ ક્ષેત્રની વૃદ્ધિને વિસ્તાર આધારિત પ્રોત્સાહન આપવાનું છે, સંભવિત વિસ્તારોમાં વાંસ હેઠળના ક્ષેત્રમાં આવરણ વધારવા માટે યોગ્ય પ્રજાતિઓ છે. વાંસ અને વાંસ આધારિત હસ્તકલાઓના માર્કેટિંગને પ્રોત્સાહન આપવું; પરંપરાગત શાણપણ અને આધુનિક વૈજ્ઞાનિક જ્ઞાનના મિશ્રણ દ્વારા તકનીકોને પ્રોત્સાહન, વિકાસ, પ્રસારણ અને કુશળ અને અકુશળ વ્યક્તિઓ, ખાસ કરીને બેરોજગાર યુવાનો માટે રોજગારની તકો ઉભી કરે છે. • રાજ્ય અને જિલ્લા કક્ષાની વાંસ વિકાસ એજન્સીની સ્થાપના કરવામાં આવી છે.
ચેર સંરક્ષણ	• રાજ્યના ચેર આવરણ 14 જિલ્લાઓમાં અને 4 પ્રદેશોમાં, કચ્છ, કચ્છનો અખાત અને સૌરાષ્ટ્ર તેમજ દક્ષિણ ગુજરાત (ખંભાતના અખાત- ડુમ્મસ અને ઉભરાટ વિસ્તારો સહિત) માં કરવામાં આવે છે. • રાજ્યના સૌથી વધુ મેંગ્રોવ કવર કચ્છમાં છે (71.5%). કચ્છનો અખાત (15.6 %) સૌરાષ્ટ્ર (0.3%) અને દક્ષિણ ગુજરાત, ખંભાતના અખાત અને ડુમ્મસ અને ઉભરાટ સહિતના વિસ્તારોમાં 12.6 % ચેર છે. • ગુજરાતમાં 15 ચેરની પ્રજાતિઓ છે. આ 15 પ્રજાતિઓમાંથી, *એવિસેન્નીયા મરિના* સૌથી પ્રબળ પ્રજાતિ છે જે ગુજરાતના કુલ મેંગ્રોવના 97 % હિસ્સો ધરાવે છે. • ગુજરાતમાં ચેર વાવેતર 3 મોડેલથી કરવામાં આવે છે (i) સીધું બીજ વાવેતર (ii) રેઈઝ બેડ પદ્ધતિથી વાવેતર અને (iii) ફિશબોન ચેનલ વાવેતર
લીયોજીન પ્રોજેક્ટ	• ગુજરાત વન વિભાગના વન્યજીવ પાંખ અને GSBTM ના સંયુક્ત સહયોગથી Wildlife Genomics and DNA banking ની સ્થાપના કરવામાં આવી છે. • ગુજરાતની અગત્યની વન્યપ્રાણી પ્રજાતિના જીનનો સંગ્રહ કરવો.
જૈવવિવિધતા સંરક્ષણ અને ગ્રામીણ આજીવિકા સુધારણા પ્રોજેક્ટ (BCRLIP)	• સંરક્ષણ અને આજીવિકા માટે પરંપરાગત જ્ઞાનને મુખ્ય પ્રવાહમાં લાવવા. • સંરક્ષણ અને આજીવિકા આયોજનને ટેકો આપવા માટે માહિતી તકનીક અને સંદેશાવ્યવહાર નેટવર્ક સ્થાપિત કરાવી. • જ્ઞાનનાના અંતરને ઘટાડવા માટે લેન્ડસ્કેપ સ્તરે સંશોધન અને અભ્યાસ કરવો.

GFD,2020

Eminent Environmentalists of Gujarat

Babulal Dahiya	Farmer of Satna and working to conserve indigenous seeds and promoting conservation of biodiversity in Satpura hills. Padma Shri Awardee.
Bhavbhuti Parasharya	Renowned ornithologist
Digvirendrasinh Solanki	Maharaja of Vansda and Instrumental in establishment of Vansda National Park.
Jitendra Patel	Gujarat has its first Green Ambassador with an environmentalist from Mehsana Jitendra Patel being chosen for the great honour.
Kartikeya Sarabhai	Environment conservationist and founder-director of Centre for Environment Education (CEE). Padma Shri awardee.
Lavkumar Khachar	Started nature education camp.
Raju Vyas	Renowned herpetologist

ગુજરાતના જાણીતા પર્યાવરણવિદો

બાબુલાલ દહિયા	સત્નાનો ખેડૂત અને સ્વદેશી બિયારણને બચાવવા અને સાતપુડા પર્વતોમાં જૈવવિવિધતાના સંરક્ષણને પ્રોત્સાહન આપવાનું કામ કરે છે. પદ્મશ્રી એવોર્ડ વિજેતા.
ભવભૂતિ પારાશર્ય	પ્રખ્યાત પક્ષીવિદ.
દિગ્વીરેન્દ્રસિંહ સોલંકી	વાંસદા રાષ્ટ્રીય ઉદ્યાનની સ્થાપનામાં વાંસદાના મહારાજાનો સિંહફાળો છે.
જીતેન્દ્ર પટેલ	ગુજરાતના પ્રથમ ગ્રીન એમ્બેસેડર છે. મહેસાણાના પર્યાવરણવિદ્ જીતેન્દ્ર પટેલને મહાન સન્માન માટે પસંદ કરવામાં આવ્યા છે.
કાર્તિકેય સારાભાઇ	જાણીતા પર્યાવરણ સંરક્ષણવાદી અને સેન્ટર ફોર એન્વાયર્નમેન્ટ એજ્યુકેશન (CEE) ના સ્થાપક. પદ્મ શ્રી વિજેતા.
લવકુમાર ખાચર	પ્રકૃતિ શિક્ષણ શિબિર શરૂ કરનાર જાણીતા પર્યાવરણવિદ.
રજુ વ્યાસ	પ્રખ્યાત હર્પેટોલોજિસ્ટ

Important Trees of Gujarat | ગુજરાતના મહત્વના વૃક્ષો

Saag	*Tectona grandis*
Sisam	*Dalbergia latifolia*
Mahudo	*Madhuca indica*
Khair	*Acacia catechu*
Chandan	*Santalum album*
Vad	*Ficus benghalensis*
Pipdo	*Ficus religiosa*
Timru	*Diospyros melanoxylon*
Simado	*Bombex ceiba*
Sadad	*Terminalia tomentosa*
Red Sendar	*Pterocarpus santalinus*
Kilai	*Albizia procera*
Sevan	*Gmelina arborea*
Biyo	*Pterocarpus marsupium*

Rohan	*Soymida Febrifuga*
Dhavdo	*Anogeissus latifolia*
Kalam	*Mitragyna parvifolia*
Haldu	*Adina cordifolia*
Harde	*Terminalia chebula*
Khadyo	*Sterculia urens*
Rohido	*Tecomella undulata*
Rayan	*Manilkara hexandra*

Important Wild Mammals of Gujarat | ગુજરાતના મહત્વપૂર્ણ જંગલી સસ્તન પ્રાણીઓ

Common Name	Scientific Name	Estimated Population
Asiatic Lion	*Panthera leo persica*	674
Asiatic Wild Ass	*Equus hemionus khur*	6082
Asiatic Wild Dog	*Cuon alpinus*	–
Sloth Bear	*Melursus ursinus*	–
Grey Wolf	*Canis lupus pallipes*	343
Blackbuck	*Antilope cervicapra*	
Caracal	*Caracal caracal*	–
Four-Horned Antilope	*Tetracerus quadricornis*	–
Sambar	*Rusa unicolour*	–
Honey Badger	*Mellivora capensis*	–
Leopard	*Panthera pardus fusca*	1395
Pangolin	*Manis crassicaudata*	–

GFD 2020

Important Sea Mammals of Gujarat | ગુજરાતના મહત્વપૂર્ણ સમુદ્ર સસ્તન પ્રાણીઓ

Blue Whale	*Balaenoptera musculus*
Indo Pacific Dolphin	*Sous Chinensis*
Dugong	*Dugong dugon*
Finless Porpoise	*Delphinoidea Phocoenidae*

GFD 2020

Important Birds of Gujarat | ગુજરાતના મહત્વના પક્ષીઓ

Greater Flamingo	*Phoenicopterus roseus*
Great Indian Bustard	*Ardeotis nigriceps*
Forest Owlet	*Athene blewitti*
Macqueen's Bustard	*Chlamydotis macqueenii*
Lesser Florican	*Sypheotides indicus*
Oriental dwarf Kingfisher	*Ceyx erithaca*
Asian Openbill	*Anastomus oscitans*

Black necked Vulture	*Coragyps atratus*
Black-necked Stork	*Ephippiorhynchus asiaticus*
Black-rumped flameback	*Dimopium benghalense*
Brown Hawk Owl	*Ninox scutulata*
Dalmatian pelican	*Pelecanus crispus*
Greater racket-Tailed Drongo	*Dicrurus paradiseus*
Indian Grey Hornbill	*Ocyceros birostris*
Indian Nightjar	*Caprimulgus asiaticus*
Indian Pitta	*Pitta brachyura*
Indian Skimmer	*Rynchops albicollis*
Jacobin Cuckoo	*Clamator jacobinus*
King Vulture	*Sarcoramphus papa*
Lesser Flamingo	*Phoeniconaisas minor*
Malabar Trongo	*Harpactes fasciatus*
Peregrine Falcons	*Falco peregrinus*
Sarus Cranes	*Grus antigone*
Sociable Lapwing	*Vanellus gregarious*
White backed Vulture	*Gyps africanus*
White Stork	*Ciconia ciconia*
White-rumped Shama	*Copsychus malabaricus*

GFD 2020

Important Reptiles of Gujarat | ગુજરાતના મહત્વના સરિસૃપો

Bamboo pit Viper	*Trimeresurus stejńegeri*
Bronzeback tree Snake	*Dendrelaphis tristis*
Common Krait	*Bungarus caeruleus*
Green Keelback	*Macropisthodon plumbicolor*
Green sea Turtle	*Chelonia nydus*
Gujarat Gecko	*Hemidactylus gujratensis*
Indian Chameleon	*Chemaeleo zeylanicus*
Indian rock Python	*Python molurus*
Indian Star Tortoise	*Geochelone elegans*
Leatherback Sea Turtle	*Dermochelys coriacca*
Marsh Crocodile	*Crocodylus palustris*
Olive Ridley Turtle	*Lepidochelys olivacea*
Rat Snake	*Ptyas mucosa*
Red Sand Boa	*Eryx johnii*
Russel's Viper	*Daboia russelii*
Saw Scaled Viper	*Echis carinatus*
Spectacled Cobra	*Naja naja*

Important Fishes of Gujarat | ગુજરાતની મહત્વની માછલીઓ

Puffer Fish	*Takifugu oblongus*
Whale Shark	*Rhincodon typus*
Atlantic Salmon	*Salmo salar*
Rohu Fish	*Labeo rohita*
Tuna Fish	*Thunnini*
Catla Fish	*Catla catla*
Pomfret Fish	*Bramidae*
Bombay Duck	*Harpadon nehereus*
Llish (Hilsa)	*Tenualosa ilisha*
Mullet	*Mugilidae*

Extinct Species of Gujarat | ગુજરાતની લુપ્ત જાતો

Name	Remarks
Asian Elephant	–
Cheetah	–
Dhole	Recently rediscovered in 2020
Gaur	–
Giant Squirrel	–
Tiger	Recently rediscovered in 2020

Reserved Trees of Gujarat | ગુજરાતનાં અનામત વૃક્ષો

Teak	*Tectona grandis*
Sisham	*Dalbergia latifolia*
Mahudo	*Madhuca indica*
Khair	*Acacia catechu*
Chandan	*Santalum album*

GFD 2020

Who gives the permission to cut the unreserved trees in Gujarat

ગુજરાતમાં અનારક્ષિત વૃક્ષોને કાપવાની પરવાનગી કોણ આપે છે

Municipality Area	Authorized officer by the Municipal Commissioner
Urban Development Area	Chief Executive Officer
Nagarpalika Area	Chief Officer
Rural Areas	Mamlatdar

CMS species found in Gujarat | ગુજરાતમાં જોવા મળતી માઇગ્રેટરી પ્રજાતિઓ

	MAMMALS	
1.	Finless Porpoise	*Neophocaena phocaenoides*
2.	Indo-Pacific Hump-backed Dolphin	*Sousa chinensis*
3.	Asiatic Wild Ass	*Equus hemionus*
4.	Humpback Whale	*Megaptera novaeangliae*
5.	Killer Whale, Orca	*Orcinus orca*
6.	Blue Whale	*Balaenoptera musculus*
7.	Dugong	*Dugong dugon*
8.	Lion	*Panthera leo*
9.	Leopard	*Panthera pardus*
	BIRDS	
1.	Eurasian Spoonbill	*Platalea leucorodia*
2.	Crab Plover	*Dromas ardeola*
3.	Demoiselle Crane	*Anthropoides virgo*
4.	Griffon Vulture	*Gyps fulvus*
5.	Steppe Eagle	*Aquila nipalensis*
6.	Egyptian Vulture	*Neophron percnopterus*
7.	Amur Falcon	*Falco amurensis*
8.	Greater Flamingo	*Phoenicopterus ruber*
9.	Pied Harrier	*Circus melanoleucos*
10.	Common Sandpiper	*Tringa hypoleucos*
11.	Bee-Eater	*Merops apiaster*
12.	Common Pochard	*Aythya ferina*
13.	Eurasian Curlew	*Numenius arquata*
14.	Ruddy Shelduck	*Tadorna ferruginea*
15.	Black Stork	*Ciconia nigra*
16.	Shikra	*Accipiter badius*
17.	Black Kite	*Milvus migrans*
18.	Merlin	*Falco columbarius*
19.	Common Crane	*Grus grus*
20.	Lesser Flamingo	*Phoenicopterus minor*
21.	Glossy Ibis	*Plegadis falcinellus*
22.	Siberian Crane	*Leucogeranus leucogeranus*
23.	Pied Avocet	*Recurvirostra avosetta*
24.	Sarus Crane	*Antigone Antigone*
25.	MacQueen's Bustard	*Chlamydotis macqueenii*

REPTILES		
1.	Green Turtle	*Chelonia mydas*
2.	Ridley Turtle, Olive Ridley Turtle	*Lepidochelys olivacea*
3.	Hawksbill Turtle	*Eretmochelys imbricata*
FISHES		
1.	Shortfin Mako Shark	*Isurus oxyrinchus*
2.	Whale Shark	*Rhincodon typus*
3.	Pelagic Thresher Shark	*Alopias pelagicus*
4.	Common Thresher Shark	*Alopias vulpinus*
5.	Scalloped Hammerhead Shark	*Sphyrna lewini*
6.	Great hammerhead Shark	*Sphyrna mokarran*
7.	Narrow Sawfish	*Anoxypristis cuspidata*
8.	Large-tooth Sawfish	*Pristis pristis*
9.	Oceanic Whitetip Shark	*Carcharhinus longimanus*

GFD 2020

Chapter 2
પ્રકરણ ૨

India's Forest Profile and Forest Resources
ભારતના વનોની રૂપરેખા અને દેશની વન સંપદા

India Facts

Geographical Area (GA)	32,87,469 sq km
Recorded Forest Area (RFA) of India	7,76,419 sq km
% of RFA of GA	23.34 %
Forest cover of India	7,12,249 sq km
% of Forest Cover of GA	21.67 %
State with the highest Forest Cover	Mizoram (85.41%)
UT with the highest Forest Cover	Lakshadweep (90.33 %)
State with the lowest Forest Cover	Haryana (3.62%)
UT with the lowest Forest Cover	Ladakh (1.47 %)

ISFR 2019

ભારત હકીકતો

ભૌગોલિક ક્ષેત્ર (GA)	32,87,469 ચોરસ કિ.મી.
ભારતનો રેકોર્ડ કરેલ વન વિસ્તાર (RFA)	7,76,419 ચોરસ કિ.મી.
RFA નો GA %	23.34 %
ભારતનું વન આવરણ	7,12,249 ચોરસ કિ.મી.
GA નો વન આવરણ %	21.67 %
સૌથી વધુ વન આવરણ ધરાવતું રાજ્ય	મિઝોરમ (85.41%)
સૌથી ઓછું વન આવરણ ધરાવતો કેન્દ્રશાસિત પ્રદેશ	લક્ષદ્વીપ (90.33%)
સૌથી વધુ વન આવરણ ધરાવતું રાજ્ય	હરિયાણા (3.62%)
સૌથી ઓછું વન આવરણ ધરાવતો કેન્દ્રશાસિત પ્રદેશ	લદાખ (1.47%)

ISFR 2019

Who's who

Minister for Environment, Forest & Climate Change	Prakash Javadekar
Minister of State for Environment, Forest & Climate Change	Babul Supriyo
Secretary, Environment, Forest & Climate Change	Rameshwar Prasad Gupta
Director General of Forests	Sanjay Kumar
Additional Director General of Forest for Wildlife	Soumitra Dasgupta

MoEFCC 2020

કોણ શું છે

મંત્રી, પર્યાવરણ, વન અને ક્લાઈમેટ ચેન્જ	પ્રકાશ જાવડેકર
રાજ્ય મંત્રી, પર્યાવરણ, વન અને ક્લાઈમેટ ચેન્જ	બાબુલ સુપ્રિયો
સચિવ, પર્યાવરણ, વન અને ક્લાઈમેટ ચેન્જ	રામેશ્વર પ્રસાદ ગુપ્તા
ભારતના વન મહાનિદેશક	સંજયકુમાર
વન્યપ્રાણીસૃષ્ટિ ના અધિક વન મહાનિદેશક	સૌમિત્ર દાસગુપ્તા

MoEFCC 2020

Forest Cover in the States/UTs in India | ભારતમાં રાજ્યો / કેન્દ્રશાસિત કેન્દ્રો માં વન આવરણ

State/UT	% of Geographical Area
Andhra Pradesh	17.88
Arunachal Pradesh	79.63
Assam	79.63
Bihar	7.76
Chhattisgarh	41.13
Delhi	13.18
Goa	60.43
Gujarat	7.57
Haryana	3.62
Himachal Pradesh	27.72
UT of J&K	39.66
UT of Ladakh	1.47
Jharkhand	29.62
Karnataka	20.11
Kerala	54.42
Madhya Pradesh	25.14
Maharashtra	16.50
Manipur	75.46
Meghalaya	76.33

State/UT	% of Geographical Area
Mizoram	85.41
Nagaland	75.31
Punjab	3.67
Odisha	33.15
Rajasthan	4.86
Sikkim	47.10
Tamil Nadu	20.27
Telangana	18.36
Tripura	73.68
Uttar Pradesh	6.15
Uttarakhand	45.44
West Bengal	19.04
A & N Islands	81.74
Chandigarh	19.32
Dadra & Nagar Haveli	42.16
Daman & Diu	18.46
Lakshadweep	90.33
Puducherry	10.70

ISFR 2019

Forest types of India | ભારતના વન પ્રકારો

5 Major groups	16 Types
I. Tropical forest	Wet evergreen forest Semi- evergreen forest Moist deciduous forest Littoral and swampy forest Dry deciduous forest Thorn forest Dry evergreen forest
II. Montane sub-tropical forest	Subtropical broad-leaved forest Subtropical pine forest Subtropical dry evergreen forest
III. Montane temperate forest	Montane wet temperate forest Himalayan moist forest Himalayan dry forest
IV. Sub-alpine forest	Sub-alpine forest
V. Alpine scrub	Moist alpine
	Dry alpine

Champion and Seth 1968

Forest Types of India with Regional Climate and Speciation

પ્રાદેશિક હવામાન અને વિશિષ્ટ પ્રજાતિ સાથે ભારતના વન પ્રકાર

	Forest type	Distribution	Climatic condition	Plant species
1. Tropical wet evergreen forest				
1A.	Southern Tropical Wet Evergreen Forest	Andamans, Kerala, Kanataka, Maharashtra, TN	MAT (Mean Annual Temperature): 27°C MARF (Mean Annual Rainfall): 2000-1000 mm	**Top Canopy** - *Dipterocarpus, Mesua, Hopea, Artocarpus, Mangifera, Syzygium* **Undergrowth:** *Canes, Bamboos, Palms* **Ground Cover:** *Strobilanthes, Epiphytes, Ferns, Orchids*
1B.	Northern Tropical Wet Evergreen Forest	North East India, West Bengal, Assam, Orissa	MAT: 22-26°C MARF: 2500- 6400 mm	**Over wood:** *Dipterocarpus, Shorea Artocarpus, Syzygium, Mesua, Michelia.* **Middle storey:** *Vatica, Garcinia, Dendrocalamus, Melocanna, Bambusa* **Under growth:** *Clerodendron Ixora, Canes*
2. Tropical semi evergreen forest				
2A.	Southern Tropical Semi Evergreen Forest	Western Ghats, Karnataka, Kerala, Andamans	MAT: 26°C MRF: 2000 3000 mm	**Over wood:** *Kyia. Terminalia, Dipterocarpus, Hopea* **Middle storey:** *Schleichera oleasa, Eleocarpus, Mallotus* **Under wood:** *Strobilanthus Clerodendron, lxora*
2B.	Northern Tropical Semi Evergreen Forest	Assam, West Bengal, Coastal parts of Orissa	MAT: 24°C MARF: 1500-3000 mm	**Over wood**: *Dipterocarpus, Syzygium, Cinnamomum. Michelia, Artocarpus, Sal* **Middle storey:** *Michelias Polyalthia* **Under wood:** *Bamboos, Arundinacea, Deridrocalamus, Hamiltoni, Melocanna*
3. Tropical moist deciduous forest				
3A.	Andamans Moist Deciduous Forest	Andaman Islands	MAT: 26°C MARF: 3000 mm	**Over wood:** *Pterocarpus dalbergiordes, Terminalia* **Middle storey:** *Lannea, Cinnamomum, Bambos* **Under growth:** *Murraya, Ixora, Mallotus*

	Forest type	Distribution	Climatic condition	Plant species
3B.	South Indian Tropical Moist Deciduous Forest	Gujarat, Karnataka, Maharashtra, TN, M.P, Kerala	MAT: 24-27°C MARF: 1200-3000 mm	**Over wood:** *Teak, Grewia, Lagerstroemia, Terminalia tomentosa, Adina* **Middle Storey:** *Dillenia, Kydia, Dendrocalamus* **Undergrowth:** Clerodendron
3C.	North Indian Tropical Moist Deciduous forest	U.P., Bihar, Orissa, M.P., Bengal, Assam	MAT:21-26°C MARF:1000-2000 mm	**Over wood:** *Sal. Schima wallichi and Terminalia, tomentosa, Michelia champaca* **Middle storey:** *Dendrocalamus hamiltoni syzygium*
4. Littoral and swamp forest				
4A.	Littoral Forest	Coastal region	MAT: 26-29°C MARF: 760-5000 mm	**Over wood:** *Casuarina, Manilkara, Calophyllum* **Middle survey:** *Pongamia, Barringtonia*
4B.	Tidal Swamp Forest	Estuaries of rivers	—	*Rhizophora, Avicennia, Ceriops. Xylocarpus*
4C.	Tropical Fresh Water Swamp Forest	Himalayan sub montane forest, Kerala	—	*Myristica* spp.
4D.	Tropical Seasonal Swamp Forest	Along Ganga & Brahmaputra rivers		**Over wood:** *Altingia, Barringtonia, Syzygium* spp.
4E.	Tropical Riparian Fringing Forest	Along the banks of the largest streams in hilly tracts	—	*Terminalia arjuna, Lagerstromia speciosa, Terminalia myriocarpa.*
5. Tropical dry deciduous forest				
5A.	Southern Tropical Dry Deciduous Forest	M.P., Gujarat, Maharashtra, A.P., Karnataka, T.N.	MAT: 23 -29°C MARF: 750-1900 mm	**Over wood:** *Teak, Terminalia, Anogeissus latifolia, Diospyros*
5B.	Northern Tropical Dry Deciduous Forest	H.P., U.P., Bihar, Orissa W.B., M.P., Rajasthan	MAT: 24-27°C MARF: 750-1300 mm	**Over wood:** *Sail, Anogeissus, Diospyros, Adina, Schleichera, Lannea, Acacia catechu.*
6. Tropical thorn forest				
6A.	Southern Tropical Thorn Forest	Dry peninsular India, M.P.	MAT: 24-28°C MARF: 460-950 mm	*Acacia catechu, Acacia nilotica Albizzia, Prosopis, Acacia planiferans*

Forest type		Distribution	Climatic condition	Plant species
6B.	Northern Tropical Thorn Forest	Semi-arid areas of Punjab, Rajasthan, Gujarat, U.P., M.P.	MAT: 25-27°C MARF: 750-1250 mm	*Prosopis, Acacia* spp.
7. Tropical dry evergreen forest				
7.	Tropical Dry Evergreen Forest	Coastal belt from Tirunelveli to Nellore	MAT:28 -29°C MARF: 870-1270 mm	**Over wood:** *Azadirachta indica Manilkara hexandara, Canthium, Syzygium*
8. Sub-tropical broad leaved hill forest				
8A.	Southern sub - tropical broad leaved hill forest	South Indian Hills	MAT:17-22°C MARF:1500-6600 mm	*Wigthtiana, Canthium, Syzygium cumimi, Manilkara haxendra, Mangifera*
8B.	Northern sub-tropical broad leaved hill forest	Eastern Himalayas and Khasi hills	MAT: 16-20°C MARF:2100-11400 mm	*Castonopis, Schima, Quercus*
9. Sub-tropical pine forest				
9.	Sub-tropical pine Forest	Central Himalayas, Nagar and Manipur Hills	MAT: 15-22°C MARF: 1000-3000 mm	*Chir. Deodar, Kail, Quercus incana, Rhododendron*
10. Sub-tropical dry evergreen forest				
10.	Sub-Tropical Dry Evergreen Forest	Himalayas	—	*Olea cuspidata, Acacia modesta*
11. Montane wet temperate forest				
11A.	Southern Montane Wet Temperate Forest	Hills of Tamil Nadu and Kerala	MAT: 14-17°C MARF:1300-6000 mm	*Lagerstroemia, Syzygium, Rhododendron, Michelia*
11B.	Northern Montane Wet Temperate Forest	Eastern Himalayas	MAT: 12°C MARF: 1900-4000 mm	*Machilus, Michelia, Magnolia, Cinnamomum, Quercus, Acer* spp.

Forest type		Distribution	Climatic condition	Plant species
12. Himalayan moist temperate forest				
12.	Himalayan Moist Temperate Forest	Himalayas	MAT:13-16°C MARF: 1500-3300 mm	*Querces incana Q dilatata, chir pine*
13. Himalayan dry temperate forest				
13.	Himalayan dry temperate forest	Kashmir, H.P.,U.P., Sikkim, Arunachal Pradesh	MAT: 6 - 17°C MARF:80-800 mm	*Pinus gerardiana, Blue pine, Q. ilex, Deodar, J. macropoda*
14. Sub alpine forest				
14.	Sub alpine forest	Himalayas	MAT: 2 - 6°C MARF:10-55 mm	*Abies spectabilis, Abies densa, Juniperus, Spruce*
15.	Moist alpine scrub	Above 3400 m altitude	—	*Rhododendron, birch* and *other deciduous trees*
16. Dry alpine scrub				
16.	Dry alpine scrub	Xerophytic condition	—	*Juniperus wallichiana, J.communes, Artimisia*

Champion and Seth 1968

Statutory bodies of MoEFCC, Govt of India

ભારત સરકારના પર્યાવરણ, વન અને ક્લાઈમેટ ચેન્જ મંત્રાલયની વૈધાનિક સંસ્થાઓ

Authority/ Board	HQ	Year of establishment	Head	Function
Central Zoo Authority (CZA)	New Delhi	1992	S. P. Yadav	Controlling Zoos of India
National Biodiversity Authority (NBA)	Chennai	2003	V. B. Mathur	Implement India's Biological Diversity Act
National Green Tribunal (NGT)	New Delhi	2010	Justice Adarsh Kumar Goel	Protection of environment and conservation of forests of India
National Tiger Conservation Authority (NTCA)	New Delhi	2005	S. P. Yadav	Statutory authority to Project Tiger
Central Pollution Control Board (CPCB)	New Delhi	1974	Shiv Das Meena	Control of pollution in India

Authority/ Board	HQ	Year of establishment	Head	Function
National Afforestation & Eco-Development Board (NAEB)	New Delhi	1992	Pankaj Asthana	Promote afforestation in India
Wildlife Crime Control Bureau (WCCB)	New Delhi	2006	Tilotama Varma	Control wildlife crime in India

MoEFCC 2020

Autonomous bodies of MoEFCC, Govt of India

ભારત સરકારના પર્યાવરણ, વન અને ક્લાઈમેટ ચેન્જ મંત્રાલયની સ્વાયત સંસ્થાઓ

Organization	HQ	Year of establishment	Head	Function
G.B. Pant National Institute of Himalayan Environment (GBPNIHESD)	Almora	1988	R. S. Rawal	Conduct research and development in Indian Himalayan Region
Indian Council of Forestry Research and Education (ICFRE)	Dehradun	1986	A. K. Rawat	Forestry education and research in India
Indian Institute of Forest Management (IIFM)	Bhopal	1982	Pankaj Shrivastava	Work on professional forestry management.
Indian Plywood Industries Research and Training Institute (IPIRTI)	Bengaluru	1962	Shakti Singh Chauhan	Research and Training on wood
Wildlife Institute of India (WII)	Dehradun	1982	Dhananjay Mohan	Scientific knowledge on wildlife resources of India

MoEFCC 2020

Organizations of MoEFCC, Govt of India
ભારત સરકારના પર્યાવરણ, વન અને ક્લાઈમેટ ચેન્જ મંત્રાલયની સંસ્થાઓ

Organization	HQ	Year of establishment	Director	Function
Botanical Survey of India (BSI)	Kolkata	1890	A. A. Mao	Document plant resources of India
Forest Survey of India (FSI)	Dehradun	1981	Subhash Ashutosh	Assessing forest cover of India, Prepare State of Forest Report biennially, Monitor changes in forests
Indira Gandhi National Forest Academy (IGNFA)	Dehradun	1938	Bharat Jyoti	Impart Forestry Training to Forest Officers
National Centre for Sustainable Coastal Management (NCSCM)	Chennai	2011	Ramesh Ramachandran	Research in the areas of coastal zone management
Sálim Ali Centre for Ornithology and Natural History (SACON)	Coimbatore	1990	S. Sankar	Education and research in Ornithology
Zoological Survey of India (ZSI)	Kolkata	1961	Kailash Chandra	Work on faunal diversity of India
Society of Integrated Coastal Management (SICOM)	New Delhi	2010	R. P. Gupta	Mapping the coastline of India
CPR Environmental Education Centre (CPREEC)	Chennai	1989	Nanditha Krishna	Disseminate environment education and awareness.
Centre of Environment Education (CEE)	Ahmedabad	1984	Kartikeya Sarabhai	Environment education for sustainable development

MoEFCC 2020

Indian Council of Forestry Research and Education (ICFRE) Institutes and Research Centre

ભારતીય વાનિકી સંશોધન અને શિક્ષણ પરિષદ અંતર્ગતની સંસ્થાઓ અને સંશોધન કેન્દ્રો

ICFRE established in 1878

HQ: Dehradun

Director General: A. K. Rawat

ICFRE Institute	HQ	Year of Establishment	Director
Arid Forest Research Institute (AFRI)	Jodhpur	1987	M. R. Baloch
Forest Research Institute (FRI)	Dehradun	1906	Arun Singh Rawat
Himalayan Forest Research Institute (HFRI)	Shimla	1988	S. S. Samant
Institute of Forest Biodiversity (IFB)	Hyderabad	2012	D. Jayaprasad
Institute of Forest Genetics and Tree Breeding (IFGTB)	Coimbatore	1988	S. Murugesan
Institute of Forest Productivity (IFP)	Ranchi	1993	Nitin Kulkarni
Institute of Wood Science and Technology (IWST)	Bangalore	1988	Madan Prasad Singh
Rain Forest Research Institute (RFRI)	Jorhat, Assam	1988	R.S.C. Jayaraj
Tropical Forest Research Institute (TFRI)	Jabalpur	1988	G. Rajeshwar Rao
Forest Research Centre			
FRC for Bamboo & Rattan (FRC-BR)	Aizawl	2004	
FRC for Coastal Ecosystem (FRC-CE)	Visakhapatnam	2016	
FRC for Eco-Rehabilitation (FRC-ER)	Prayagraj	1992	
FRC for Livelihood Extension (FRC-LE)	Agartala	2012	
FRC for Skill Development (FRC-SD)	Chhindwara	1995	

ICFRE 2020

Forestry organizations providing specialized education and training

વિશિષ્ટ શિક્ષણ અને તાલીમ પૂરી પાડતી વાનિકી સંસ્થાઓ

No.	Institute	Location
1.	Indian Institute of Forest Management (IIFM)	Bhopal
2.	Indian Plywood Industries Research and Training Institute (IPIRTI)	Bangalore
3.	Institute of Paper Technology (IPT)	Saharanpur
4.	Wildlife Institute of India (WII)	Dehradun

Universities, Colleges and Departments offering Forestry, Agroforestry and Wildlife Education in India

ભારતમાં વનવિદ્યા, એગ્રોફોરેસ્ટ્રી અને વન્યજીવ શિક્ષણ પ્રદાન કરતી યુનિવર્સિટીઓ, કોલેજો અને વિભાગો

1.	ASPEE College of Horticulture and Forestry	Navsari, Gujarat
2.	College of Forestry & Hill Agriculture, Tehri Garhwal of Uttarakhand University of Horticulture and Forestry	Uttarakhand
3.	College of Forestry and Environment, Allahabad of Sam Higginbottom Institute of Agriculture Technology and Sciences (Formerly Allahabad Agricultural Institute)	Allahabad, UP
4.	College of Forestry and Hill Agriculture, Ranichauri of G.B. Pant University of Agriculture and Technology	Uttarakhand
5.	College of Forestry, Akola of Dr. Panjabrao Deshmukh Krishi Vidyapeeth	Maharashtra
6.	College of Forestry, Dapoli of Dr. Balasaheb Sawant Konkan Krishi Vidyapeeth	Maharashtra
7.	College of Forestry, Nauni, Solan of Dr.Y.S. Parmar University of Horticulture and Forestry	Himachal Pradesh
8.	College of Forestry, Ponnampet of University of Agricultural Sciences	Bangalore, Karnataka
9.	College of Forestry, Sirsi of University of Agricultural Sciences	Dharwad, Karnataka
10.	College of Forestry, Thrissur of Kerala Agricultural University	Kerala
11.	College of Horticulture and Forestry, Pasighat, Arunachal Pradesh of Central Agricultural University	Arunachal Pradesh
12.	College of Horticulture and Forestry, Jhalawar of Maharana Pratap University of Agriculture and Technology	Rajasthan
13.	Department of Agroforestry, Jhansi of Bundelkhand University	Uttar Pradesh
14.	Department of Forestry and Natural Resources, Ludhiana of Punjab Agricultural University	Ludhiana, Punjab
15.	Department of Forestry of Kumaun University	Uttarakhand
16.	Department of Forestry, Bhubaneswar of Orissa University of Agriculture and Technology	Orissa
17.	Department of Forestry, Cooch Behar of Uttar Banga Krishi Vishwavidyalaya	West Bengal
18.	Department of Forestry, Hisar of CCS Haryana Agricultural University	Haryana
19.	Department of Forestry, Kanpur of C.S. Azad University of Agriculture and Technology	Kanpur, UP
20.	Department of Forestry, Nirjuli of North Eastern Regional Institute of Science and Technology (NERIST) Deemed University	Arunachal Pradesh

21.	Department of Forestry, Raipur of Indira Gandhi Agricultural University	Chhattisgarh
22.	Department of Forestry, Srinagar (Garhwal) of H.N.B. Garhwal University	Uttarakhand
23.	Department of Forestry, Wildlife and Environment Science, Bilaspur of Guru Ghasidas Vishwavidyalaya University	Chhattisgarh
24.	Department of Wildlife Science of Aligarh Muslim University	UP
25.	Division of Agroforestry of Sher-e-Kashmir University of Agricultural Sciences and Technology	Jammu
26.	Dolphin Institute of Biomedical & Natural Sciences	Dehradun, Uttarakhand
27.	Doon College of Agriculture Science and Technology	Dehradun, Uttarakhand
28.	Faculty of Forestry, Kanke, Ranchi of Birsa Agricultural University	Ranchi
29.	Faculty of Forestry, Sher-e-Kashmir University of Agricultural Sciences and Technology	Srinagar, Kashmir
30.	Forest College Research Institute	Mettupalayam of Tamil Nadu
31.	Forest Research Institute-Deemed University	Dehradun, Uttarakhand
32.	School of Studies (Forestry & Wildlife), Bastar of Bastar University	Chhattisgarh

Forest Administration in India

Designation	In charge of
Director General of Forests	Country
PCCF & Head of the Forest Force	State
Principal Chief Conservator of Forests	Wing
Chief Conservator of Forests	Circle/ wing
Conservator of Forests	Circle
Deputy Conservator of Forests	Division
Assistant Conservator of Forests	Sub division
Range Forest Officer	Range
Forester/ Forest Officer	Round
Beat Guard	Beat

ભારતમાં વન વહીવટ

હોદ્દો	હવાલો
વન મહાનિદેશક	દેશ
અગ્ર મુખ્ય વન સંરક્ષક અને રાજ્યના વન દળના વડા	રાજ્ય
અગ્ર મુખ્ય વન સંરક્ષક	પાંખ
મુખ્ય વન સંરક્ષક	વર્તુળ / પાંખ

હોદ્દો	હવાલો
વન સંરક્ષક	વર્તુળ
નાયબ વન સંરક્ષક	વિભાગ
સહાયક વન સંરક્ષક	પેટા વિભાગ
પરિક્ષેત્ર વન અધિકારી	પરિક્ષેત્ર
વનપાલ/ વન અધિકારી	રાઉન્ડ
બીટ ગાર્ડ	બીટ

Forest Service in India | ભારતમાં વન સેવા

Gazetted Forest Service	Non-Gazetted Forest Service
Indian Forest Service, Class I	Executive Branch: Forest Guards, Foresters, Range Forest Officers (Survey), Surveyors, Drivers.
Gujarat Forest Service Class I and II	Ministerial Branch: Head Clerks, Accountants, Statistical Assistants, Clerks and Stenographers.
Assistant Conservator of Forests, Class II	Class IV: Peons, Zoo Keepers, Cleaners and Watchmen etc.
Range Forest Officer Class II	

Forest Department field official's Uniform | વન વિભાગના ક્ષેત્રીય અધિકારીનો ગણવેશ

Range Forest Officer	3 Stars on shoulders	Khakhi P-Cap
Forester	2 Stars on shoulders	Blue Cap
Beat Guard	1 Stars on shoulders	Blue Cap

Endemic Species of India | ભારતની સ્થાનિક પ્રજાતિઓ

Taxa	Species	Region
Mammals	Asiatic Lion	Gir, Gujarat
	Asiatic Wild Ass	Kutch, Gujarat
	Indian Giant Squirrel	Satpuda, MP
	Jungle Stripped Squirrel	Western Ghats
	Lion-tailed Macaque	Western Ghats
	Malabar Civet	Kerala
	Nilgiri Tahr	Nilgiri hills, TN
	Pygmy Hog	Assam
Birds	Forest Owlet	Western Ghat
	Great Indian Bustard	Rajasthan, Gujarat
	Jerdon's Courser	AP, Telangana
	Nicobar Megapod	Nicobar Island
	Nicobar Parakeet	Nicobar Island

Taxa	Species	Region
Reptiles	Andaman Cobra	Andaman Island
	Assam Keelback	Assam
	Darjeeling Kukri Snake	Darjeeling, WB
	Nilgiri Forest Lizard	Nilgiri hills
Amphibia	Malabar Toad	Western Ghat
	Nagaland Toad	Nagaland
Butterfly	Nilgiri Tiger	Nilgiri hills
	Andaman Mormon	Andaman Island

Reputed Indian awards in Environment
પર્યાવરણ ના ક્ષેત્રમાં પ્રતિષ્ઠિત ભારતીય પુરસ્કારો

Award	
Amrita Devi Bishnoi Wild Life Protection Award	MoEFCC, GoI
B.P. Pal National Environment Fellowship Award for Biodiversity	MoEFCC, GoI
Desert Ecology Fellowship	MoEFCC, GoI
Dr. Salim Ali National Wild Life Fellowship Award	MoEFCC, GoI
ICFRE award for excellence in Forestry	ICFRE, Dehradun
India Biodiversity Award	MoEFCC, GoI
Indira Gandhi Paryavaran Purashkar	MoEFCC, GoI
Indira Priyadarshini Vrikshamitra Awards	MoEFCC, GoI
Janaki Ammal National Award for Taxonomy	MoEFCC, GoI
Janaki Ammal National Award for Taxonomy	MoEFCC, GoI
MEDINI Puraskar Yojana	MoEFCC, GoI
National Award for Best ENVIS Center	MoEFCC, GoI
National Award for Prevention of Pollution	MoEFCC, GoI
Padma Awardees Honored for Environmental Protection Activities (Padma Vibhushan, Padma Bhushan and Padma Shri)	Govt. of India
Paryavaran Aur Van Mantralaya Vishisht Vaigyanik Puraskar	MoEFCC, GoI
Pitambar Pant National Environment Fellowship Award	MoEFCC, GoI
Rajiv Gandhi Environment Award for Clean Technology	MoEFCC, GoI
Rajiv Gandhi Wildlife Conservation Award	MoEFCC, GoI
Shri Kailash Sakhla National Wildlife Fellowship Award	MoEFCC, GoI

MoEFCC 2020

Eco Sensitive Zones in India

Starting year	2011
Governed by	MoEFCC, Govt of India
Objective	The purpose of declaring Eco-sensitive zones around National parks and Sanctuaries is to create some kind of "Shock Absorber" for the protected areas. They would also act as a transition zone from areas of high protection to areas involving lesser protection. Notification of Eco Sensitive Zones not only strengthens the corridors and buffers around the Protected Area but also minimize the negative effect of anthropogenic pressure, industrialization and unplanned development around the area.
Prohibited activities	Commercial mining, setting of saw mills and industries causing pollution, commercial use of firewood and major hydro-power projects, are prohibited in such areas. It also prohibits tourism activities like flying over protected areas in an aircraft or hot air balloon, and discharge of effluents and solid waste in natural water bodies or terrestrial areas.
Regulated activities	Felling of trees, drastic change in agriculture systems and commercial use of natural water resources, including groundwater harvesting and setting up of hotels and resorts, are the activities regulated in the areas.
Permitted and promoted activities	Activities permitted in the areas include ongoing agriculture and horticulture practices by local communities, rainwater harvesting, organic farming, adoption of green technology and use of renewable energy sources.
Area	In case where sensitive corridors, connectivity and ecological important patches, crucial for landscape linkage, are even beyond 10 Kms width, these should be included in the eco-sensitive zone.
Official document	Zonal Master Plan is the guiding document for the entire eco-sensitive zone.

ભારતમાં ઇકો સેન્સિટિવ ઝોન

પ્રારંભ વર્ષ	2011
સંચાલિત	પર્યાવરણ, વન અને ક્લાઈમેટ ચેન્જ મંત્રાલય, ભારત સરકાર
ઉદ્દેશ્ય	રાષ્ટ્રીય ઉદ્યાનો અને અભયારણ્યોની આસપાસના ઇકો સેન્સેટીવ ઝોન જાહેર કરવાનો હેતુ સંરક્ષિત વિસ્તારો માટે અમુક પ્રકારના "શોક એબ્સોર્બર" બનાવવાનો છે. તેઓ ઉચ્ચ સંરક્ષણવાળા ક્ષેત્રોથી ઓછા રક્ષણ ધરાવતા વિસ્તારોમાં સંક્રમણ ક્ષેત્ર તરીકે પણ કાર્ય કરશે. ઇકો સેન્સિટિવ ઝોનના જાહેરનામાથી સૂચનાથી સંરક્ષિત ક્ષેત્રની આજુબાજુના કોરિડોર અને બફર્સને માત્ર મજબૂત કરવામાં આવે છે, પરંતુ તે આજુબાજુના માનવ દબાણ, ઔદ્યોગિકરણ અને આયોજિત વિકાસની નકારાત્મક અસરને પણ ઘટાડે છે.

પ્રતિબંધિત પ્રવૃત્તિઓ	આવા ક્ષેત્રોમાં વાણિજ્યિક ખાણકામ, લાકડાની મિલ અને ઉદ્યોગોનું પ્રદૂષણ, લાકડાંનો વ્યાપારી ઉપયોગ અને મુખ્ય હાઇડ્રો-પાવર પ્રોજેક્ટ્સ પ્રતિબંધિત છે. તે વિમાન અથવા ગરમ હવાના બલૂનમાં સુરક્ષિત વિસ્તારોમાં ઉડાન અને કુદરતી જળ સંસ્થાઓ અથવા પાર્થિવ વિસ્તારોમાં પ્રવાહી અને નક્કર કચરાના વિસર્જન જેવી પર્યટન પ્રવૃત્તિઓને પણ પ્રતિબંધિત કરે છે.
નિયમિત પ્રવૃત્તિઓ	વૃક્ષોનું ભંગાણ, કૃષિ પ્રણાલીમાં ધરખમ ફેરફાર અને ભૂગર્ભ જળ સંચય અને હોટલ અને રિસોર્ટની સ્થાપના સહિત કુદરતી જળ સંસાધનોનો વ્યાપારી ઉપયોગ, તે વિસ્તારોમાં નિયમનકારી પ્રવૃત્તિઓ છે.
પ્રોત્સાહિત પ્રવૃત્તિઓ	આ ક્ષેત્રોમાં મંજૂરી આપવામાં આવતી પ્રવૃત્તિઓમાં સ્થાનિક સમુદાયો દ્વારા કૃષિ અને બાગાયતી પદ્ધતિઓ, વરસાદી પાણીનો સંગ્રહ, સજીવ ખેતી, જૈવિક તકનીકી અપનાવવા અને નવીનીકરણીય ઉર્જા સ્ત્રોતોનો ઉપયોગ સામેલ છે.
વિસ્તાર	સંવેદનશીલ કોરિડોર, કનેક્ટિવિટી અને ઇકોલોજીકલ અગત્યના પેચો, લેન્ડસ્કેપ જોડાણ માટે નિર્ણાયક, 10 કિલોમીટરથી પણ વધુ છે, તેવા વિસ્તારોને ઇકો સેન્સેટિવ ઝોનમાં સામેલ કરવા જોઈએ.
સત્તાવાર દસ્તાવેજ	ઝોનલ માસ્ટર પ્લાન એ સમગ્ર ઇકો-સેન્સેટીવ ઝોન માટે માર્ગદર્શક દસ્તાવેજ છે.

Important Marine Projects in India / ભારતમાં મહત્વપૂર્ણ દરિયાઈ પ્રોજેક્ટ્સ

Project	Starting Year
Ganga Action Plan (GAP)	1986
Coastal Ocean Monitoring and Prediction System (COMAPS)	1991
Land Ocean Interactions in the Coastal Zone (LOICZ)	1995
Integrated Coastal and Marine Area Management (ICMAM)	1998
Society of Integrated Coastal Management (SICOM)	2010

Chapter 3

પ્રકરણ 3

Protected Areas of Gujarat

ગુજરાતના રક્ષિત વિસ્તારો

Protected Areas of Gujarat | ગુજરાતના રક્ષિત વિસ્તારોની રૂપરેખા

Protected Area	Nos	Area (sq km)	% area to the total GA of India
National Park	4	479.67	0.24 %
Wildlife Sanctuary	23	16629.37	8.47 %
Conservation Reserve	1	227.0	0.12 %
Community Reserve	0	0	0
Total	**28**	**17336.04**	**8.83%**

National Parks of Gujarat | ગુજરાતના રાષ્ટ્રીય ઉદ્યાનો

NP	Dist.	Year	Area (sq km)	Major wildlife
Blackbuck National park	Bhavnagar	1976	34.08	Blackbuck, Grey wolf, Lesser florican, Harrier
Gir National Park	Junagadh	1965	258.70	Asiatic lion, Leopard, Four-horned antelope, Spotted deer, Hyena, Blue bull
Marine National Park	Jamnagar	1982	162.89	Sponges, Jelly fish, Octopus, Dugongs, Star fish, Olive Ridley sea turtle
Vansda National Park	Navsari	1979	23.99	Leopard, Spotted deer, Four-horned antelope, Rusty spotted cat, Forest-owlet

GFD 2020

Wildlife Sanctuaries of Gujarat | ગુજરાતના વન્યપ્રાણી અભ્યારણ્યો

WLS	Dist.	Year	Area (sq km)	Major Wildlife
Balaram Ambaji WLS	Bansaskatha	1989	542.08	Sloth Bear, Leopard, Rhesus Macaque, Blue bull, Fox, Hyena
Barda WLS	Porbandar	1979	192.3	Leopard, Blue bull, Hyena, Wild boar, Wolf, Birds

WLS	Dist.	Year	Area (sq km)	Major Wildlife
Gaga WLS	Dwarka	1988	3.33	Great Indian Bustard, Wolf, Jackal
Gir WLS	Junagadh	1965	1153.40	Asiatic Lion, Leopard, Four-horned antelope, Spotted Deer, Hyena, Chinkara
Girnar WLS	Junagadh	2008	178.87	Asiatic Lion, Leopard, Four-horned antelope, Spotted Deer, Hyena, Chinkara
Hingolgadh Nature Education Sanctuary	Rajkot	1980	6.54	Blue bull, Chinkara, Wolf, Fox, Birds
Jambughoda WLS	Panchmahal	1990	130.38	Sloth Bear, Leopard, Rhesus Macaque, Jungle Cat, Wolf, Hyena
Jessore Sloth Bear Sanctuary	Bansaskatha	1978	180.66	Sloth Bear, Leopard, Rhesus Macaque, Spotted Deer, Hyena
Khijadiya Bird Sanctuary	Jamnagar	1981	6.05	Indian Skimmer, Ibises, Painted Stork, Cormorants
Kutch Bustard Sanctuary	Kutch	1992	2.03	Great Indian Bustard, Lesser Florican, Macqueen's Bustard
Kutch desert WLS	Kutch	1968	7506.22	Chinkara, Hyena, Fox, Flamingo, Pelicans and Water Fowls
Marine Sanctuary	Jamnagar, Dwarka, Gulf of Kutch	1980	295.03	Sponges, Jelly Fish, Octopus, Star Fish, Dolphin
Mitiyala WLS	Amreli	2004	18.22	Asiatic Lion, Leopard, Four-horned antelope, Spotted Deer, Hyena, Chinkara
Nal Sarovar Bird Sanctuary	Ahmedabad, Surendranagar	1969	120.82	Flamingo, Pelicans, Coot, Ducks, Waders, Storks, Herons and Waterfowl
Narayan Sarovar Sanctuary	Kutch	1981	444.23	Chinkara, Desert Cat, Hyena, Fox, Jackal
Pania WLS	Amreli	1989	39.63	Asiatic Lion, Leopard, Four-horned antelope, Spotted Deer, Hyena, Chinkara, Pangolin, Blue bull
Porbandar Bird Sanctuary	Porbandar	1988	0.09	Flamingo, Pelicans, Coot, Ducks, Waders, Storks, Herons and Waterfowl
Purna WLS	Dang	1990	160.84	Leopard, Rhesus Macaque, Four-horned antelope, Spotted Deer, Hyena
Rampara WLS	Morbi	1988	15.01	Blue bull, Chinkara, Wolf, Fox, Jackal, Birds

WLS	Dist.	Year	Area (sq km)	Major Wildlife
Ratanmahal Sloth Bear Sanctuary	Dahod	1982	55.65	Sloth Bear, Leopard, Rhesus Macaque, Jungle Cat, Civet Cat, Hyena
Shoolpaneshwar WLS	Narmada	1982	607.70	Sloth Bear, Leopard, Rhesus Macaque, Four-horned antelope, Spotted Deer, Hyena, Chinkara, Pangolin
Thol Bird Sanctuary	Mehsana	1988	6.99	Sarus Cranes, Geese, Flamingos
Wild Ass Sanctuary	Kutch	1973	4953.70	Wild Ass, Chinkara, Hyena, Blackbuck, Flamingo, Pelicans, Other Water Fowl

GFD 2020

Conservation Reserve of Gujarat | ગુજરાતના કન્ઝર્વેશન રિઝર્વ

CR	Dist.	Year	Area (sq km)	Major wildlife
Chhari Dhand Conservation Reserve	Kutch	2008	227.0	Grey Wolf, Chinkara Caracal, Desert Cat, Desert Foxe, Flamingo, Storks, Steppe Eagle, Montagu's Harrier, Peregrine Falcon,

GFD 2020

Biosphere Reserve of Gujarat | ગુજરાતના બાયોસ્ફિયર રિઝર્વ

Biosphere Reserves are governed under UNESCO's MAB Programme

BR	Dist.	Year	Area (sq km)	Major wildlife
Kachchh Biosphere Reserve	Kutch	2008	12454.0	Grey Wolf, Desert Fox, Blue bull, Chinkara Sarus Crane, Ducks, Pelicans and Flamingoes

GFD 2020

Chapter 4

પ્રકરણ ૪

Protected Areas of India

ભારતના રક્ષિત વિસ્તારો

Protected Areas of India | ભારતના રક્ષિત વિસ્તારો ની રૂપરેખા

Protected Area	Nos	Area (sq km)	% area to the total GA of India
National Park	101	40,564	1.23 %
Wildlife Sanctuary	553	119,776	3.64 %
Conservation Reserve	86	3858.25	0.12 %
Community Reserve	163	833.34	0.03%
Total	**903**	**165,031.59**	**5.02 %**

National Wildlife Database, WII 2020

Biosphere Reserves | બાયોસ્ફિયર રિઝર્વ

Starting Year	1971
Total BRs World	714 (129 countries)
Total BRs India	18
Governed by	UNESCO Man and the Biosphere (MAB)
First BR	Nilgiri BR, Karnataka - 1986
Latest	Panna, MP – 2011
Biggest	Kutchh, Gujarat - 12,454 sq km
Smallest	Dibru-Saikhowa, Assam - 765 sq km

ENVIS, WII 2020

Biosphere Reserves of India | ભારતના બાયોસ્ફિયર રિઝર્વ

No.	Biosphere Reserve	Year	State
1	Nilgiri	1986	Tamil Nadu, Kerala and Karnataka
2	Nanda Devi	1988	Uttarakhand
3	Nokrek	1988	Meghalaya
4	Great Nicobar	1989	A&N Islands
5	Gulf of Mannar	1989	Tamil Nadu
6	Manas	1989	Assam

No.	Biosphere Reserve	Year	State
7	Sunderbans	1989	West Bengal
8	Simlipal	1994	Orissa
9	Dibru-Saikhowa	1997	Assam
10	Dehang-Dibang	1998	Arunachal Pradesh
11	Pachmarhi	1999	Madhya Pradesh
12	Khangchendzonga	2000	Khangchendzonga hills and Sikkim
13	Agasthyamalai	2001	Kerala
14	Achanakamar -Amarkantak	2005	M.P. and Chhattishgarh
15	Kachchh	2008	Gujarat
16	Cold Desert	2009	Himachal Pradesh
17	Seshachalam Hills	2010	Andhra Pradesh
18	Panna	2011	Madhya Pradesh

ENVIS, WII 2020

National Park Facts | રાષ્ટ્રીય ઉદ્યાન તથ્યો

World's first National Park	Yellow Stone National Park, USA - 1872
India's first National Park	Jim Corbett National Park, Uttarakhand- 1936
World's largest National Park	Greenland's National Park- 9,72,000 sq km
India's largest National Park	Hemis National Park, Ladakh - 3,350 sq km
India's smallest National Park	Fossil National Park, MP - 0.27 sq km
India's State with highest National Parks	Madhya Pradesh (10)

As on Dec 2020

National Parks of India | ભારતના રાષ્ટ્રીય ઉદ્યાનો

No.	National Park	Year of establishment	Area (sq km)	District
Andaman & Nicobar Islands				
1	Campbell Bay NP	1992	426.23	Nicobar
2	Galathea Bay NP	1992	110.00	Nicobar
3	Mahatama Gandhi Marine (Wandoor) NP	1983	281.50	Andaman
4	Rani Jhansi Marine NP	1996	320.06	Andaman
5	Saddle Peak NP	1987	32.54	Andaman
6	Mount Harriett NP	1987	46.62	Andaman
Andhra Pradesh				
1	Papikonda NP	2008	1012.86	East & West Godavari
2	Rajiv Gandhi (Rameswaram) NP	2005	2.40	Kadapa
3	Sri Venkateswara NP	1989	353.62	Chittoor & Cuddapah

No.	National Park	Year of establishment	Area (sq km)	District
Arunachal Pradesh				
1	Mouling NP	1986	483.00	Upper Siang
2	Namdapha NP	1983	1807.82	Changlang
Assam				
1	Dibru-Saikhowa NP	1999	340.00	Tinsukia & Dibrugarh
2	Kaziranga NP	1974	858.98	Golaghat, Nagaon & Sonitpur
3	Manas NP	1990	500.00	Barpeta & Bongaigaon
4	Nameri NP	1998	200.00	Sonitpur
5	Rajiv Gandhi Orang NP	1999	78.81	Darrang & Sonitpur
Bihar				
1	Valmiki NP	1989	335.65	West Champaran
Chhattisgarh				
1	Guru Ghasidas (Sanjay) NP	1981	1440.71	Surguja & Koria
2	Indravati (Kutru) NP	1982	1258.37	Dantewada
3	Kanger Valley NP	1982	200.00	Bastar
Goa				
1	Mollem NP	1992	107.00	North Goa
Gujarat				
1	Vansda NP	1979	23.99	Navsari
2	Blackbuck (Velavadar) NP	1976	34.53	Bhavnagar
3	Gir NP	1975	258.70	Junagadh
4	Marine NP	1982	162.89	Jamnagar
Haryana				
1	Kalesar NP	2003	46.82	Yamuna Nagar
2	Sultanpur NP	1989	1.43	Gurgaon
Himachal Pradesh				
1	Great Himalayan NP	1984	754.40	Kullu
2	Inderkilla NP	2010	104.00	Kullu
3	Khirganga NP	2010	710.00	Kullu
4	Pin Valley NP	1987	675.00	Lahul & Spiti
5	Simbalbara NP	2010	27.88	Sirmaur
Jammu & Kashmir				
1	City Forest (Salim Ali) NP	1992	9.00	Srinagar

No.	National Park	Year of establishment	Area (sq km)	District
2	Dachigam NP	1981	141.00	Srinagar & Pulwama
3	Hemis NP	1981	3350.00	Leh
4	Kishtwar NP	1981	425.00	Kishtwar & Doda
Jharkhand				
1	Betla NP	1986	226.33	Latehar
Karnataka				
1	Anshi NP	1987	417.34	Uttara Kannada
2	Bandipur NP	1974	874.20	Mysore & Chamarajanagar
3	Bannerghatta NP	1974	260.51	Bangalore
4	Kudremukh NP	1987	600.32	Dakshin Kannada, Udipi & Chikmagalur
5	Nagarahole (Rajiv Gandhi) NP	1988	643.39	Kodagu & Mysore
Kerala				
1	Anamudi Shola NP	2003	7.50	Idukki
2	Eravikulam NP	1978	97.00	Idukki
3	Mathikettan Shola NP	2003	12.82	Idukki
4	Pambadum Shola NP	2003	1.32	Idukki
5	Periyar NP	1982	350.00	Idukki & Quilon
6	Silent Valley NP	1984	89.52	Palakkad
Madhya Pradesh				
1	Bandhavgarh NP	1968	448.85	Umaria & Katni
2	Fossil NP	1983	0.27	Mandla
6	Indira Priyadarshini Pench NP	1975	292.85	Seoni & Chhindwara
3	Kanha NP	1955	940.00	Mandla, Balaghat & Dindori
4	Madhav NP	1959	375.22	Shivpuri
5	Panna NP	1981	542.67	Panna & Chhatarpur
7	Sanjay NP	1981	466.88	Sidhi
8	Satpura NP	1981	585.17	Hoshangabad
9	Van Vihar NP	1979	4.45	Bhopal
Maharashtra				
1	Chandoli NP	2004	317.67	Sangli, Satara, Kolhapur, Ratnagiri
2	Gugamal NP	1975	361.28	Amravati

No.	National Park	Year of establishment	Area (sq km)	District
3	Nawegaon NP	1975	133.88	Bhandara (Gondia)
4	Pench (Jawaharlal Nehru) NP	1975	257.26	Nagpur
5	Sanjay Gandhi (Borivilli) NP	1983	86.96	Thane & Mumbai
6	Tadoba NP	1955	116.55	Chandrapur
Manipur				
1	Keibul-Lamjao NP	1977	40.00	Bishnupur
Meghalaya				
1	Balphakram NP	1985	220.00	South Garo Hills
2	Nokrek Ridge NP	1986	47.48	East Garo Hills
Mizoram				
1	Murlen NP	1991	100.00	Champhai
2	Phawngpui Blue Mountain NP	1992	50.00	Lawngtlai
Nagaland				
1	Intanki NP	1993	202.02	Dimapur
Odisha				
1	Bhitarkanika NP	1988	145.00	Kendrapara
2	Simlipal NP	1980	845.70	Mayurbhanj
Rajasthan				
1	Mukundra Hills NP	2006	200.54	Kota & Chittourgarh
2	Desert NP	1992	3162.00	Barmer & Jaisalmer
3	Keoladeo Ghana NP	1981	28.73	Bharatpur
4	Ranthambhore NP	1980	282.00	Sawai Madhopur
5	Sariska NP	1992	273.80	Alwar
Sikkim				
1	Khangchendzonga NP	1977	1784.00	North Sikkim
Tamil Nadu				
1	Guindy NP	1976	2.82	Chennai
2	Gulf of Mannar Marine NP	1980	6.23	Ramanathpuram & Tuticorin
3	Indira Gandhi (Annamalai) NP	1989	117.10	Coimbatore
4	Mudumalai NP	1990	103.23	Nilgiris
5	Mukurthi NP	1990	78.46	Nilgiris
Telangana				
1	Kasu Brahmananda Reddy NP	1994	1.43	Hyderabad

No.	National Park	Year of establishment	Area (sq km)	District
2	Mahaveer Harina Vanasthali NP	1994	14.59	Ranga Reddy
3	Mrugavani NP	1994	3.60	Ranga Reddy
Tripura				
1	Clouded Leopard NP	2007	5.08	West Tripura
2	Bison (Rajbari) NP	2007	31.63	South Tripura
Uttar Pradesh				
1	Dudhwa NP	1977	490.00	Lakhimpur-Kheri
Uttarakhand				
1	Corbett NP	1936	520.82	Nainital & Pauri Garhwal
2	Gangotri NP	1989	2390.02	Uttarkashi
3	Govind NP	1990	472.08	Uttarkashi
4	Nanda Devi NP	1982	624.60	Chamoli
5	Rajaji NP	1983	820.00	Dehradun, Pauri Garhwal & Haridwar
6	Valley of Flowers NP	1982	87.50	Chamoli
West Bengal				
1	Buxa NP	1992	117.10	Jalpaiguri
2	Gorumara NP	1992	79.45	Jalpaiguri
3	Jaldapara NP	2014	216.51	Jalpaiguri
4	Neora Valley NP	1986	159.89	Darjeeling
5	Singalila NP	1986	78.60	Darjeeling
6	Sunderban NP	1984	1330.10	North & South 24-Paraganas

ENVIS, WII 2020

State-wise break up of National Parks | રાજ્યો પ્રમાણે રાષ્ટ્રીય ઉદ્યાનોનો વિસ્તાર

State / UT	State Area (sq km)	No. of NPs	PA Area (sq km)	% of State Area
Andhra Pradesh	160229	3	1368.88	0.85
Arunachal Pradesh	83743	2	2,290.82	2.74
Assam	78438	5	1,977.79	2.52
Bihar	94163	1	335.65	0.36
Chhattisgarh	135191	3	2,899.08	2.14
Goa	3702	1	107.00	2.89
Gujarat	196022	4	480.12	0.24

State / UT	State Area (sq km)	No. of NPs	PA Area (sq km)	% of State Area
Haryana	44212	2	48.25	0.11
Himachal Pradesh	55673	5	2,271.38	4.08
Jharkhand	79714	1	226.33	0.28
Karnataka	191791	5	2,795.76	1.46
Kerala	38863	6	558.16	1.44
Madhya Pradesh	308245	10	3,657.26	1.19
Maharashtra	307713	6	1,273.60	0.41
Manipur	22327	1	40.00	0.18
Meghalaya	22429	2	267.48	1.19
Mizoram	21081	2	150.00	0.71
Nagaland	16579	1	202.02	1.22
Odisha	155707	2	990.70	0.64
Punjab	50362	0	0.00	0.00
Rajasthan	342239	5	3,947.07	1.15
Sikkim	7096	1	1,784.00	25.14
Tamil Nadu	130058	5	307.85	0.24
Telangana	114840	3	19.62	0.02
Tripura	10486	2	36.71	0.35
Uttar Pradesh	240928	1	490.00	0.20
Uttarakhand	53483	6	4,915.02	9.19
West Bengal	88752	6	1,981.65	2.23
Andaman & Nicobar	8249	6	1,216.95	14.75
Chandigarh	114	0	0.00	0.00
Dadra & Nagar Haveli	491	0	0.00	0.00
Daman & Diu	112	0	0.00	0.00
Delhi	1483	0	0.00	0.00
Jammu & Kashmir	163090	3	575.00	0.36
Ladakh	59146	1	3350.00	5.66
Lakshadweep	32	0	0.00	0.00
Puducherry	480	0	0.00	0.00
TOTAL	**32,87,263**	**101**	**40,564**	**1.23**

NWLD, WII 2020

Wildlife Sanctuaries | વન્યપ્રાણી અભ્યારણ્યો

India's first Wildlife Sanctuary	Periyar (1950), Kerala
India's largest Wildlife Sanctuary	Kachchh Desert Wildlife Sanctuary (7506.22 sq km), Gujarat
India's smallest Wildlife Sanctuary	Goose Island (0.01 sq km), A & N Rowe Island (0.01 sq km), A & N
State with highest Wildlife Sanctuaries	Maharashtra (48)
UT with highest Wildlife Sanctuaries	Andaman & Nicobar (96)

As on Dec 2020

Wildlife Sanctuaries of India | ભારતના વન્યપ્રાણી અભ્યારણ્યો

No.	WLS	Year of establishment	Area (Sq km)
Andaman & Nicobar Islands			
1.	Arial Island	1977	0.05
2.	Bamboo Island	1977	0.05
3.	Barren Island	1977	8.10
4.	Battimalv Island	1977	2.23
5.	Belle Island	1977	0.08
6.	Bennett Island	1977	3.46
7.	Bingham Island	1977	0.08
8.	Blister Island	1977	0.26
9.	Bluff Island	1977	1.14
10.	Bondoville Island	1977	2.55
11.	Brush Island	1977	0.23
12.	Buchanan Island	1977	9.33
13.	Chanel Island	1977	0.13
14.	Cinque Islands	1977	9.51
15.	Clyde Island	1977	0.54
16.	Cone Island	1977	0.65
17.	Curlew (B.P.) Island	1977	0.16
18.	Curlew Island	1977	0.03
19.	Cuthbert Bay	1997	5.82
20.	Defence Island	1977	10.49
21.	Dot Island	1977	0.13
22.	Dottrell Island	1977	0.13
23.	Duncan Island	1977	0.73
24.	East Island	1977	6.11
25.	East of Inglis Island	1977	3.55
26.	Egg Island	1977	0.05

No.	WLS	Year of establishment	Area (Sq km)
27.	Elat Island	1977	9.36
28.	Entrance Island	1977	0.96
29.	Galathea Bay	1997	11.44
30.	Gander Island	1977	0.05
31.	Girjan Island	1977	0.16
32.	Goose Island	1977	0.01
33.	Hump Island	1977	0.47
34.	Interview Island	1977	133.87
35.	James Island	1977	2.10
36.	Jungle Island	1977	0.52
37.	Kwangtung Island	1987	0.57
38.	Kyd Island	1977	8.00
39.	Landfall Island	1977	29.48
40.	Latouche Island	1977	0.96
41.	Lohabarrack	1977	22.21
42.	Mangrove Island	1977	0.39
43.	Mask Island	1977	0.78
44.	Mayo Island	1977	0.10
45.	Megapode Island	1977	0.12
46.	Montogemery Island	1977	0.21
47.	Narcondam Island	1977	6.81
48.	North Brother Island	1977	0.75
49.	North Island	1977	0.49
50.	North Reef Island	1977	3.48
51.	Oliver Island	1977	0.16
52.	Orchid Island	1977	0.10
53.	Ox Island	1977	0.13
54.	Oyster Island-I	1977	0.08
55.	Oyster Island-II	1977	0.21
56.	Paget Island	1977	7.36
57.	Parkinson Island	1977	0.34
58.	Passage Island	1977	0.62
59.	Patric Island	1977	0.13
60.	Peacock Island	1977	0.62
61.	Pitman Island	1977	1.37
62.	Point Island	1977	3.07
63.	Potanma Islands	1977	0.16
64.	Ranger Island	1977	4.26

No.	WLS	Year of establishment	Area (Sq km)
65.	Reef Island	1977	1.74
66.	Roper Island	1977	1.46
67.	Ross Island	1977	1.01
68.	Rowe Island	1977	0.01
69.	Sandy Island	1977	1.58
70.	Sea Serpent Island	1977	0.78
71.	Shark Island	1977	0.60
72.	Shearme Island	1977	7.85
73.	Sir Hugh Rose Island	1977	1.06
74.	Sisters Island	1977	0.36
75.	Snake Island-I	1977	0.73
76.	Snake Island-II	1977	0.03
77.	South Brother Island	1977	1.24
78.	South Reef Island	1977	1.17
79.	South Sentinel Island	1977	1.61
80.	Spike Island-I	1977	0.42
81.	Spike Island-II	1977	11.70
82.	Stoat Island	1977	0.44
83.	Surat Island	1977	0.31
84.	Swamp Island	1977	4.09
85.	Table (Delgarno) Island	1977	2.29
86.	Table (Excelsior) Island	1977	1.69
87.	Talabaicha Island	1977	3.21
88.	Temple Island	1977	1.04
89.	Tillongchang Island	1977	16.83
90.	Tree Island	1977	0.03
91.	Trilby Island	1977	0.96
92.	Tuft Island	1977	0.29
93.	Turtle Islands	1977	0.39
94.	West Island	1977	6.40
95.	Wharf Island	1977	0.11
96.	White Cliff Island	1977	0.47
Andhra Pradesh			
1.	Coringa	1978	235.7
2.	Gundla Brahmeswaram	1990	1194.00
3.	Kambalkonda	2002	71.90
4.	Kaundinya	1990	357.60
5.	Kolleru	1953	308.55

No.	WLS	Year of establishment	Area (Sq km)
6.	Krishna	1989	194.81
7.	Nagarjunasagar Srisailam	1978	1401.81
8.	Nellapattu	1976	4.59
9.	Pulicat Lake	1980	500.00
10.	Rollapadu	1988	6.14
11.	Sri Lankamalleswaram	1988	464.42
12.	Sri Penusila Narasimha	1997	1030.85
13.	Sri Venkateswara	1985	172.35
14.	D' Ering Memorial	1978	190.00
15.	Dibang	1991	4149.00
16.	Eagle Nest	1989	217.00
17.	Itanagar	1978	140.30
18.	Kamlang	1989	783.00
19.	Kane	1991	31.00
20.	Mehao	1980	281.5
21.	Pakhui	1977	861.95
22.	Sessa Orchid	1989	100.00
23.	Tale Valley	1995	337.00
24.	Yorde-Rabe-Supse	1996	397
Arunachal Pradesh			
1.	D' Ering Memorial	1978	190.00
2.	Dibang	1991	4149.00
3.	Eagle Nest	1989	217.00
4.	Itanagar	1978	140.30
5.	Kamlang	1989	783.00
6.	Kane	1991	31.00
7.	Mehao	1980	281.5
8.	Pakhui	1977	861.95
9.	Sessa Orchid	1989	100.00
10.	Tale Valley	1995	337.00
11.	Yorde-Rabe-Supse	1996	397
Assam			
1.	Amchang	2004	78.64
2.	Barail	2004	326.24
3.	Barnadi	1980	26.22
4.	Bherjan-Borajan-Padumoni	1999	7.22
5.	Burachapori	1995	44.06

No.	WLS	Year of establishment	Area (Sq km)
6.	Chakrasila	1994	45.57
7.	Diparbeel	1989	4.14
8.	Dihing Patkai	2004	111.19
9.	East Karbi Anglong	2000	221.81
10.	Garampani	1952	6.05
11.	Hollongapar Gibbon	1997	20.98
12.	Laokhowa	1972	70.11
13.	Marat Longri	2003	451.00
14.	Nambor	2000	37.00
15.	Nambor-Doigurung	2003	97.15
16.	Pabitora	1987	38.83
17.	Pani-Dihing	1995	33.93
18.	Sonai-Rupai	1998	220.00
Bihar			
1.	Barela Jheel Salim Ali	1997	1.96
2.	Bhimbandh	1976	681.99
3.	Gautam Budha	1976	138.34
4.	Kaimur	1978	1342.00
5.	Kanwarjheel	1989	63.11
6.	Kusheshwar Asthan	1994	29.17
7.	Nagi Dam	1987	1.92
8.	Nakti Dam	1987	3.33
9.	Rajgir	1978	35.84
10.	Udaypur	1978	8.87
11.	Valmiki	1978	545.15
12.	Barela Jheel Salim Ali	1997	1.96
13.	Bhimbandh	1976	681.99
Chandigarh			
1.	City Bird	1998	0.03
2.	Sukhna Lake	1986	25.98
Chhattisgarh			
1.	Achanakmar	1975	551.55
2.	Badalkhol	1975	104.45
3.	Barnawapara	1976	244.66
4.	Bhahramgarh	1983	138.95
5.	Bhoramdev	2001	351.24
6.	Gomardha	1975	277.82

No.	WLS	Year of establishment	Area (Sq km)
7.	Pamed	1985	262.12
8.	Semarsot	1978	430.35
9.	Sitanadi	1974	553.36
10.	Tamorpingla	1978	608.51
11.	Udanti	1983	237.27
Dadra & Nagar Haveli, Daman & Diu, Delhi			
1.	Dadra & Nagar Haveli	2000	92.16
2.	Fudam	1991	2.18
3.	Asola Bhati	1992	27.82
Goa			
1.	Mollem	1967	133.00
2.	Bondla	1969	7.95
3.	Chorao Island	1988	1.78
4.	Cotigao	1968	85.65
5.	Madei	1999	208.48
6.	Netravali	1999	211.05
Gujarat			
1.	Balaram Ambaji	1989	542.08
2.	Barda	1979	192.31
3.	Great Indian Bustard	1992	3.33
4.	Gir	1965	1153.42
5.	Girnar	2008	178.8
6.	Hingolgarh	1980	6.54
7.	Jambughoda	1990	130.38
8.	Jessore	1978	180.66
9.	Kachchh Desert	1986	7506.22
10.	Khijadia	1981	6.05
11.	Lala Great Indian Bustard	1995	2.03
12.	Gulf of Kachchh	1980	295.03
13.	Mitiyala	2004	18.22
14.	Nal Sarovar	1969	120.82
15.	Narayan Sarovar	1995	442.91
16.	Paniya	1989	39.63
17.	Porbandar Lake	1988	0.09
18.	Purna	1990	160.84
19.	Rampara Vidi	1988	15.01
20.	Ratanmahal	1982	55.65

No.	WLS	Year of establishment	Area (Sq km)
21.	Shoolpaneswar	1982	607.7
22.	Thol Lake	1988	6.99
23.	Wild Ass	1973	4953.71
Haryana			
1.	Abubshehar	1987	115.30
2.	Bhindawas	1986	4.12
3.	Bir Shikargarh	1987	7.67
4.	Chhilchila	1986	0.29
5.	Kalesar	1992	54.06
6.	Khaparwas	1991	0.83
7.	Morni Hills	2004	48.83
8.	Nahar	1987	2.11
Himachal Pradesh			
1.	Bandli	1962	32.11
2.	Chail	1976	16.00
3.	Chandratal	2007	38.56
4.	Churdhar	1985	55.52
5.	Daranghati	1962	171.5
6.	Dhauladhar	1994	982.86
7.	Gamgul Siahbehi	1962	108.40
8.	Kais	1954	12.61
9.	Kalatop-Khajjair	1958	17.17
10.	Kanawar	1954	107.29
11.	Khokhan	1954	14.94
12.	Kibber	1992	2220.12
13.	Kugti	1962	405.49
14.	Lippa Asrang	1962	31.00
15.	Majathal	1954	30.86
16.	Manali	1954	29.00
17.	Nargu	1962	132.3731
18.	Pong Dam Lake	1982	207.59
19.	Sangla Valley	1989	304.00
20.	Renukaji	1964	4.00
21.	Rupi Bhaba	1982	503.00
22.	Sainj	1994	90.00
23.	Sechu Tuan Nala	1962	390.29
24.	Shikari Devi	1962	29.94

No.	WLS	Year of establishment	Area (Sq km)
25.	Shimla Catchment	1958	10.00
26.	Talra	1962	46.48
27.	Tirthan	1992	61.00
28.	Tundah	1962	64.00
Jammu & Kashmir			
1.	Baltal-Thajwas	1987	210.50
2.	Gulmarg	1987	180.00
3.	Hirapora	1987	110.00
4.	Hokarsar Lake	1992	13.75
5.	Jasrota	1987	25.75
6.	Lachipora	1987	80.00
7.	Limber	1987	26.00
8.	Nandini	1981	33.34
9.	Overa-Aru	1981	425.00
10.	Rajparian	2002	20.00
11.	Ramnagar Rakha	1981	31.5
12.	Surinsar Mansar	1981	55.5
13.	Trikuta	1981	31.77
Jharkhand			
1.	Dalma	1976	193.22
2.	Gautam Budha	1986	121.14
3.	Hazaribagh	1976	186.25
4.	Koderma	1985	177.35
5.	Lawalong	1978	211.00
6.	Mahauaduar	1976	63.25
7.	Palamau	1976	752.94
8.	Palkot	1990	182.83
9.	Parasnath	1984	49.33
10.	Topchanchi	1978	12.82
11.	Udhwa Lake	1991	5.65
Karnataka			
1.	Adichunchunagiri	1981	0.84
2.	Arabithittu	1985	13.50
3.	Attiveri	1994	2.22
4.	Bhadra	1974	492.46
5.	Bhimgad	2010	190.42
6.	Biligiri Rangaswami Temp	1987	539.52

No.	WLS	Year of establishment	Area (Sq km)
7.	Brahmagiri	1974	181.29
8.	Cauvery	1987	1027.53
9.	Chincholi	2012	134.88
10.	Dandeli	1987	886.41
11.	Daroji Bear	1992	82.72
12.	Ghataprabha	1974	29.79
13.	Gudavi Bird Sanctuary	1989	0.73
14.	Gudekote Sloth Bear	2013	38.48
15.	Jogimatti	2015	100.48
16.	Kappathagudda	2015	244.15
17.	Malai Mahadeshwara	2013	906.19
18.	Melkote Temple	1974	49.82
19.	Mookambika	1974	370.37
20.	Nugu	1974	30.32
21.	Pushpagiri	1987	102.96
22.	Ramadevara Betta Vulture	2012	3.46
23.	Ranebennur	1974	119.00
24.	Ranganathittu	1940	0.67
25.	Rangayyanadurga	2012	77.24
26.	Sharavathi Valley LTM	—	930.16
27.	Shettihalli	1974	395.6
28.	Someshwara	1974	314.25
29.	Talakaveri	1987	105.01
30.	Thimlapura	2016	50.86
31.	Yadahalli Chinkara	2015	96.36
Kerala			
1.	Aralam	1984	55.00
2.	Chimmony	1984	85.00
3.	Chinnar	1984	90.44
4.	Chulannur	2007	3.42
5.	Idukki	1976	70.00
6.	Kottiyoor	2011	30.38
7.	Kurinjimala	2006	32.00
8.	Malabar	2010	74.22
9.	Mangalavanam Bird	2004	0.03
10.	Neyyar	1958	128.00
11.	Parambikulam	1973	285.00
12.	Peechi-Vazhani	1958	125.00

No.	WLS	Year of establishment	Area (Sq km)
13.	Peppara	1983	53.00
14.	Periyar	1950	427.00
15.	Shendurney	1984	100.32
16.	Thattekadu	1983	25.00
17.	Wayanad	1973	344.44
Ladakh & Lakshadweep			
1.	Wayanad	1973	344.44
2.	Changthang	1987	4000.00
3.	Karakoram	1987	5000.00
Madhya Pradesh			
1.	Ralamandal	1989	2.34
2.	Sailana	1983	12.96
3.	Veerangna Durgavati	1997	23.97
4.	Son Gharial	1981	41.8
5.	Orcha	1994	44.91
6.	Ken Gharial	1981	45.20
7.	Narsingarh	1978	59.19
8.	Panna (Gangau)	1979	68.14
9.	Phen	1983	110.74
10.	Pench	1975	118.47
11.	Kheoni	1982	122.7
12.	Karera	1981	202.21
13.	Panpatha Ws	1983	245.84
14.	Singhori	1976	287.91
15.	Palpur	1981	344.68
16.	Sardarpur	1983	348.12
17.	Sanjay Dubri	1975	364.59
18.	Gandhi Sagar	1981	368.62
19.	Pachmarhi	1977	417.78
20.	National Chambal	1979	435.00
21.	Bagdara	1978	478.00
22.	Bori	1977	485.72
23.	Ghatigaon	1981	511.00
24.	Ratapani	1978	823.84
25.	Noradehi	1984	1194.67

No.	WLS	Year of establishment	Area (Sq km)
Maharashtra			
1.	Amba Barwa	1997	127.11
2.	Andhari	1986	509.27
3.	Aner Dam	1986	82.94
4.	Bhamragarh	1997	104.38
5.	Bhimashankar	1985	130.78
6.	Bor	1970	61.10
7.	Chaprala	1986	134.78
8.	Deulgaon-Rehkuri	1980	2.17
9.	Dhyanganga	1997	205.23
10.	Gautala	1986	260.61
11.	Ghodazari	—	159.00
12.	Great Indian Bustard	1979	366.73
13.	Isapur	—	37.80
14.	Jaikwadi	1986	341.05
15.	Kalsubai	1986	361.71
16.	Karanja Sohal	2000	18.32
17.	Karnala	1968	4.48
18.	Katepurna	1988	73.63
19.	Koka	—	100.14
20.	Koyna	1985	423.35
21.	Lonar	2000	1.17
22.	Malvan Marine	1987	29.12
23.	Mansingdeo	2010	182.59
24.	Mayureshwar Supe	1997	5.15
25.	Melghat	1985	778.75
26.	Nagzira	1970	152.81
27.	Naigaon Peacock	1994	29.89
28.	Nandur-Madhameshwar	1986	100.12
29.	Narnala Bird	1997	12.35
30.	Nawegaon	2012	122.76
31.	New Bor	2012	60.70
32.	Gangewadi New Great Indian Bustard	1979	1.98
33.	New Nagzira	2012	151.33
34.	Painganga	1986	324.62
35.	Phansad	1986	69.79
36.	Pranhita	—	420.06
37.	Radhanagari	1958	351.16

No.	WLS	Year of establishment	Area (Sq km)
38.	Sagareshwar	1985	10.87
39.	Sudhagad	—	77.13
40.	Tamhini	—	49.62
41.	Tansa	1970	304.81
42.	Thane Creek Flamingo	2015	16.91
43.	Tipeshwar	1997	148.63
44.	Tungareshwar	2003	85.00
45.	Umred-Kharngla	2012	189.3
46.	Wan	1997	211.00
47.	Yawal	1997	177.52
48.	Yedsi Ramlin Ghat	1997	22.38
Manipur			
1.	Khongjaingamba Ching	2016	0.412
2.	Yangoupokpi-Lokchao	1989	184.4
Meghalaya			
1.	Baghmara Pitcher Plant	1984	.0.02
2.	Narpuh WLS	2014	59.9
3.	Nongkhyllem	1981	29.00
4.	Siju	1979	5.18
Mizoram			
1.	Buhvum	—	94.00
2.	Dampa	1985	500.00
3.	Khawnglung	1991	35.00
4.	Lengteng	1999	60.00
5.	Ngengpui	1992	110.00
6.	Pualreng	2004	50.00
7.	Tawi	1978	35.75
8.	Thorangtlang	2002	50
9.	Tokalo	2007	250.00
Nagaland			
1.	Fakim	1980	6.41
2.	Puliebadze	1980	9.23
3.	Rangapahar	1986	4.70
Odisha			
1.	Badarma	1962	304.03
2.	Baisipalli	1981	168.35
3.	Balukhand Konark	1984	71.72

No.	WLS	Year of establishment	Area (Sq km)
4.	Bhitarkanika	1975	525.00
5.	Chandaka	1982	175.79
6.	Chilka (Nalaban)	1987	15.53
7.	Debrigarh	1985	346.91
8.	Gahirmatha	1997	1435.00
9.	Hadgarh	1978	191.06
10.	Kapilash	2011	125.50
11.	Karlapat	1992	147.66
12.	Khalasuni	1982	116.00
13.	Kothagarh	1981	399.50
14.	Kuldiha	1984	272.75
15.	Lakhari Valley	1985	185.87
16.	Nandankanan	1979	14.16
17.	Satkosia-Gorge	1976	745.52
18.	Simlipal	1979	1354.30
19.	Sunabeda	1988	500.00
Puducherry			
1.	Oussudu	2008	3.90
Punjab			
1.	Abohar	1988	186.50
2.	Bir Aishvan	1952	2.64
3.	Bir Bhadson	1952	10.23
4.	Bir Bunerheri	1952	6.50
5.	Bir Dosanjh	1952	5.18
6.	Bir Gurdialpura	1977	6.20
7.	Bir Mehaswala	1952	1.23
8.	Bir Motibagh	1952	6.40
9.	Harike Lake	1982	86.00
10.	Jhajjar Bacholi	1998	1.16
11.	Kathlaur Kushlin	2007	7.58
12.	Nangal	2007	3.40
13.	Takhni-Rehampur	1992	3.82
14.	Abohar	1988	186.50
Rajasthan			
1.	Bandh Baratha	1985	199.50
2.	Bassi	1988	138.69
3.	Bhensrodgarh	1983	229.14

No.	WLS	Year of establishment	Area (Sq km)
4.	Darrah	1955	80.75
5.	Jaisamand	1955	52.00
6.	Jamwa Ramgarh	1982	300.00
7.	Jawahar Sagar	1975	153.41
8.	Kela Devi	1983	676.38
9.	Kesarbagh	1955	14.76
10.	Kumbhalgarh	1971	608.58
11.	Mount Abu	1960	326.10
12.	Nahargarh	1980	50.00
13.	National Chambal	1979	274.75
14.	Phulwari Ki nal	1983	692.68
15.	Ramgarh Vishdhari	1982	252.79
16.	Ramsagar	1955	34.40
17.	Sajjangarh	1987	5.19
18.	Sariska	1955	219.00
19.	Sawai Man Singh	1984	103.25
20.	Sawaimadhopur	1955	131.3
21.	Shergarh	1983	98.71
22.	Sitamata	1979	422.94
23.	Tal Chhapper	1971	7.19
24.	Todgarh Raoli	1983	495.27
25.	Van Vihar	1955	25.60
Sikkim			
1.	Barsey Rhododendron	1998	104.00
2.	Fambong Lho	1984	51.76
3.	Kitam	2005	6.00
4.	Kyongnosla Alpine	1977	31.00
5.	Maenam	1987	35.34
6.	Pangolakha	2002	128.00
7.	Shingba	1984	43.00
Tamil Nadu			
1.	Cauvery North	2014	504.33
2.	Chitrangudi	1989	0.48
3.	Gangaikondam Spotted Deer	2013	2.88
4.	Anamalai	1976	841.49
5.	Kalakad	1976	223.58
6.	Kanjirankulam	1989	1.04

No.	WLS	Year of establishment	Area (Sq km)
7.	Kanyakumari	2002	457.78
8.	Karaivetti	1999	4.54
9.	Karikili	1998	0.61
10.	Kodaikanal	2013	608.95
11.	Koonthankulam-Kadankulam	1994	1.29
12.	Megamalai	—	269.11
13.	Melasalvanoor-Keelaselvanoor	1998	5.93
14.	Mudumalai	1942	217.76
15.	Mundanthurai	1977	567.38
16.	Nellai	2015	356.73
17.	Oussudu Lake Bird	2015	3.32
18.	Point Calimere	1967	17.26
19.	Pulicat	1980	153.67
20.	Sakkarakottai Bird	2012	2.30
21.	Sathyamangalam	2008	1411.61
22.	Srivilliputhur	1988	485.20
23.	Theerthangal Bird	2010	0.29
24.	Udayamarthandapuram Lake	1991	0.45
25.	Vaduvoor	1991	1.28
26.	Vedanthangal	1936	0.30
27.	Vellanadu	1987	16.41
28.	Vellode	1997	0.77
29.	Vettangudi	1977	0.38
Telangana			
1.	Nagarjunasagar Srisailam	1978	2166.28
2.	Eturnagaram	1953	806.15
3.	Kawal	1965	892.23
4.	Kinnerasani	1977	635.41
5.	Lanja Madugu Sivaram	1978	29.81
6.	Manjeera Crocodile	1978	20.00
7.	Pakhal	1952	860.00
8.	Pocharam	1952	130.00
9.	Pranahita	1980	136.02
Tripura			
1.	Gumti	1988	389.54
2.	Roa	1988	0.858
3.	Sepahijala	1987	13.45
4.	Trishna	1988	163.08

No.	WLS	Year of establishment	Area (Sq km)
Uttar Pradesh			
1.	Bakhira	1982	28.94
2.	Nawabganj	1990	2.25
3.	Chandraprabha	1976	78.00
4.	Dr. Bhimrao Ambedkar Bird	2003	4.27
5.	Hastinapur	1972	2073.00
6.	Surha Tal	1989	34.32
7.	Kaimur	1982	500.73
8.	Katarniyaghat	1977	400.09
9.	Kishanpur	1979	227.00
10.	Lakh Bahosi	1979	80.24
11.	Mahavir Swami	1984	5.41
12.	National Chambal	1979	635.00
13.	Okhla	1990	4.00
14.	Parvati Aranga	1977	10.84
15.	Patna	1990	1.09
16.	Pilibhit	2014	602.798
17.	Ranipur	1987	230.31
18.	Saman	1987	5.26
19.	Samaspur	1987	7.99
20.	Sandi	1990	3.09
21.	Shekha Bird Sanctuary		0.25
22.	Sohagibarwa	1987	428.2
23.	Sohelwa	1988	452.47
24.	Sur Sarovar	1991	4.03
25.	Turtle	1990	7.00
26.	Vijai Sagar	1990	2.62
Uttarakhand			
1.	Askot Musk Deer	1986	600.00
2.	Binsar	1988	47.07
3.	Govind Pashu Vihar	1955	485.89
4.	Kedarnath	1972	975.20
5.	Mussoorie	1993	10.82
6.	Nandhaur	2012	269.96
7.	Sonanadi	1987	301.18

No.	WLS	Year of establishment	Area (Sq km)
West Bengal			
1.	Ballavpur	1977	2.02
2.	Bethuadahari	1980	0.67
3.	Bibhutibhusan	1980	0.64
4.	Buxa	1986	267.92
5.	Chapramari	1976	9.60
6.	Narendrapur	1982	0.07
7.	Haliday Island	1976	5.95
8.	Jorepokhri Salamander	1985	0.04
9.	Lothian Island	1976	38.00
10.	Mahananda	1976	158.04
11.	Pakhi Bitan Bird Sanctuary	2016	14.09
12.	Raiganj	1985	1.30
13.	Ramnabagan	1981	0.14
14.	Sajnakhali	1976	362.40
15.	Senchal	1976	38.88
16.	West Sunderban	2013	556.45

NWLD, WII 2020

State-wise break up of Wildlife Sanctuaries

રાજ્યો પ્રમાણે વન્યપ્રાણી અભ્યારણ્યોનો વિસ્તાર

State/ Union Territory	State Area (sq km)	No. of WLS	WLS Area (sq km)	% of State Area
Andhra Pradesh	160229	13	5942.23	3.71
Arunachal Pradesh	83743	11	7487.75	8.94
Assam	78438	18	1840.14	2.35
Bihar	94163	12	2901.67	3.08
Chhattisgarh	135191	11	3760.29	2.78
Goa	3702	6	647.91	17.50
Gujarat	196022	23	16618.42	8.48
Haryana	44212	8	233.21	0.53
Himachal Pradesh	55673	28	6116.10	10.99
Jharkhand	79714	11	1955.82	2.45
Karnataka	191791	31	7517.69	3.92
Kerala	38863	17	1928.24	4.96
Madhya Pradesh	308245	25	7158.42	2.32
Maharashtra	307713	48	7592.31	2.47
Manipur	22327	2	184.81	0.83

State/ Union Territory	State Area (sq km)	No. of WLS	WLS Area (sq km)	% of State Area
Meghalaya	22429	4	94.10	0.42
Mizoram	21081	9	1184.75	5.62
Nagaland	16579	3	20.34	0.12
Odisha	155707	19	7094.65	4.56
Punjab	50362	13	326.60	0.65
Rajasthan	342239	25	5592.38	1.63
Sikkim	7096	7	399.10	5.62
Tamil Nadu	130058	29	6157.12	4.73
Telangana	114840	9	5675.91	4.94
Tripura	10486	4	566.93	5.41
Uttar Pradesh	240928	26	5829.20	2.42
Uttarakhand	53483	7	2690.12	5.03
West Bengal	88752	16	1456.21	1.64
Andaman & Nicobar	8249	96	389.39	4.72
Chandigarh	114	2	26.01	22.82
Dadra & Nagar Haveli	491	1	92.16	18.77
Daman & Diu	112	1	2.19	1.96
Delhi	1483	1	27.82	1.88
Jammu & Kashmir	163090	13	1243.11	0.76
Ladakh	59146	2	9000.00	15.22
Lakshadweep	32	1	0.01	0.03
Puducherry	480	1	3.90	0.81
TOTAL	**3287263**	**553**	**119757**	**3.64**

NWLD, WII 2019

Conservation Reserves | કન્ઝર્વેશન રિઝર્વ

India's first Conservation Reserve	Asan Wetland, Uttrakhand (2005)
India's largest Conservation Reserve	Garpanchkot, West Bengal (1340.34 sq km)
India's smallest Conservation Reserve	Tiruppadaimarathur, Tamil Nadu (0.03 sq km)

As on Dec 2020

Conservation Reserves of India | ભારતના કન્ઝર્વેશન રિઝર્વ્સ

No.	Conservation Reserve	State	Area (sq km)	Year of Establishment
1	Ajas	Jammu & Kashmir	48.00	2008
2	Ajas (WL)	Jammu & Kashmir	1.00	2008
3	Anjneri	Maharashtra	5.69	2017
4	Ankasamudra Birds	Karnataka	0.9826	2017

No.	Conservation Reserve	State	Area (sq km)	Year of Establishment
5	Asan Wetland	Uttarakhand	4.44	2005
6	Bahu	Jammu & Kashmir	19.75	2008
7	Bankapur Peacock	Karnataka	0.56	2006
8	Basur Amruth Mahal Kaval	Karnataka	7.37	2011
9	Beas River	Punjab	N.A.	2017
10	Bedthi	Karnataka	57.30	2011
11	Bhorkada (Bhorgad)	Maharashtra	3.49	2008
12	Bir Bara Ban	Haryana	4.19	2007
13	Bir Jhunjhunu	Rajasthan	10.47	2012
14	Bisalpur	Rajasthan	48.31	2008
15	Boodh Karbu	Ladakh	12.00	2008
16	Brain-Nishat	Jammu & Kashmir	15.75	2008
17	Chatlam, Pampore (WL)	Jammu & Kashmir	0.25	2008
18	Chharidhandh Wetland	Gujarat	227.00	2008
19	Darlaghat	Himachal Pradesh	0.67	2013
20	Deul	West Bengal	10.50	2017
21	Garpanchkot	West Bengal	1340.34	2017
22	Gharana (WL)	Jammu & Kashmir	0.75	2008
23	Gogelo	Rajasthan	3.58	2012
24	Gudha Bisnoiyan	Rajasthan	2.31	2011
25	Hijli	West Bengal	15.50	2017
26	Hokera (Ramsar Site) (WL)	Jammu & Kashmir	13.75	2008
27	Hornbill	Karnataka	52.50	2011
28	Hygam (WL)	Jammu & Kashmir	7.25	2008
29	Jawahar Tunnel	Jammu & Kashmir	18.00	2008
30	Jawai Bandh Leopard	Rajasthan	19.79	2013
31	Jayamangali Blackbuck	Karnataka	3.23	2007
32	Jhilmi Jheel	Uttarakhand	37.84	2005
33	Jor Beed Gadwala Bikaner	Rajasthan	56.47	2008
34	Khanagund	Jammu & Kashmir	15.00	2008
35	Khetri Bansyal	Rajasthan	70.18	2019
36	Khimber/Dara/Sharazbal	Jammu & Kashmir	34.00	2008
37	Khiram	Jammu & Kashmir	15.75	2008
38	Khonmoh	Jammu & Kashmir	67.00	2008
39	Khrew Con R	Jammu & Kashmir	50.25	2008
40	Kolamarka	Maharashtra	180.72	2013
41	Kukarian (WL)	Jammu & Kashmir	24.25	2008
42	Magadi Kere	Karnataka	0.54	2015

No.	Conservation Reserve	State	Area (sq km)	Year of Establishment
43	Malgam (WL)	Jammu & Kashmir	4.50	2008
44	Mamdapur	Maharashtra	54.46	2014
45	Manibugh (WL)	Jammu & Kashmir	4.50	2008
46	Melapura Bee Eater Bird	Karnataka	0.0318	2015
47	Mirgund (WL)	Jammu & Kashmir	4.00	2008
48	Muktai Bhavani	Maharashtra	122.74	2014
49	Mukutmanipur	West Bengal	43.70	2017
50	Naganari	Jammu & Kashmir	22.25	2008
51	Naina Devi Himalayan Bird	Uttarakhand	111.92	2015
52	Nanga (WL)	Jammu & Kashmir	15.25	2008
53	Narkara (WL)	Jammu & Kashmir	3.25	2008
54	Norrichain (WL)	Ladakh	2.00	2008
55	Panyar	Jammu & Kashmir	10.00	2008
56	Pargwal (WL)	Jammu & Kashmir	49.25	2008
57	Pawalgarh	Uttarakhand	58.25	2012
58	Puttenahalli Lake Birds	Karnataka	0.15	2015
59	Rakh Sarai Amanat Khan	Punjab	4.95	2010
60	Ranjit Sagar	Punjab	18.65	2017
61	Ropar Wetland	Punjab	2.11	2017
62	Rotu	Rajasthan	0.73	2012
63	Sabu	Ladakh	15.00	2008
64	Sangral-Asa Chak (WL)	Jammu & Kashmir	7.00	2008
65	Saraswati	Haryana	44.53	2007
66	Shakambhari	Rajasthan	131.00	2012
67	Shallabugh (WL)	Jammu & Kashmir	16.00	2008
68	Shalmale Riparian Bio-system	Karnataka	4.89	2012
69	Shikargah	Jammu & Kashmir	15.50	2008
70	Shilli	Himachal Pradesh	1.49	1963
71	Naina Devi	Himachal Pradesh	17.01	1982
72	Sling Dong Fairreanum Orchid	Sikkim	0.06	2008
73	Suchindrum-Theroor-Managudi	Tamil Nadu	4.85	2015
74	Sudhmahadev	Jammu & Kashmir	142.25	2008
75	Sundha Mata	Rajasthan	117.49	2010
76	Tekonia	West Bengal	5.87	
77	Thein	Jammu & Kashmir	19.00	2008
78	Thimlapura	Karnataka	17.38	
79	Thungabhadra Otter	Karnataka	N.A.	2015
80	Tiruppadaimarathur	Tamil Nadu	0.03	2005

No.	Conservation Reserve	State	Area (sq km)	Year of Establishment
81	Toranmal	Maharashtra	93.42	
82	Tsomoiri (Ramsar Site) (WL)	Ladakh	120.00	2008
83	Umdeganj Bird	Rajasthan	2.72	2012
84	Ummathur	Karnataka	6.08	—
85	Wangat/Chatergul	Jammu & Kashmir	12.00	2008
86	Zaloora, Harwan	Jammu & Kashmir	25.25	2008

NWLD, WII 2020

State-wise break up of Conservation Reserves | રાજ્યો પ્રમાણે કન્ઝર્વેશન રિઝર્વ નો વિસ્તાર

State / UT	No. of Conservation Reserve	Area (sq km)	% of State Area
Gujarat	1	227.00	0.116
Haryana	2	48.72	0.110
Himachal Pradesh	3	19.17	0.034
Karnataka	13	151.02	0.079
Maharashtra	6	460.52	0.150
Punjab	4	25.71	0.051
Rajasthan	11	463.06	0.135
Sikkim	1	0.06	0.001
Tamil Nadu	2	4.88	0.004
Uttarakhand	4	212.45	0.397
West Bengal	5	1415.91	1.595
Jammu & Kashmir	30	680.75	0.417
Ladakh	4	149.00	0.252
TOTAL	**86**	**3858.25**	**0.117**

NWLD, WII 2020

Community Reserves | કોમ્યુનિટી રિઝર્વ

India's largest Community Reserve	Benreu (30 sq km) Longtoker (30 sq km) Wakchin Chingla (30 sq km)
India's smallest Community Reserve	Dangkipara (0.025 sq km)

As on Dec 2020

Community Reserves of India | ભારતના કોમ્યુનિટી રિઝર્વ્સ

No.	Community Reserves	State	Area (sq km)	Year of Establishment
1	Aimekhi (Japu)	Nagaland	2.05	—
2	Akumen	Nagaland	3.96	—
3	Aokangching (Pangsha Old)	Nagaland	3.50	—
4	Aruakgre	Meghalaya	1.00	2014
5	Atoizu	Nagaland	4.00	—
6	Baladingre	Meghalaya	0.50	2013
7	Balsri Adingi	Meghalaya	0.456	—
8	Bandarigre	Meghalaya	0.67	2013
9	Benreu	Nagaland	30.00	—
10	Bhumbak	Nagaland	6.50	—
11	Bonchu	Nagaland	9.05	—
12	Chandigre	Meghalaya	0.37	2013
13	Chemekong	Nagaland	29.175	—
14	Chenggni	Meghalaya	1.74	—
15	Chimanpara	Meghalaya	0.102	—
16	Chimitap	Meghalaya	22.8	—
17	Chingmelen	Nagaland	10.00	—
18	Chishilimi	Nagaland	3.50	—
19	D. Khel, Kohima Village	Nagaland	3.00	—
20	Dallenggittim	Meghalaya	0.2217	—
21	Dambuk Allonge	Meghalaya	36.6	—
22	Dambuk Jongkol	Meghalaya	2.648	—
23	Dangkipara	Meghalaya	0.025	2014
24	Daribokgre	Meghalaya	1.73	2013
25	Deukwaram	Nagaland	11.20	—
26	Dihoma	Nagaland	2.00	—
27	Dumitdikgre	Meghalaya	0.70	2013
28	Dura Kalkgre	Meghalaya	0.60	2013
29	Eman Asakgre	Meghalaya	0.30	2013
30	Ena Leu (Kingniu)	Nagaland	20.10	—
31	Ethungya Tongti Project (Vikhum)	Nagaland	23.00	—
32	Gariphema	Nagaland	2.65	—
33	Ghotovi	Nagaland	1.30	—
34	Gokagre	Meghalaya	0.1789	—
35	Hakchang	Nagaland	932	—

No.	Community Reserves	State	Area (sq km)	Year of Establishment
36	Halwa Ambeng	Meghalaya	0.941	—
37	Hekheshe	Nagaland	1.85	—
38	Hovishe Arkha (Yezami)	Nagaland	5.10	—
39	Hukphang	Nagaland	3.00	—
40	Jaksongram	Meghalaya	0.555	—
41	Jingru Luyong (Alisopur)	Nagaland	15.00	—
42	Jirang Com R	Meghalaya	2.00	2014
43	Jotsoma	Nagaland	5.00	—
44	Ka Khloo Langdoh Kur Pyrtuh	Meghalaya	0.154	2014
45	Ka Khloo Pohblai Mooshutia	Meghalaya	0.335	2014
46	Ka Khloo Thangbru Umsymphu	Meghalaya	0.196	2014
47	Ka Lum Luwe	Meghalaya	0.141	
48	Kadalundi Vallikkunnu	Kerala	1.50	2007
49	Kanjang	Nagaland	1.00	—
50	Kekhazong (Hurong)	Nagaland	4.50	—
51	Keshopur Chhamb	Punjab	3.40	2007
52	Kezoma	Nagaland	2.65	—
53	Khekiye	Nagaland	2.50	—
54	Khloo Amrawan	Meghalaya	1.29	2015
55	Khloo Blai Chyrmang Sein Raij Kongwasan Chyrmang Kwai	Meghalaya	0.07	
56	Khloo Blai Ka Raij U Landoh longlang	Meghalaya	0.15	2016
57	Khloo Blai Kongwasan Khloo Blai Chyrmang	Meghalaya	0.07	2014
58	Khloo Blai Sein Raij Tuber	Meghalaya	0.965	2014
59	Khonoma	Nagaland	2.65	—
60	Khrieyalienuomaiko	Nagaland	2.65	—
61	Khrokhropfu - Lephori	Nagaland	6.15	—
62	Khudei	Nagaland	4.80	—
63	Khutur	Nagaland	4.89	—
64	Khwuma Khel Jotsoma	Nagaland	3.00	—
65	Kidema	Nagaland	2.65	—
66	Kigwema	Nagaland	2.65	—
67	Kikruma	Nagaland	1.100	—
68	Kilo Old	Nagaland	2.00	—
69	Kitmadamgre	Meghalaya	0.70	2014
70	Kiyelho	Nagaland	3.00	—

No.	Community Reserves	State	Area (sq km)	Year of Establishment
71	Kokkare Bellur	Karnataka	3.12	2007
72	Kpoh Eijah Com R	Meghalaya	0.17	2014
73	Lalwan	Punjab	12.67	2007
74	Lawbah Com R	Meghalaya	2.10	2014
75	Liangmain (Wui)	Nagaland	25.00	—
76	Litem	Nagaland	1.60	—
77	Lizuto	Nagaland	2.50	—
78	Longra	Nagaland	2.275	—
79	Longtang	Nagaland	5.80	—
80	Longtoker	Nagaland	30.00	—
81	Lotnagar	Meghalaya	0.456	—
82	Lotovi	Nagaland	1.00	—
83	Lotsu	Nagaland	10.00	—
84	Lum Jusong Com R	Meghalaya	0.70	2014
85	Lumkohkriah Com R	Meghalaya	6.11	2014
86	Luzaphuhu	Nagaland	14.00	—
87	Mandalgre	Meghalaya	0.50	2013
88	Matchirampat	Meghalaya	0.217	—
89	Mezoma	Nagaland	2.85	—
90	Miewsyiar Com R	Meghalaya	0.87	2014
91	Mikadogre	Meghalaya	0.01	2013
92	Molungkimong	Nagaland	10.00	—
93	Mongalgre	Meghalaya	0.20	2014
94	Morakjo	Nagaland	6.50	—
95	Mpai Namci	Nagaland	20.00	—
96	Naltoqa (Yezami)	Nagaland	5.25	—
97	Namen Min (Khar)	Nagaland	15.00	—
98	Nerhema Perazatsa	Nagaland	20.00	—
99	Nerhema Yaoke	Nagaland	20.00	—
100	Nian	Nagaland	2.00	—
101	Nikwatgre	Meghalaya	0.495	—
102	Noksen	Nagaland	1.00	—
103	Nongsangu Com R	Meghalaya	1.00	2014
104	Nongumiang	Meghalaya	0.31	2003
105	Old Jalukie	Nagaland	3.70	—
106	Phudja—ud	Meghalaya	1.20	2014

No.	Community Reserves	State	Area (sq km)	Year of Establishment
107	Piphema "A"	Nagaland	1.00	
108	Piphema "B"	Nagaland	2.80	—
109	Pukhato	Nagaland	1.40	—
110	Raid Nongbri	Meghalaya	0.70	2014
111	Raid Nonglyngdoh/ Pdah Kyndeng	Meghalaya	0.75	2014
112	Rangkang	Nagaland	5.15	—
113	Resu Haluapra	Meghalaya	0.50	2014
114	Rewak Daburam	Meghalaya	0.183	—
115	Rewak Watregittim	Meghalaya	0.098	—
116	Rongalgre	Meghalaya	0.165	—
117	Rongcheng	Meghalaya	2.356	—
118	Rongma Paromgre	Meghalaya	0.62	2013
119	Rongma Rekmangre	Meghalaya	1.92	2013
120	Ryngibah	Meghalaya	0.80	2014
121	Ryngud	Meghalaya	5.22	2014
122	Sakalgre	Meghalaya	1.22	2013
123	Sakhabama	Nagaland	2.50	—
124	Sangdak	Nagaland	5.09	—
125	Sasatgre	Meghalaya	0.60	2013
126	Scaly—Mopungchuket	Nagaland	15.00	—
127	Sekai (Konya)	Nagaland	18.00	—
128	Selbalgre	Meghalaya	0.20	2013
129	Sele Aoyang Khanshu (Tuensang P. Khel)	Nagaland	4.90	—
130	Shitsuasulu (Akhoya)	Nagaland	10.00	—
131	Siju Duramong—I	Meghalaya	0.768	—
132	Siju Duramong—II	Meghalaya	25.1	—
133	Siswan	Punjab	12.95	2017
134	Sitap	Nagaland	1.50	—
135	Taidang	Meghalaya	1.214	—
136	Tamlu	Nagaland	2.00	—
137	Teyozwu Saku (P. Khel of Viswema)	Nagaland	5.00	—
138	Thamushui (Chingmei)	Nagaland	15.00	—
139	Thangkharang	Meghalaya	1.11	2014
140	Thokpara	Meghalaya	0.30	—
141	Tokugha	Nagaland	1.85	—
142	Tsecu (Piphewa Old)	Nagaland	5.00	—

No.	Community Reserves	State	Area (sq km)	Year of Establishment
143	Tsekhwelu	Nagaland	8.00	—
144	Tsiepama	Nagaland	3.325	—
145	Tsuruhu	Nagaland	2.70	—
146	Tuophema Village	Nagaland	2.50	—
147	Tutheze	Nagaland	1.62	—
148	Tutim (Kengjong)	Nagaland	10.00	—
149	Umsum Pitcher Plant	Meghalaya	0.40	2014
150	Upper Dosogre	Meghalaya	0.19691	—
151	Uza Zuzong (Langkok)	Nagaland	1.87	—
152	Viswema	Nagaland	2.65	—
153	Viyixe	Nagaland	2.65	—
154	Wakchin Chingla	Nagaland	30.00	—
155	Waromung	Nagaland	3.00	—
156	Yali	Nagaland	14.00	—
157	Yangpi	Nagaland	3.0007	—
158	Yaongyimchen	Nagaland	8.00	—
159	Yehemi	Nagaland	7.49	—
160	Yongshei	Nagaland	1.50	—
161	Yonyu	Nagaland	4.80	—
162	Zhanuolie (Chichewa)	Nagaland	0.51	—
163	Zurutongtchu Project Land Owner Committee	Nagaland	10.00	—

NWLD, WII 2020

State-wise break up of Community Reserves | રાજ્યો પ્રમાણે કોમ્યુનિટી રિઝર્વનો વિસ્તાર

Name of State & Union Territory	State Area (sq km)	No. of Community Reserve	Area (sq km)	% of State Area
Karnataka	1,91,791	1	3.12	0.002
Kerala	3,88,63	1	1.50	0.004
Meghalaya	22,429	65	136.97	0.610
Nagaland	16,579	93	662.73	3.997
Punjab	50,362	3	29.02	0.058
	TOTAL	**163**	**833.34**	**0.03**

NWLD, WII 2020

Marine Protected Areas | દરિયાઇ રક્ષિત વિસ્તારો

India's first Marine Protected Area	Point Calimere, Tamil Nadu (1967)
India's largest Marine Protected Area	Sajnakhali, West Bengal (1435 sq km)
India's smallest Marine Protected Area	Kadalundi Vallikkunnu Com R, Kerala (1.50 sq km)

As on Dec 2020

Marine Protected Areas in peninsular India | દ્વીપકલ્પિક ભારતમાં દરિયાઇ રક્ષિત વિસ્તારો

No.	Name of MPA	State	Category	Area	Year of establishment
1	Coringa	Andhra Pradesh	Sanctuary	235.7	1978
2	Krishna	Andhra Pradesh	Sanctuary	194.81	1989
3	Pulicat Lake	Andhra Pradesh	Sanctuary	500	1980
4	Dadra & Nagar Haveli	Dadra & Nagar Haveli	Sanctuary	92.16	2000
5	Fudam	Daman & Diu	Sanctuary	2.18	1991
6	Chorao Island	Goa	Sanctuary	1.78	1988
7	Marine (Gulf of Kachchh)	Gujarat	National Park	162.89	1995
8	Khijadia	Gujarat	Sanctuary	6.05	1981
9	Marine (Gulf of Kachchh)	Gujarat	Sanctuary	295.03	1980
10	Kadalundi Vallikkunnu Com R	Kerala	Community Reserve	1.50	2007
11	Malvan Marine	Maharashtra	Sanctuary	29.12	1987
12	Thane Creek Flamingo	Maharashtra	Sanctuary	16.905	2015
13	Bhitarkanika	Odisha	National Park	145	1998
14	Bhitarkanika	Odisha	Sanctuary	672	1975
15	Chilka (Nalaban)	Odisha	Sanctuary	15.53	1987
16	Gahirmatha	Odisha	Sanctuary	1435	1997
17	Balukhand Konark	Odisha	Sanctuary	71.72	1984
18	Gulf of Mannar Marine	Tamil Nadu	National Park	6.23	1980
19	Point Calimere	Tamil Nadu	Sanctuary	172.6	1967
20	Pulicat Lake	Tamil Nadu	Sanctuary	153.67	1980
21	Sundarbans	West Bengal	National Park	1330.1	1984
22	West Sundarbans	West Bengal	Sanctuary	556.45	2013
23	Haliday Island	West Bengal	Sanctuary	5.95	1976
24	Sajnakhali	West Bengal	Sanctuary	2091.12	1976
25	Lothian Island	West Bengal	Sanctuary	38	1976

excluding small island MPAs of Andaman and Nicobar Islands

ENVIS, WII 2020

Chapter 5

પ્રકરણ ૫

India's unique affinity with Forests and Environment

વન અને પર્યાવરણ સાથેના ભારતના અનન્ય સંબંધો

National Symbols of India

Category	Animal	Year	Remarks
National Emblem	Lion	1950	Lion capital of Sarnath was adopted as the national emblem of India It has four lions, a bull, galloping horse and the wheel.
National animal	Tiger	1973	The tiger was declared as national animal of India in 1973 in the same year project tiger was started. Before that, the Asiatic lion was the national animal.
National heritage animal	Asian Elephant	2010	—
National aquatic animal	Ganges river Dolphin	2009	—

ભારતનાં રાષ્ટ્રીય પ્રતીકો

પ્રકાર	પ્રાણી	વર્ષ	વિશેષ નોંધ
ભારતનું રાજ્ય ચિહ્ન	સિંહ	1950	સારનાથની સિંહ પ્રતિકૃતિ ભારતના રાષ્ટ્રીય પ્રતીક તરીકે અપનાવવામાં આવી હતી. તેમાં ચાર સિંહો, એક બળદ, ઘોડો અને ચક્ર છે.
રાષ્ટ્રીય પ્રાણી	વાઘ	1973	1973 માં વાઘને ભારતનો રાષ્ટ્રીય પ્રાણી તરીકે ઘોષિત કરવામાં આવ્યો હતો તે જ વર્ષમાં વાઘ પરિયોજના ની શરૂઆત કરવામાં આવી હતી. તે પહેલા એશિયાટિક સિંહ ભારતનું રાષ્ટ્રીય પ્રાણી હતું.
રાષ્ટ્રીય ધરોહર પ્રાણી	એશિયન હાથી	2010	—
રાષ્ટ્રીય જળચર પ્રાણી	ગંગા નદીની ડોલ્ફીન	2009	—

Historic events of Forests and Environment of India

1800	The first step in Indian Forestry began in West Coast. A commission was appointed to enquire the availability of teak in Malabar forests.
1805	A forest Committee was constituted to enquire into the capacity of forests and the status of proprietary rights in them. As a result of the Forest Committee's report, a proclamation was made declaring 'royalty' rights over teak trees in the South and prohibiting unauthorized felling of teak.
1806	The Government of Madras appointed Captain Watson of the Police as the first conservator of forests. It was basically to organize the production of teak and other timber supply from the West Coast suitable for the Navy for ship building.
1823	The post of Conservator of Forests was abolished.
1842	Conolly, Colllector of Malabar, initiated action that resulted in the world-famous first teak plantation in the country at Nilambur (Kerala). This was the first step towards the conservation forestry. This work was consolidated by the tireless effort of his assistant Sub-Conservator, Chathu Menon.
1847	Gibson was appointed as Conservator of Forests by Bombay Govt.
1855	The Charter of Indian Forests was promulgated on August 3 for the first time and Lord Dalhouise, the Governor General of India, established the Indian Forest Department.
1856	Cleghorn was appointed as the first regular conservator of forests in Madras Presidency.
1858	Exotic Acacia and Eucalyptus species were introduced and resulted in the large-scale plantations in Nilgiris.
1861	Publication of book Forest and Gardens of South India authored by Cleghorn.
1861	Govt. constituted the Central Provinces as a Chief commissionership. Temple was appointed as first Chief Commissioner and Captain person was appointed as Superintendent of Forests.
1864	Lord Canning appointed Sir Dietrich Brandis (a German forester) as first Inspector General of Forests to the Govt. of India, a post he held from 1864 to 1883. He is the founder of Scientific Forestry in India. There was a shift from the exploitative principles towards conservation and better management of forest resources (systematic forest management).
1865	First Indian Forest Act enacted.
1867	Imperial Forest Service was constituted.
1869	Reorganization of Forest department and Forest services and forest officers classified into Conservators, Deputy Conservators and Assistant Conservators.
1865-1870	Planting of Casuarina in the coastal areas of Madras for providing fuel needs of railways.
1871	The forest department was placed under the newly established department of revenue and agriculture, itself under the umbrella of the Home Department.
1874	Publication of book Forest Flora of the North West and Central India authored by Sir Dietrich Brandis.
1875	The Indian Forester was founded.
1878	Formation of Indian Forest school at Dehradun for training forest rangers of all the provinces.

1878	A revised Indian Forest Act came into existence. For the first time forests were classified into Reserve and Protected forests.
1871 to 1990	Compilation of working plans and forest settlements.
1881	Publication of book Manual of Indian timbers authored by James Sykes Gamble.
1882	Madras Forest Act was passed.
1885	Forest training commenced at Cooper's Hill in England.
1889-1896	Publication of book Manual of Forestry in five volumes authored by Sir Wilhelm Schlich.
1891	Having realized the importance of a multi-tier forest administration in the federal and provincial Govt. for effective management of forest resources the British Indian Govt. constituted Provincial Forest Service and Executive & Subordinate Services, which were quite similar to the present-day forest administrative hierarchy.
1894	First National Forest Policy.
1906	On June 5, Forest Research Institute (FRI) was established as Imperial Forest Research Institute and College with six research disciplines silviculture, forest working plans, forest zoology, forest botany, forest economics and forest chemistry.
1906	Publication of book Indian Trees authored by Sir Dietrich Brandis.
1906	Appointment of R. S. Troup as first Imperial Forest Economist.
1907	Appointment of Chief Conservator of Forests for each province.
1909	R. S. Troup was appointed as first regular Silviculturist and served from 1909 to 1919. He laid the foundation of systemic research in the field of silviculture.
1909	Royal Commission on Decentralization recommendations stated it was not considered necessary that Inspector General should be as adviser over every working plan and it was considered sufficient that chief conservator of forests will do the final scrutiny and check the programmes.
1910	Board of Forestry, comprising of representative chief conservators under the presidentship of the Inspector General of Forests was set up, which met once in three years to discuss the forestry programmes.
1912	Madras Forest College was started in Coimbatore for training forest rangers of South India.
1917	The appointment of Chief Conservators of Forests began.
1921	Publication of book Silviculture of Indian Trees in three volumes authored by R. S. Troup.
1922	40 % of the Imperial Forest Services was reserved for Indians.
1927	New Indian Forest Act enacted.
1935	The subject of "Forestry" which was managed by Federal Govt. until then, was transferred to the "Provincial List" by the Govt. of India Act, 1935. National Character of forest administration was considerably diluted and forests became a provincial subject.
1938	Indian Forest College (IFC) was established at Dehradun and the officers recruited to the Superior Forest Service by the province/states were trained there.
1950	Central Board of Forestry was constituted.
1952	New National Forest Policy.
1952	Indian Board of Wildlife was formed.

1966	The Indian Forest Service, one of the three All India Services, was constituted under the All India Services Act, 1951 by the Govt. of India.
1972	The Indian Wildlife (Protection) Act was enacted.
1973	Project Tiger, now renamed as the Nation Tiger Conservation Authority, was launched.
1980	Forest (Conservation) Act was enacted.
1981	Establishment of Forest Survey of India (FSI).
1982	Establishment of Wildlife Institute of India (WII).
1983	First National Wildlife Action Plan (NWAP) was adopted.
1985	Creation of Ministry of Environment and Forests (MoFF) as a separate ministry.
1986	Establishment of Indian Council of Forestry Research and Education (ICFRE).
1988	Current National Forest Policy.
1990	Launch of Joint Forest Management Programme.
1992	Project Elephant was launched.
1999	Formulation of National Forestry Action Programme (NFAP).
2002-2016	New Wildlife Action Plan has been adopted.
2003	National Forest Commission was set up.
2006	The Scheduled Tribes and other Traditional Forest Dwellers (Recognition of Forest Rights) Act was enacted.
2011	First Indian Forest Congress was organized.
2014	National Agroforestry Policy was launched.
2017-2031	India's third Wildlife Action Plan formulated and implemented.
2018	All India Tiger estimation conducted by MoEFCC. Estimated population of tigers in India is 2967 (2603-3346).
2020	Asian Elephant, Indian Bustard and Bengal Florican were declared as "Endangered Migratory Species" at the CMS COP 13.

ભારતમાં વન અને પર્યાવરણના ઐતિહાસિક ઘટનાક્રમ

1800	ભારતીય વનીકરણનું પ્રથમ પગલું પશ્ચિમ દરિયાકાંઠે શરૂ થયું. માલાબાર જંગલોમાં સાગની ઉપલબ્ધતાની તપાસ માટે એક કમિશનની નિમણૂક કરવામાં આવી હતી.
1805	જંગલોની ક્ષમતા અને તેમાં માલિકીના હકની સ્થિતિની તપાસ માટે વન સમિતિની રચના કરવામાં આવી હતી. વન સમિતિના અહેવાલને પરિણામે, દક્ષિણમાં સાગના વૃક્ષો ઉપર 'રોયલ્ટી' અધિકાર જાહેર કરવા અને સાગના અનધિકૃત કાપણી પર પ્રતિબંધ મૂકવાની જાહેરાત કરવામાં આવી
1806	મદ્રાસ સરકારે પોલીસના કપ્તાન વોટસનને પ્રથમ વન સંરક્ષક તરીકે નિયુક્ત કર્યા. તે મૂળભૂત રીતે જહાજ બનાવવા માટે નૌકાદળ માટે યોગ્ય વેસ્ટ કોસ્ટથી સાગ અને અન્ય લાકડાનો પુરવઠો ઉત્પાદનને ગોઠવવાનું હતું.
1823	વન સંરક્ષકનું પદ રદ કરાયું હતું.

1842	માલબારના કલેક્ટર કોનોલીના પ્રયત્નોને લીધે નિલમ્બુરના (કેરળ) ખાતે દેશમાં જગપ્રસિદ્ધ પ્રથમ સાગ વાવેતર પરિણમ્યું હતું. સંરક્ષણ વનીકરણ તરફનું આ પ્રથમ પગલું હતું. આ કાર્ય તેમના સહાયક સબ-કન્ઝર્વેટર, ચથુ મેનનના અથાક પ્રયત્નોથી એકીકૃત કરવામાં આવ્યું હતું.
1847	ગિબ્સનને બોમ્બે સરકાર દ્વારા વન સંરક્ષક તરીકે નિયુક્ત કરવામાં આવ્યા હતા.
1855	ભારતીય જંગલોના સનદની પહેલી વાર 3 ઓગસ્ટના રોજ જાહેરાત કરવામાં આવી અને ભારતના ગવર્નર જનરલ લોર્ડ ડેલહાઉસીએ ભારતીય વન વિભાગની સ્થાપના કરી.
1856	ક્લેગહોર્નને મદ્રાસ પ્રેસિડેન્સીમાં પ્રથમ વન સંરક્ષક તરીકે નિયુક્ત કરવામાં આવ્યા હતા
1858	બહારની પ્રજાતિઓ એકેસિયા અને નીલગિરી દાખલ કરવામાં આવી હતી અને પરિણામે નીલગિરિસમાં મોટા પાયે વાવેતર થયું.
1861	ક્લેગહોર્ન દ્વારા રચિત 'દક્ષિણ ભારતનાં વન અને બગીચા' પુસ્તકનું પ્રકાશન
1861	સરકારે ચીફ કમિશનરશીપ તરીકે કેન્દ્રીય પ્રાંતની રચના કરી. ટેમ્પલ ની પ્રથમ ચીફ કમિશનર તરીકે નિમણૂક કરવામાં આવી હતી અને કેપ્ટન પરસનને વન અધિક્ષક તરીકે નિયુક્ત કરવામાં આવ્યા હતા.
1864	લોર્ડ કેનિંગે સર ડાયટ્રિચ બ્રાંડિસ (એક જર્મન ફોરેસ્ટર) ની સરકાર માટે વનના પ્રથમ મહાનિરીક્ષક તરીકે નિમણૂક કરી. ભારતનું આ પદ તેમણે 1864 થી 1883 દરમિયાન શોભાવ્યું હતું. તે ભારતમાં વૈજ્ઞાનિક વનીકરણના સ્થાપક છે. વન સંસાધનો (વ્યવસ્થિત વન વ્યવસ્થાપન) ના સંરક્ષણ અને વધુ સારી વ્યવસ્થાપન તરફના શોષણકારી સિદ્ધાંતોમાંથી ફેરફાર થયો.
1865	પ્રથમ ભારતીય વન અધિનિયમ ઘડ્યો.
1867	વન સેવાની રચના કરવામાં આવી હતી.
1869	વન વિભાગ અને વન સેવાઓ અને વન અધિકારીઓનું પુનર્ગઠન સંરક્ષકો, નાયબ સંરક્ષકો અને સહાયક સંરક્ષકોમાં વર્ગીકૃત કરવામાં આવ્યું છે.
1865-1870	રેલવેની બળતણ જરૂરિયાતો પૂરી પાડવા માટે મદ્રાસના દરિયાકાંઠાના વિસ્તારોમાં શરૂનું વાવેતર કરવામાં આવ્યું.
1871	વન વિભાગને મહેસૂલ અને કૃષિ વિભાગના નવા સ્થાપિત વિભાગ હેઠળ મૂકવામાં આવ્યો હતો, તે ગૃહ વિભાગ હેઠળ હતો.
1874	સર ડાયેટ્રીક બ્રાન્ડિસ દ્વારા રચિત ઉત્તર પશ્ચિમ અને મધ્ય ભારતના ફોરેસ્ટ ફ્લોરા પુસ્તકનું પ્રકાશન.
1875	ઇન્ડિયન ફોરેસ્ટરની સ્થાપના કરવામાં આવી હતી.
1878	બધા પ્રાંતના વન રેન્જરોને તાલીમ આપવા માટે દહેરાદૂન ખાતે ભારતીય વન શાળાની રચના.
1878	એક સુધારેલો ભારતીય વન અધિનિયમ અસ્તિત્વમાં આવ્યો. પ્રથમ વખત જંગલોને રિઝર્વ અને પ્રોટેક્ટેડ જંગલોમાં વર્ગીકૃત કરવામાં આવી હતી.
1871-1990	કાર્યકારી યોજનાઓ અને વન વસાહતોનું સંકલન.

1881	જેમ્સ સાઇકસ ગેમ્બલ દ્વારા રચિત મેન્યુઅલ ઇન્ડિયન ટિમ્બર્સ પુસ્તકનું પ્રકાશન
1882	મદ્રાસ ફોરેસ્ટ એક્ટ પસાર થયો
1885	ઇંગ્લેન્ડના કૂપર હિલ ખાતે વન તાલીમ શરૂ થઈ
1889-1896	સર વિલ્હેમ શ્લિચ દ્વારા રચિત પાંચ ભાગમાં મેન્યુઅલ ઓફ ફોરેસ્ટ્રી પુસ્તકનું પ્રકાશન
1891	સંઘીય અને પ્રાંતીય સરકારમાં મલ્ટી-ટાયર ફોરેસ્ટ એડમિનિસ્ટ્રેશનનું મહત્ત્વ સમજ્યા પછી વન સંસાધનોના અસરકારક સંચાલન માટે બ્રિટીશ ભારતીય સરકાર પ્રાંતીય વન સેવા અને કારોબારી અને ગૌણ સેવાઓ રચાયેલ જે આજકાલના વન વહીવટી વંશવેલો સમાન હતા.
1894	પ્રથમ રાષ્ટ્રીય વન નીતિ.
1906	5 જૂન ના રોજ, વન સંશોધન સંસ્થા (FRI) સ્થાપના કરી હતી. વન સંશોધન સંસ્થા અને કોલેજ છ સંશોધન શાખાઓમાં વન સંવર્ધન, વન કાર્યકારી યોજનાઓ, વન પ્રાણીશાસ્ત્ર, વન વનસ્પતિશાસ્ત્ર, વન અર્થશાસ્ત્ર અને જંગલ રસાયણશાસ્ત્ર સાથે શરૂ થઈ હતી.
1906	સર ડાયેટ્રીક બ્રાન્ડિસ દ્વારા રચિત ઇન્ડિયન ટ્રીઝ પુસ્તકનું પ્રકાશન.
1906	પ્રથમ શાહી વન અર્થશાસ્ત્રી તરીકે આર. એસ. ટ્રુપની નિમણૂક.
1907	દરેક પ્રાંત માટે મુખ્ય વન સંરક્ષકની નિમણૂક.
1909	આર. એસ. ટ્રુપની પ્રથમ નિયમિત સિલ્વીકલ્ચરિસ્ટ તરીકે નિમણૂક કરવામાં આવી હતી અને 1909 થી 1919 દરમિયાન સેવા આપી હતી. તેમણે સિલ્વીકલ્ચર ક્ષેત્રે પ્રણાલીગત સંશોધનનો પાયો નાખ્યો.
1909	વિકેન્દ્રિયકરણની ભલામણો અંગેના રોયલ કમિશને જણાવ્યું હતું કે દરેક કાર્યકારી યોજનામાં ઇન્સ્પેક્ટર જનરલ સલાહકાર હોવા જોઈએ તે જરૂરી માનવામાં આવતું ન હતું અને તે પૂરતું માનવામાં આવતું હતું કે જંગલોના મુખ્ય સંરક્ષક અંતિમ ચકાસણી કરશે અને કાર્યક્રમોની તપાસ કરશે.
1910	વન મહાનિરીક્ષકની અધ્યક્ષતા હેઠળ પ્રતિનિધિ મુખ્ય સંરક્ષકોની બનેલી વન મંડળની સ્થાપના કરવામાં આવી હતી, જે વનીકરણ કાર્યક્રમો અંગે ચર્ચા કરવા ત્રણ વર્ષમાં એકવાર મળી હતી.
1912	દક્ષિણ ભારતના વન રેન્જરોને તાલીમ આપવા માટે કોઈમ્બતુરમાં મદ્રાસ ફોરેસ્ટ કોલેજ શરૂ કરવામાં આવી હતી.
1917	મુખ્ય વન સંરક્ષકોની નિમણૂક શરૂ થઈ.
1921	આર. એસ. ટ્રુપ દ્વારા રચિત ત્રણ ભાગમાં સિલ્વીકલ્ચર ઓફ ઇન્ડિયન ટ્રીઝ પુસ્તકનું પ્રકાશન.
1922	શાહી વન સેવાઓનો 40% હિસ્સો ભારતીયો માટે અનામત હતો.
1927	નવો ભારતીય વન અધિનિયમ ઘડ્યો.
1935	ભારતીય અધિનિયમ, 1935 હેઠળ "વનવિદ્યા" નો વિષય જેનું સંચાલન ફેડરલ સરકાર દ્વારા કરવામાં આવ્યું હતું સરકાર દ્વારા "પ્રાંતીય સૂચિ" માં સ્થાળાંતરિત કરવામાં આવ્યું હતું. વન વહીવટનું રાષ્ટ્રીય કદ નોંધપાત્ર રીતે ઓછું થયું હતું અને 'જંગલો' પ્રાંતીય વિષય બન્યો હતો.

1938	ભારતીય વન કોલેજ (IFC) ની સ્થાપના દહેરાદૂન ખાતે કરવામાં આવી હતી અને પ્રાંત / રાજ્યો દ્વારા સુપિરિયર ફોરેસ્ટ સર્વિસમાં અધિકારીઓને ત્યાં તાલીમ આપવામાં આવી હતી.
1950	સેન્ટ્રલ ફોરેસ્ટ્રી બોર્ડની રચના કરવામાં આવી હતી.
1952	નવી રાષ્ટ્રીય વન નીતિ.
1952	ભારતીય વન્યપ્રાણી મંડળ (IBWL) ની રચના કરવામાં આવી હતી.
1966	ભારતીય વન સેવા, ત્રણ અખિલ ભારતીય સેવાઓમાંથી એક, સરકાર દ્વારા ઓલ ઈન્ડિયા સર્વિસિસ એક્ટ, 1951 હેઠળ બનાવવામાં આવી હતી.
1972	ભારતીય વન્યજીવ (સંરક્ષણ) અધિનિયમ ઘડાયો.
1973	પ્રોજેક્ટ ટાઇગર, જે હવે નેશન ટાઇગર કન્સર્વેઝન ઓથોરિટી તરીકે નામ આપવામાં આવ્યું છે, શરૂ કરવામાં આવ્યું હતું.
1980	વન (સંરક્ષણ) અધિનિયમ ઘડાયો.
1981	ભારતીય વન સર્વેક્ષણ (FSI) ની સ્થાપના .
1982	ભારતીય વન્યજીવ સંસ્થા (WII) ની સ્થાપના
1983	પ્રથમ રાષ્ટ્રીય વન્યજીવ એક્શન પ્લાન (NWAP) અપનાવવામાં આવ્યો.
1985	એક અલગ મંત્રાલય તરીકે પર્યાવરણ અને વન મંત્રાલય (MoEF) ની રચના.
1986	ભારતીય વન સંશોધન અને શિક્ષણ પરિષદ (ICFRE)ની સ્થાપના .
1988	વર્તમાન રાષ્ટ્રીય વન નીતિ.
1990	સહભાગી વન વ્યવસ્થાપન (JFM) કાર્યક્રમનો પ્રારંભ.
1992	પ્રોજેક્ટ હાથીની શરૂઆત કરવામાં આવી હતી
1999	રાષ્ટ્રીય વનીકરણ એક્શન પ્લાન (NFAP) ની રચના.
2002-2016	નવો રાષ્ટ્રીય વન્યજીવ એક્શન પ્લાન અપનાવવામાં આવ્યો.
2003	રાષ્ટ્રીય વન પંચની સ્થાપના કરવામાં આવી હતી.
2006	અનુસૂચિત જનજાતિ અને અન્ય પરંપરાગત વન નિવાસીઓ (વન અધિકારની માન્યતા) અધિનિયમ ઘડાયો.
2011	પ્રથમ ભારતીય વન કોંગ્રેસનું આયોજન કરવામાં આવ્યું હતું.
2014	રાષ્ટ્રીય એગ્રોફોરેસ્ટ્રી નીતિ શરૂ કરવામાં આવી હતી.
2017-2031	ભારતનો ત્રીજો વાઇલ્ડલાઇફ એક્શન પ્લાન ઘડાયો અને અમલમાં આવ્યો.
2018	MoEFCC દ્વારા હાથ ધરવામાં આવેલ ઓલ ઇન્ડિયા ટાઇગર એસ્ટિમેશન. ભારતમાં વાઘની અંદાજિત વસ્તી 2967 (2603-3346) છે.
2020	CMP COP 13 માં એશિયન એલિફન્ટ, ઇન્ડિયન બસ્ટાર્ડ અને બંગાળ ફ્લોરિકનને "જોખમમાં મૂકાયેલ સ્થળાંતર પ્રજાતિઓ" તરીકે જાહેર કરવામાં આવ્યા.

Eminent Environmentalists of India

Almrita Patel	Environmentalist - She has been actively associated with causes such as protecting the Gir Lions, saving Ulsoor Lake, solid waste management, and building low cost homes.
Anil Prakash Joshi	Environmental Activist and Botanist impacted over 10,000 villages in Himalayan region through his Himalayan Environmental Studies & Conservation Organization (HESCO). The Govt. Of India awarded him Padma Shri in 2006 and Padma Bhushan in 2020.
Anup Sah	Anup Sah is a nature photographer and author from the Himalayan state, Uttarakhand.
Anupam Mishra	India's Water Guru who had dedicated his life efforts to the conservation of water. Anupam Mishra was a Gandhian environmentalist and the recipient of the 1996 Indira Gandhi Paryavaran Puraskar.
Bano Haralu	Journalist and now environment conservationist whose 'Friends of the Amur Falcon' campaign is widely accepted and impacted.
Chandi Prasad Bhatt	Padma Bhusan award in 2005, Chandi Bhatt is in fact one of India's first modern environmentalist. He is known for his work on subaltern social ecology and is also the recipient of the Gandhi Peace Prize.
Chewang Motup Goba	Founder of Rimo Expedition. He is Mountaineer, Entrepreneur and Climate Change Activist in Ladakh. Padma Shri Awardee.
Claude Alvares & Norma Alvares	Famous Alvares environmentalist couple in India. Norma Alvares 2002 Padma Shri awardee has initiated numerous environmental conservation projects
Himmata Ram Bhambhu	Farmer and environmentalist who has developed a unique integrated agro-forestry plantation and biodiversity conservation centre and he has created green patch in desert area of Nagaur district which has become a unique rehabilitation center for peacocks. Padma Shri Awardee.
Indira Gandhi	Formulated the Wildlife Protection Act. Also, she was an active crusader of the Save the Tiger campaign.
Jadav Payeng	Known as 'Forest man of India'. From Assam and created 550 hectare long man-made forest all by himself.
Jamuna Tudu	Jamuna is nicknamed Lady Tarzan for her courage and ferocity. She is fighting against the timber mafia to protect forests in Jarkhand.
John C. Jacob	John C. Jacob was amongst the pioneers of the environmental movement in the state of Kerala. He started the Society for Environment Education, Kerala (SEEK). He played a pivotal role in the movement against the Silent Valley project in South Kerala.
Kallen Pokkudan	Founder of the Mangrove School that attempts to educate people about the ecological importance of mangroves.
Kalyan Singh Rawat	Environmentalist and Founder President of Himalayan Vanya Jeev Sansthan – famous for initiating Maiti movement. Padma Shri Awardee.
Kinkri Devi	Activist and environmentalist, best known for waging a war on illegal mining and quarrying in her native state of Himachal Pradesh.

Kushal Konwar Sarma	He devoted his life to the conservation of Asian Elephants. Popularly known as Hati Doctor. Dr. Sarma is a member of the steering Committee, Project Elephant of the MoEFCC. Padma Shri Awardee.
M. C. Mehta	He is India's famous environmental lawyer.
M. S. Swaminathan	Known as 'Father of green revolution'.
Madhav Gadgil	A well-known ecologist in India.
Maneka Gandhi	Animal right activist.
Marimuthu Yoganathan	known as 'Tree man of India' Planted more than 3 lakh saplings in 30 years.
Medha Patkar	Famous for Narmada Bachao Andolan
Mike Pandey	A film maker specializing in films about wildlife and the environment. Pandey is the first Asian producer / director to win the Wildscreen Panda Award, also known as the Green Oscar.
Pawar Popatrao Bhaguji	Model Sarpanch of Hiware Bazaar village in Ahmedanagar district who transformed it from a drought-prone village into a green role model of Gram Swaraj.
Pradip Krishen	Worked on jungle and urban tree.
Purnima Barman	Whitley awardee, which is known as the Green Oscars, Barman was conferred the award in recognition of her efforts towards the conservation of the Greater Adjutant Stork and its habitat. Her work with the local people has led to the formulation of the 'Hargila Army'.
Rahibai Soma Popere	An Indian farmer and conservationist. Padma Shri Awardee.
Rajendra Singh	Known as 'Watermen of India' and aorked on efficient water management and harvesting.
Robin Banerjee	Founder of the Kaziranga Wildlife Society and active regarding issues concerning the national park.
Romulus Whitaker	Herpetologist. Established Madrash Crocodile Bank. Padma Shre Awardee.
Saalumarada Thimmakka	Known as 'Mother of trees. Received Padma Shri for planting more than 8,000 trees from her own expenditure.
Salim Ali	Known as 'Bird man of India'. Created Bharatpur Bird sanctuary and protected Silent Valley National Park.
Soumitra Abdulali	Worked on sand mining and environmental hazards.
Sugatha Kumari	Government of India awarded Padma Shri for her efforts in environmental conservation and afforestation.
Sundaram Verma	50,000 trees in the arid Shekawati region of Rajasthan using a water-saving technique – dryland agroforestry. Padma Shri Awardee.
Sundarlal Bahuguna	Known for 'Chipko movement'. Green peace worker.
Tulsi Gowda	Environmentalist from the state of Karnataka, she has planted 100,000 saplings and looks after them.
Usha Chaumar	Work as manual scavenger, and the forefront of Environmental Sanitation. Padma Shri Awardee.

Vandana Shiva	Founder of Navdanya, a movement for biodiversity conservation and farmers' rights and also the founder and director of the Research Foundation for Science, Technology and Natural Resource Policy.

ભારતના જાણીતા પર્યાવરણવિદો

અલમ્રિતા પટેલ	પર્યાવરણવિદ્. તે ગીર સિંહોનું રક્ષણ, અલ્સોર તળાવને બચાવવા, ઘન કચરાનું વ્યવસ્થાપન અને ઓછા ખર્ચે મકાનો બનાવવા જેવી પ્રવૃત્તિઓ સાથે સક્રિયપણે સંકળાયેલી છે.
અનિલ પ્રકાશ જોશી	પર્યાવરણીય કાર્યકર અને વનસ્પતિશાસ્ત્રીએ તેની હિમાલયન પર્યાવરણીય અધ્યયન અને સંરક્ષણ સંગઠન (HESCO) દ્વારા હિમાલય ક્ષેત્રના 10,000 થી વધુ ગામોને અસર કરી. સરકાર ભારત તરફથી તેમને 2006 માં પદ્મ શ્રી અને 2020 માં પદ્મ ભૂષણથી નવાજવામાં આવ્યા હતા.
અનુપ સાહ	હિમાલય રાજ્ય, ઉત્તરાખંડના એક પ્રકૃતિ ફોટોગ્રાફર અને લેખક છે.
અનુપમ મિશ્રા	ભારતના પાણી ગુરુ. પાણી સંરક્ષણ માટે પોતાના જીવનને અર્પણ કર્યું છે. અનુપમ મિશ્રા ગાંધીવાદી પર્યાવરણવિદને 1996 માં ઇન્દિરા ગાંધી પર્યાવરણ પુરસ્કાર મળ્યો હતો.
બાનો હરાલુ	પત્રકાર અને હવે પર્યાવરણ સંરક્ષણકાર જેમનો "ફ્રેન્ડ્સ ઓફ અમુર ફાલ્કન" અભિયાન બહોળી પ્રસિદ્ધિ અને સ્વીકૃતિ પામેલ છે.
ચંદી પ્રસાદ ભટ્ટ	2005 માં પદ્મ ભૂષણ એવોર્ડ, ચંદી ભટ્ટ હકીકતમાં ભારતના પહેલા આધુનિક પર્યાવરણવિદ છે. તેઓ સબલટર્ન સામાજિક ઇકોલોજી પરના તેમના કાર્ય માટે જાણીતા છે અને ગાંધી શાંતિ પુરસ્કાર મેળવનાર પણ છે.
ચેવાંગ મોટોપ ગોબા	રિમો અભિયાનના સ્થાપક. તે લદ્દાખમાં પર્વતારોહક, ઉદ્યોગસાહસિક અને ક્લાયમેટ ચેન્જ એક્ટિવિસ્ટ છે. પદ્મ શ્રી એવોર્ડ વિજેતા.
ક્લાઉડ અલ્વેરેસ અને નોર્મા અલ્વેરેસ	ભારતમાં પ્રખ્યાત એલ્વેર્સ પર્યાવરણવાદી દંપતી. નોર્મા અલ્વેરેસ 2002 પદ્મ શ્રી એવોર્ડ ના લીધે અસંખ્ય પર્યાવરણીય સંરક્ષણ પ્રોજેક્ટ શરૂ થયા છે
હિંમતા રામ ભંભુ	ખેડૂત અને પર્યાવરણવાદી કે જેમણે એક અનોખી સંકલિત કૃષિ-વાનિકી વાવેતર અને જૈવવિવિધતા સંરક્ષણ કેન્દ્ર વિકસાવ્યું છે અને તેમણે નાગૌર જિલ્લાના રણ વિસ્તારમાં લીલો રંગ બનાવ્યો છે જે મોર માટે એક અનન્ય પુનર્વસન કેન્દ્ર બની ગયું છે. પદ્મ શ્રી એવોર્ડ વિજેતા.
ઇન્દિરા ગાંધી	વન્યપ્રાણી સંરક્ષણ કાયદો ઘડયો. તે વાઘ બચાવો અભિયાનની સક્રિય ક્રૂસેડર હતી.
જાદવ પાયેંગ	'ભારતના ફોરેસ્ટ મેન' તરીકે જાણીતા. આસામના અને જાતે જ 550 હેક્ટર વિસ્તારનું માનવસર્જિત વન બનાવ્યું,
જમુના તુડુ	જમુનાને તેની હિંમત માટે 'લેડી ટારઝન' ઉપનામ આપવામાં આવ્યું છે. તે ઝારખંડમાં જંગલોની સુરક્ષા માટે ટીમ્બર માફિયા સામે લડી રહી છે.
જ્હોન સી જેકબ	કેરળ રાજ્યમાં પર્યાવરણીય ચળવળના પ્રણેતાઓમાં સામેલ હતા. તેમણે સોસાયટી ફોર એન્વાયર્નમેન્ટ એજ્યુકેશન, કેરળ (SEEK) ની શરૂઆત કરી.

	દક્ષિણ કેરળમાં સાયલેન્ટ વેલી પ્રોજેક્ટ વિરુદ્ધના આંદોલનમાં તેમણે મુખ્ય ભૂમિકા ભજવી હતી.
ક્લેન પોક્કુદાન	મેંગ્રોવ સ્કૂલના સ્થાપક કે જે લોકોને મેંગ્રોવના ઇકોલોજીકલ મહત્વ વિશે જાગૃત કરવા પ્રયાસ કરે છે.
કલ્યાણસિંહ રાવત	પર્યાવરણવાદી અને હિમાલયન વન્યજીવ સંસ્થાના સ્થાપક પ્રમુખ. મૈતી આંદોલન શરૂ કરવા માટે પ્રખ્યાત. પદ્મ શ્રી એવોર્ડ વિજેતા.
કિંકરી દેવી	કાર્યકર અને પર્યાવરણવાદી, હિમાચલ પ્રદેશમાં ગેરકાયદેસર ખાણકામ વિરુદ્ધ ચળવળ માટે જાણીતા છે.
કુશાલ કોનવર શર્મા	એશિયન હાથીઓના સંરક્ષણ માટે તેમનું જીવન સમર્પિત કર્યું છે. હાથી ડોક્ટર તરીકે લોકપ્રિય. ડો શર્મા પ્રોજેક્ટ પર્યાવરણ અને વન મંત્રાલય અંતર્ગત એલિફન્ટ સ્ટીઅરીંગ સમિતિના સભ્ય છે. પદ્મ શ્રી એવોર્ડ વિજેતા.
એમ.સી.મહેતા	તે ભારતના પ્રખ્યાત પર્યાવરણ ક્ષેત્રના વકીલ છે.
એમ.એસ.સ્વામિનાથન	'ગ્રીન ક્રાંતિના પિતા' તરીકે જાણીતા
માધવ ગાડગીલ	ભારતના એક જાણીતા ઇકોલોજીસ્ટ.
મેનકા ગાંધી	પ્રાણી અધિકાર કાર્યકર.
મરીમુથુ યોગનાથન	'ટ્રી મેન ઓફ ઈન્ડિયા' તરીકે ઓળખાય છે. 30 વર્ષમાં 3 લાખથી વધુ રોપાઓ ઉછેર્યા.
મેધા પાટકર	નર્મદા બચાવો આંદોલન માટે પ્રખ્યાત.
માઇક પાંડે	વન્યજીવન અને પર્યાવરણ વિશેની ફિલ્મોમાં વિશેષતા આપતો એક ફિલ્મ નિર્માતા. તેઓ 'વાઇલ્ડસ્ક્રીન પાંડા એવોર્ડ' જીતનાર પ્રથમ એશિયન નિર્માતા/ દિગ્દર્શક છે, જે ગ્રીન ઓસ્કાર તરીકે પણ ઓળખાય છે.
પવાર પોપટરાવ ભગુજી	અહમદનગર જિલ્લાના હાયવેર બજાર ગામના મોડેલ સરપંચ જેમણે તેને દુષ્કાળગ્રસ્ત ગામથી ગ્રામ સ્વરાજના રોલ મોડેલમાં પરિવર્તિત કર્યું.
પ્રદિપ ક્રિશેન	જંગલ અને શહેરી વૃક્ષ પર કામ કર્યું.
પૂર્ણિમા બર્મન	ગ્રેટર એડજ્યુટન્ટ સ્ટોર્ક અને તેના રહેઠાણના સંરક્ષણ તરફના પ્રયત્નોને લીધે ગ્રીન ઓસ્કાર તરીકે જાણીતા 'વ્હિટલી એવોર્ડ' થી સન્માનિત. સ્થાનિક લોકો સાથેના તેમના કાર્યને લીધે 'હરગીલા આર્મી' બનાવવામાં આવી છે.
રાહીબાઈ સોમા પોપરે	ભારતીય ખેડૂત અને સંરક્ષણવાદી. પદ્મ શ્રી એવોર્ડ વિજેતા.
રાજેન્દ્ર સિંહ	'ભારતના વોટરમેન' તરીકે ઓળખાય છે અને કાર્યક્ષમ પાણી વ્યવસ્થાપન પર કામ કર્યું છે.
રોબિન બેનર્જી	કાઝીરંગા વાઇલ્ડલાઇફ સોસાયટીના સ્થાપક અને રાષ્ટ્રીય ઉદ્યાનને લગતા મુદ્દાઓ વિશે ખાસ સક્રિય હતા.
રોમ્યુલસ વ્હાઇટેકર	હર્પેટોલોજિસ્ટ. મદ્રાસ મગર બેંકની સ્થાપના કરી. પદ્મ શ્રી એવોર્ડ વિજેતા.
સાલુમરાદા થિમ્મક્કા	ઝાડની માતા તરીકે ઓળખાય છે. પોતાના ખર્ચથી 8,000 થી વધુ વૃક્ષો વાવવા બદલ ભારત સરકર દ્વારા પદ્મ શ્રી એનાયત કરવામાં આવેલ છે.

સાલીમ અલી	'બર્ડમેન ઓફ ઇન્ડિયા' તરીકે ઓળખાય છે. ભરતપુર પક્ષી અભયારણ્ય બનાવ્યું અને સાયલેન્ટ વેલી રાષ્ટ્રીય ઉદ્યાનનું રક્ષણ કર્યું.
સૌમિત્રા અબ્દુલઅલી	રેતી ખનન અને પર્યાવરણીય જોખમો પર કામ કર્યું છે.
સુગાથ કુમારી	પર્યાવરણીય સંરક્ષણ અને વનીકરણમાં પ્રયત્નો માટે ભારત સરકાર દ્વારા પદ્મ શ્રી એનાયત કરવામાં આવેલ છે.
સુંદરમ વર્મા	પાણી બચાવવાની તકનીકનો ઉપયોગ કરીને રાજસ્થાનના શુષ્ક શેકાવાટી ક્ષેત્રમાં 50,000 વૃક્ષો - ડ્રાયલેન્ડ એગ્રોફોરેસ્ટ્રી તકનીક દ્વારા વાવેતર કર્યું. પદ્મ શ્રી એવોર્ડ વિજેતા.
સુંદરલાલ બહુગુણા	'ચિપકો આંદોલન' માટે જાણીતા છે. પર્યાવરણ માટેના શાંતિ કાર્યકર.
તુલસી ગૌડા	કર્ણાટક રાજ્યના પર્યાવરણવિદ, તેમણે 100,000 રોપાઓ વાવ્યા છે અને તેમની સંભાળ રાખે છે.
ઉષા ચૌમાર	પર્યાવરણીય સ્વચ્છતા માટે મેન્યુઅલ સ્કેવેંજર તરીકે કામ કરે છે. પદ્મ શ્રીએવોર્ડ વિજેતા.
વંદના શિવા	નવદાન્યની સ્થાપના કરી છે, જૈવવિવિધતા સંરક્ષણ અને ખેડુતોના હકો માટેના આંદોલન અને વિજ્ઞાન, ટેકનોલોજી અને પ્રાકૃતિક સંસાધન નીતિના સંશોધન ફાઉન્ડેશનના સ્થાપક અને નિયામક.

India's First

Year	First	Remarks
1806	First conservator of forest	Watson
1854	First zoo in India	Calcutta zoo, West Bengal
1856	First regular conservator of forest	Cleghorn
1864	First Inspector general of forest	Dietrich Brandis
1865	First national forest Act	Indian forest Act
1867	First special Assistant Conservator of Forest	Ribben Trop and W.Schlich
1875	First issue of Indian forester	Powel and Schlich
1878	First director of the forest school, Dehradun	F.Bailey
1894	First forest policy	Indian forest policy
1895	First working plan of India	Nilambur Forest Working Plan
1898	First Wildlife Sanctuary of India	Vedamthngal Wildlife Sanctuary, Madras Precidency
1901	First forest Entomologist	Stebbing
1906	First Chief Conservator of Forest	Ruther
1906	First president of Forest Research Institute and College, Dehradun	S. Eardly Wilmot
1909	First Silviculturist of India	Troup
1936	First National Park	Hailey National Park
1949	First Indian Inspector General of Forest	M.D. Chaturvedi

Year	First	Remarks
1973	First Forest Movement	Chipko movement by Sunderlal Bahuguna
1986	First Biosphere Reserve in India	Nilgiri Biosphere Reserve
2011	First Indian Forest Congress organized	In New Delhi
2014	First National Agroforest Policy	MoEFCC, Govt of Inida

ભારતનું પ્રથમ

વર્ષ	પ્રથમ	વિશેષ નોંધ
1806	પ્રથમ વન સંરક્ષક	વોટસન
1854	ભારતમાં પ્રથમ પ્રાણી સંગ્રહાલય	કલકત્તા ઝૂ, પશ્ચિમ બંગાળ
1856	વનનો પ્રથમ નિયમિત સંરક્ષક	ક્લગોર્ન
1864	વનના પ્રથમ મહાનિરીક્ષક	ડાયટ્રિચ બ્રાંડિસ
1865	પ્રથમ રાષ્ટ્રીય વન અધિનિયમ	ભારતીય વન અધિનિયમ
1867	વનના પ્રથમ વિશેષ સહાયક સંરક્ષક	રિબબેન ટ્રોપ અને ડબલ્યુ. સ્ક્લિચ
1875	ઇન્ડિયન ફોરેસ્ટર પ્રથમ અંક	પોવેલ અને શ્લિચ
1878	દહેરાદૂન ફોરેસ્ટ સ્કૂલના પ્રથમ ડિરેક્ટર	એફ.બેલે
1894	પ્રથમ વન નીતિ	ભારતીય વન નીતિ
1895	ભારતની પ્રથમ કાર્યકારી યોજના	નિલામ્બુર વન કાર્યકારી યોજના
1898	ભારતનું પ્રથમ વન્યપ્રાણી અભયારણ્ય	વેદમથંગલ વન્યપ્રાણી અભયારણ્ય, મદ્રાસ પૂર્વશાળા
1901	પ્રથમ વન એન્ટોમોલોજિસ્ટ	સ્ટેબિંગ
1906	પ્રથમ વન સંરક્ષક	રૂથર
1906	વન સંશોધન સંસ્થા અને કોલેજના પ્રથમ પ્રમુખ, દહેરાદૂન	એસ. એર્ડલી વિલ્મોટ
1909	ભારતના પ્રથમ સિલ્વીકલ્ચરિસ્ટ	ટ્રોઉપ
1936	પ્રથમ રાષ્ટ્રીય ઉદ્યાન	હેલી નેશનલ પાર્ક
1949	પ્રથમ ભારતીય વન મહાનિરીક્ષક	એમ. ડી. ચતુર્વેદી
1973	પ્રથમ વન ચળવળ	સુંદરલાલ બહુગુણા દ્વારા ચિપકો આંદોલન
1986	ભારતમાં પ્રથમ બાયોસ્ફિયર રિઝર્વ	નીલગિરી બાયોસ્ફિયર રિઝર્વ
2011	પ્રથમ ભારતીય વન કોંગ્રેસનું આયોજન	નવી દિલ્હીમાં
2014	પ્રથમ રાષ્ટ્રીય એગ્રોફોરેસ્ટ નીતિ	પર્યાવરણ, વન અને ક્લાઈમેટ ચેન્જ મંત્રાલય, ભારત સરકાર

India's ... | ભારતનું ...

Longest life span tree	*Cedrus deodara* (704 years) and *Dalbergia latifolia* (600 years)
Biggest tree by girth	*Cedrus deodara* (1646 cm) and *Shorea robusta* (1782 cm)
Largest fruit bearing tree	*Artocarpus heterophyllus* (Jackfruit)
Most valuable root parasite	*Santalum album* (Chandan)
Tallest grass	*Bambusa arundinacea*
Biggest leaf	*Tectona grandia* and *Dillenia indica*

World's... | વિશ્વનું...

Tallest tree	*Seqoia sempervirens* (Red Wood tree)
Oldest living fossil tree	*Gingo biolba* (Maiden Hair tree)
Longest life span	*Seqoia sempervirens* (4000 years) in California
Largest tree crown	*Ficus bengalensis* (Banyan tree)
Oldest living tree	*Pinus lengaeva* (Bristle cone pine)
Largest fruit bearing	*Artocarpus heterophyllus* (Jackfruit)
Largest seed	Double coconut
Largest leaf	*Raphia regalis*
Largest palm	*Raffia* palm
Largest family	Asteraceae (Sunflower family)
Fastest growing plants	*Macrocystis pyrifera* (18 inch/day)
Smallest growing plants	Wolffia species
Largest woody plants genera	Euphorbia (species- 1600)

National Forest Institutes | ભારતની વન સંસ્થાઓ

Institute	Name in full	Location	State
AFRI	Arid Forest Research Institute	Jodhpur	Rajasthan
BIOTRIM	Biotechnological Centre for tree Improvement	Trupathy	Andhra Pradesh
BSI	Botanical Survey of India	Kolkata	West Bengal
CAZRI	Central Arid Zone Research Institute	Jodhpur	Rajasthan
CEMDE	Centre for Environmental Management of Degraded Ecosystem	New Delhi	Delhi
CERT	Centre for Ecological Research and Training	Bengaluru	Karnataka
CES	Centre for Ecological Sciences	Bengaluru	Karnataka
CME	Centre of Mining Environment	Dhanbad	Jharkhand
CPCB	Central Pollution Control Board	New Delhi	Delhi
CSSRI	Central Soil Salinity Institute	Karnal	Haryana
CZA	Central Zoo Authority	Delhi	Delhi
DFE	Directorate of forest education	Dehradun	Uttarakhand

Institute	Name in full	Location	State
FRC	Forest Research Centre	Hyderabad	Andhra Pradesh
FRI	Forest Research Institute	Dehradun	Uttarakhand
GSI	Geological Survey of India	Kolkata	West Bengal
HFRI	Himalayan Forest Research Institute	Shimla	Himachal Pradesh
ICFRE	Indian Council of Forest Research and Education	Dehradun	Uttarakhand
IFGTB	Institute of Forest Genetics and Tree Breeding	Coimbatore	Tamil Nadu
IFP	Institute of Forest Productivity	Ranchi	Jharkhand
IGNFA	Indira Gandhi National Forest Academy	Dehradun	Uttarakhand
IIB	Indian Institute of Biodiversity	Itanagar	Arunachal Pradesh
IIFM	Indian Institute of Forest Management	Bhopal	Madhya Pradesh
IISS	Indian Institute of Soil Science	Bhopal	Madhya Pradesh
IPIRITI	Indian Plywood Industries Research and Training Centre	Bengaluru	Karnataka
IWST	Institute of Wood Science and technology	Bengaluru	Karanataka
NBSS & LUP	National Bureau of Soil Survey and Land Use Planing	Nagpur	Maharashtra
NBWL	National Board of Wildlife	New Delhi	Delhi
NEERI	National Environmental Engineering Research Institute	Nagpur	Maharshtra
NRCAF	National Research Centre for Agroforestry	Jhansi	Uttar Pradesh
NRCMAP	National Research Centre for Medicinal & Aromatic Plants	Anand	Gujarat
NRCO	National Research Centre for Oil palm	Eluru	Andhra Pradesh
NRCO	National Research Centre for Orchids	Gangtok	Sikkim
NRSA	National Remote Sensing Agency	Hyderabad	Andhra Pradesh
NZP	National Zoological Park	New Delhi	Delhi
RFRI	Rain Forest Research Institute	Jorhat	Assam
SACON	Salim All Centre for Ornithology and Natural History	Coimbatore	Tamil Nadu
SFRC	State Forest Service College	Dehradun	Uttarakhand
TBGRI	Tropical Botanic Garden and Research Institute	Thiruvananthapuram	Kerala
TFRI	Tropical Forest Research Institute	Jabalpur	Madhya Pradesh
WII	Wildlife Institute of India	Dehradun	Uttarakhand
ZSI	Zoological survey of India	Kolkata	West Bengal

Important National Environment NGOs
મહત્વપૂર્ણ રાષ્ટ્રીય પર્યાવરણ બિનસરકારી સંસ્થાઓ

NGO	HQ	Establishment Year	Head
Ashoka Trust for Research in Ecology and Environment (ATREE)	New Delhi	1996	Nitin Pandit
Bombay Natural History Society (BNHS)	Mumbai	1883	Bivash Pandav
Centre for Environment Education (CEE)	Ahmedabad	1984	Karikeya Sarabhai
Centre For Environmental Studies (CES)	Bhubaneswar	2000	Mona Sharma
Centre for Science and Environment (CSE)	New Delhi	1980	Sunita Narain
Kalpavriksh	Pune	1979	Ashish Kothari
M S Swaminathan Research Foundation (MSSRF)	Chennai	1988	Madhura Swaminathan
Nature Conservation Foundation (NCF)	Mysuru	1996	Charudutt Mishra
Satpuda Foundation	Amravati - Maharashtra.	2012	Prakash Muralidhar Laddha
The Energy and Resource Institute (TERI)	New Delhi	1974	Ajay Mathur
Trade Record Analysis of Flora and Fauna In Commerce (TRAFFIC)	New Delhi	1976	Shekhar Niraj
Wildlife Protection Society of India (WPSI)	New Delhi	1994	Belinda Wright
Wildlife Trust of India (WTI)	Noida	1998	Vivek Menon
World Wide Fund for Nature (WWF-India)	New Delhi	1969	Ravi Singh
Zoo Outreach Organization (ZOO)	Coimbatore	1985	Sanjay Molur

ENVIS 2020

Chapter 6

પ્રકરણ ૬

Ecology and Biodiversity
પરિસ્થિતિ વિજ્ઞાન અને જૈવવિવિધતા

Zoogeographic regions of the world | વિશ્વના ઝૂજીઓગ્રાફિક વિસ્તારો

Realm	Area
Palaearctic	North Asia, whole of Europe, Mediterranean Africa
Nearctic	North America above tropic, Greenland, Maxican Plateau
Ethiopian	Africa south of Sahara, Madagascar & South of Arabi
Oriental	India, Ceylon, Burma, Indochina, Siam, Malaya, Sumatra, Java, Philippine, Celiba & South Island
Australian	Australia, New Zealand & Tasmania
Neotropical	South America, Central America, West Indies, South Maxico

Wallace 1876

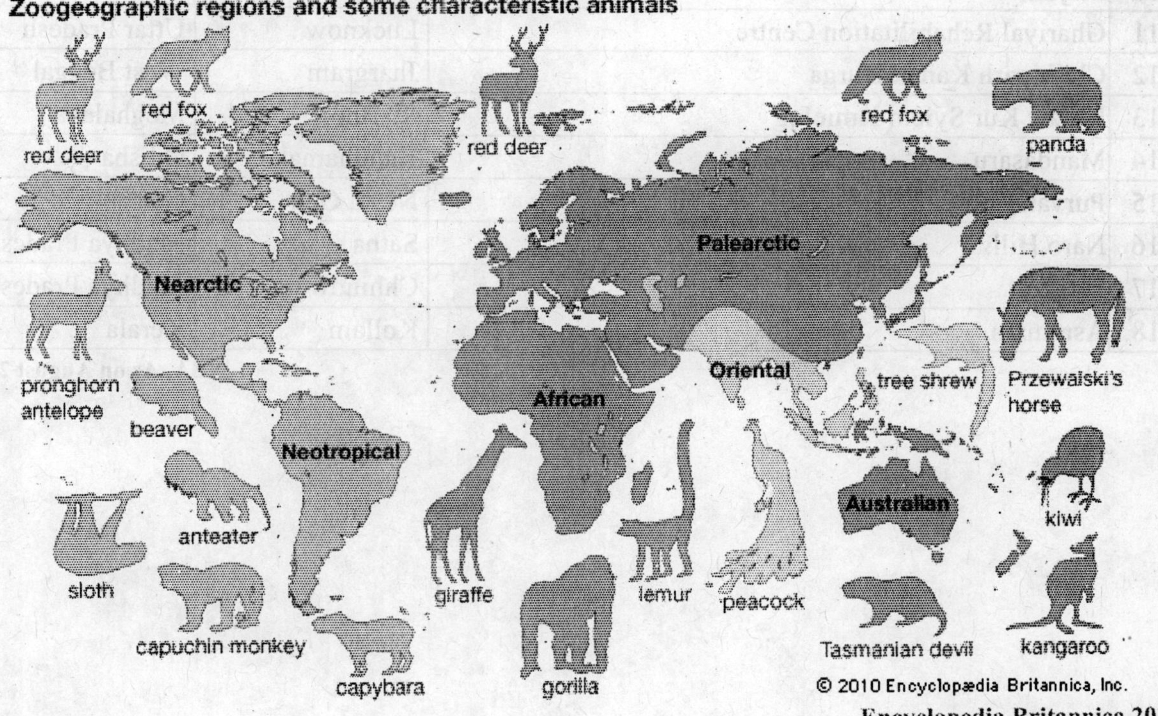

Encyclopedia Britannica 2010

Biodiversity Heritage Sites (BHS) of India

ભારતની જૈવવિવિધતા હેરિટેજ સાઇટ્સ

First BHS	Nallur Tamarind Grove, Karnataka
Year	2007
Total BHS	18
Governed by	National Biodiversity Authority (NBA), Chennai
Largest	Maujuli, Assam
Smallest	Glory of Allapalli, Maharashtra

No.	Name of the Site	District	State
1	Nallur Tamarind Grove	Bengaluru	Karnataka
2	Hogrekan	Chikmagalur	Karnataka
3	University of Agricultural Sciences, GKVK Campus	Bengaluru	Karnataka
4	Ambaraguda	Shimoga	Karnataka
5	Glory of Allapalli	Gadchiroli	Maharashtra
6	Tonglu BHS under the Darjeeling Forest Division	Darjeeling	West Bengal
7	Dhotrey BHS under the Darjeeling Forest Division	Darjeeling	West Bengal
8	Dialong Village	Tamenglong	Manipur
9	Ameenpur lake	Sangareddy	Telangana
10	Majuli	Majuli	Assam
11	Ghariyal Rehabilitation Centre	Lucknow	Uttar Pradesh
12	Chilkigarh Kanak Durga	Jhargram	West Bengal
13	Khlaw Kur Syiem KmieIng	Ri-Bhoi	Meghalaya
14	Mandasaru	Kandhamal	Odisha
15	Purvatali Rai	North Goa	Goa
16	Naro Hills	Satna	Madhya Pradesh
17	Patalkot	Chhindwara	Madhya Pradesh
18	Asramam	Kollam	Kerala

NBA, As on August 2020

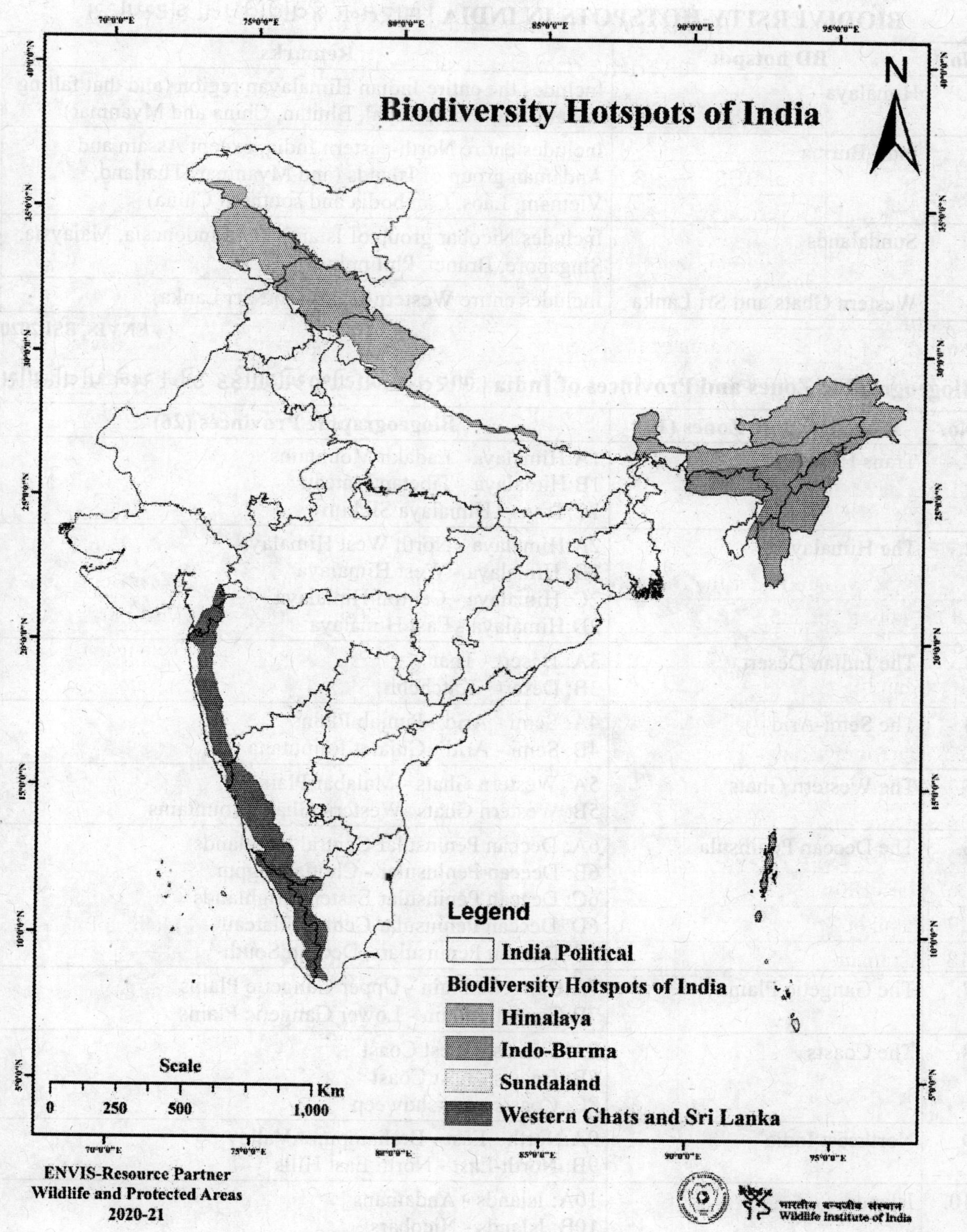

WII, As on August 2020

BIODIVERSITY HOTSPOTS IN INDIA | ભારતના જૈવવિવિધતા હોટસ્પોટ્સ

No.	BD hotspot	Remarks
1	Himalaya	Includes the entire Indian Himalayan region (and that falling in Pakistan, Tibet, Nepal, Bhutan, China and Myanmar)
2	Indo-Burma	Includes entire North-eastern India, except Assam and Andaman group of Islands (and Myanmar, Thailand, Vietnam, Laos, Cambodia and southern China)
3	Sundalands	Includes Nicobar group of Islands (and Indonesia, Malaysia, Singapore, Brunei, Philippines)
4	Western Ghats and Sri Lanka	Includes entire Western Ghats (and Sri Lanka)

ENVIS, BSI 2020

Biogeographic Zones and Provinces of India | ભારતના બાયોજીઓગ્રાફિક ઝોન અને પ્રોવીન્સીસ

No.	Biogeographic Zones (10)	Biogeographic Provinces (26)
1.	Trans Himalaya	1A:Himalaya- Ladakh Mountains 1B:Himalaya - Tibetan Plateau 1C:Trans - Himalaya Sikkim
2.	The Himalaya	2A:Himalaya - North West Himalaya 2B: Himalaya - West Himalaya 2C: Himalaya - Central Himalaya 2D:Himalaya - East Himalaya
3.	The Indian Desert	3A: Desert - Thar 3B: Desert - Katchchh
4.	The Semi-Arid	4A: Semi - Arid - Punjab Plains 4B: Semi - Arid - Gujarat Rajputana
5.	The Western Ghats	5A: Western Ghats - Malabar Plains 5B: Western Ghats -Western Ghats Mountains
6.	The Deccan Peninsula	6A: Deccan Peninsular Central Highlands 6B: Deccan Peninsular - Chotta Nagpur 6C: Deccan Peninsular Eastern Highlands 6D: Deccan Peninsular Central Plateau 6E: Deccan Peninsular - Deccan South
7.	The Gangetic Plains	7A:Gangetic Plain - Upper Gangetic Plains 7B:Gangetic Plain - Lower Gangetic Plains
8.	The Coasts	8A: Coasts - West Coast 8B: Coasts - East Coast 8C: Coasts - Lakshdweep
9.	Northeast India	9A:North - East - Brahamputra Valley 9B: North-East - North East Hills
10.	Islands	10A: Islands - Andamans 10B: Islands - Nicobars

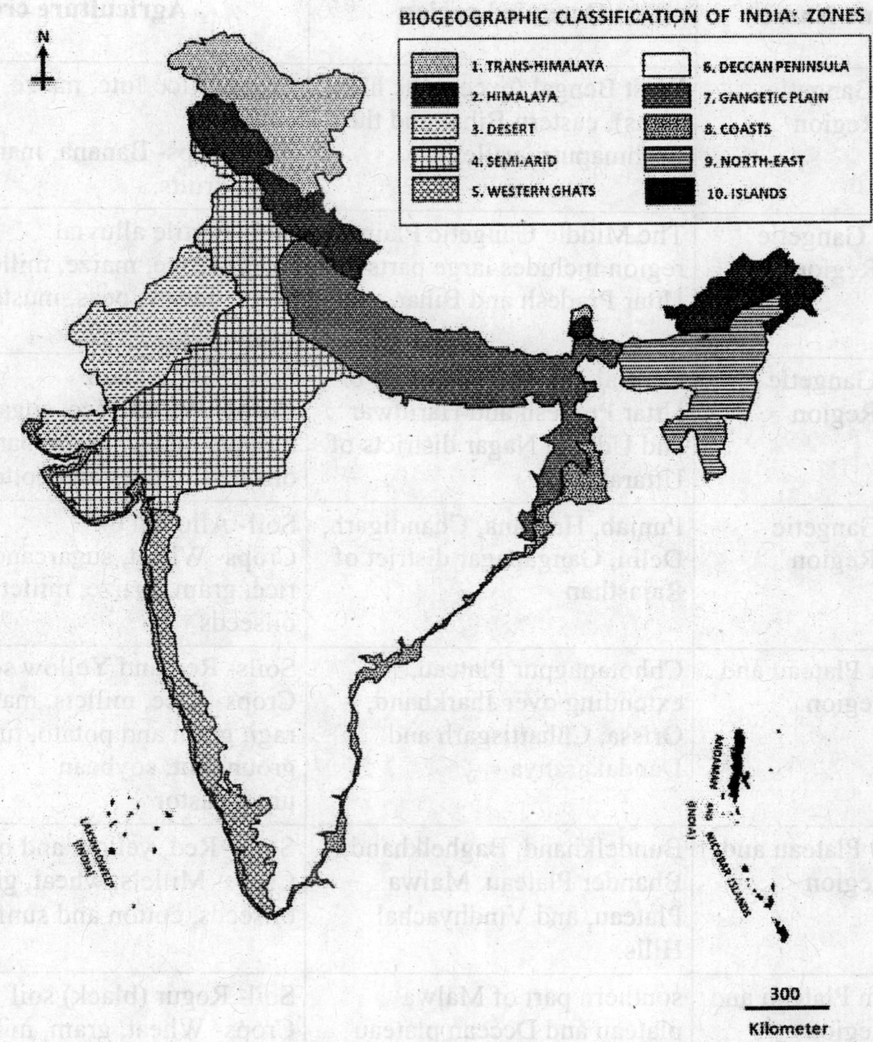

Rodgers *et al.*, 2000

Agroclimatic Zones of India | ભારતના એગ્રોક્લાઈમેટિક ઝોન

Zone No	Zone Name	Occupied region	Agriculture crops
1	Western Himalayan Region	Jammu and Kashmir, Himachal Pradesh, and the hill region of Uttarakhand.	Crops- Rice, maize Fruit crops- Apple, peaches, apricot, pears, cherry, almond, litchis, walnut, Saffron
2	Eastern Himalayan Region	Arunachal Pradesh, the hills of Assam, Sikkim, Meghalaya, Nagaland, Manipur, Mizoram, Tripura, and the Darjeeling district of West Bengal	Soil- Red-brown, Jhumming (shifting cultivation) prevails in the hilly areas. Crops- Rice, maize, potato, tea Fruit crops- Pineapple, litchi, oranges and lime.

Zone No	Zone Name	Occupied region	Agriculture crops
3	Lower Gangetic Plains Region	West Bengal (except the hilly areas), eastern Bihar and the Brahmaputra valley	Crops- Rice Jute, maize, potato, and pulses. Fruit crops- Banana, mango and citrus fruits.
4	Middle Gangetic Plains Region	The Middle Gangetic Plain region includes large parts of Uttar Pradesh and Bihar	Soil- Fertile alluvial. Crops- Rice, maize, millets, wheat, gram, barley, peas, mustard and potato
5	Upper Gangetic Plains Region	Central and Western parts of Uttar Pradesh and Haridwar and Udham Nagar districts of Uttarakhand	Soil- Sandy loam Crops- Wheat, rice, sugarcane, millets, maize, gram, barley, oilseeds, pulses and cotton
6	Trans-Gangetic Plains Region	Punjab, Haryana, Chandigarh, Delhi, Ganganagar district of Rajasthan	Soil- Alluvial Crops- Wheat, sugarcane, cotton, rice, gram, maize, millets, pulses and oilseeds
7	Eastern Plateau and Hills Region	Chhotanagpur Plateau, extending over Jharkhand, Orissa, Chhattisgarh and Dandakaranya	Soils- Red and Yellow soil Crops- Rice, millets, maize, oilseeds, ragi, gram and potato, tur, groundnut, soybean urad, castor
8	Central Plateau and Hills Region	Bundelkhand, Baghelkhand, Bhander Plateau, Malwa Plateau, and Vindhyachal Hills.	Soils- Red, yellow and black. Crops- Millets, wheat, gram, oilseeds, cotton and sunflower
9	Western Plateau and Hills Region	southern part of Malwa plateau and Deccan plateau (Maharashtra)	Soil- Regur (black) soil Crops- Wheat, gram, millets, cotton, pulses, groundnut, oilseeds, sugarcane, rice, jowar, bajra Fruit crops- Oranges, grapes, bananas, ber, pomegranate, mango and guava
10	Southern Plateau and Hills Region	Interior Deccan and includes parts of southern Maharashtra, the greater parts of Karnataka, Andhra Pradesh, and Tamil Nadu	Crops- Millets, oilseeds, and pulses are grown. Coffee, tea, cardamom and spices
11	East Coast Plains and Hills Region	Coromandal and northern Circar coasts of Andhra Pradesh and Orissa.	Soils- Alluvial, loam and clay Crops- Rice, jute, tobacco, sugarcane, maize, millets, groundnut and oilseeds, pepper and cardamom

Zone No	Zone Name	Occupied region	Agriculture crops
12	West Coast Plains and Ghat Region	Malabar and Konkan coastal plains and the Sahyadris.	Soils- Laterite and alluvial Crops- Rice, coconut, oilseeds, sugarcane, millets, pulses and cotton
13	Gujarat Plains and Hills	The hills and plains of Kathiawar, and the fertile valleys of Mahi and Sabarmati rivers.	Soils- Regur & alluvium. Crops- Groundnut, cotton, rice, millets, oilseeds, wheat and tobacco are the main crops. It is an important oilseed producing region.
14	Western Dry Region	Extending over Rajasthan, West of the Aravallis	Crops- Bajra, jowar, moth, wheat and gram Fruit crops- Water melon, guava and date palm,
15	The Islands Region	Andaman-Nicobar and Lakshadweep	Soil- Sandy loam Crops- Rice, maize, millets, pulses, arecanut, turmeric and cassava. Nearly half of the cropped area is under coconut

Planning Commission 2020

Agro-climatic zones of India

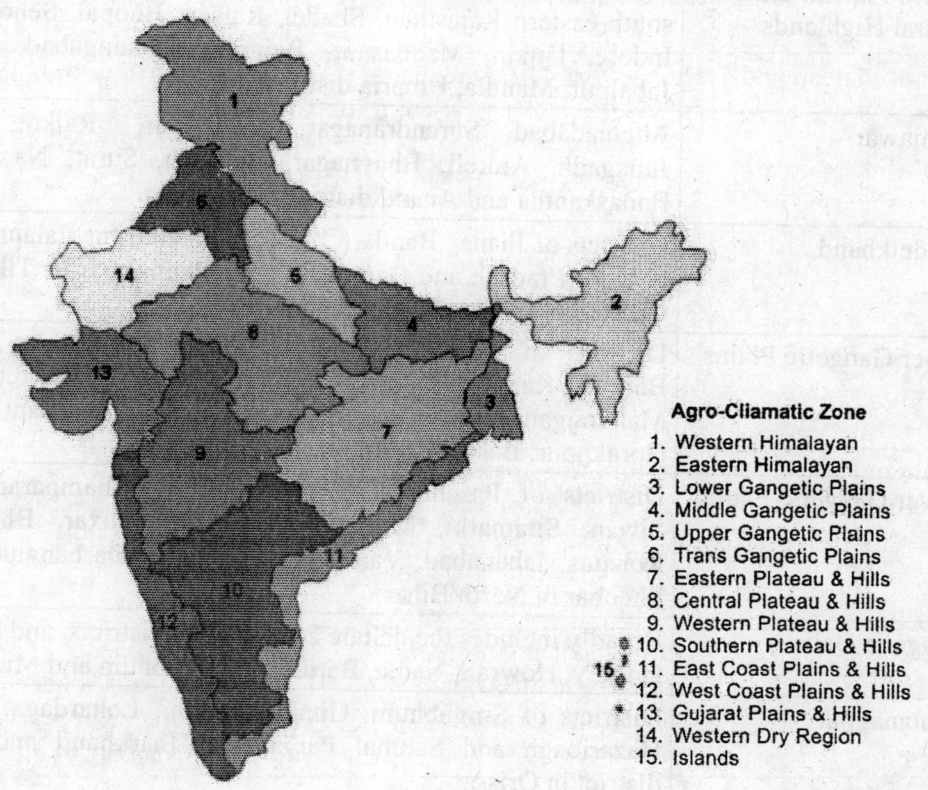

Agro-biodiversity Hotspots of India | ભારતના કૃષિ જૈવવિવિધતા હોટસ્પોટ્સ

No.	Hotspot Region	Area Covered
1.	Cold Desert	Western Himalyas covering Ladakh and Kargil. Upper reaches of Lahuai-Spiti districts of Himachal Pradesh.
2.	Western Himalayan	Districts of Srinagar,Anantnag,Udhamput,Riasi,Kathna in Jammu & Kashmir,all the districts of Himachal Pradesh except the cold arid region and all the districts of Uttarakhand.
3.	Eastern Himalayan	All the districts of Arunachal Pradesh, Sikkim and Darjeeling district of West Bengal.
4.	Brahmaputra Valley	Dhubri, Kokrajhar, Bongaigaon, Barpeta, Nalbari, Goalpara, Kamrup, Golaghat, Darrang, Morigaon, Nagaon, Sonitpur, Jorhat, Lakhimpur, Sibsagar, Dibrugarh, Dhemaji and Tinsukia.
5.	Khasia-Jaintia-Garo Hills	All the seven districts of Meghalaya,i.e. East Garo Hills,West Garo Hills, South Garo Hills, East Khasi Hills, West Khasi Hills, Jaintia Hills and Ri-Bhoi.
6.	North-Eastern Hills	All the districts of Manipur, Mizoram, Nagaland, Tripura and the adjoining Cachar and North Cachar districts of Asom.
7.	Arid Western	Sikar, Nagaur, Pali, Hanumangarh, Ganganagar, Jalore, Sirohi, Jodhpur, parts of Jaisalmer and Bikaner, Udaipur, Dungarpur, Churu,and Jhunjhunun districts of Rajastahan.
8.	Malwa Plateau and Central Highlands	Malwa plateau, Central highlands,the Mewar plateau and semi- arid south-eastern Rajasthan. Shadel, Raisen, Bhopal, Sehore, Shajapur, Indore, Ujjain, Mandasaur, Rajgarh Hoshangabad, Narsinghpur, Jabalpur, Mandla, Umaria districts.
9.	Kathiawar	Ahemadabad, Surendranagar, Jamnagar, Rajkot, Porbandar, Junagadh, Amreli, Bhavnagar, Bharuch, Surat, Navsari, Valsad, Banaskantha and Anand districts of Gujarat.
10.	Bundelkhand	Districts of Jhansi, Banda, Chitrakoot, Hamirpur, Jalaun and Lalitpur in Uttar Pradesh and Damoha, Datia, Panna, Sagar, Tikamagarh and Chattarpur in Madhya Pradesh.
11.	Upper Gangetic Plains	Districts of Hardoi, Sitapur, Barabanki, Lucknow, Unnao, Rae Bareilly, Kanpur, Kannuj of Central Uttar Pradesh and the districts of Maharajganj, Sidharatnagar, Kushinagar, Deoria, Sant Kabir Nagar, Gorakhpur, Basti of North-eastern Uttar Pradesh.
12.	Lower Gangetic Plains	Districts of Paschim Champaran, Purbi Champaran, Gopalganj, Siwan, Sitamarhi, Muzaffarpur, Saran, Buxar, Bhojpur, Patna, Rohatas, Jahanabad, Vaishali, Samastipur, Darbangha, Madhubani, Sheohar in North Bihar.
13.	Gangetic Delta	Broadly includes the deltaic 24-Parganas districts, and the districts of Hoogly, Howrah, Nadia, Bardhaman, Birbhum and Murshidabad.
14.	Chotanagpur	Districts of Singhbhum, Gumla, Ranchi, Lohardaga, Palamau and Hazaribagh and Santhal Pargana in Jharkhand and Mayurbhanj district in Orissa.

No.	Hotspot Region	Area Covered
15.	Bastar	Districts of Bastar, Bilaspur, Durg, Jashpur, Kabirdham, Kanker, Kirba, Koria, Mahasamund, Kondaigaon, and Rajnangoan of Chattisgarh.
16.	Koraput	Districts of Malkangiri, Sonabeda,Jeypore,Koraput, Nabrangpur, Kalahandi, Bolangir, Rayagada of Orissa and districts of north eastern Andhra Pradesh i.e. Srikakulam, Vijanagaram and Vizagapatnam.
17.	Southern Eastern Ghats	Districts of Chittoor, Ananthapur, Cuddapah and`Kurnoolin Andhra
18.	Kaveri	Districts of Chengalput, South Arcot, North Arcot, Thiuvannamalai, Tiruchirapalli, Pudukottai, Thiruarur, Vellore, Kanchipuram, Dharmapuri, Salem, Namakkal, Karur and Dindigal.
19.	Deccan	Districts of Jalna, Hingoli, Parbhani, Beed, Nanded, latur, Osmanabad, Solapur, Sangli, Gondia, Gadchiroli in Maharashtra, districts of Adilabad, Karimnagar, Warangal and Khamman in Andhra Pradesh and districts of Bidar and Gulbarga in Karnataka.
20.	Konkan	Coastal districts of Thane,Raigad,Ratnagiri,Sindhudurg and part of Sahyadridistricts of Pune,Satara and Kolhapur of Maharashtra, all the districts of Goa and Uttar Kannda district of Karnataka.
21.	Malabar	Districts of Kasargod, Kannur,Wayanad, Kozikode, Malappuram, Palakkad, Thrissur, Idukki, Ernakulam, Alappuza, Kollam, Kottayam, Pathanamthitta & Thiruvananthpuram in Kerala, Udhagamandalam (Nilgiri) and Kanyakumari districts of Tamil Nadu and districts of Dakshin Kannada, Kodagu and Udipi in Karnataka.
22.	Islands	Andaman and Nicobar Islands and Lakshadweep

NBA 2020

Ecological Lifeforms | ઇકોલોજીકલ લાઇફફોર્મ્સ

Symbol	Life Form	Description
Ch	Chamaephytes	Small shrubs, herbs and prostrate plants, in which perennating buds are above ground, but at less than 25 cm height. Predominating at high altitudes and latitudes
Cr	Cryptophytes	Perennating buds under water or below the soil surfaces as in hydrophytes, marsh plants, or terrestrial plants with underground rhizome, tuber etc. Predominating in zones characterized by a long dry season.
H	Hemicryptophytes	Biennial or perennial herbs and grasses in which perennating buds are just beneath or in soil surface covered with snow or little during unfavorable periods. Predominating in cold temperate zones
P	Phanerophytes	Tree or shrubs, large perennial herbs, woody lianas, epiphytes, system succulent in which perennating buds are well above ground (above 25 cm from ground surface) of twigs or branches. Predominating in tropical flora.

Symbol	Life Form	Description
Th	Therophytes	Usually annual herbs, which survive unfavourable period through seeds or spores. Predominating in cold and hot deserts.

Raunkiaer 1934

Mangrove Zonation in India | ભારતમાં ચેરના ઝોનેશન

Zone	Type	Area	Example
Proximal Zone	Front mangroves	Towards sea water	*Avicennia* and *Sonneratia*
Middle Zone	Mid mangroves	Core zone	*Ceriops* and *Bruguiera*
Distal Zone	Back mangroves	Towards land region	*Heritiera* and *Xylocarpus*

Grassland Types of India | ભારતના ઘાસના મેદાનના પ્રકાર

Type	Remarks
Himalayan grasslands	Extends up to snowline. At lower altitudes, grasslands are present in patches along with coniferous or broadleaved forests.
Terai-Duars grasslands	About 25 km wide stretches of Terai-Duars grasslands are located in the low-lying waterlogged areas and marshes at the base of the Himalayas.
Semi-arid grasslands	Usually found in the semi desert conditions. The semi-arid plains of western India, central India and the Deccan.
Shola grasslands	The Shola grasslands are the patches of undulating grasslands on hill-slopes between the Shola forests on the Western Ghats. Nilgiri and Annamalai ranges above 2000 m from sea level.

Important Grass Species of India | ભારતની મહત્વની ઘાસની જાતો

Grassland type	Dominant species	Regions
Sehima Dichanthium	*Sehima sulcanum*	Maharashtra, M.P., South-west U.P. and parts of Tamilnadu and Karnataka
Dichanthium Cenchrus	*Dichanthium Cenchrus*	Plains of Punjab. Haryana. Dellhi, Rajasthan, Saurashtra, Eust M.P. costal Maharashtra and Tamilnadu
Phragmites Saccharum	*Pharagmites karka* *Saccharum spontaneum* *Imperata cylindrica* *Bothriochloa pertusa* *Bothriochloa odorata*	Taral areas of north U.P. Bihar, Bengal, Assam, and Tamilnadu (Sundarbans and Kaveri delta)
Bothriochloa	*Bothriochloa odorata*	Lonavala tract of Maharashtra
Cymbopogon	*Cymbopogon*	Western Ghats, Vindhyas, Satpura, Aravali, Chhota Nagapur
Arundinella	*Arundinella nepalensis* *Themeda anathera*	Western Ghats, Nilgiris, lower Himalayas (Assam to Kashmir)

Grassland type	Dominant species	Regions
Deyeuxia- Arundinella	*Deyeuxia, Arundinella brachypodium, Bromus, Festuca*	Upper Himalayas between 6,200 ft to 10,000 ft from Assam, Bengal, through U.P. to Punjab and H.P.
Deschampsia- Deyeuxia	*Deyeuxia, Deschampsia, Poa, Stipa. Glyceria, Festuca*	Himalayas (above 8,300ft) Kashmir

State Symbols in India | ભારતમાં રાજ્ય ચિહ્નો

State/UT	Animal	Bird	Tree	Flower
Andhra Pradesh	Blackbuck	Rose ringed Parakeet	Neem	Jasmine
Arunachal Pradesh	Mithun	Great hornbill	Hollong	Lady slipper orchid
Assam	One-horned rhinoceros	White-winged wood duck	Hollong	Foxtail orchids
Bihar	Ox	House sparrow	Peepal	Kachnar
Chhattisgarh	Wild buffalo	Hill myna	Sal	Indian lurel
Delhi	Nilgai	House sparrow	Gulmohar	Alfalafa
Goa	Gaur	Flame throated bulbul	Matti	—
Gujarat	Asiatic lion	Greater flamingo	Mango	Marigold
Haryana	Blackbuck	Black francolin	Peepal	Lotus
Himachal Pradesh	Snow leopard	Western tragopan	Deodar	Rhododendron
Jammu & Kashmir	Hangul	Black necked crane	Chinar	Lotus
Jharkhand	Elephant	Koel	Sal	Palash
Karnataka	Elephant	Indian roller	Sandal	Lotus
Kerala	Elephant	Great hornbill	Coconut	Golden shower tree
Madhya Pradesh	Swamp deer	Asian paradise flycatcher	Banyan	Parrot tree
Maharashtra	Giant squirrel	Yellow-footed green pigeon	Mango	Jarul
Manipur	Sangai	Mrs. Hume's pheasant	Toon	Shirui lily
Meghalaya	Clouded leopard	Hill myna	White teak	Lady slipper orchid
Mizoram	Serow	Mrs. Hume's pheasant	Iron wood	Red vanda
Nagaland	Mithun	Blyth's tragopan	Alder	Rhododendron
Odisha	Sambar	Indian roller	Banyan	Lotus
Puducherry	Squirrel	Asian koel	Vilva tree	Cannonball

State/UT	Animal	Bird	Tree	Flower
Punjab	Blackbuck	Northern goshawk	Sheesham	—
Rajasthan	Camel	Great Indian bustard	Khejri	Rohira
Sikkim	Red panda	Blood pheasant	Rhododendron	Noble orchid
Tamil Nadu	Nilgiri tahr	Emerald dove	Palmyra palm	Kandhal
Telangana	Deer	Indian roller	Jammi	Tangedu
Tripura	Phayre's langur	Green imperial pigeon	Agar	Nageshwar
Uttarakhand	Musk deer	Himalayan monal	Burans	Brahm kamal
Uttar Pradesh	Swamp deer	Sarus crane	Ashok	Brahm kamal
West Bengal	Fishing cat	White-breasted kingfisher	Chatian	Shephali

ENVIS, FRI 2019

Important Trees of India | ભારતના મહત્વના વૃક્ષો

Common Name	Botanical name	Family	Alternate Name
Aam	*Mangefera indica*	Anacardaceae	—
Almond	*Prunus amygdalus*	Rosaceae	—
Amaltas	*Cassia fistlua*	Caesalpinacea	Indian laburnum
Aonla	*Emblica officinalis*	Euphorbiaceae	—
Apricoat	*Prunus armoniaca*	Rosaceae	Khubani
Aritha	*Sapindus emarginatus*	Euphorbiaceae	—
Arjun	*Terminalia arjuna*	Combretaceae	—
Ashok	*Ashoka indica*	Legumniosae	—
Aus. Wattle	*Acacia auriculoformis*	Mimoseae	—
Babul	*Acacia nilotica*	Mimoseae	—
Baheda	*Terminalia bellerica*	Combretaceae	—
Bakain	*Melia azadirache*	Meliaceae	—
Bamboo	*Bambusa bamboos*	Poaceae	Green Gold
Bel	*Aegle marmelos*	Rutaceae	God Fruit
Ber	*Zizyhus mauritiana*	Rhamnaceae	Poor man's apple
Bergad	*Ficus benghalensis*	Moraceae	—
Bijasal	*Pterocapus marsupium*	Dipterocaarpaceae	Bengal Kino
Black popular	*Populus nigra*	Salicaceae	—
Blue pine	*Pinus wallichiana*	Pineaceae	—
Bottle brush	*Callistimon viminalis*	Myrtaceae	—
Casuarina	*Casuarina equisitifolia*	Casuarinaceae	Beef Wood
Chandan	*Santalum album*	Santalaceae	Sandal wood

Common Name	Botanical name	Family	Alternate Name
Charoli	*Buchanania lanzan*	Caesalpinaceae	Chironzi
Chilgoza pine	*Pinus gerardiana*	Pineaceae	—
Chinar	*Platinum orientalis*	Platanaceae	—
Chir pine	*Pinus roxburghii*	Pineaceae	—
Cupress	*Cupresses torulosa*	Cupreseaceae	—
Cycas	*Cycus revoluta*	Cycadaceae	Sago palm
Deodar	*Cedrus deodara*	Pineaceae	—
Devil tree	*Alostonia scholaris*	Legumniosae	Devil's Tree
Dhak	*Butea monosperma*	Legumniosae	Forest flame
Dhonkra	*Anogeissus pendula*	Combretaceae	—
Dhudhi	*Wrightia tinctoria*	Apocynaceae	Pala indigo plant
Elm	*Ulmus species*	Ulmaceae	—
Eugenia	*Eugenia oojensis*	Myrtaceae	—
Ghamar	*Gmelia arborea*	Lamiaceae	White teak
Gular	*Ficus glomeruta*	Moraceae	—
Gulmohar	*Delonix regia*	Malvaceae	—
Gurjan	*Dipterocarpus turbinatus*	Dipterocarpaceae	—
Gutel	*Grewia tilifolia*	Malvaceae	Dhaman
Haldu	*Adina cordifoia*	Rubiaceae	—
Harad	*Terminalia chebula*	Combretaceae	—
Imli	*Tamarindus indica*	Caesalpiniaceae	—
Indian kopak	*Ceiba patendra*	Bombaceae	White Semal
Indian laurel	*Terminalia tomentosa*	Combretaceae	—
Indian trumpet	*Oroxylum indicum*	Bignoniaceae	Midnight horror
Jamun	*Szygigium cuminii*	Myrtaceae	—
Kachnar	*Bauhania variegata*	Caesalpinaceae	—
Kadamb	*Anthocephalus cinensis*	Rubiaceae	—
Kadaya	*Caryota urens*	Arecaceae	Fishtail palm
Kaith	*Limonia acidica*	Rutaceae	—
Kakooni	*Cochlosprmum religiosum*	Bixaceae	Silk cotton tree
Kanju	*Holoptelia integrifolia*	Ulmaceae	Indian elm
Karanj	*Pongamia pinnata*	Leuminosae	—
Kesiya pine	*Pinus kassiya*	Pineaceae	—
Khair	*Acacia catechu*	Mimoseae	—
Khejra	*Prosopis juliflora*	Leuminosae	—
Khejri	*Prosopis cineraria*	Leuminosae	—
Khirni	*Manilkara hexandra*	Sapotaceae	—

Common Name	Botanical name	Family	Alternate Name
Kindal tree	*Terminalia peniculata*	Combretaceae	—
Kurchi	*Holorhhinea antidycentrica*	Apocynaceae	Kutuj
Kusum	*Schleria oleosa*	Sapindaceae	Ceylon oak
Lendi crape	*Lagerstomia parviflora*	Lytraceae	—
Mahaneem	*Ailanthus excelsa*	Simarubiaceae	Tree of Heaven
Mahua	*Madhuca indica*	Sapotaceae	Butter tree
Malawar wood	*Hopea parviflora*	Dipterocarpaceae	Malawar wood
Marwar teak	*Tecomella undulata*	Bignoniaceae	Marwar teak
Mysoore gum	*Eucalyptus tereticornis*	Myrtaceae	—
Neem	*Azadirachta indica*	Meliaceae	—
Oak	*Quercus rubra*	Fagaceae	Oak tree
Pepal	*Ficus religiosa*	Moraceae	—
Pink cedar	*Acrocarpus fraxinifolius*	Legumniosae	—
Popular	*Populus ciliata*	Salicaceae	Himalayan poplar
Popular	*Populus deltoides*	Salicaceae	—
Ratanjot	*Jatropha curcus*	Euphorbiaceae	Biodiesal plant
Rose wood	*Dalbergia latifolia*	Paplionaceae	Rose wood
Rubinia	*Robinia pseudoacacia*	Fabaceae	—
Safed donkra	*Anogeissus latifolia*	Combretaceae	Axle wood
Safeda	*Eucalyptus globulus*	Myrtaceae	—
Saffron	*Crocus sativus*	Iridaceae	—
Sal	*Shorea robusta*	Dipterocaarpaceae	—
Salix	*Salix alba*	Salicaceae	White willow
Sehjana	*Moringa oleifera*	Moringaceae	Drumstick tree
Sehtoot	*Morus alba*	Moraceae	Mulberry
Semal	*Bombax ceiba*	Bombacaceae	Kapok
Sequoia	*Sequoia semperviranes*	Pineaceae	—
Shisham	*Dalbergia sisso*	Paplionaceae	Biradi
Silver fir	*Abies pindrew*	Pinacea	—
Spruce	*Picea smithiana*	Pineaceae	—
Sterculia	*Sterculia urens*	Sterculaceae	—
Subabul	*Leucaena leucocephala*	Mimoseae	—
Teak	*Tectona grandis*	Verbenaceae	Queen wood
Tendu	*Diospyrus melanoxylon*	Ebenaceae	Bedi leaf
Toon	*Toona ciliata*	Meliaceae	Red cedar
Walnut	*Juglans regia*	Juglandaceae	—

ENVIS, BSI 2020

Multiple Choice Questions | હેતુલક્ષી પ્રશ્નો

1. એક પ્રજાતિ નો અર્થ શું થાય છે?
A. સમાન દેખાતા ઇન્ડિવિજ્યુઅલ
B. આનુવંશિક રૂપે સમાન ઇન્ડિવિજ્યુઅલ
C. ઇન્ડિવિજ્યુઅલ જે પ્રજનન કરી શકે
D. ઇન્ડિવિજ્યુઅલ કે જેઓ સાથે રહી શકે છે

2. મનુષ્ય અને ચિમ્પાન્ઝીઝના જનીનોમાં કેટલી સમાનતા છે?
A. 60% B. 70%
C. 80% D. 98%

3. જીવતંત્રમાં કોષને આધારે વર્ગીકરણની પદ્ધતિ કોણે ઘડી છે?
A. કાર્લ વોઇઝ
B. કેરોલસ લિનાઇઅસ
C. કાર્લ લેન્ડસ્ટીનર
D. ડી અર્ન્સ્ટ હેક્કેલ

4. ગુજરાત રાજ્ય જૈવવિવિધતા એક્શન પ્લાન (GSBAP) ની શરૂઆત ક્યારે કરવામાં આવી હતી?
A. 2008 B. 2014
C. 2016 D. 2018

5. ગુજરાત સ્ટેટ એક્શન પ્લાન ઓન ક્લાઈમેટ ચેન્જ (GSAPCC) ની શરૂઆત ક્યારે કરવામાં આવી હતી?
A. 2008 B. 2014
C. 2016 D. 2018

6. ભારતમાં જૈવવિવિધતા લક્ષ્યાંક ક્યારે અપનાવવામાં આવ્યો હતો?
A. 2002 B. 2004
C. 2005 D. 2007

1. **A species means**
A. Similar looking individuals
B. Genetically identical individuals
C. Individual who can interbred
D. Individuals who can live together

2. **The genes of humans and chimpanzees have similarity upto**
A. 60 % B. 70 %
C. 80 % D. 98 %

3. **Who devised a system of classification based on the cellular organization of organism?**
A. Carl Woese
B. Carolus Linnaeus
C. Carl Landsteiner
D. D Ernst Haeckel

4. **Gujarat State Biodiversity Action Plan (GSBAP) was launched in**
A. 2008 B. 2014
C. 2016 D. 2018

5. **Gujarat State Action Plan on Climate Change (GSAPCC) was launched in**
A. 2008 B. 2014
C. 2016 D. 2018

6. **In India, the Biodiversity Target was adopted in**
A. 2002 B. 2004
C. 2005 D. 2007

7. રાષ્ટ્રીય જૈવવિવિધતા સત્તામંડળ (NBA) નું મુખ્ય મથક ક્યાં છે ?
A. મુંબઈ B. ચેન્નાઈ
C. હૈદરાબાદ D. નવી દિલ્હી

8. રાષ્ટ્રીય જૈવવિવિધતા બોર્ડ (NBB) ક્યાં આવેલું છે ?
A. બેંગલુરુ B. નવી દિલ્હી
C. કોલકાતા D. ગુજરાત

9. __________ ભારતમાં બાયોડાયવર્સીટી હોટસ્પોટ તરીકે જાણીતું છે.
A. સુંદરવન ડેલટા
B. પશ્ચિમ ઘાટ
C. પૂર્વી ઘાટ
D. ગેન્ગેટીક પ્લેન

10. સક્શેશનની સેરલ કમ્યુનિટી કઈ છે ?
A. પાયોનિયર કમ્યુનિટી
B. ક્લાઈમેક્સ કમ્યુનિટી
C. પાયોનિયર અને ક્લાઈમેક્સ કમ્યુનિટી વચ્ચેના તબક્કા
D. ઉપરોક્ત કંઈ નથી

11. ભેજવાળા પ્રદેશમાં સક્શેશન શું છે?
A. લિથોસેર B. સામોસેરે
C. મેસોસેરે D. સેરુલા

12. __________ ઘાસના મેદાન એ એક પ્રકારનું પ્રતિક્રિયાશીલ સક્શેશન છે.
A. પમ્પાસ B. સવાના
C. વેલ્ડેટ્સ D. ડોન

13. કયા પ્રકારનાં ઈન્ટરેક્શનમાં એક અથવા બન્ને જાતિના સભ્યને નુકસાન થાય છે?
A. નેગેટિવ ઈન્ટરેક્શન
B. પોઝીટીવ ઈન્ટરેક્શન
C. ન્યુટ્રલ ઈન્ટરેક્શન
D. કંઈ નહીં

7. National Biodiversity Authority (NBA) headquarter is in
A. Mumbai B. Chennai
C. Hyderabad D. New Delhi

8. National Biodiversity Board (NBB) is situated in
A. Bengaluru B. New Delhi
C. Kolkata D. Gujarat

9. __________ are known as hotspot of biodiversity in India
A. Sunderbans Deltas
B. Western Ghats
C. Eastern Ghats
D. Gangetic Plain

10. Seral communities in succession are
A. Pioneer communities
B. Climax communities
C. Stages between pioneer and climax communities
D. None of the above

11. Succession in moistened region is
A. Lithosere B. Psamosere
C. Mesosere D. Serula

12. __________ grassland is a type of retrogressive succession
A. Pampas B. Savana
C. Veldts D. Down

13. The type of interaction in which member of one or both the interacting species are harmed is
A. Negative interaction
B. Positive interaction
C. Neutral interaction
D. None

14. રફ્લેસિયા શું છે?

A. થડ પરનું સંપૂર્ણ પરોપજીવી
B. મૂળ પરનું સંપૂર્ણ પરોપજીવી
C. થડ પરનું આંશિક પરોપજીવી
D. મૂળ પરનું આંશિક પરોપજીવી

15. તેની પોતાની પ્રજાતિઓ દ્વારા ખાવામાં આવેલ સજીવને શું કહેવામાં આવે છે?

A. મેક્રોફેગી B. કોમેન્સાલિઝમ
C. ઓટોફેગી D. કેનાબિલિઝમ

16. સજીવો જે વિસ્તારમાં જીવે છે તેને શું કહે છે?

A. નિશ B. હેબીટાટ
C. સ્ટ્રાટા D. ટ્રોફિક

17. નિવાસસ્થાન ના પેટા વિભાગ ને શું કહે છે ?

A. માઇક્રો ક્લાઈમેટ B. મેક્રો ક્લાઈમેટ
C. મેગા ક્લાઈમેટ D. સબ હેબીટાટ

18. એ. જી. ટાંસલીએ કયો શબ્દ આપ્યો?

A. ઇકોલોજી
B. ઇકોટોન
C. ઇકોસિસ્ટમ
D. ઇકોસિસ્ટમ ઇકોલોજી

19. __________ એ ઇકોસિસ્ટમ ઇકોલોજીના પિતા તરીકે ઓળખાય છે.

A. ગ્રિનલ B. રીટર
C. ઇ. પી. ઓડમ D. એલ્ટન

20. વિશ્વ જૈવવિવિધતા દિવસ ક્યારે ઉજવવામાં આવે છે?

A. 21 ફેબ્રુઆરી
B. 12 સપ્ટેમ્બર
C. 11 જુલાઈ
D. 22 મે

14. Rafflesia is a

A. Total stem parasite
B. Total root parasite
C. Partial root parasite
D. Partial stem parasite

15. Organism eaten by its own species is called

A. Macrophagy B. Commensalism
C. Autophagy D. Cannabilism

16. Physical area where an organism lives

A. Niche B. Habitat
C. Strata D. Trophic

17. Sub-division of habitat is called-

A. Micro climate B. Macro climate
C. Mega climate D. Sub habitat

18. A. G. Tansley gave the term

A. Ecology
B. Ecotone
C. Ecosystem
D. Ecosystem ecology

19. ______________ is the Father of ecosystem ecology

A. Grinnel B. Reiter
C. E. P. Odum D. Elton

20. World Biodiversity Day is celebrated on

A. 21 February
B. 12 September
C. 11 July
D. 22 May

21. ઇકોસિસ્ટમના અજૈવિક ઘટકો કયા છે?
A. માઇક્રોબ B. હવા
C. પાણી D. B અને C બન્ને

22. ઇકોસિસ્ટમના ઓટોટ્રોપ એટલે
A. પ્રોડ્યુસર
B. ટ્રાંસડ્યુસર
C. કન્વર્ટર
D. ઉપરોક્ત તમામ

23. ઇકોસિસ્ટમના મુખ્ય વિઘટનકર્તા કયા છે?
A. સુક્ષ્મસજીવો B. છોડ
C. પ્રાણીઓ D. ઉપરના બધા

24. ઇકોસિસ્ટમમાં છોડના પરોપજીવીઓ કયા છે?
A. પ્રાથમિક ઉત્પાદકો
B. પ્રાથમિક ઉપભોક્તાઓ
C. દ્વિતીય ઉત્પાદકો
D. દ્વિતીય ઉપભોક્તાઓ

25. પ્રકૃતિમાં કેટલા પ્રકારની આહાર શૃંખલા ઉપ્લબ્દ્ધ છે?
A. 2 B. 3
C. 4 D. 5

26. કઈ આહાર શૃંખલા વિઘટનકર્તાઓ દ્વારા મૃત કાર્બનિક પદાર્થોના વિઘટનથી પ્રારંભ થાય છે?
A. શિકારી આહાર શૃંખલા
B. ચરાઈ આહાર શૃંખલા
C. સેપ્રોફાઇટિક આહાર શૃંખલા
D. ડેટ્રિટસ આહાર શૃંખલા

27. ઇકોસિસ્ટમના જુદા જુદા ટ્રોફિક સ્તર પર ઇકોલોજીકલ પરિમાણોની ગ્રાફિકલ રજૂઆત શું છે?
A. ટ્રોફિક્સ B. આહાર જાળ
C. આહાર શૃંખલા D. પિરામિડ્સ

21. Abiotic components of ecosystem include
A. Microbe B. Air
C. Water D. Both B and C

22. Autotrophs of an ecosystem are called
A. Producer
B. Transducer
C. Converter
D. All of the above

23. The main decomposers in an ecosystem are
A. Micro-organisms B. Plants
C. Animals D. All of above

24. Plant parasites in an ecosystem are
A. Primary producers
B. Primary consumers
C. Secondary producers
D. Secondary consumers

25. __________ number of food chains are present in nature
A. 2 B. 3
C. 4 D. 5

26. The food chain begins with decomposition of dead organic matter by decomposers is
A. Predatory food chain
B. Grazing food chain
C. Saprophytic food chain
D. Detritus food chain

27. Graphical representation of ecological parameters at different trophic level in ecosystem
A. Trophics B. Food web
C. Food chain D. Pyramids

28. ભારતની સંસદે કયા વર્ષમાં જૈવવિવિધતા બિલ પસાર કર્યું?

A. 2001 B. 2002
C. 2005 D. 2007

29. પ્રકાશસંશ્લેષણ દરમિયાન છોડ દ્વારા ચોક્કસ સમયગાળા દરમિયાન એકમ દીઠ ઉત્પાદિત બાયોમાસની માત્રાને શું કહે છે?

A. પ્રાથમિક ઉત્પાદકતા
B. દ્વિતીય ઉત્પાદકતા
C. ચોખ્ખી ઉત્પાદકતા
D. કુલ ઉત્પાદકતા

30. ઊર્જા પિરામિડ હંમેશા____________ હોય છે.

A. ઊંધા B. સીધા
C. અનિયમિત D. રેખીય

31. લિન્ડેમને કયો સિદ્ધાંત આપ્યો?

A. 10% ઉર્જાનો નિયમ
B. ઇકોસિસ્ટમની વિભાવના
C. ઇકોલોજીકલ પિરામિડની વિભાવના
D. ઉપર્યુક્તમાંથી કોઈ નહિ

32. ઇકોસિસ્ટમમાં ચોક્કસ સમયે ચોક્કસ ક્ષેત્રમાં જીવંત કાર્બનિક પદાર્થોની કુલ માત્રાને શું કહે છે?

A. સ્ટેન્ડિંગ ફોર્મ
B. સ્ટેન્ડિંગ સ્ટેટ
C. સ્ટેન્ડિંગ ક્રોપ
D. સ્ટેન્ડિંગ મેટર

33. આપણા દેશમાં સુરખાબની એક માત્ર જાણીતી બ્રીડિંગ કોલોની ક્યાં મળી આવે છે?

A. થાર રણ
B. લદ્દાખ રણ
C. કચ્છનો મોટો રણ
D. શીત રણ

28. In which year Indian parliament passed biodiversity bill?

A. 2001 B. 2002
C. 2005 D. 2007

29. The amount of biomass produced per unit area over a time period by plants during photosynthesis

A. Primary productivity
B. Secondary Productivity
C. Net Productivity
D. Gross Productivity

30. Pyramid of energy is always

A. Inverted B. Upright
C. Irregular D. Linear

31. Lindeman gave

A. 10% law of energy
B. Concept of ecosystem
C. Concept of ecological pyramid
D. None of the above

32. Total amount of living organic matter present in particular area at particular time in an ecosystem is

A. Standing form
B. Standing state
C. Standing crop
D. Standing matter

33. The only known breeding colony of Greater Flamingos in our country is found in

A. The Thar desert
B. Ladakh desert
C. The Great Rann of Kutch
D. Cold desert

34. બાયોમાસની ગણતરી કરવા માટેનું સૂત્ર શું છે?

A. બાયોમાસ = વોલ્યુમ × ગ્રેવિટી

B. બાયોમાસ = વેઇટ × સ્પેસિફિક ગ્રેવિટી

C. બાયોમાસ = વોલ્યુમ × વેઇટ

D. બાયોમાસ = વોલ્યુમ × સ્પેસિફિક ગ્રેવિટી

35. નીચેનામાંથી કઈ ઇકોસિસ્ટમમાં સંખ્યાના પિરામિડ ઊંધા છે?

A. પોન્ડ ઈકોસિસ્ટમ

B. ટ્રી ઈકોસિસ્ટમ

C. ગ્રાસલેન્ડ ઈકોસિસ્ટમ

D. ફોરેસ્ટ ઈકોસિસ્ટમ

36. બાયોમાસ ઉત્પાદનના દર ને શું કહે છે?

A. શ્વસન

B. પ્રકાશસંશ્લેષણ

C. ઉત્પાદકતા

D. પરિવર્તન

37. રિયો ડી જાનેરો (1992) ની અર્થ સમિટ શાને માટે હતી?

A. બાયોસ્ફિયર રિઝર્વની સ્થાપના

B. જૈવવિવિધતાનું સંરક્ષણ

C. રેડ લીસ્ટનું સંકલન

D. વાયુ પ્રદૂષણમાં ઘટાડો

38. પ્રાથમિક ઉત્પાદકતાને કેવી રીતે વિભાજિત કરી શકાય છે?

A. GPP B. NPP

C. TPP D. A અને B બન્ને

39. મહત્તમ ઉત્પાદકતા પ્રતિ યુનિટ ક્યાં જોવા મળે છે?

A. સમશીતોષ્ણ જંગલ

B. સુકા પાનખર જંગલ

C. ઉષ્ણકટિબંધીય વરસાદનું જંગલ

D. રણ

34. Formula to calculate biomass is

A. Biomass= Volume × gravity

B. Biomass= Weight × specific gravity

C. Biomass= Volume × weight

D. Biomass= Volume × specific gravity

35. In which of the following ecosystem the pyramid of number is inverted?

A. Pond ecosystem

B. Tree ecosystem

C. Grassland ecosystem

D. Forest ecosystem

36. The rate of biomass production is known as

A. Respiration

B. Photosynthesis

C. Productivity

D. Transformation

37. Earth summit of Rio de Janeiro (1992) was for

A. Establishment of biosphere reserves

B. Conservation of biodiversity

C. Compilation of Red list

D. Reduction in air pollution

38. Primary productivity can be divided into

A. GPP B. NPP

C. TPP D. Both A and B

39. Maximum productivity per unit area is found in

A. Temperate forest

B. Dry deciduous forest

C. Tropical rain forest

D. Desert

40. ઠંડા પ્રદેશના સસ્તન પ્રાણીઓની પૂંછડીઓ, કાન, અંગો, મોઢું એ નાના હોય છે અને ગરમ પ્રદેશમાં મોટા હોય છે. આ નિયમને શું કહે છે ?

A. બર્ગરનો નિયમ
B. એલેન્સનો નિયમ
C. બર્ગમેન નિયમ
D. જોર્ડન નિયમ

40. The tails, ears, limbs, snout of mammals are smaller in colder region and larger in warmer region. This rule is known as

A. Berger rule
B. Allens rule
C. Bergman rule
D. Jordan rule

41. વન અને બરફ માટે અલ્બીડો મૂલ્ય કેટલું છે?

A. 5-10% અને 80%
B. 20-30% અને 80%
C. 50% અને 30%
D. 18% અને 30%

41. Albido value for forest and snow are respectively

A. 5-10% & 80%
B. 20-30% & 80%
C. 50% & 30%
D. 18% & 30%

42. વર્ષ દરમિયાન નીચા તાપમાને છોડ ક્યાં ઉગે છે?

A. મેગાથર્મ્સ B. મેસોથર્મ્સ
C. માઇક્રોથર્મ્સ D. હેકીસ્ટોથર્મ્સ

42. Plants growing at low temperature throughout the year

A. Megatherms B. Mesotherms
C. Microtherms D. Hekistotherms

43. ________ પ્રજાતિઓની પ્રવૃત્તિ સમુદાયની રચના નક્કી કરે છે.

A. કી પ્રજાતિઓ
B. એપેક્સ પ્રજાતિઓ
C. એડજ જાતિઓ
D. ઉપર્યુક્તમાંથી કોઈ નહિ

43. The activities of ________ species determine the structure of the community.

A. Key species
B. Apex species
C. Edge species
D. None of the above

44. ઇકોલોજી શબ્દ કોણે આપ્યો?

A. રીટર B. ઇ. હેક્કેલ
C. ઇ. પી. ઓડમ D. ડાર્વિન

44. The term Ecology was coined by

A. Reiter B. E. Haeckel
C. E. P. Odum D. Darwin

45. કોઈ ચોક્કસ વિસ્તારમાં જોવા મળતી પ્રજાતિઓ ને શું કહે છે?

A. કી સ્ટોન પ્રજાતિઓ
B. એપેક્સ પ્રજાતિઓ
C. એડજ પ્રજાતિઓ
D. સ્થાનિક જાતિઓ

45. The species found at a particular area is called

A. Key stone species
B. Apex species
C. Edge species
D. Endemic species

46. એસ્ચ્યુઅરીઝ ક્યાં જોવા મળે છે?
A. નદી સમુદ્રમાં પ્રવેશ કરે
B. સરોવરો નદીમાં પ્રવેશ કરે
C. જ્યાં નદીઓ ભેગી થાય
D. જ્યાં ઝરણાં ભેગા થાય

47. સરેરાશ દરિયાઇ સપાટીથી ઊંચાઈને શું કહે છે?
A. રેખાંશ B. અક્ષાંશ
C. અલ્ટીટ્યૂડ D. પોલ

48. વાતાવરણ કેટલા સ્તરમાં વહેંચાયેલું છે?
A. 4 B. 5
C. 6 D. 7

49. ટ્રોપોસ્ફિયરમાં ઊંચાઈ વધવાની સાથે તાપમાનમાં ઘટાડો કયા દરે થાય છે?
A. 1000 મીટર દીઠ 4.5°C
B. 1000 મીટર દીઠ 5.5°C
C. 1000 મીટર દીઠ 6.5°C
D. 1000 મીટર દીઠ 7.5°C

50. બાયોસ્ફીયર બીજા કયા નામે ઓળખાય છે ?
A. હાઇડ્રોસ્ફિયર B. એટમોસ્ફિઅર
C. ઇકોસ્ફિયર D. B અને C બન્ને

51. નીચેનામાંથી શું બાયોસ્ફિયરનો ઘટક નથી?
A. હાઇડ્રોસ્ફિયર
B. લિથોસ્ફીયર
C. એટમોસ્ફિઅર
D. ઉપરના બધા

52. આહાર શૃંખલામાં બિન-વિઘટનક્ષમ પ્રદુષકો ની સૌથી વધુ સાંદ્રતા ક્યાં આવે છે?
A. પ્રાથમિક ઉપભોક્તા
B. દ્વિતીય ઉપભોક્તા
C. તૃતીય ઉપભોક્તા
D. ટોચના ઉપભોક્તા

46. **Estuaries are the regions where**
A. river enters into the ocean
B. lakes enter into the river
C. Where to rivers merge
D. Where streams merge

47. **Height of a place from mean sea level is**
A. Longitude B. Latitude
C. Altitude D. Pole

48. **Atmosphere is divided into ______ layers**
A. 4 B. 5
C. 6 D. 7

49. **In troposphere temperature decrease with increase in altitude at a rate of**
A. 4.5°C per 1000 meter
B. 5.5°C per 1000 meter
C. 6.5°C per 1000 meter
D. 7.5°C per 1000 meter

50. **Biosphere is also known as**
A. Hydrosphere B. Atmosphere
C. Ecosphere D. Both B and C

51. **Which one of the following is not a component of Biosphere?**
A. Hydrosphere
B. Lithosphere
C. Atmosphere
D. All of above

52. **In a food chain, the highest concentration of non-biodegradable pollutants occurs in**
A. Primary consumer
B. Secondary consumer
C. Tertiary consumer
D. Top consumer

53. કાર્બનિક પદાર્થો ની વધુ માત્રા અને ઓક્સિજનની ઓછી માત્રાવાળા તળાવને શું કહે છે?

A. યુટ્રોફિક તળાવ
B. ઓલિગોટ્રોફિક તળાવ
C. ડિસ્ટ્રોફિક તળાવ
D. ઉપર્યુક્તમાંથી કોઈ નહિ

54. પ્રેરીઝ ઘાસના મેદાન ક્યાં જોવા મળે છે?

A. દક્ષિણ અમેરિકા
B. ઉત્તર અમેરિકા
C. યુરોપ અને એશિયા
D. ઓસ્ટ્રેલિયા

55. સજીવની ક્રિયાપ્રતિક્રિયા અને જીવસૃષ્ટિના આંતર સંબંધ અભ્યાસ ને શું કહે છે?

A. પર્યાવરણ વિજ્ઞાન
B. પરિસ્થિતિ વિજ્ઞાન
C. પ્રાણીશાસ્ત્ર
D. A અને B બન્ને

56. એશિયામાં સ્થિત સૌથી મોટું ખારુ તળાવ ક્યાં છે?

A. દાલ સરોવર B. વુલર સરોવર
C. ચિલ્કા સરોવર D. માનસબલ સરોવર

57. છાયાવાળા નિવાસસ્થાન ને પસંદ કરનાર વનસ્પતિને શું કહે છે?

A. મેસોફાઇટ્સ B. સાયોફાઇટ્સ
C. એપિફાઇટ્સ D. હેલિઓફાઇટ્સ

58. 'ચીપ્કો મૂવમેન્ટ' અને 'ઇકોલોજી કાયમી અર્થતંત્ર છે' જેવા સૂત્રો કોણે આપ્યા હતા?

A. મહાત્મા ગાંધી
B. જવાહરલાલ નહેરુ
C. સાલીમ અલી
D. સુંદરલાલ બહુગુણા

53. Lake with high amount of organic materials and little amount of oxygen is

A. Eutrophic lake
B. Oligotrophic lake
C. Dystrophic lake
D. None of the above

54. Prairies grassland found in

A. South America
B. North America
C. Europe and Asia
D. Australia

55. The study of interaction and inter-relationship of organisms with their environment

A. Environmental Sciences
B. Ecology
C. Zoology
D. Both A and B

56. The largest brackish water lake situated in Asia is

A. Dal lake B. Woolar lake
C. Chilka lake D. Manasbal lake

57. Plants that prefer Shady habitats are known as

A. Mesophytes B. Sciophytes
C. Epiphytes D. Heliophytes

58. Who gave the slogan of 'Chipko Movement 'Ecology is Permanent economy'?

A. Mahatama Gandhi
B. Jawaharlal Nehru
C. Salim Ali
D. Sunderlal Bahuguna

59. શેલ્ફોર્ડના લૉ ઓફ ટોલરન્સના નિયમ અનુસાર, જે સજીવો વિસ્તીર્ણ પર્યાવરણીય સહનશીલતા દર્શાવે છે કે તેઓ
A. વધુ વસ્તી સંખ્યા સાથે વ્યાપક રીતે વિસ્તરેલ હોય છે
B. વધુ વસ્તી સંખ્યા સાથે માર્યાદિત રીતે વિસ્તરેલ હોય છે
C. ઓછી વસ્તી સંખ્યા સાથે વ્યાપક રીતે વિસ્તરેલ હોય છે
D. ઓછી વસ્તી સંખ્યા સાથે માર્યાદિત રીતે વિસ્તરેલ હોય છે

59. According to Shelford's Law of Tolerance, the organisms wide environmental factor tolerance limit show
A. Wide distribution with high population size
B. Narrow distribution with high population size
C. Wide distribution with low population size
D. Narrow distribution with low population size

60. વૃક્ષ પર રહેતું એક ઓર્કિડ શું છે ?
A. સહજીવન B. સહભોજિતા
C. પરસ્પરવાદ D. પરોપજીવીકરણ

60. An orchid living on a tree exhibit
A. Symbiosis B. Commensalism
C. Mutualism D. Parasitism

61. મહત્તમ ઊર્જા કોને આવશ્યક છે ?
A. દ્વિતીય ઉપભોક્તા
B. પ્રાથમિક ઉપભોક્તા
C. પ્રાથમિક ઉત્પાદક
D. વિઘટન કરનાર

61. Maximum energy is required by
A. Secondary consumer
B. Primary consumer
C. Primary producer
D. Decomposer

62. આહાર શૃંખલામાં જુદા જુદા તબક્કાઓમાં ઊર્જાપ્રવાહના ગુણોત્તર ને શું કહે છે?
A. ઇકોલોજીકલ કેપેસીટી
B. ઇકોલોજીકલ એસિમિલેશન
C. ઇકોલોજીકલ પોટેન્શિયલ
D. ઇકોલોજીકલ એફિસિઅન્સી

62. The ratio in energy flow at different points in a food chain is known as
A. Ecological capacity
B. Ecological assimilation
C. Ecological potential
D. Ecological efficiency

63. લેમુરની પ્રજાતિ ક્યા ક્ષેત્રમાં સ્થાનિક છે ?
A. સેશેલ્સ આઇલેન્ડ
B. ગાલાપાગોસ આઇલેન્ડ
C. મેડાગાસ્કર
D. ન્યુ કેલેડોનીયા

63. Species of Lemur are endemic to which of the following area
A. Seychelles Island
B. Galapagos Island
C. Madagascar
D. New Caledonia

64. ઊર્જાપિરામિડ હંમેશા સીધા હોય છે; જેનો અર્થ થાય કે

A. માંસભક્ષકોની ઊર્જારૂપાંતર કાર્યક્ષમતા શાકાહારીઓ કરતાં વધુ સારી છે

B. ઉત્પાદકોમાં સૌથી ઓછી ઊર્જારૂપાંતર કાર્યક્ષમતા હોય છે

C. બધા ટ્રોફિક સ્તરોમાં ઊર્જારૂપાંતર કાર્યક્ષમતા સમાન છે

D. શાકાહારીની ઊર્જારૂપાંતર કાર્યક્ષમતા માંસભક્ષકો કરતાં વધુ સારી છે

64. The pyramid of energy is always upright means

A. The energy conversion efficiency of carnivores is better than herbivores

B. Producers have the lowest energy conversion efficiency

C. Energy conversion efficiency is the same in all trophic levels

D. The energy conversion efficiency of herbivores is better than carnivores

65. આનુવંશિક વિતરણ ની વિશાળ શ્રેણીવાળી પ્રજાતિઓ સ્થાનિક વસ્તીમાં વિકસિત થાય છે જેને _____ કહે છે.

A. ઇકોટાઇપ B. કોમ્યુનિટી

C. ઇકોસિસ્ટમ D. બાયોમ

65. Species with a wide range of genetic distribution evolve into a local population is called

A. Ecotype B. Community

C. Ecosystem D. Biome

66. તળાવના કયા ઝોનમાં ફાયટોપ્લેન્કટોન વિપુલ પ્રમાણમાં હોય છે?

A. લેટોરલ ઝોન B. લીમનેટીક ઝોન

C. બેન્થિક ઝોન D. પ્રોફેન્ડલ ઝોન

66. __________ zone of lake has phytoplanktons in abundance

A. Littoral zone B. Limnetic zone

C. Benthic zone D. Profundal zone

67. 'K' સિલેક્ટેડ પ્રજાતિ કઈ છે?

A. *એસ્પરગિલસ* B. ઘાસ

C. માનવ D. વરુ

67. Which one is a 'K' selected species?

A. *Aspergillus* B. Grass

C. Human D. Wolf

68. કઈ વિશિષ્ટતા સજીવોને તેના નિવાસસ્થાનની પ્રતિકૂળ પરિસ્થિતિઓમાં ટકાવી રાખે છે?

A. અનુકૂલન

B. સમાયોજન

C. દશાનુકૂલન

D. સુધારો

68. The feature that allows organisms to survive in adverse conditions of its habitat is called

A. Adaptation

B. Adjustment

C. Acclimatization

D. Amendment

69. સજીવોની વિશાળ વિવિધતાને _______ કહેવામાં આવે છે

A. વિવિધતા

B. જૈવવિવિધતા

C. વિતરણ

D. વસ્તી

69. Wide variety of living organisms is called

A. Diversity

B. Biodiversity

C. Distribution

D. Population

70. ગીધ ________ છે.
A. સર્વભક્ષી
B. વિઘટનકર્તા
C. માંસભક્ષક
D. સફાઇ કામદારો

70. **Vultures are**
A. Omnivorous
B. Decomposers
C. Carnivores
D. Scavengers

71. સૌથી નાનું કૃત્રિમ ઇકોસિસ્ટમ પસંદ કરો કે જે લાંબા સમય સુધી ટકી રહ્યું છે.
A. ફોલ્સમ બોટલ
B. ફોલ્સમ તળાવ
C. ફોલ્સમ સરોવર
D. B અને C બન્ને

71. **Select the smallest artificial ecosystem that has sustained for a long period**
A. Folsom bottle
B. Folsom pond
C. Folsom lake
D. Both B and C

72. પ્રજાતિના જૂથ કે જે જૈવિક અને અજૈવિક સ્રોતોનો એક જ રીતે ઉપયોગ કરે છે તેને શું કહે છે?
A. બાયોમ B. કોમ્યુનિટી
C. ગિલ્ડ D. એકડસ

72. **The group of species that exploits the biotic and abiotic resources in same way are known as**
A. Biome B. Community
C. Guild D. Ecads

73. ઇકોસિસ્ટમ્સનો સમૂહ ________ કહેવામાં આવે છે.
A. ઇકોસ્ફિયર
B. બાયોસ્ફીયર
C. એટમોસ્ફિયર
D. A અને B બન્ને

73. **The set of ecosystems is called**
A. Ecosphere
B. Biosphere
C. Atmosphere
D. Both A and B

74. પ્રકાશ, પોષક તત્વો અને સ્પેસ માટેની સ્પર્ધા કઈ પરિસ્થિતિમાં સૌથી વધુ તીવ્ર હોય છે.
A. જુદી જુદી નિશમાં ઉછરેલા નજીકથી સંબંધિત સજીવો
B. એક જ નિશમાં ઉછરેલા નજીકથી સંબંધિત સજીવો
C. સમાન હેબીટાટમાં ઉછરેલા દૂરથી સંબંધિત સજીવો
D. જુદી જુદી નિશમાં ઉછરેલા દૂરથી સંબંધિત સજીવો

74. **Competition for light, nutrients and space is most severe between**
A. closely related organism growing in different niches
B. closely related organisms growing in the same niche
C. distantly related organisms growing in the same habitat
D. distantly related organisms growing in different niches

75. ઇકોસિસ્ટમ વિશે નીચેના નિવેદનોમાંથી કોઈ એક પસંદ કરો જે સાચું છે.
A. પ્રાથમિક ઉપભોક્તાઓ ઉત્પાદકો પર ઓછામાં ઓછા નિર્ભર છે
B. પ્રાથમિક ઉપભોક્તાઓની સંખ્યા ઉત્પાદકોથી વધારે છે
C. ઉત્પાદકો પ્રાથમિક ઉપભોક્તાઓ કરતાં વધુ છે
D. ઉત્પાદકો સૌથી વધુ અને સૌથી શક્તિશાળી છે

76. ઘાસના મેદાનો/વૃક્ષ નિવાસનતંત્ર માં સંખ્યાકીય પિરામિડ કેવું હોય છે?
A. હંમેશાં ઉંધા
B. હંમેશાં સીધા
C. ક્યારેક ઉંધા
D. સંખ્યાકીય પિરામિડ હોતું જ નથી

77. પર્યાવરણમાં પરિવર્તન સાથે પરિવર્તનનો પ્રતિકાર કરવાની જૈવિક પ્રણાલીઓની વૃત્તિ ______ તરીકે ઓળખાય છે
A. હોમિઓસ્ટેસિસ
B. બફરિંગ
C. A અને બB બન્ને
D. ઉપર્યુક્તમાંથી કોઈ નહિ

78. બાયોસ્ફિયરમાં શેનો સમાવેશ થાય કે જે એક ઉચ્ચ સંકલિત અને ક્રિયાપ્રતિક્રિયાત્મક ક્ષેત્રનું પ્રતિનિધિત્વ કરે છે ?
A. હાઇડ્રોસ્ફિયર
B. લિથોસ્ફીયર
C. એટમોસ્ફિયર
D. ઉપરોક્ત તમામ

75. Select one of the following statements about ecosystem that is correct
A. primary consumers are least dependent upon producers
B. primary consumers out-number producers
C. producers are more than primary consumers
D. producers are the largest and most powerful

76. Pyramid of numbers in a grassland/tree ecosystem
A. Always inverted
B. Always upright
C. Sometimes inverted
D. There is no pyramid of numbers

77. Tendency of biological systems to resist change with change in environment is known as
A. Homeostasis
B. Buffering
C. Both A and B
D. None of the above

78. Biosphere represents a highly integrated and interacting zone comprising of
A. Hydrosphere
B. Lithosphere
C. Atmosphere
D. All of the above

79. સમાન પ્રકારના સજીવોની વસ્તીના સરવાળાથી ______ રચાય છે
A. કોલોની બનાવતા એકમો
B. પ્રજાતિઓ
C. જીનસ
D. સમુદાય

79. **The sum of populations of the same kind of organisms constitute**
A. Colony forming units
B. Species
C. Genus
D. Community

80. નિવાસસ્થાનની અંદર એક પ્રજાતિની વસ્તીની વિપુલતા ______ છે.
A. નિવાસ ઘનતા
B. પ્રજાતિઓની ઘનતા
C. વસ્તી ઘનતા
D. નિશ ઘનતા

80. **The abundance of a species population within its habitat is**
A. habitat density
B. species density
C. population density
D. niche density

81. નીચેનામાંથી સૌથી સ્થિર ઇકોસિસ્ટમ કઈ છે?
A. મહાસાગર B. રણ
C. પર્વત D. વન

81. **Which of the following is the most stable ecosystem?**
A. Ocean B. Desert
C. Mountain D. Forest

82. જીવન દરમિયાન સજીવની આસપાસની અથવા તેને અસર કરતી દરેક વસ્તુ સામૂહિક રીતે તેના ______ તરીકે ઓળખાય છે
A. વાતાવરણ B. આસપાસ
C. પર્યાવરણ D. ઇકોસ્ફિયર

82. **Everything that surrounds or affects an organism during its life time is collectively known as its**
A. Atmosphere B. Surroundings
C. Environment D. Ecosphere

83. ઇકોલોજીના ______ મુખ્ય સ્તર છે.
A. 5 B. 4
C. 6 D. 3

83. **There are ______ main levels of organisation of ecology**
A. 5 B. 4
C. 6 D. 3

84. સંબંધિત સ્વતંત્રતા અને કદને આધારે સમુદાયોને ______ માં વિભાજિત કરવામાં આવે છે .
A. મુખ્ય અને ગૌણ સમુદાય
B. ઉચ્ચ અને નીચા સમુદાય
C. મહત્વપૂર્ણ અને તુચ્છ સમુદાય
D. આશ્રિત અને સ્વતંત્ર સમુદાય

84. **On the basis of size and degree of relative independence communities are divided into**
A. Major and minor community
B. High and low community
C. Important and trivial community
D. Dependent and independent community

85. ______________ સારી રીતે સજીવ તેના પર્યાવરણ માં અસ્તિત્વમાં સૌથી મહત્વપૂર્ણ નિર્ધારકો છે.

A. જૈવિક પરિબળો
B. અજૈવિક પરિબળો
C. તાપમાન અને વરસાદ
D. ઊર્જા અને સબસ્ટ્રેટમ

86. ______________ જંગલો દરિયાઇ અને પાર્થિવ ઇકોસિસ્ટમ વચ્ચેના ઇકોટોનનું પ્રતિનિધિત્વ કરે છે.

A. સમશીતોષ્ણ
B. ઉષ્ણકટિબંધીય
C. બોરિયલ
D. ચેર

87. બાયોમની લાક્ષણિકતા શું છે ?

A. વનસ્પતિ અને પ્રાણી જીવન
B. આબોહવા
C. માટીના પ્રકારો
D. ઉપરોક્ત તમામ

88. બાયોસ્ફીયર ક્યાં નથી હોતા?

A. ઉત્તર ધ્રુવની ચરમસીમા પર
B. દક્ષિણ ધ્રુવની ચરમસીમા પર
C. A અને B બન્ને
D. ઉપર્યુક્તમાંથી કોઈ નહિ

89. જયારે એક પ્રજાતિને નુકસાન થાય છે, ત્યારે બીજીને અસર ન થાય તેવા સંબંધને શું કહેવાય છે?

A. અમ્મેન્સાલિઝમ
B. સહભોજિતા
C. તટસ્થતા
D. પરસ્પરવાદ

85. ______________ are the most important determinants of where and how well an organism exists in its environment

A. Biotic factors
B. Abiotic factors
C. Temperature and rainfall
D. Energy and substratum

86. ______________ forests represent an ecotone between marine and terrestrial ecosystem.

A. Temperate
B. Tropical
C. Boreal
D. Mangrove

87. Biomes are characterized by

A. Vegetation and animal life
B. Climate
C. General soil type
D. All of the above

88. Biosphere is absent at

A. extremes of the North pole
B. extremes of the South pole
C. Both A and B
D. None of the above

89. One species is harmed, the other is unaffected is

A. Ammensalism
B. Commensalism
C. Neutralism
D. Mutualism

90. હાઇડ્રોલોજિક ચક્ર એ પૃથ્વી-વાતાવરણ પ્રણાલીમાં પાણીનું સતત પરિભ્રમણ છે જે __________ દ્વારા ચાલે છે.

A. હાઇડ્રોલોજિકલ ઊર્જા

B. પરમાણુ ઊર્જા

C. સૌર ઊર્જા

D. પવન ઊર્જા

91. નેશનલ ગંગા રિવર બેસિન ઓથોરિટીના અધ્યક્ષ કોણ હોય છે?

A. રાષ્ટ્રપતિ B. પ્રધાનમંત્રી

C. મુખ્યમંત્રી D. ઉપરાષ્ટ્રપતિ

92. પરંપરાગત માનવ જીવન સાથે જૈવવિવિધતાના સંરક્ષણ માટેની સૌથી મહત્વપૂર્ણ વ્યૂહરચના __________ ની સ્થાપના છે.

A. વનસ્પતિ ઉદ્યાનો

B. રાષ્ટ્રીય ઉદ્યાનો

C. બાયોસ્ફિયર રિઝર્વ

D. વન્યપ્રાણી અભયારણ્ય

93. ચેમ્પિયન અને શેઠ મુજબ ભારતના જંગલોનું __________ પ્રકારમાં વર્ગીકરણ કરવામાં આવે છે.

A. 12 B. 14

C. 16 D. 18

94. દરિયાઇ નિવાસનતંત્રો નું તેમના મીઠાના જથ્થાને આધારે કેટલા પ્રકારમાં વર્ગીકરણ કરવામાં આવે છે?

A. તાજા પાણી અને દરિયાઇ ઇકોસિસ્ટમ

B. તાજા પાણી, મરીન અને બ્રેકિશ વોટર ઇકોસિસ્ટમ

C. તાજા અને ખારા પાણીના ઇકોસિસ્ટમ

D. દરિયાઈ નિવસનતંત્રો મીઠાના જથ્થાને આધારે વર્ગીકૃત નથી

90. The hydrologic cycle is the continuous circulation of water in the Earth-atmosphere system which is driven by

A. hydrological energy

B. nuclear energy

C. solar energy

D. wind energy

91. National Ganga River Basin Authority is chaired by the

A. President B. Prime Minister

C. Chief Minister D. Vice President

92. The most important strategy for the conservation of biodiversity together with traditional human life is the establishment of

A. botanical gardens

B. national parks

C. biosphere reserves

D. wildlife sanctuaries

93. Forest types in India are classified by Champion and Seth into __________ types

A. 12 B. 14

C. 16 D. 18

94. Aquatic ecosystems are classified based on their salt content into

A. Freshwater and Marine ecosystem

B. Freshwater, Marine and Brackish water ecosystem

C. Fresh and Saline water ecosystem

D. Aquatic ecosystems are not classified of the basis of salt content

95. જૈવવિવિધતા હેરિટેજ સાઇટ અમિનપુર તળાવને જૈવવિવિધતા અધિનિયમ હેઠળ દેશમાં પ્રથમ જૈવવિવિધતા હેરિટેજ સાઇટનો દરજ્જો ક્યારે મળ્યો?

A. 2002 B. 2004
C. 2006 D. 2008

96. ભારતે તેની રાષ્ટ્રીય જૈવવિવિધતા એક્શન પ્લાન (NBAP) ને અપડેટ કરીને શું ક્રિયાન્વિત કર્યું?

A. 2014 માં 8 રાષ્ટ્રીય જૈવવિવિધતા લક્ષ્યાંક
B. 2014 માં 12 રાષ્ટ્રીય જૈવવિવિધતા લક્ષ્યાંક
C. 2012 માં 8 રાષ્ટ્રીય જૈવવિવિધતા લક્ષ્યાંક
D. 2012 માં 12 રાષ્ટ્રીય જૈવવિવિધતા લક્ષ્યાંક

97. હિમાલયની પર્વત ઇકોસિસ્ટમ્સ અને જૈવિક સંસાધનો પર પર્યટનની અસરોને કારણે ચિંતાનો વિષય છે કારણ કે

A. હિમાલયની ઉચ્ચ જૈવવિવિધતા અને પર્યાવરણીય સંવેદનશીલતા છે
B. હિમાલયની ઊંચાઈ વધુ છે
C. હિમાલયમાં પ્રદૂષણ વધુ છે
D. હિમાલયનું કઠોર વાતાવરણ છે

98. જૈવવિવિધતાને ત્રણ સ્તરે આંકવામાં આવે છે, તે કયા છે?

A. જિનેટિક, મોર્ફોલોજિકલ અને ઈકોસિસ્ટમ
B. મોર્ફોલોજિકલ, સ્પીસીસ અને ઈકોસિસ્ટમ
C. જિનેટિક, સ્પીસીસ અને ઈકોસિસ્ટમ
D. જિનેટિક, મોર્ફોલોજિકલ અને સ્પીસીસ

99. ઇકોસિસ્ટમ્સ વચ્ચેની વિવિધતાની તુલના જે સામાન્ય રીતે ઇકોસિસ્ટમ્સ વચ્ચેની પ્રજાતિઓની માત્રામાં ફેરફાર તરીકે માપવામાં આવે છે તેને શું કહે છે?

A. આલ્ફા વિવિધતા B. બીટા વિવિધતા
C. ગામા વિવિધતા D. ડેલ્ટા વિવિધતા

95. First Biodiversity Heritage Site Ameenpur Lake gets the status of the first Biodiversity Heritage Site in the country under the biodiversity act

A. 2002 B. 2004
C. 2006 D. 2008

96. India updated its National Biodiversity Action Plan (NBAP) and developed

A. 8 National Biodiversity Targets in 2014
B. 12 National Biodiversity Targets in 2014
C. 8 National Biodiversity Targets in 2012
D. 12 National Biodiversity Targets in 2012

97. The impacts of tourism on Himalayan mountain ecosystems and biological resources are of great concern because of

A. High biodiversity and environmental sensitivity of the Himalayas
B. Altitude of the Himalayas
C. Pollution in the Himalayas
D. Harsh climate of the Himalayas

98. Biodiversity is considered to exist at three levels

A. Genetic, Morphological and Ecosystem
B. Morphological, Species and Ecosystem
C. Genetic, Species and Ecosystem
D. Genetic, Morphological and Species

99. Comparison of diversity between ecosystems, usually measured as the change in amount of species between the ecosystems

A. Alpha diversity B. Beta diversity
C. Gamma diversity D. Delta diversity

100. ક્યાં કોરલ રીફ જોવા મળતા નથી?

A. આંદામાન અને નિકોબાર દ્વીપ સમૂહ

B. કચ્છનો અખાત

C. સુંદરવન

D. મન્નારનો અખાત

100. Which of the following do not have coral reefs?

A. Andaman and Nicobar Islands

B. Gulf of Kachchh

C. Sunderbans

D. Gulf of Mannar

Answers | જવાબો

Que.	Ans.	Que.	Ans.	Que.	Ans.	Que.	Ans.	Que.	Ans.	Que.	Ans.	Que.	Ans.
1.	C	**16.**	B	**31.**	A	**46.**	A	**61.**	C	**76.**	B	**91.**	B
2.	D	**17.**	A	**32.**	C	**47.**	C	**62.**	D	**77.**	A	**92.**	C
3.	A	**18.**	C	**33.**	C	**48.**	B	**63.**	C	**78.**	D	**93.**	C
4.	C	**19.**	C	**34.**	D	**49.**	C	**64.**	D	**79.**	B	**94.**	B
5.	B	**20.**	D	**35.**	B	**50.**	C	**65.**	A	**80.**	D	**95.**	A
6.	A	**21.**	D	**36.**	C	**51.**	D	**66.**	B	**81.**	A	**96.**	B
7.	B	**22.**	D	**37.**	B	**52.**	D	**67.**	C	**82.**	C	**97.**	A
8.	B	**23.**	A	**38.**	D	**53.**	A	**68.**	A	**83.**	C	**98.**	C
9.	B	**24.**	B	**39.**	C	**54.**	C	**69.**	B	**84.**	A	**99.**	B
10.	C	**25.**	B	**40.**	B	**55.**	B	**70.**	D	**85.**	B	**100.**	C
11.	C	**26.**	C	**41.**	A	**56.**	C	**71.**	A	**86.**	D		
12.	B	**27.**	D	**42.**	C	**57.**	B	**72.**	C	**87.**	D		
13.	A	**28.**	B	**43.**	A	**58.**	D	**73.**	B	**88.**	C		
14.	B	**29.**	A	**44.**	B	**59.**	A	**74.**	B	**89.**	A		
15.	D	**30.**	B	**45.**	D	**60.**	B	**75.**	C	**90.**	C		

Chapter 7
પ્રકરણ ૭

Environment and Conservation
પર્યાવરણ અને સંરક્ષણ

Forests of the World | વિશ્વમાં વન વિસ્તાર

Total Forests on Earth	4.06 billion ha
% of total land area	31 %
Highest net gain of forest area (2010-2020)	Asia
Highest net loss of forest area (2010-2020	Africa
Annual rate of deforestation (2015-2020)	10 million ha

GFRA 2020

વિશ્વના જંગલો

પૃથ્વી પરના કુલ જંગલો	4.06 અબજ હેક્ટર
કુલ જમીન ક્ષેત્રનો %	31%
દસ વર્ષમાં વન વિસ્તારનો સૌથી વધુ વધારો નોંધાવતો ખંડ (2010-2020)	એશિયા
દસ વર્ષમાં વન વિસ્તારનો સૌથી વધુ ઘટાડો નોંધાવતો ખંડ (2010-2020)	આફ્રિકા
વાર્ષિક વનવિનાશનો દર (2015-2020)	10 મિલિયન હેક્ટર

GFRA 2020

Forest area in the World | વિશ્વમાં વન વિસ્તાર

Country	% of Forest area in the World
Russian Federation	20
Brazil	12
Canada	9
USA	8
China	5
Rest of the world	46

GFRA 2020

Important Days of Forest and Environment Conservation

વન અને પર્યાવરણ સંરક્ષણના મહત્વપૂર્ણ દિવસો

Month	Date	Day	Year of Starting	Agency
February	2	World Wetland Day	1971	Wetlands International
	10	World Pulses Day	2019	UN
	27	International Polar Bear Day	2011	Polar Bears International
	28	National Science Day	1987	National Council for Science and Technology
March	3	World Wildlife Day	2013	UN
	14	International Day of Action for Rivers	1997	International Rivers
	20	World Sparrow Day	2010	Nature Forever Society
	21	International Day of Forests	2012	UN
	21	World Wood Day	2014	World Wood Day Foundation
	22	World Water & Sanitation Day	1993	UN
	23	World Meteorological Day	1961	WMO
April	7	World Health Day	1950	WHO
	10	World Atmosphere Day	1974	UN
	18	World Heritage Day	1982	UNESCO/ ICOMOS
	21	World fish migration Day	2014	FFSG-IUCN, WWF
	22	World Earth Day	1970	UNESCO
May	Second Saturday of May	World Migratory Bird Day	2006	UNEP
	11	National Technology Day	1999	Govt. of India
	14	Endemic Bird Day	2010	eBird
	20	International Endangered Species Day	2006	US Senate
	22	World Biodiversity Day	2001	UN
	23	World Turtle Day	2000	American Tortoise Rescue
June	5	World Environment Day	1972	UN
	5	International Day for the Fight against Illegal, Unreported and Unregulated Fishing	2018	UN
	7	World Food Safety Day	2019	UN

Month	Date	Day	Year of Starting	Agency
June	08	World Oceans Day	1992	UNCED
	09	Coral Triangle Day	2012	Coral Triangle Initiative on Coral Reefs, Fish
	15	Global Wind Day	2007	World Wind Energy Council (GWEC) and WindEurope (WE)
	17	World Day to Combat Desertification	1995	UN
	21	World Giraffe Day	2014	Giraffe Conservation Federation / Montgomery Zoo
	29	International Day of the Tropics	2016	UN
July	1st week of July	Van Mahotsava	1950	Govt of India
	3	World Seabird Day	2017	World Seabird Unioun
	11	World Population Day	1990	UNDP
	26	International Day for the Conservation of the Mangrove Ecosystem	2016	UNESCO
	29	International Tiger Day	2010	Saint Petersburg Tiger Summit
August	9	International Day of the World's Indigenous Peoples	1995	UN
	10	World Lion day	2013	Big Cat Initiative and National Geographic
	12	World Elephant Day	2012	The Elephant Reintroduction Foundation
	Third Saturday in August	National Honey Bee Day	2009	USA
	29	International Day against Nuclear Tests	2009	UN
September	07	International Day of Clean Air for Blue Skies	2020	UN
	08	World Clean-up Day	2018	Let's Do It World (LDIW)
	16	International Day for the Preservation of the Ozone Layer	1994	UN
	18	World Water Monitoring Day	2003	America's Clean Water Foundation (ACWF)

Month	Date	Day	Year of Starting	Agency
September	21	Zero Emissions Day	2008	Ken Wallace
	22	World Rhino Day	2010	WWF South Africa
	26	World Environment Health Day	2011	International Federation of Environmental Health (IFEH)
	26	International Day for the Total Elimination of Nuclear Weapons	2013	UN
	27	World Tourism Day	1980	UNWTO
	Last Sunday of September	World River Day	2005	UN
October	2nd to 8th October	Wildlife Week	1957	NBWL
	First Monday of October	World Habitat Day	1986	UN
	04	World Animal Day	1929	Naturewatch Foundation first World Animal Day on 24 March, 1925 at the Sport Palace in Berlin, Germany. The date for the same was moved to 4 October in 1929.
	06	World Wildlife Day	2013	CITES
	second Saturday in May and October	World Migratory Bird Day	2006	Agreement on the Conservation of African-Eurasian Migratory Waterbirds (AEWA), CMS, UNEP
	13	International Day for Disaster Risk Reduction	1989	UN
	16	World Food Day	1979	UN
	22	World Energy Day	2012	World Energy Forum
	24	International Day of Climate Action		
November	6	International Mangrove Day	2016	UNESCO
	6	International Day for Preventing the Exploitation of the Environment in War and Armed Conflict	2001	UNEP
	10	World Science Day for Peace and Development	2002	UNESCO

Month	Date	Day	Year of Starting	Agency
December	5	World Soil Day	2014	UN
	11	International Mountain Day	2003	UN
	14	National Energy Conservation Day	1991	The Bureau of Energy Efficiency (BEE), under Ministry of Power
	19	International Day for the Preservation of the Ozone Layer	1995	UN

ENVIS, WII 2020
UN 2020

International Years observed by UN to promote Conservation

સંરક્ષણને પ્રોત્સાહન આપવા યુ.એન. દ્વારા આંતરરાષ્ટ્રીય વર્ષોની ઉજવણી

World Population Year	1974
International Year of the Ocean	1998
International Year of Mountains	2002
International Year of Ecotourism	2002
International Year of Freshwater	2003
International Year of Deserts and Desertification	2006
International Polar Year	2007–2009
International Year of Planet Earth	2008
International Year of Sanitation	2008
International Year of Natural Fibres	2009
Year of the Gorilla	2009
International Year of Biodiversity	2010
International Year of Forests	2011
International Year of Sustainable Energy for All	2012
International Year of Water Cooperation	2013
International Year of Family Farming	2014
International Year of Soils	2015
International Year of Pulses	2016
International Year of Sustainable Tourism for Development	2017
International Year of Plant health	2020
International Year of Fruits and Vegetables	2021

UN 2020

International Decades observed by UN to promote Conservation

સંરક્ષણને પ્રોત્સાહન આપવા યુ.એન. દ્વારા આંતરરાષ્ટ્રીય દાયકાઓની ઉજવણી

International Drinking Water Supply and Sanitation Decade	1981-1990
International Decade for Natural Disaster Reduction	1990-1999
Decade of the World's Indigenous People	1994-2004
International Decade for Action, "Water for Life"	2005-2015
United Nations Decade on Biodiversity	2011-2020
United Nations Decade of Sustainable Energy for All	2014-2024
International Decade for Action "Water for Sustainable Development"	2018-2028
United Nations Decade of Ocean Science for Sustainable Development	2021-2030
United Nations Decade on Ecosystem Restoration	2021-2030

UN 2020

Important Revolutions | મહત્વપૂર્ણ ક્રાંતિઓ

Revolution	Focus area
Black	Petroleum Production
Blue	Fish Production
Golden	Fruit Production
Green	Food grain Production
Grey	Fertilizer Production
Pink	Prawn Production
Red	Meat/Tomato Production
Round	Potato Production
Silver	Egg/Poultry Production
White	Milk Production
Yellow	Oil seeds Production

Multiple Choice Questions | હેતુલક્ષી પ્રશ્નો

1. કોરલ રીફ અને દરિયાઇ જીવન માટે જોખમી એવા ઝેરી સનસ્કીનો ને સત્તાવાર રીતે પ્રતિબંધ મૂકનાર વિશ્વનો પ્રથમ દેશ કયો છે?
A. ભારત B. ફિલિપાઇન્સ
C. પલાઉ D. ઓસ્ટ્રેલિયા

2. ભેજવાળાપ્રદેશના શંકુદ્રુપ જંગલો હેઠળ બરછટ ટેક્ષ્ચરને શું કહેવાય છે ?
A. ઇનસેપ્ટિસોલ્સ B. વર્ટિસોલ્સ
C. ઓક્સિસોલ્સ D. સ્પોડોસોલ્સ

3. કઈ કંપનીએ તાજેતરમાં જાહેરાત કરી હતી કે તેઓ 2030 સુધીમાં તેમનો અડધો કાર્બન ઉત્સર્જન કાપશે?
A. ફેસબુક B. ગુગલ
C. માઈક્રોસોફ્ટ D. રિલાયન્સ

4. કયો એવોર્ડ પર્યાવરણનો નોબેલ પ્રાઈઝ તરીકે ઓળખાય છે?
A. ટાઇલર પ્રાઈઝ
B. વૈશ્વિક પર્યાવરણ પ્રાઈઝ
C. ચેમ્પિયન્સ ઓફ અર્થ
D. હેન્ઝ એવોર્ડ

5. તાજેતરમાં પ્રકાશિત વર્લ્ડ વાઇલ્ડલાઇફ ક્રાઇમ રિપોર્ટ 2020 મુજબ વિશ્વનું સૌથી વધુ શિકાર થતું સસ્તન પ્રાણી કયું છે?
A. વાઘ B. પેંગોલિન
C. ચિત્તો D. હાથી

6. કયું પ્રાણી ક્લાઈમેટ ચેન્જની અસરોને લીધે લુપ્ત થઈ શકે છે?
A. ધ્રુવીય રીંછ B. પેંગ્વિન
C. સ્નો લેપર્ડ D. સીલ

1. **The first country in the world to officially ban the toxic sun screens that are hazardous to coral reefs and marine life of seas**
A. India B. Philippines
C. Palau D. Australia

2. **Coarse textured deposits under coniferous forests of humid region consists of**
A. Inceptisols B. Vertisols
C. Oxisols D. Spodosols

3. **The company that recently announced that they will cut their half carbon emission by 2030**
A. Facebook B. Google
C. Microsoft D. Reliance

4. **Which award is known as the noble prize of environment**
A. Tyler prize
B. Global environment prize
C. Champions of earth
D. Heinz award

5. **The recently released World Wildlife Crime Report 2020 revealed that the world's most trafficked mammal is**
A. Tiger B. Pangolin
C. Leopard D. Elephant

6. **The animal which might become extinct due to impacts of climate change**
A. Polar Bear B. Penguin
C. Snow leopard D. Seal

7. રાષ્ટ્રીય મહાસાગર અને વાતાવરણીય વહીવટ (NOAA) દ્વારા તાજેતરમાં કરવામાં આવેલા એક અભ્યાસ મુજબ કયા દેશના કોરલ રીફ્સમાં નોંધપાત્ર ઘટાડો થયો છે?'

A. ભારત B. યુનાઈટેડ સ્ટેટ્સ
C. મલેશિયા D. ઓસ્ટ્રેલિયા

8. ભારતની સર્વોચ્ચ અદાલતે તાજેતરમાં રાજસ્થાનના ડેઝર્ટ નેશનલ પાર્કની કઈ પ્રજાતિને બચાવવા માટે પાવરલાઇનો ભૂગર્ભમાં મૂકવાનો ચુકાદો આપ્યો?

A. રણનું શિયાળ
B. ગીધ
C. ગ્રેટ ઈન્ડિયા બસ્ટાર્ડ
D. હેરિયર્સ

9. એક્ઝોટીક સ્પીસીસ ________ તરીકે પણ ઓળખાય છે

A. ફ્લેગશીપ સ્પીસીસ
B. અમ્બ્રેલા સ્પીસીસ
C. એલીયન સ્પીસીસ
D. ઇન્ડિકેટર સ્પીસીસ

10. રેટર ______ ના પિતા તરીકે ઓળખાય છે.

A. વનવિદ્યા
B. પરિસ્થિતિ વિજ્ઞાન
C. વનસ્પતિશાસ્ત્ર
D. બાગાયત

11. ઓટોઈકોલોજી ________ નો અભ્યાસ છે.

A. જીવસૃષ્ટિ સાથે જીવંત સ્વરૂપોની ક્રિયાપ્રતિક્રિયા
B. જૈવિક અને અજૈવિક પરિબળોની ક્રિયાપ્રતિક્રિયા
C. એક પ્રજાતિનો તેના પર્યાવરણ સાથે સંબંધ
D. વિવિધ જાતિઓના જૂથનો તેમના પર્યાવરણ સાથે સંબંધ

7. Which country's coral reefs have degraded significantly, as per a recent study by National Oceanic and Atmospheric Administration (NOAA)?

A. India B. United States
C. Malaysia D. Australia

8. Supreme court of India recently passed judgement to put powerlines underground to save which species of Desert National Park, Rajasthan?

A. Desert fox
B. Vultures
C. Great India Bustard
D. Harriers

9. Exotic species are also known as

A. Flagship species
B. Umbrella species
C. Alien species
D. Indicator species

10. Reiter is the father of

A. Forestry
B. Ecology
C. Botany
D. Horticulture

11. Autoecology is the study of

A. Interaction of living forms with ecosystem
B. Interaction of biotic and abiotic factors
C. relationship of single species with its environment
D. relationship of the group of different species with their environment

12. પ્રિસેન્ટિવ પ્રજાતિઓ __________ તરીકે પણ ઓળખાય છે.
   A. સર્વદેશીય પ્રજાતિઓ
   B. ભયગ્રસ્ત જાતિઓ
   C. દુર્લભ પ્રજાતિઓ
   D. સ્થાનિક જાતિઓ

13. ઝૂપ્લાંકટોન્સ __________ હોય છે.
   A. માઇક્રોસ્કોપિક, ઇનએક્ટિવ ફ્લોટિંગ વનસ્પતિ
   B. માઇક્રોસ્કોપિક, ઇનએક્ટિવ ફ્લોટિંગ પ્રાણીઓ
   C. માઇક્રોસ્કોપિક, એક્ટિવ ફ્લોટિંગ પ્રાણીઓ
   D. માઇક્રોસ્કોપિક, એક્ટિવ ફ્લોટિંગ વનસ્પતિ

14. બેંથોનિક સજીવો __________ છે.
   A. સમુદ્રના સેડેન્ટરી સજીવો
   B. સમુદ્ર સપાટી પરના સજીવ
   C. જમીનના નિષ્ક્રિય સજીવ
   D. જમીનની સપાટી પરના સજીવ

15. એસ્ટિવેશન શબ્દ __________ માટે વપરાય છે.
   A. શિયાળાની ઊંઘ
   B. ઉનાળાની ઊંઘ
   C. શુષ્ક સમયગાળામાં લાંબા સમય સુધી નિષ્ક્રિયતા
   D. બંને A અને B

16. પર્યાવરણમાં બે ઇકોસિસ્ટમ્સ અથવા સજીવના સમુદાયો વચ્ચેની જાતિઓની રચનામાં ક્રમશ: સતત ફેરફાર__________ તરીકે ઓળખાય છે.
   A. ઇકો ક્લાઇન
   B. ઇકોટાઇપ
   C. ઇકોટોન
   D. કો-ઇકોટોન

12. Precentive species is also known as
   A. Cosmopolitan species
   B. Endangered species
   C. Rare species
   D. Endemic species

13. Zooplanktons are
   A. Microscopic, inactive floating plants
   B. Microscopic, inactive floating animals
   C. Microscopic, active floating animals
   D. Microscopic, active floating plants

14. Benthonic organisms are
   A. Sedentary organisms of sea
   B. Surface organisms of sea
   C. Sedentary organisms of land
   D. Surface organisms of land

15. Aestivation is the term used for
   A. Winter sleep
   B. Summer sleep
   C. Prolonged dormancy in dry period
   D. Both A and B

16. A gradual, continuous change in the species composition between two ecosystems or communities of organisms across an environmental gradient is known as
   A. Ecocline
   B. Ecotype
   C. Ecotone
   D. Co-ecotone

17. __________ એ એક એવી સ્થિતિ છે જે પ્રતિકૂળ પર્યાવરણીય પરિસ્થિતિઓના નિયમિત અને રિકરિંગ પીરિયડના પ્રતિભાવ તરીકે વિલંબ પ્રસ્થાપિત કરે છે.

A. મિટિગેશન

B. ડાયપોઝ

C. બિહેવિયરલ ડોર્મન્સી

D. ફિઝિયોલોજીકલ ડોર્મન્સી

17. **__________ physiological state is the delay in development in response to regularly and recurring periods of adverse environmental conditions.**

A. Mitigation

B. Diapause

C. Behavioural dormancy

D. Physiological dormancy

18. છીછરા પાણીથી ભરાયેલાની ચાણવાળા વિસ્તાર ________ તરીકે ઓળખાય છે

A. તળાવો

B. ઘાસના મેદાનો

C. જલપ્લાવિત વિસ્તારો

D. મેદાનો

18. **Low lying areas covered with shallow water are known as**

A. Ponds

B. Grasslands

C. Wetlands

D. Grounds

19. જળચર છોડ અથવા પ્રાણીઓ કે જે સક્રિય રીતે તરણ કરવામાં સક્ષમ છે તેને ________ કહેવામાં આવે છે.

A. ફાયટોપ્લાંકટોન

B. નેક્ટોન્સ

C. ઝૂપ્લાંકટોન

D. પ્લેન્કટોન્સ

19. **Those aquatic plants or animals which are capable of swimming actively are called-**

A. Phytoplankton

B. Nektons

C. Zooplankton

D. Planktons

20. તળાવના કાંઠે મહત્તમ મૂળિયા ધરાવતી વનસ્પતિ સાથે જોવા મળતા વિસ્તારને __________ ઝોન તરીકે ઓળખવામાં આવે છે.

A. પ્રોફેન્ડલ ઝોન B. લિમિનેટિક ઝોન

C. લીટોરલ ઝોન D. બેન્થિક ઝોન

20. **The zone which is found at bank of lake with maximum rooted vegetation is known as**

A. Profundal zone B. Limnetic zone

C. Littoral zone D. Benthic zone

21. હોમ રેન્જમાં પ્રજનન માટે ચિહ્નિત કરેલ ચોક્કસ જગ્યા __________ તરીકે ઓળખાય છે.

A. બાઉન્ડ્રી B. હેબીટાટ

C. ટેરીટરી D. નિશ

21. **In home range particular space marked by an individual for breeding is known as**

A. Boundary B. Habitat

C. Territory D. Niche

2. પ્રાણીની વર્તણુક જે બીજાને લાભ કરે તેને શું કહેવાય છે?

A. એમેન્સાલિઝમ

B. પરસ્પરવાદ

C. સહભોજિતા

D. ઉલ્ટ્રઇઝમ

22. Behavior of an animal that benefits another at its own expense is known as

A. Amensalism

B. Mutualism

C. Commensalism

D. Altruism

3. પ્રકાશની જરૂરિયાત મુજબ, ઊંડું તળાવ _____ ઝોનમાં વહેંચાયેલું છે.

A. 6 B. 5

C. 4 D. 3

23. According to the need of light, a deep lake is divided into _______ zones.

A. 6 B. 5

C. 4 D. 3

4. વિવિધ આબોહવાની સ્થિતિમાં રહેતા સમાન પ્રજાતિના સભ્યોમાં કેટલાક આનુવંશિક રીતે જોવા મળતી ભિન્નતા __________તરીકે ઓળખાય છે.

A. ઇકોલોજીકલ રેસ B. એકેડસ

C. બ્રીડ D. A અને C બંને

24. Members of same species living in different climatic conditions have some genetically variations known as

A. Ecological races B. Ecades

C. Breed D. Both A and C

5. લાઇકેન અને માઇકોરિઝા ______________ નાં ઉદાહરણો છે

A. પ્રોટોકોપરેશન

B. સહભોજિતા

C. સિમ્બાયોટિક એસોસિએશન

D. અલ્ટરૂઈઝમ

25. Lichens and Mycorrhizae are the examples of

A. Protocooperation

B. Commensalism

C. Symbiotic associations

D. Altruism

6. ઉજ્જડ વિસ્તાર પર પ્લાન્ટ કોમ્યુનિટીનો વિકાસ_____________ છે.

A. બોટનિકલ સક્શેશન

B. બાયોટિક સક્શેશન

C. ઇકોલોજીકલ સક્શેશન

D. એબિઓટિક સક્શેશન

26. Development of plant community on barren area is

A. Botanical succession

B. Biotic succession

C. Ecological succession

D. Abiotic succession

7. રાષ્ટ્રીય વાઘ સંરક્ષણ સત્તામંડળ (NTCA) નું મુખ્ય મથક ક્યાં આવેલું છે?

A. દહેરાદૂન B. અમદાવાદ

C. હૈદરાબાદ D. નવી દિલ્હી

27. Headquarter of National Tiger Conservation Authority (NTCA) is located in

A. Dehradun B. Ahmedabad

C. Hyderabad D. New Delhi

28. વન સંરક્ષણ અધિનિયમ _____ વર્ષમાં લાગુ કરાયો હતો.

A. 1982 B. 1980

C. 1986 D. 1984

28. Forest Conservation Act was enacted in the year

A. 1982 B. 1980

C. 1986 D. 1984

29. રામસાર કન્વેન્શન શાને લાગતું છે ?

A. પર્વતો

B. નદીઓ

C. જલપ્લાવિત વિસ્તારો

D. રણ

29. Ramsar Convention refers to the conservation of

A. Mountains

B. Rivers

C. Wetlands

D. Deserts

30. ભારતમાં વન્યપ્રાણી સંરક્ષણ ચળવળનો _____ દ્વારા પ્રારંભ કરાયો હતો

A. BNHS B. IUCN

C. WWF D. CITES

30. First wildlife conservation movement in India was started by

A. BNHS B. IUCN

C. WWF D. CITES

31. તેમના મૂળ સ્થાને વનસ્પતિ અને પ્રાણીસૃષ્ટિનું સંરક્ષણ શું છે?

A. Ex-situ સંરક્ષણ

B. જર્મપ્લાઝમ સંરક્ષણ

C. *In-situ* સંરક્ષણ

D. નેટિવ સંરક્ષણ

31. _____ is the conservation of flora and fauna at their native place

A. Ex-situ conservation

B. Germplasm conservation

C. *In-situ* conservation

D. Native conservation

32. વન્યપ્રાણી સંરક્ષણ અધિનિયમનું કયું શેડ્યૂલ દુર્લભ અને જોખમી જાતિઓના સંરક્ષણ માટે છે?

A. શેડ્યૂલ I

B. શેડ્યૂલ II

C. શેડ્યૂલ III

D. શેડ્યૂલ VI

32. Which schedule of WPA deals with Conservation of rare and endangered species

A. Schedule I

B. Schedule II

C. Schedule III

D. Schedule VI

33. રિયો ડી જનેરો ની અર્થ સમિટનો મુખ્ય હેતુ શો હતો?

A. બાયોસ્ફિયર રિઝર્વની સ્થાપના

B. રેડ લીસ્ટનું સંકલન

C. જૈવવિવિધતાનું સંરક્ષણ

D. ગ્રીન-હાઉસ ગેસિસ ઘટાડવા

33. The main motive of Earth summit of Rio de Janeiro was

A. Establishment of biosphere reserves

B. Compilation of Red list

C. Conservation of biodiversity

D. Reduced green-house gasses

34. મૂળ ગુણવત્તાને અસર કર્યા વિના શુષ્ક સ્વરૂપમાં લીલા ઘાસચારાનું સંરક્ષણ શું છે?
A. હે
B. સોલીડ સ્ટેટ ફર્મેન્ટેશન
C. સિલેજ
D. ફોડર

34. The conservation of green forage into dry form without affecting the quality of original material
A. Hay
B. Solid state fermentation
C. Silage
D. Fodder

35. રિમોટ સેન્સિંગ શબ્દ કોણે આપ્યો?
A. પિશારોથ રામ પિશારોતિ
B. શ્રીમતી એવલિન પ્રુટ
C. જેસી બોઝ
D. મણિકનંદન વી સી

35. The term remote sensing was coined by
A. Pisharoth Rama Pisharoty
B. Ms. Evelyn Pruitt
C. J. C. Bose
D. Maniknandan V. C.

36. આગના નિર્માણ માટે કયા પરિબળો જવાબદાર છે?
A. હવામાન
B. દાહક પદાર્થો
C. ટોપોગ્રાફી
D. ઉપરોક્ત તમામ

36. Which is/are the factors responsible for make up the fire environment
A. Weather
B. Inflammable materials
C. topography
D. All of the above

37. વિશ્વ પર્યાવરણ દિવસ ક્યારે ઉજવવામાં આવે છે?
A. 13 ડિસેમ્બર B. 5 જૂન
C. 12 એપ્રિલ D. 27 જુલાઈ

37. World environment day is celebrated on
A. 13 December B. 5 June
C. 12 April D. 27 July

38. રાષ્ટ્રીય પર્યાવરણ સંરક્ષણ અધિનિયમ ક્યારે લાગુ કરાયો?
A. 1983 B. 1986
C. 1987 D. 1989

38. National Environment Protection Act was enacted
A. 1983 B. 1986
C. 1987 D. 1989

39. સ્ટોકહોમમાં માનવ પર્યાવરણ પર પરિષદ ક્યારે યોજાઇ હતી?
A. 1969 B. 1972
C. 1975 D. 1978

39. Conference on Human Environment was held in Stockholm in the year
A. 1969 B. 1972
C. 1975 D. 1978

40. જુદી જુદી પ્રજાતિના જૂથોના પર્યાવરણ સાથેના સંબંધોના અભ્યાસને શું કહે છે?
A. સિનેકોલોજી B. ઇકોલોજી
C. ઑટેકોલોજી D. ફિઝિયોલોજી

40. Study of relationship of group of different species with their environment is called
A. Synecology B. Ecology
C. Autecology D. Physiology

41. પર્યાવરણનું સૌથી નાનું માળખાકીય અને કાર્યાત્મક એકમ ______ છે.

A. વાતાવરણ B. સમુદાય

C. નિવસનતંત્ર D. પ્રજાતિ

41. **Smallest structural and functional unit environment is**

A. Atmosphere B. Community

C. Ecosystem D. Species

42. દરેક જીવતંત્રમાં પર્યાવરણીય પરિબળોના સંદર્ભમાં સહનશીલતાની ન્યૂનતમ અને મહત્તમ મર્યાદા હોય છે અને તે સાથે કેન્દ્રીય મહત્તમ શ્રેણી જેમાં સજીવો વિપુલ પ્રમાણમાં જોવા મળે છે, આ મહત્તમ શ્રેણી ______ તરીકે ઓળખાય છે

A. મહત્તમ વિપુલતાનો ઝોન

B. મહત્તમ વૃદ્ધિનો ઝોન

C. મહત્તમ સ્વીકૃતિનો ઝોન

D. મહત્તમ સહનશીલતાનો ઝોન

42. **Every organism has minimum and maximum limit of tolerance with respect to environmental factors along with central optimum range in which organisms are found abundant this optimum range is known as**

A. optimum zone of abundance

B. optimum zone of growth

C. optimum zone of acceptance

D. optimum zone of tolerance

43. દરેક જીવતંત્રમાં પર્યાવરણીય પરિબળોના સંદર્ભમાં સહનશીલતાની ન્યૂનતમ અને મહત્તમ મર્યાદા હોય છે અને તે સાથે કેન્દ્રીય મહત્તમ શ્રેણી જેમાં સજીવો વિપુલ પ્રમાણમાં જોવા મળે છે આ નિયમ ______ ના નિયમ તરીકે ઓળખાય છે.

A. ડાર્વિનનો નિયમ

B. શેલ્ફોર્ડનો નિયમ

C. એલનનો નિયમ

D. ઉપર્યુક્તમાંથી કોઈ નહિ

43. **Every organism has minimum and maximum limit of tolerance with respect to environmental factors along with central optimum range in which organisms are found abundant; this law is known as**

A. Darwin law

B. Shelford law

C. Allen law

D. None of the above

44. પૃથ્વીની આસપાસનો એક પારદર્શક વાયુયુક્ત કવર છે જેને ______ કહે છે.

A. આબોહવા

B. પર્યાવરણ

C. વાતાવરણ

D. હવા

44. **A transparent gaseous envelope surrounding the earth is**

A. Climate

B. Environment

C. Atmosphere

D. Air

45. પર્યાવરણનો મુખ્ય ઊર્જા સ્રોત કયો છે?

A. સૌર ઊર્જા B. પરમાણુ ઊર્જા

C. હાઈડ્રો ઊર્જા D. રાસાયણિક ઊર્જા

45. **The chief energy source of environment is**

A. Solar energy B. Nuclear energy

C. Hydro energy D. Chemical energy

46. સંસાધનો કે જેમના ઉપયોગ પછી પર્યાવરણમાં પુનઃ પ્રાપ્ત થતા નથી તેમને કેવા સંસાધનો કહેવાય છે ?

A. પુનઃપ્રાપ્ય સંસાધનો

B. માનવસર્જિત સંસાધનો

C. પુનઃઅપ્રાપ્ય સંસાધનો

D. કાયમી સંસાધનો

47. પ્રાણીઓ કે જેમના શરીરનું તાપમાન વાતાવરણના તાપમાન સાથે બદલાય છે, તેઓ કયા નામે ઓળખાય છે ?

A. યુરીથર્મલ B. સ્ટેનોથર્મલ

C. ઇક્ટોથર્મલ D. B અને C બંને

48. રાસાયણિક અવરોધકોના ઉત્પાદન દ્વારા એક છોડની બીજા છોડ પાર સીધી અથવા આડકતરી અસરોને ________ કહે છે.

A. એમેન્સાલિઝમ B. સિમ્બાયોસીસ

C. એલેલોપેથી D. કમેન્સલીઝમ

49. પર્યાવરણીય પરિબળો સામે વધુ સારી સુરક્ષા માટે વિન્ડ બ્રેકનું વાવેતર કઈ દિશામાં કરવું જોઈએ?

A. ઉત્તર પશ્ચિમ B. ઉત્તર-પૂર્વ

C. દક્ષિણ-પૂર્વ D. ઉત્તર-દક્ષિણ

50. ફેનોટાઇપિક વેરિએશનમાં ________ શામેલ છે

A. એડિટિવ વેરિએશન

B. નોન-એડિટિવ વેરિએશન

C. એન્વાયરલમેન્ટલ વેરિએશન

D. ઉપરોક્ત તમામ

51. એસિડિક પાણીમાં સક્સેશન ________ તરીકે ઓળખાય છે

A. હાઇડ્રોસેર B. હલોસેર

C. ઓક્ષલોસેર D. ક્ષેરોસેર

46. Resources which are not replaced in the environment after their utilization are called

A. Renewable resources

B. Man-made resources

C. Non-renewable resources

D. Permanent resources

47. Animals whose body temperature varies with the changes in the temperature are known as

A. Eurythermal B. Stenothermal

C. Ectothermal D. Both B and C

48. The direct or indirect effects of one plant to another through the production of chemical inhibitors in the environment is called

A. Amensalism B. Symbiosys

C. Allelopathy D. Commensalism

49. For better protection against environmental factors, wind break should be planted in

A. North-West B. North-East

C. South-East D. North-South

50. Phenotypic variations include

A. Additive variations

B. Non-additive variations

C. Environmental variations

D. All of the above

51. Succession in acidic water is known as

A. Hydrosere B. Halosere

C. Oxalosere D. Xerosere

52. વનીકરણ એ શાનો એક પ્રકાર જ છે ?
A. પ્રાયમરી સક્સેશન
B. સેકન્ડરી સક્સેશન
C. ટર્શરી સક્સેશન
D. એલોજેનિક સક્સેશન

53. પ્રજાતિઓ જે વિપુલ પ્રમાણમાં થાય છે અને તેઓનો સમય ઇકોટોનમાં વિતાવે છે, તેને _______ પ્રજાતિઓ કહે છે.
A. કી સ્ટોન પ્રજાતિઓ
B. પાયોનિયર પ્રજાતિઓ
C. એડ્જ પ્રજાતિઓ
D. ક્લાઇમેક્ષ પ્રજાતિઓ

54. ઇકોસિસ્ટમનો ઓટોટ્રોફ્સ _______ તરીકે ઓળખાય છે ?
A. પ્રોડ્યુશર્સ B. ટ્રાંસડ્યુસર્સ
C. કન્વર્ટરર્સ D. ઉપરોક્ત તમામ

55. માંડલા પ્લાન્ટ ફોસિલ રાષ્ટ્રીય ઉદ્યાન ક્યાં આવેલો છે ?
A. ઉત્તરાખંડ
B. ગુજરાત
B. મધ્યપ્રદેશ
C. હિમાચલ પ્રદેશ

56. સંરક્ષણ માટેનો "ઓપરેશન ઓલિવા" શરૂ કરનાર સંસ્થા કઈ છે?
A. ભારતીય કોસ્ટગાર્ડ
B. ભારતીય સેના
C. ભારતીય નૌકાદળ
D. બોર્ડર સિક્યુરિટી ફોર્સ

57. પી એમ 10 અને પીએમ 2.5 સેમ્પલર્સનો ઉપયોગ શેમાં થાય છે?
A. પર્યાવરણ પ્રદૂષણ
B. ખોરાકમાં ભેળસેળ
C. પેટ્રોલમાં ભેળસેળ
D. દૂધમાં ભેળસેળ

52. Afforestation is a type of
A. Primary succession
B. Secondary succession
C. Tertiary succession
D. Allogenic succession

53. Species which occur most abundantly and spend their time in ecotone are known as
A. Key stone species
B. Pioneer species
C. Edge species
D. Climax species

54. The autotrophs of an ecosystem are known as
A. Producers B. Transducers
C. Converters D. All of the above

55. The Mandla Plant Fossils National Park (MPNP) is located in
A. Uttarakhand
B. Gujarat
B. Madhya Pradesh
C. Himachal Pradesh

56. The organisation who has launched the conservation project "Operation Oliva"
A. Indian Coast Guard
B. Indian Army
C. Indian Navy
D. Boarder Security Force

57. PM10 and PM2.5 Samplers are used for
A. Environment Pollution
B. Food adulteration
C. Adulterated Petrol
D. Milk adulteration

58. ભારતની પ્રથમ ઝીરો એનર્જી બિલ્ડિંગનું નિર્માણ ક્યાં કરવામાં આવ્યું છે?

A. ગાંધીનગર B. કોલકાતા

C. નવી દિલ્હી D. મુંબઈ

58. India's first net zero energy building has been constructed in

A. Gandhinagar B. Kolkata

C. New Delhi D. Mumbai

59. જલપ્લાવિત નિવાસનતંત્ર શું છે?

A. યુટ્રોફિકેશન દ્વારા માર્ક થયેલ

B. ઘાસના મેદાનોથી ઘેરાયેલા

C. જંગલોથી ઘેરાયેલા

D. કંઈ નહીં

59. Wetland ecosystem is

A. marked by eutrophication

B. surrounded by grasslands

C. surrounded by forests

D. None

60. પન્ના બાયોસ્ફિયર રિઝર્વ ક્યાં આવેલું છે?

A. સિક્કિમ B. મધ્યપ્રદેશ

C. છત્તીસગઢ D. હિમાચલ પ્રદેશ

60. Panna Biosphere Reserve is located in

A. Sikkim B. Madhya Pradesh

C. Chhattisgarh D. Himachal Pradesh

61. નીચેનામાંથી કયું એસ્ટ્યુરી વિશે સાચું છે?

A. તે સમુદ્રના પાણી અને નદીના પાણીના મિશ્રણ દ્વારા અંકિત થયેલ ઇકોસિસ્ટમ છે

B. તે હંમેશા ખૂબ ઓછી ખારાશ દ્વારા અંકિત થયેલ છે

C. બંને

D. કંઈ નહીં

61. Which of the following is true about Estuary?

A. It is the ecosystem marked by the mixing of sea water and river water

B. They are always marked by very low salinity

C. Both

D. None

62. ______________ નો ઉપયોગ જીવસૃષ્ટિની ગુણવત્તાની તુલના કરવા માટે થાય છે.

A. વિઘટનકર્તા

B. શિકારી

C. બાયોઇન્ડિકેટર

D. જૈવ-ઉપચારક

62. __________ is an organism used to gauge the quality of an ecosystem.

A. Decomposers

B. Predator

C. Bioindicator

D. Bio-remediator

63. ભારતમાં જાહેર કરાયેલ પ્રથમ પર્યાવરણીય સંરક્ષણ અધિનિયમ કયો હતો?

A. જળ અધિનિયમ

B. વાયુ અધિનિયમ

C. પર્યાવરણ અધિનિયમ

D. અવાજ પ્રદૂષણ નિયમ

63. The first of the major environmental protection act to be promulgated in India was

A. Water Act

B. Air Act

C. Environment Act

D. Noise Pollution Rule

64. ગ્રીન ક્લાઇમેટ ફંડ માટે કયુ નિવેદન યોગ્ય છે.

A. મેક્સિકોમાં કોન્ફરન્સ ઓફ ધ પાર્ટીઝ (COP 16) ની સ્થાપના કરી હતી

B. UNFCCC હેઠળ નાણાકીય મિકેનિઝમ છે

C. તે ભારતમાં NABARD દ્વારા સંચાલિત છે

D. ઉપરોક્ત તમામ

64. Which statement/s is/are correct for Green Climate Fund

A. was established at Conference of the Parties (COP 16) in Mexico

B. is the financial mechanism under the UNFCCC

C. is managed by NABARD in India

D. All of the above

65. ઝૂપ્લાંકટોન્સ શામાં વર્ગીકૃત કરી શકાય?

A. પ્રાથમિક ઉત્પાદકો

B. પ્રાથમિક ઉપભોક્તાઓ

C. ડેટ્રિવાર્સ

D. દ્વિતીય ઉપભોક્તાઓ

65. Zooplanktons can be categorized in

A. Primary producers

B. Primary consumers

C. Detrivores

D. Secondary consumers

66. સમગ્ર વિશ્વમાં માછલીના ઉત્પાદનમાં ભારતનો ક્રમ કેટલો છે ?

A. 2 B. 8

C. 12 D. 14

66. What is the rank of India in fish production all over the world?

A. 2 B. 8

C. 12 D. 14

67. વાતાવરણને સ્વચ્છ અને લીલા રાખવા માટેની રીતોમાં ___________ શામેલ છે

A. ઓછી ઊર્જાનો ઉપયોગ

B. પર્યાવરણ મૈત્રીપૂર્ણ ટેકનોલોજી

C. વન્યજીવન રક્ષણ

D. ઉપરોક્ત તમામ

67. Methods to keep environment clean and green include

A. Use of less energy

B. environment friendly technology

C. protecting wildlife

D. All of the above

68. _________ને રીડ્યુસ, રીયુઝ અને રીસાયકલ પ્રક્રિયાને અનુસરીને સુરક્ષિત કરી શકાય છે

A. પર્યાવરણ B. ખોરાક

C. વાતાવરણ D. પ્રાણવાયુ

68. _________ can be conserved by following reduce, reuse and recycle process

A. environment B. food

C. climate D. oxygen

69. ભાવિ ઉપયોગ માટે વરસાદી પાણીનો સંગ્રહ ___________ તરીકે ઓળખાય છે

A. રેઇન વોટર કલેક્શન

B. રેઇન વોટર હાર્વેસ્ટિંગ

C. રેઇન ડિગીંગ

D. રેઇન વોટર પમ્પીંગ

69. Collection of rainwater for future use is known as

A. Rainwater collection

B. Rainwater harvesting

C. Rain digging

D. Rain water pumping

70. વરસાદી પાણીના સંગ્રહની ________ પદ્ધતિઓ છે.

A. એક B. બે

C. ત્રણ D. ચાર

71. પાણીની આસપાસનું ક્ષેત્ર જેમાં પાણી તેના પર આધીન હોય તેને ____________ કહે છે.

A. રેઇન વોટર હાર્વેસ્ટિંગ

B. વોટર પમ્પિંગ

C. વોટર સાયકલ

D. વોટરશેડ

72. યોગ્ય રીતે મેળ ખાતી જોડીને ઓળખો

A. બેઝલ કન્વેનશન - જૈવવિવિધતા સંરક્ષણ

B. મોન્ટ્રીયલ પ્રોટોકોલ - ગ્લોબલ વોર્મિંગ

C. ક્યોટો પ્રોટોકોલ - ક્લાઈમેટ ચેન્જ

D. રામસર કન્વેનશન - ભૂગર્ભ જળ પ્રદૂષણ

73. *ટ્રાઇકોડર્મા હર્ઝિયનમ* ____________ માટે વપરાય છે

A. જમીન સુધારણા

B. દુષિત જમીનોનો નો જૈવિક ઉપાય

C. માટી દ્વારા થતા છોડના પેથોજેન્સનું જૈવિક નિયંત્રણ

D. છોડમાં જીન ટ્રાન્સફર

74. પ્રકૃતિમાં પડેલા થડના વિઘટનનો ધીમો દર તેમના ____________ કારણે હોય છે.

A. ઓછી ભેજ

B. ઓછો સેલ્યુલોઝ

C. ઓછો નાઇટ્રોજન

D. તેમની આજુબાજુમાં એનારોબિક વાતાવરણ

70. There are ________ methods of rainwater harvesting

A. One B. Two

C. Three D. Four

71. An area surrounding a body of water in which that body of water is subject to is known as

A. Rainwater harvesting

B. Water pumping

C. Water cycle

D. Watershed

72. Identify the correctly matched pair

A. Basal Convention – Biodiversity Conservation

B. Montreal Protocol - Global warming

C. Kyoto protocol – Climatic change

D. Ramsar Convention – Ground water pollution

73. *Trichoderma harzianum* is used for:

A. Reclamation of wastelands

B. Bioremediation of contaminated soils

C. Biological control of soil-borne plant pathogens

D. Gene transfer in higher plants

74. The slow rate of decomposition of fallen logs in nature is due to their

A. Low moisture content

B. Low cellulose content

C. Poor nitrogen content

D. Anaerobic environment around them

75. હવામાનની આગાહી દ્વારા ________ કરવામાં આવે છે
A. પર્યાવરણવાદીઓ
B. હવામાનશાસ્ત્રીઓ
C. ક્લાઇમેટોલોજિસ્ટ્સ
D. ઉપર્યુક્તમાંથી કોઈ નહિ

76. પાણી ઇકોસિસ્ટમમાં BODના સૂચક ________ છે.
A. ટ્રાઉટ B. ડાફનીયા
C. ગાંબુસિયા D. લાઇકેન

77. ભારતમાં સૌ પ્રથમવાર કયા રાજ્યએ પોલિથીન પેક પર પ્રતિબંધ મૂકવાનો પ્રસ્તાવ મૂક્યો હતો?
A. હિમાચલ પ્રદેશ
B. ઉત્તરપ્રદેશ
C. કર્ણાટક
D. મહારાષ્ટ્ર

78. પૃથ્વીની સપાટીની સરેરાશ ઘનતા કેટલી છે?
A. 5.5 ગ્રામ/સે.મી.^{3} B. 6.5 ગ્રામ/સે.મી.^{3}
C. 15 ગ્રામ/સે.મી.^{3} D. 21 ગ્રામ/સે.મી.^{3}

79. ભારતમાં આવેલી ________ વનસ્પતિની પ્રજાતિ ભારતીય જળસંગ્રહમાં તેની તીવ્ર વૃદ્ધિને લીધે મુશ્કેલી થઈ રહી છે.
A. *આઇકોર્નિયા* B. *પિસ્ટિયા*
C. *સિંચોના* D. *નક્સવોમિકા*

80. ________ નો ઉપયોગ *E-situ* પ્લાન્ટ સંરક્ષણ માટે થતો નથી.
A. બીજ બેંકો
B. જનીન બેંકો
C. શિફટિંગ કલ્ટીવેશન
D. બોટનિકલ ગાર્ડન

75. Forecasting of weather is done by
A. Environmentalists
B. Meteorologists
C. Climatologists
D. None of the above

76. What is the indicator of BOD in a water ecosystem
A. Traught B. Daphnia
C. Gambusia D. Lichen

77. Which state proposed a ban on polythene packing for the first time in India
A. Himachal Pradesh
B. Uttar Pradesh
C. Karnataka
D. Maharashtra

78. What is the average density of Earth's surface?
A. 5.5 g/cm^3 B. 6.5 g/cm^3
C. 15 g/cm^3 D. 21 g/cm^3

79. ________ plant species introduced in India appeared to be a trouble due to its heavy growth in the Indian water bodies.
A. *Eichhornia* B. *Pistia*
C. *Cinchona* D. *Nux vomica*

80. ________ is not used for *Ex-situ* plant conservation
A. Seed banks
B. Field gene banks
C. Shifting cultivation
D. Botanical Gardens

81. __________ *in-situ* સંરક્ષણ હેઠળ સામેલ નથી.
A. રાષ્ટ્રીય ઉદ્યાન
B. બોટનિકલ ગાર્ડન
C. વન્યપ્રાણી અભયારણ્ય
D. બાયોસ્ફિયર રિઝર્વ

82. નીચેનામાંથી કયું ક્ષેત્ર યાકની જાતિનું મૂળ સ્થાન છે?
A. હિમાચલ પ્રદેશ
B. લદાખ
C. કાશ્મીર
D. ઉત્તરાખંડ

83. સપ્રોફાઇટ્સ શું છે ?
A. ઝાડની છાલમાંથી સેપ ખાનાર
B. અન્ય સજીવો ખાનાર
C. મૃત અથવા સડેલા કાર્બનિક પદાર્થો ખાનાર
D. ઉપર્યુક્તમાંથી કોઈ નહિ

84. એબીસોપેલેજિક નો અર્થ થાય કે
A. 1000 મી થી વધુ ઊંડાણોમાં રહેતા પાણીના સજીવો
B. 2000 મી થી વધુ ઊંડાણોમાં રહેતા પાણીના સજીવો
C. 3000 મી થી વધુ ઊંડાણોમાં રહેતા પાણીના સજીવો
D. 4000 મી થી વધુ ઊંડાણોમાં પર રહેતા પાણીના સજીવો

85. સજીવ દ્વારા પ્રકાશનું ઉત્પાદન __________ તરીકે ઓળખાય છે.
A. બાયોલ્યુમિનેશન્સ
B. બાયોલાઇટિંગ
C. બાયોફ્લોરોશન્સ
D. બાયોલેન્સ

81. __________ is not included under *In-situ* conservation.
A. National Park
B. Botanical Gardens
C. Wildlife Sanctuary
D. Biosphere Reserve

82. Which of the following region is a native place of the species of Yak?
A. Himachal Pradesh
B. Laddakh
C. Kashmir
D. Uttarakhand

83. Saprophytes are
A. feed on the sap from tree bark
B. feed on other living beings
C. feed on dead or decayed organic matter
D. None of the above

84. Abyssopelagic include
A. Organisms living at water depths greater than 1000 m
B. Organisms living at water depths greater than 2000 m
C. Organisms living at water depths greater than 3000m
D. Organisms living at water depths greater than 4000 m

85. The production of light by living organisms is known as
A. Bioluminescence
B. Biolighting
C. Biofluorescence
D. Biolens

86. કયો દેશ 'લેન્ડ ઓફ વિન્ડમીલ' તરીકે જાણીતો છે ?

A. ઝિમ્બાબ્વે B. નેધરલેન્ડ
C. ફ્રાન્સ D. નોર્વે

87. એનિમોલોજિસ્ટ શું અભ્યાસ કરે છે?

A. ઇલેક્ટ્રોન B. રેતી
C. પવન D. અણુઓ

88. ઉત્તરપૂર્વ ભારતમાં સૌથી મોટું તાજા પાણીનું સરોવર કયું છે?

A. ગુરુડોંગમાર સરોવર
B. ત્સોંગ્મો સરોવર
C. ડિપોર બીલ
D. લોકટક સરોવર

89. વિશ્વનો એકમાત્ર ફ્લોટિંગ પાર્ક , જે ઉત્તર પૂર્વ ભારતમાં સ્થિત છે.

A. કાંચનઝંઘા રાષ્ટ્રીય ઉદ્યાન
B. કીબુલ લમજાઓ રાષ્ટ્રીય ઉદ્યાન
C. ડિબ્રુ-સાઈખોવા રાષ્ટ્રીય ઉદ્યાન
D. માનસ રાષ્ટ્રીય ઉદ્યાન

90. નીચા તાપમાને વધતા સુક્ષ્મજીવાણુઓ ____________ તરીકે ઓળખાય છે.

A. થર્મોફિલ્સ B. મેસોફિલ્સ
C. સાયકોફાઇલ્સ D. થર્મોટ્રોફિક

91. પૃથ્વીના વાતાવરણનો સૌથી બાહ્ય સ્તર ____________ છે.

A. ટ્રોપોસ્ફિઅર B. એક્સ્પોયર
C. મેસોસ્ફિયર D. આયનોસ્ફિયર

92. UNEP એટલે

A. United Nations Environmental Process
B. United Nations Environmental Policy
C. United Nations Environmental Protection
D. United Nations Environmental Programme

86. **Which country is known as the 'land of the windmills'?**

A. Zimbabwe B. The Netherlands
C. France D. Norway

87. **What does an anemologist study?**

A. Electrons B. Sand
C. Wind D. Atoms

88. **The largest freshwater lake in Northeast India is**

A. Gurudongmar Lake
B. Lake Tsongmo
C. Dipor Beel
D. Loktak Lake

89. **The only floating park in the world, located in North East India is**

A. Khangchendzonga National Park
B. Keibul Lamjao National Park
C. Dibru-Saikhowa National Park
D. Manas National Park

90. **The microbes that grow in low temperatures are known as**

A. Thermophiles B. Mesophiles
C. Psychrophiles D. Thermotrophic

91. **The outermost layer of Earth's atmosphere is**

A. Troposphere B. Exosphere
C. Mesosphere D. Ionosphere

92. **UNEP stands for**

A. United Nations Environmental Process
B. United Nations Environmental Policy
C. United Nations Environmental Protection
D. United Nations Environmental Programme

93. વેટલેન્ડમાં __________ પાણી હોઈ શકે છે.
A. બ્રેકિશ B. તાજું
C. સ્થિર D. ઉપરોક્ત તમામ

94. મહત્તમ ચેર વિસ્તાર ક્યાં મળી આવે છે?
A. ગુજરાત B. ઓડિશા
C. પશ્ચિમ બંગાળ D. તામિલનાડુ

95. એક વિસ્તારમાંથી વનસ્પતિ દૂર કરી નવી વનસ્પતિના વિકાસને __________ કહેવામાં આવે છે
A. પ્લાન્ટેશન B. એફોરેસ્ટેશન
C. રિફોરેસ્ટેશન D. ડીફોરેસ્ટેશન

96. ______ એ ઇકોલોજીકલ નિશ શબ્દ આપ્યો.
A. ગ્રિનલ B. મોહિત હુસેન
C. એલન D. પી. ઓડમ

97. કાળી ક્રાંતિ __________ સાથે સંબંધિત છે
A. દૂધ ક્રાંતિ
B. કૃષિ પાકની ક્રાંતિ
C. બાયોફ્યુઅલ / જાટ્રોપા ક્રાંતિ
D. જમીન ક્રાંતિ

98. ભારતનું પ્રથમ ઓર્ગેનિક રાજ્ય કયું છે?
A. સિક્કિમ B. મેઘાલય
C. મિઝોરમ D. કેરળ

99. 'રિંગ ઓફ ફાયર' એ __________ સાથે જોડાયેલું છે.
A. ગ્લોબલ વોર્મિંગ B. જ્વાળામુખી
C. પેટ્રોલિયમ D. ગરમ સમુદ્ર પ્રવાહો

100. કયા જંગલોને પૃથ્વીના ફેફસાં કહે છે?
A. એમેઝોન
B. ઓસ્ટ્રેલિયન જંગલો
C. અર્ધ પાનખર જંગલો
D. સદાબહાર જંગલો

93. Water in a wetland can be
A. Brackish B. Fresh
C. Static D. All of the above

94. The maximum mangrove cover is found in
A. Gujarat B. Odisha
C. West Bengal D. Tamil Nadu

95. Removal of vegetation from an area and development of newly vegetation is called
A. Plantation B. Afforestation
C. Reforestation D. Deforestation

96. __________ gave the Ecological Niche term
A. Grinnel B. Mohit Husain
C. Allen D. P. Odum

97. Black revolution is related with
A. Milk revolution
B. Agriculture crop revolution
C. Biofuel/Jatropa revolution
D. Soil revolution

98. First organic state in India is
A. Sikkim B. Meghalaya
C. Mizoram D. Kerala

99. 'Ring of fire' is associated with the distribution of
A. Global warming B. Volcanoes
C. Petroleum D. Warm sea currents

100. The forests which are called the lungs of the earth?
A. Amazon
B. Australian forests
C. Semi deciduous forests
D. Evergreen forests

Answers | જવાબો

Que.	Ans.	Que.	Ans.	Que.	Ans.	Que.	Ans.	Que.	Ans.	Que.	Ans.	Que.	Ans.
1.	C	**16.**	A	**31.**	C	**46.**	C	**61.**	A	**76.**	B	**91.**	B
2.	D	**17.**	B	**32.**	A	**47.**	D	**62.**	C	**77.**	A	**92.**	D
3.	C	**18.**	C	**33.**	C	**48.**	C	**63.**	A	**78.**	A	**93.**	D
4.	A	**19.**	B	**34.**	A	**49.**	D	**64.**	D	**79.**	A	**94.**	C
5.	B	**20.**	C	**35.**	B	**50.**	D	**65.**	B	**80.**	C	**95.**	C
6.	A	**21.**	C	**36.**	D	**51.**	C	**66.**	A	**81.**	B	**96.**	A
7.	B	**22.**	D	**37.**	B	**52.**	A	**67.**	D	**82.**	B	**97.**	C
8.	C	**23.**	D	**38.**	B	**53.**	C	**68.**	A	**83.**	C	**98.**	A
9.	C	**24.**	D	**39.**	B	**54.**	D	**69.**	B	**84.**	D	**99.**	B
10.	B	**25.**	C	**40.**	A	**55.**	B	**70.**	B	**85.**	A	**100.**	A
11.	C	**26.**	C	**41.**	C	**56.**	A	**71.**	D	**86.**	B		
12.	D	**27.**	D	**42.**	D	**57.**	A	**72.**	C	**87.**	C		
13.	B	**28.**	B	**43.**	B	**58.**	C	**73.**	B	**88.**	D		
14.	A	**29.**	C	**44.**	C	**59.**	A	**74.**	C	**89.**	B		
15.	D	**30.**	A	**45.**	A	**60.**	B	**75.**	B	**90.**	C		

Chapter 8

પ્રકરણ ૮

Forest Soils

વન્ય ભૂમિ

Forest Soil Profile | વન્ય ભૂમિનું પાર્શ્વચિત્ર

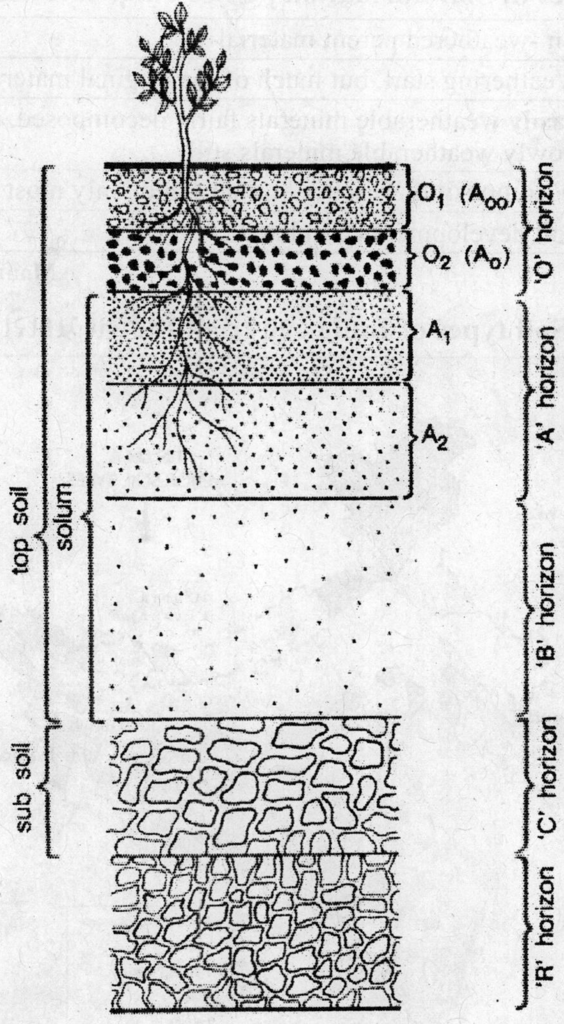

Sharma 2013

Soil Formation | ભૂમિ નિર્માણ

S = f (cl, r, o, p, t)	
S	soil properties (organic matter content, pH, soil texture etc.)
cl	region climate
o	potential biota
r	topography
p	parent matter
t	time

Hans Jenny 1941

Stages of Soil Formation | ભૂમિ નિર્માણના તબક્કાઓ

I	Initial stage	Un -weathered parent material
II	Juvenile Stage	Weathering start, but much of the original material still un-weathered
III	Verile Stage	Easily weatherable minerals fairly decomposed, clay contain increased and slowly weatherable minerals still.
IV	Senile stage	Decomposition reaches at final stage; only most resistant minerals survive
V	Final stage	Soil development completed

Mohr Edward and Van Baren 1954

Soil types of India | ભારતની માટીનાં પ્રકારો

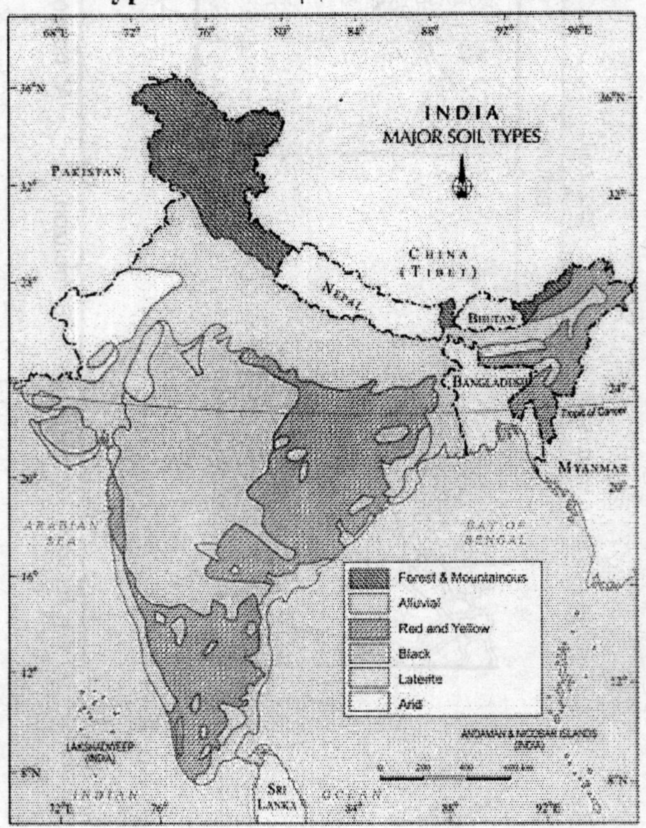

Soils of India | ભારતની વિભિન્ન માટી

No.	Soil	Region
1	Alluvial soil	Rajasthan, Haryana, Punjab, Uttar Pradesh, Valley of Krishna, Kaveri, Brahmaputra, Godavri, Narmada, Tapi and Surma rivers
2	Black cotton soil	Madhya Pradesh, Gujarat, Maharashtra, West part of Andhra Pradesh, Plateau of Deccan and Tamil Nadu
3	Red soil	Chennai, Mysore, part of Uttar Pradesh (Mirzapur, Hamirpur, Jhansi), Bihar (Santhal Pargana, Chhotanagpur), West Bengal and eastern half part of Rajasthan, Madhya Pradesh, Andhra Pradesh and Odisha
4	Laterite soil	Hill tops of South, Hills of Rajasthan, Plain part of Odisha, Maharashtra, Malabar and Central India
5	Desert soil	North – west part of India i.e. Rajasthan, South Punjab and in west district of Uttar Pradesh
6	Tarai soil	Tarai area ofd Punjab, Uttar Pradesh, Bihar, west Bengal and Himalaya
7	Saline and sodic soil	Saline soil : Punjab, Uttar Pradesh, Rajasthan, Madhya Pradesh and Bihar; Sodic Soil : Gujarat, Maharashtra and Karnataka
8	Organic soil	Tarai of Himalaya, middle part of Bihar, North East part of Uttar Pradesh, Kerala, Cochin and Sundarban (West Bengal)
9	Forest soil	Jammu & Kashmir, Uttarakhand, Himachal Pradesh, Bihar, Madhya Pradesh, Maharasthtra, Kerala and North East region
10	Podzole soil	Himalaya mountain and in some part of south India
11	Skeletal soil	Ladakh area of Himalaya Mountain and Vindhyan region of central India

Physical, Chemical and Biological properties of Soil
માટીના ભૌતિક, રાસાયણિક અને જૈવિક ગુણધર્મો

Physical properties	Soil colour, texture, structure, porosity, density, consistence, temperature, and air
Chemical properties	Soil pH, EC, total carbon, Cation Exchange Capacity (CEC), Available N, P, K, Ca, Mg, Mn, Fe, Zn and Cu
Biological properties	Soil Organic matter, soil organisms

Acidic Soil | એસિડિક જમીન

Strongly acidic soils	pH < 4
Moderately acidic soils	pH 4 - 6
Slightly acidic soil	pH 6 – 7

Trees for different types of Soils | વિવિધ પ્રકારની જમીન માટેના વૃક્ષો

Area/ Soil types	Trees
Acidic soils	*Ailanthus ultisima* *Albizia procera* *Gliricedia sepium* *Gmelina arborea*
Alkaline soils	*Acacia nilotica* *Acacia exelsa* *Luecaena leucocephala* *Terminalia arjuna* *Prosopis juliflora* *Salvadora persica* *Tamarindus indica*
Dry areas	*Acacia spp.* *Prosopis juliflora* *Zizyphus maurtiana*
Dry clay soils	*Acacia nilotica* *Acacia tortolis* *Acacia lebbeck*
High altitude areas	*Acacia mearnsi* *Ailanthus altisima* *Grevillea robustra*
Marshy areas	*Acacia auriculiformis* *Derris indica,* *Terminalia arjuna,* *Thespecia populnea* *Bamboo*
Saline soils	*Acacia catechu* *Acacia .nilotica* *Ailanthus exelsa* *Azadirachta indica* *Butea monosperma* *Eucalyptus hybrid* *Luecaena leucocephala* *Melia azadirach* *Prosopis juliflora,* *Parkinsonia acculeata* *Prosopis tamarugo* *Salvadora oleoides* *Salvadora persica* *Tamarindus indica* *Tamarix auriculate* *Terminalia arjuna*
Sandstone rocky soils	*Acacia senegal* *Albizia lebbek*
Sandstone rocky soils	*Boswellia serrata* *Cassia auriculata* *Cenchrus ciliaris* *Cenchrus setigerus*
Sandy soils	*Acacia tortilis* *Acacia albida* *Albizia lebbek* *Azadirachta indica,* *Dalbergia sissoo* *Leucaena leucocephala* *Melia azadearch* *Prosopis cineraria*
Shallow soils	*Acacia tortilis* *Ailathus excelsa* *Dichrostachys glomerata* *Prosopis juliflora* *Zizyphus mauritiana*
Shifting sand-dunes	*Prosopis juliflora* *Prosopis cineraria* *Acacia Senegal* *Albizia lebbek* *Tamarix articulata*
Soil conservation hedges	*Acacia catechu* *Acacia modesta* *Alnus nepalensis* *Grevillea robusta* *Leucaena leucocephala* *Pinus roxburghii* *Prosopis juliflora* *Fruit trees planted are ber, papaya dwarf mango, guava, citrus*
Waterlogged soils	*Acacia nilotica* *Albizia procera* *Anthocephalus indicus* *Butea monosperma* *Dalbergia sissoo* *Eucalyptus camaldulensis* *Eucalyptus pinnata* *Salix tetrasperma* *Syzygium cumini* *Terminalia arjuna*

Land Degradation and Improvement | જમીન અપઘટન અને સુધારણા

Component	Degradation	Improvement
Soil biology management	Overapplication of agrichemicals, industrial contamination	Introduction of biotic organisms, nitrogen fixing microorganisms
Soil nutrient and organic matter management	Alkalinization, acidification, nutrient leaching, removal of organic matter, burning of vegetative residues, nutrient depletion	Fertilization, composting, green manuring, animal manuring, flushing of saline and alkaline soils, liming of acid soils
Soil physical management	Crusting, compaction, sealing, wind erosion, water erosion, devegetation, overtillage	Soil conservation barriers (live, inert), terracing, revegetation of denuded lands, tree protection, soil decompaction, breaking up of pans, cover crops, windbreaks, soil deposition, improved tillage methods
Soil water management	Impended drainage, waterlogging, reduced waterholding capacity, reduced infiltration, salinization	Irrigation, water harvesting, field drainage, drainage of waterlogged areas, filter strips
Vegetation management	Decline in vegetative cover, decline in biodiversity, decline in species composition, decline in availability of valuable species	Increased vegetative cover, increased species diversity, increased species composition, improved availability of valuable species

Sara Scherr and Satya Yadav 1996

Different Forest Carbon Pools | જંગલના વિવિધ કાર્બન પૂલ

Pools		Description
Living Biomass	Above ground biomass	All living biomass above the soil including stems, stumps, branches, bark, seeds and foliage.
	Below ground biomass	All living biomass of live roots. Fine roots of less than 2mm diameter (country specific) are often excluded because these often cannot be distinguished from soil organic matter or litter.
Dead Organic Matter	Dead wood	Includes all non-living woody biomass not contained in the litter, either standing or lying on the ground. Dead wood also includes dead roots and stumps larger than or equal to 10cm in diameter or any other diameter used by the country.
	Litter	Includes all non-living biomass with a diameter less than a minimum diameter chosen by the country (for FSI 5 cm), lying dead, in various states of decomposition above the mineral or organic soil.
Soil	Soil organic matter	Includes organic carbon in mineral and organic soils (including peat) to a specific depth chosen by the country (for FSI 30 cm) and applied consistently through the time series.

ISFR 2019

Species for Wasteland Development | વેસ્ટલેન્ડ ડેવલપમેન્ટ માટેની પ્રજાતિઓ

Tree species	*Acacia albida, Acacia tortilis, Acacia senegal, Acacia catechu, Acacia auriculiformis, Acacia ampliceps, Acacia bivenosa, Acacia nilotica, Prosopis cineraria, Prosopis juliflora, Parkinsonia articulata, Tamarix articulata, Ziziphus mauritiana, Zizypus nummularia, Calligonum polygonoides, Cassia siamea, Albiziam lebbeck, Leucaena leucocephala, Gliricidium sepium, Sesbania sesban, Erythrina* spp., *Casuarina equisetifolia, Igna dulcis, Grewia tenax, Anogeisus rotundifoia, Euphorbia caducifolia, Tamarix aphylla, Cercidium floridum, Eucalyptus tereticornis, Eucalyptus camadulensis, Hardwickia binata, Pongamia pinnata, Tecomella undulata, Azadiracta indica, Dendrocalamus stricus, Bambusa bambos, Agave sisalana, Agave americana, Calliandra* spp., *Gmelina arborea, Grevillea robusta, Dalbergia sissoo, Paulownia* spp., *Terminalia spp., Anthocephalus* spp., *Pterocarpus* spp., *Emblica offficinalis, Aegel marmelos*
Grasses and Forage crops	*Vetiveria zizaniodes, Cynadon dactylon, Dichanthium annulatum, Panicum repens, Panicum antidotale, Panicum maximum, Cenchrus setigerus, Cenchrus ciliaris, Lasiurus sindicus, Sehima nervosum, Bracharia brizantha, Bracharia mutica, Sporobolus marginatus, S. airoides, S. helvolus, Fagonia cretica, Eragrostis spp., Taverniera cuneifolia, Barleria acanthoides, Tribulus terrestris, Brachiaria ramose, Cymbopogon jwarancusa*
Medicinal Plants	*Aloe vera, Balanites aegyptiaca, Commiphora wightii, Euphorbia antisyphilitica, Haloxylon* spp., *Cassia angustifolia*

Raj and Lal 2018

Multiple Choice Questions | હેતુલક્ષી પ્રશ્નો

1. પાનખર જંગલની જમીન કયા ઓર્ડર હેઠળ આવે છે?
   A. અલ્ફિસોલ્સ B. ગેલીસોલ્સ
   C. એન્ડીસોલ્સ D. એરિડીસોલ્સ

2. ભેજવાળા પ્રદેશના શંકુદ્રુપ જંગલો હેઠળ બરછટ ટેક્ષ્ચરના કણો શામેલ છે
   A. ઇનસેપ્ટિસોલ્સ B. વર્ટિસોલ્સ
   C. ઓક્સિસોલ્સ D. સ્પોડોસોલ્સ

3. માટીની રચના છિદ્ર-કદનું વિતરણ નક્કી કરે છે, જે શાના પર અસર કરે છે?
   A. પાણીનો પ્રવાહ અને ધોવાણની સંભાવના
   B. માઇક્રોબિયલ અને પ્રાણીસૃષ્ટિ વર્તન
   C. ઓર્ગેનિક મેટર ડાયનેમિક્સ
   D. ઉપરોક્ત તમામ

4. અલ્ફિસોલ્સ વિશ્વના બરફ મુક્ત જમીન ક્ષેત્રની આશરે _____ હદ બનાવે છે.
   A. 4% B. 10%
   C. 15% D. 7%

5. જંગલની જમીનમાં મુખ્ય લાક્ષણિકતા એ જમીનની સપાટી પર કાર્બનિક સ્તરનો વિકાસ છે, કયા નામે ઓળખવામાં આવે છે?
   A. ફોરેસ્ટ ડેસ્ક
   B. ફોરેસ્ટ ટાયર
   C. ફોરેસ્ટ ફ્લોર
   D. ફોરેસ્ટ ગ્રાઉન્ડ

6. સ્પોડોસોલ્સ શું છે?
   A. એસિડિક અને ઈન્ફર્ટાઈલ
   B. બેઈઝીક અને ઈન્ફર્ટાઈલ
   C. એસિડિક અને ફર્ટાઈલ
   D. બેઈઝીક અને ફર્ટાઈલ

1. **Under which order Deciduous forest soils come?**
   A. Alfisols B. Gelisols
   C. Andisols D. Aridisols

2. **Coarse textured deposits under coniferous forests of humid region consists of**
   A. Inceptisols B. Vertisols
   C. Oxisols D. Spodosols

3. **Soil structure determines pore-size distribution, which affects**
   A. water flow and erosion potential
   B. microbial and faunal behavior
   C. organic matter dynamics
   D. all of the above

4. **Alfisols approximately make _____ extent of world ice-free land area**
   A. 4 % B. 10 %
   C. 15 % D. 7 %

5. **The principal characteristic in forest soils is the development of an organic layer on the soil surface, usually referred as**
   A. Forest deck
   B. Forest tier
   C. Forest floor
   D. Forest ground

6. **Spodosols are**
   A. Acidic and infertile
   B. Basic and infertile
   C. Acidic and fertile
   D. Basic and fertile

7. જંગલની જમીનની સપાટીનો ઉપલો સ્તર ક્યા અક્ષર સાથે રજૂ થાય છે?
A. 'ઓ'
B. 'સી'
C. 'એફ'
D. 'એસ'

7. **The forest soil surface layer is represented with the letter**
A. 'O'
B. 'C'
C. 'F'
D. 'S'

8. છોડના બાયોમાસ ઉપરાંત, ________ વન કચરા અને જમીનમાં કાર્બનિક પદાર્થોના મહત્વપૂર્ણ પૂલનું પ્રતિનિધિત્વ કરે છે.
A. બેક્ટેરિયલ કોષો
B. ફંગલ માઇસેલિયા
C. જંગલી મશરૂમ્સ
D. A અને B બન્ને

8. **In addition to plant biomass, ________ represent an important pool of organic matter in forest litter and soil**
A. Bacterial cells
B. Fungal mycelia
C. Wild mushrooms
D. Both A and B

9. માટીની ઊંડાઈમાં વધારા સાથે, કાર્બનિક પદાર્થોની ઉપલબ્ધતા
A. વધે છે
B. સતત રહે છે
C. ઘટાડો થાય છે
D. વનસ્પતિના પ્રકાર પર આધારિત છે

9. **With increase in soil depth, the availability of organic matter**
A. Increases
B. Remains constant
C. Decreases
D. Depends on the type of vegetation

10. જંગલની જમીનોના ભૌતિક ગુણધર્મો લગભગ કાયમી ગુણધર્મો હોઈ શકે છે સિવાય કે તેમાં ફેરફાર કરવામાં આવે
A. લણણી કામગીરી
B. ખેતી સ્થળાંતર
C. દાવાનળ
D. ઉપરોક્ત તમામ

10. **Physical properties of forest soils may be almost permanent properties unless modified by**
A. Harvesting operations
B. Shifting cultivation
C. Forest fires
D. All of the above

11. તાપમાન અને વરસાદની સીધી અસર શાના પર પડે છે?
A. વન વનસ્પતિ
B. વિઘટન દર
C. માટી રચનાની પ્રક્રિયા
D. ઉપરોક્ત તમામ

11. **Temperature and precipitation regimes have a direct impact on the**
A. Forest vegetation
B. Decomposition rate
C. Soil formation process
D. All of the above

12. ઉષ્ણકટિબંધીય વન ભૂમિ કેવી છે ?
A. વધુ ક્લે અને ઓછું ઓર્ગેનિક મેટર
B. ઓછું ક્લે અને વધુ ઓર્ગેનિક મેટર
C. ઓછું ક્લે અને ઓછું ઓર્ગેનિક મેટર
D. વધુ ક્લે અને વધુ ઓર્ગેનિક મેટર

13. વનોની જમીનમાં શાનું પ્રભુત્વ હોય છે?
A. બેક્ટેરિયા B. ફૂગ
C. એક્ટિનોમિસેટ્સ D. A અને B બન્ને

14. ________ જમીનની રચનામાં સ્વતંત્ર પરિબળ નથી
A. ટોપોગ્રાફી
B. પેરન્ટ મટેરીયલ
C. ક્લાઈમેટ
D. વનસ્પતિ અને પ્રાણીસૃષ્ટિ

15. માયકોરીઝા શાનું બનેલું છે?
A. ફૂગ અને ઝાડના મૂળ
B. એક્ટિનોમિસેટ્સ અને ઝાડના મૂળ
C. ફૂગ અને શેવાળ
D. એક્ટિનોમિસેટ્સ અને ફૂગ

16. જંગલની જમીનમાં ન્યુટ્રિયન્ટ સાયક્લિંગમાં ખલેલ અને ત્યારબાદના ફેરફારોના કોના વાતાવરણીય સ્તરો પર સંભવિતપણે મોટી અસર થઈ શકે છે.
A. N_2 B. CO_2
C. CH_4 D. SO_2

17. માટીના ________ એ ઇકોસિસ્ટમ ન્યુટ્રિયન્ટ સાયક્લિંગના મુખ્ય નિયંત્રકો છે
A. જૈવિક ઘટકો B. ખનિજો
C. સુક્ષ્મસજીવો D. A અને B બન્ને

12. Tropical forest soils have
A. High clay content and lower soil organic matter
B. Low clay content and higher soil organic matter
C. Low clay content and lower soil organic matter
D. High clay content and higher soil organic matter

13. The microbial community in forest soil is dominated by
A. Bacteria B. Fungi
C. Actinomycetes D. Both A and B

14. ________ is not an independent factor in soil formation
A. Topography
B. Parent material
C. Climate
D. Flora and Fauna

15. Mycorrhizae is the association of
A. Fungi and tree roots
B. Actinomycetes and tree roots
C. Fungi and Algae
D. Actinomycetes and fungi

16. Disturbance and subsequent changes in nutrient cycling in forest soils can potentially have large impacts on atmospheric levels of
A. N_2 B. CO_2
C. CH_4 D. SO_2

17. Soil ________ are major controllers of ecosystem nutrient cycling
A. Organic Matter B. Minerals
C. Microorganisms D. Both A and B

18. જંગલની જમીનમાં મહત્તમ ઓબિટેઇડ ઘનતા ક્યાં જોવા મળે છે ?
A. શંકુદ્રુપ જંગલો
B. હાર્ડવુડ જંગલો
C. ટુન્દ્રા
D. ઘાસીયા મેદાનો

19. વન જમીનના ભૌતિક ગુણધર્મો મોટા પ્રમાણમાં શાના દ્વારા નિયંત્રિત થાય છે.
A. માટીના કણોનું કદ
B. માટીના કણોનું વિતરણ
C. માટીના કણોની ગોઠવણ
D. ઉપરોક્ત તમામ

20. વન વ્યવસ્થાપનમાં, જમીનની ઉત્પાદકતા મોટાભાગે શાની માત્રામાં માપવામાં આવે છે?
A. વન્યજીવન B. વૃક્ષ ઉત્પન્નતા
C. હ્યુમસ D. સુક્ષ્મજીવાણુઓ

21. વન વ્યવસ્થાપન પ્રવૃત્તિઓનું આયોજન કરતી વખતે વધુ ઇચ્છનીય શું છે?
A. નકારાત્મક અસરોની રોકથામ
B. અસરો આવી ગયા પછી સુધારો
C. A અને B બન્ને
D. વન્યપ્રાણી સંરક્ષણ

22. ક્લે સોઇલ કેવી જમીન તરીકે ઓળખાય છે?
A. ઓછી ટેક્ષ્ચરવાળી માટી
B. મધ્યમ ટેક્ષ્ચર માટી
C. ભારે ટેક્સચરવાળી માટી
D. નરમ ટેક્ષ્ચરવાળી માટી

23. શું વનોની જમીનના ભૌતિક ગુણધર્મોમાં સામેલ નથી ?
A. ડ્રેનેજ
B. છિદ્રાળુતા
C. સર્ફેસ હાઇડ્રોલોજી
D. પોષક સ્થિતિ

18. Maximum oribatid densities in forest soils are found in
A. Coniferous forests
B. Hardwood forests
C. Tundra
D. Grassland

19. Forest soil physical properties are largely controlled by
A. size of soil particles
B. distribution of soil particles
C. arrangement of soil particles
D. all of the above

20. In forest management, soil productivity is most often measured in volume of
A. Wildlife B. trees produced
C. Humus D. Microbes

21. When planning forest management activities what is more desirable?
A. Prevention of negative impacts
B. Correcting impacts after they have occurred
C. Both A and B
D. Conserving Wildlife

22. Clay soils are known as
A. Low textured soil
B. Medium textured soil
C. Heavy textured soil
D. Soft textured soil

23. The physical properties of forest soil does not include
A. Drainage
B. Porosity
C. Surface hydrology
D. Nutrient status

24. શું વનોની જમીનના રાસાયણિક ગુણધર્મોમાં સામેલ છે ?
A. માટીની ઘનતા
B. માટીનું ટેક્ષ્ચર
C. પીએચ
D. A અને C બન્ને

25. જમીનની સપાટી પર લાગતા ભારને પરિણામે જમીનની ઘનતામાં ________ એ વન જમીન સંકોચન છે.
A. ઘટાડો
B. નિરંતરતા
C. વધારો
D. સંપીડન

26. સૌથી વધુ મૂળભૂત ક્ષીણતાવાળી વન જમીનમાં શું ઉચ્ચ પ્રમાણ હોય છે?
A. ફાઈન સેન્ડ અને સિલ્ટ
B. જમીનમાં કાર્બનિક પદાર્થોની ઓછી માત્રા
C. ઓછી પરમેબિલિટી
D. ઉપરોક્ત તમામ

27. વન-પર્વતની જમીન ભારતના લગભગ_____ ભૂમિ ક્ષેત્ર પર કબજો કરે છે.
A. 10.42% B. 9.53%
C. 8.67% D. 6.78%

28. વન જમીનમાં શાની ઉણપ હોય છે?
A. પોટાશ B. ફોસ્ફરસ
C. લાઇમ D. ઉપરોક્ત તમામ

29. સામાન્ય રીતે મોટા પાનવાળા-સદાબહાર જંગલોની જમીન કેવી હોય છે?
A. લોમી B. ફળદ્રુપ
C. કાર્બોનેટમાં સમૃદ્ધ D. એસિડિક

30. વન જમીન શાની હાજરીથી ખૂબ સમૃદ્ધ છે?
A. હ્યુમસ B. કાર્બોનેટ
C. સુકા પદાર્થ D. નાઈટ્રેટ્સ

24. The chemical properties of forest soil include
A. Soil density
B. Soil texture
C. pH
D. Both A and C

25. Forest soil compaction is the ________ in soil density resulting from loads applied to the soil surface.
A. Decrease
B. Consistency
C. Increase
D. Compression

26. Forest soils with the highest inherent erodibility contain high proportions of
A. Fine sand and silt
B. Low amounts of soil organic matter
C. Slow permeability
D. All of the above

27. Forest-Mountain soils occupy almost _____ land area of India
A. 10.42 % B. 9.53 %
C. 8.67 % D. 6.78 %

28. Forest soils are deficient in
A. Potash B. Phosphorus
C. Lime D. All of the above

29. Soils in broad-leaved evergreen forest are generally
A. Loamy B. Fertile
C. Rich in carbonates D. Acidic

30. Forest soils are very rich in
A. Humus B. Carbonates
C. Dry matter D. Nitrates

31. હિમાલયના ક્ષેત્રમાં વન-પર્વતની જમીન ક્યાં મળતી નથી?
A. ખીણો
B. ઓછા ઢોળાવો પર
C. ઉત્તર તરફના ઢોળાવો
D. દક્ષિણ સામેના ઢોળાવો

32. સોઇલ રેસ્પિરેશસન એ CO_2 દ્વારા ઉત્પાદિત થવાને કારણે માટીમાંથી CO_2 ની મુક્તિને શાનાથી સંદર્ભિત કરે છે?
A. મૂળ
B. માટીના સૂક્ષ્મજીવો
C. કાર્બન સંયોજનોનું રાસાયણિક ઓક્સિડેશન
D. ઉપરોક્ત તમામ

33. એશી વ્હાઇટ લેયર્સવાળી હોરાઇઝન શાંનાથી ઓળખાય છે?
A. O હોરાઇઝન B. A હોરાઇઝન
C. E હોરાઇઝન D. C હોરાઇઝન

34. જંગલની જમીનોની લાક્ષણિકતાઓ સાથે શું બદલાય છે?
A. પિતૃ ખડક
B. ગ્રાઉન્ડ રૂપરેખાંકન
C. આબોહવા
D. ઉપરોક્ત તમામ

35. વન જમીનો ક્યાં જોવા મળતી નથી?
A. સહ્યાદ્રીસ
B. તરાઇ પ્રદેશ
C. દરિયાકાંઠાના વિસ્તારો
D. પૂર્વી ઘાટ

36. મોટાભાગે જંગલની જમીન ______ રંગની હોય છે.
A. કાળા B. ઘેરા ભૂરા
C. લાલ D. પીળા

31. In Himalayan region, the Forest-Mountain soils are not found in
A. Valleys
B. Less steep slopes
C. North facing slopes
D. South facing slopes

32. Soil respiration refers to the release of CO_2 from soils due to production of CO_2 by
A. Roots
B. Soil organisms
C. Chemical oxidation of carbon compounds
D. All of the above

33. Horizon with ashy white layers is known as
A. O horizon B. A horizon
C. E horizon D. C horizon

34. The characteristics of forest soils changes with
A. Parent rock
B. Ground-configuration
C. Climate
D. All of the above

35. Forest soils are not found in
A. Sahyadris
B. Terai region
C. Coastal areas
D. Eastern Ghats

36. Mostly forest soils are ______ colored.
A. Black B. Dark brown
C. Red D. Yellow

37. ખીણોમાં જંગલની જમીન કેવી હોય છે

A. લોમી-સિલ્ટી

B. બરછટ દાણાદાર

C. લોમી-દાણાદાર

D. બરછટ-સિલ્ટી

38. હિમાલયના બરફવાળા વિસ્તારોમાં જંગલની જમીન કેવી છે?

A. વધુ હ્યુમસ સાથેની બેઈઝીક જમીન

B. ઓછા હ્યુમસ સાથેની બેઈઝીક જમીન

C. ઓછા હ્યુમસ સાથેની એસિડિક જમીન

D. વધુ હ્યુમસ સાથેની એસિડિક જમીન

39. નીચલા ખીણોમાં મળતી વન જમીન કેવી છે?

A. ફળદ્રુપ અને ખનિજો સમૃદ્ધ

B. ફળદ્રુપ અને જૈવિક પર્દાર્થોથી સમૃદ્ધ

C. બિનફળદ્રુપ અને ઓછા જૈવિક પર્દાર્થો

D. ભિનફળદ્રુપ અને ઓછા ખનિજો વાળી

40. પાનખર જંગલની જમીન કેવી હોય છે?

A. E હોરાઈઝોન વિનાની

B. ક્લે કે જે સબ સોઇલમાં એકઠા થાય છે

C. E હોરાઈઝોન વાળી

D. A અને B બન્ને

41. સમશીતોષ્ણ સ્ક્લેરોફિલોસ જંગલોમાં જમીન કેવી હોય છે?

A. ખનિજોમાં નબળી

B. ખનિજોથી સમૃદ્ધ

C. ખૂબ જ ફળદ્રુપ

D. હ્યુમસથી સમૃદ્ધ

42. લાક્ષણિક સમશીતોષ્ણ પાનખર જંગલની જમીન કેવી હોય છે?

A. છીછરી જમીન B. મૂલ જમીન

C. ફળદ્રુપ જમીન D. A અને B બન્ને

37. In valleys the forest soils are

A. Loamy-silty

B. Coarse-grained

C. Loamy-grained

D. Coarse-silty

38. In the snow-bound areas of the Himalayas, forest soils are

A. Basic with high humus content

B. Basic with low humus content

C. Acidic with low humus content

D. Acidic with high humus content

39. Forest soils found in the lower valleys are

A. Fertile and rich in minerals

B. Fertile and rich in organic content

C. Infertile and low in organic content

D. Infertile and low in minerals

40. Deciduous forest soils

A. Do not have a beached E horizon

B. Have clays that accumulate in the sub soils

C. Have a beached E horizon

D. Both A and B

41. Soils in temperate sclerophyllous forests are frequently

A. poor in mineral nutrients

B. rich in mineral nutrients

C. very fertile

D. rich in humus

42. Typical temperate deciduous forest soils are

A. shallow soils B. mull soils

C. fertile soils D. Both B and C

43. સ્પોંડોસોલ્સ ઘણો વરસાદ જ્યારે રેતી સાથે જોડાઈ, પાણી ખનિજો અને કાર્બનિક પદાર્થોનું ને ઝડપથી શોષી લે ત્યારે શું બને છે
A. O હોરાઈઝોન
B. E હોરાઈઝોન
C. A હોરાઈઝોન
D. C હોરાઈઝોન

43. **Spodosols get a lot of rain, and when combined with sand, the water leaches minerals and organic matter downward, resulting in**
A. O horizon
B. E horizon
C. A horizon
D. C horizon

44. જ્યાં પેરેન્ટ મટેરીયલ રેતાળ હોય છે ત્યાં ________ લેન્ડસ્કેપ પર વર્ચસ્વ ધરાવે છે.
A. સદાબહાર વનસ્પતિ
B. બોરિયલ જંગલો
C. સમશીતોષ્ણ જંગલો
D. ઉપરોક્ત તમામ

44. **Where parent materials are sandy, ________ tends to dominate the landscape.**
A. evergreen vegetation
B. boreal forests
C. temperate forests
D. All of the above

45. ઝાડની નીચે જમીનની વનસ્પતિની સમૃદ્ધિ સામાન્ય રીતે જમીનની ફળદ્રુપતા સાથે ________.
A. ઘટે છે
B. વધે છે
C. જમીનના પ્રકાર પર આધારીત છે
D. અસર થતી નથી

45. **The richness of the ground flora beneath the trees generally ________ with soil fertility.**
A. decreases
B. increases
C. depends on soil type
D. not affected

46. સામાન્ય રીતે જંગલોમાં ઓકના વૃક્ષો કેવી જમીનમાં પ્રબળ હોય છે?
A. શેલો, સિલ્ટ જમીન
B. શેલો, રેતાળ જમીન
C. ઊંડી, ક્લે જમીન
D. ઊંડી, લોમી જમીન

46. **In forests oak trees are generally dominant on**
A. shallow, silt soils
B. shallow, sandy soils
C. deeper, clay soils
D. deeper, loamy soils

47. વનની જમીનોમાં સ્વતંત્ર કણો ભેગા થાય ત્યારે ક્યાં નામે ઓળખાય છે?
A. પ્રાથમિક કણો
B. દ્વિતીય કણો
C. તૃતીય કણો
D. ચતુર્થ કણો

47. **Individual discrete particles of forest soils when aggregate are known as**
A. primary particles
B. secondary particles
C. tertiary particles
D. quaternary particles

48. **2 મીમી થી વધુ વ્યાસના માટીના ક્યાં કણો કયા નામે ઓળખાય છે?**
A. ગ્રાવેલ્સ B. પેબલ્સ
C. કોબલ્સ D. બોલ્ડર્સ

49. **_______ જમીન આછા ટેક્ષ્ચરવાળી જમીન તરીકે ઓળખાય છે**
A. લોમી B. ક્લે
C. સિલ્ટ D. સેન્ડી

50. **એકત્રિકરણ દ્વારા ન્યુટ્રિયન્ટના ટર્નઓવરને શું અસર થઈ શકે છે?**
A. માઇક્રોબાયલ પૂર્વાનુમાનનું નિયંત્રણ
B. ઓર્ગેનિક મેટરને માઈક્રોબિયલ ડિગ્રેડેશનથી બચાવવા
C. મિનરલ્સના સંરક્ષણ દ્વારા
D. A અને B બન્ને

51. **વન જમીનો કૃષિ જમીનો કરતા કુદરતી રીતે વધુ _______ હોય છે.**
A. બેઈઝીક
B. ન્યુટ્રલ
C. એસિડિક
D. ઉપર્યુક્તમાંથી કોઈ નહિ

52. **સ્પેરોડલ જમીનની રચના કેવી હોય છે?**
A. ગ્રેન્યુલેર અને ક્રમ્બ
B. સબ એન્ગ્યુલર બ્લોકી
C. પ્રેઇઝમેટિક અને કોલુમનાર
D. એન્ગ્યુલર બ્લોકી

53. **કદના આધારે, જમીનની રચનાને _______ વર્ગોમાં વહેચવામાં આવે છે**
A. ચાર B. ત્રણ
C. પાંચ D. છ

54. **માટીના કયા કણો ઇલેક્ટ્રિકલી ચાર્જ્ડ હોય છે?**
A. સેન્ડ B. ક્લે
C. સિલ્ટ D. A અને C બન્ને

48. **Soil particles greater than 2 mm diameter are known as**
A. gravels B. pebbles
C. cobbles D. boulders

49. **_______ soils are known as light textured soils**
A. loamy B. clay
C. silt D. sandy

50. **Aggregation may affect nutrient turnover by**
A. controlling microbial predation
B. protecting organic matter from microbial degradation
C. by conserving mineral resources
D. both A and B

51. **Forest soils tend to be naturally more _______ than agricultural soils**
A. Basic
B. Neutral
C. Acidic
D. None of the above

52. **Spheroidal soil structures are**
A. granular and crumb
B. sub-angular blocky
C. prismatic and columnar
D. angular blocky

53. **On the basis of size of aggregates, soil structure is divided into _______ classes**
A. Four B. Three
C. Five D. Six

54. **Which of these soil particles are electrically charged?**
A. Sand B. Clay
C. Silt D. Both A and C

55. માટીની રચના વૃક્ષની વૃદ્ધિ માં ક્યાં અસર કરે છે?
A. મૂળના વ્યાપમાં
B. પાણી અને પોષક તત્વો લેવાની ક્ષમતામાં
C. માઇક્રોબિયલ નિયંત્રણ દ્વારા
D. A અને B બન્ને

56. જંગલની જમીનમાં બલ્ક ડેન્સિટી મુખ્યત્વે શાને કારણે ઊંડાઈ સાથે વધે છે?
A. જૈવિક ઘટકો અને એકત્રીકરણનો અભાવ
B. જૈવિક ઘટકો અને એકત્રિકરણની પૂર્તતા
C. માત્ર જૈવિક ઘટકોનો અભાવ
D. માત્ર એકત્રીકરણની પૂર્તતા

57. માટીની છિદ્રાળુતા શું સૂચવે છે.
A. ઇનફીલ્ટ્રેશન અને પર્કોલેશનની ક્ષમતા
B. કોમ્પેક્શન
C. ભેજ
D. ઘનતા

58. જંગલની જમીનમાં કોમ્પેક્શન લાવવાનું મુખ્ય પરિબળો મોટે ભાગે શું છે?
A. લાકડાની કાપણી દરમ્યાન ભારે મશીનરીનો ઉપયોગ
B. સ્થળની મિકેનિકલ તૈયારી
C. દાવાનળ
D. A અને B બન્ને

59. ઓર્ગેનિક મેટરથી સમૃદ્ધ જંગલની જમીનમાં માટીની હવામાં __________ ની ટકાવારી વધારે હોય છે
A. O_2
B. CO_2
C. CH_4
D. N_2

55. Soil structure affects tree growth by influencing
A. Root distribution
B. Ability to take up water and nutrients
C. By controlling microbial predation
D. Both A and B

56. Bulk density in forest soils tends to increase with depth primarily due to the
A. lack of organic matter and aggregation
B. Adequacy of organic matter and aggregation
C. Lack of organic matter only
D. Adequacy of aggregation only

57. Porosity of a soil is an index of its
A. Infiltration and percolation capacity
B. Compaction
C. Moisture content
D. Density

58. The main forces causing compaction of forest soils are mostly
A. Heavy machinery used during timber harvesting
B. Mechanical site preparation
C. Forest fires
D. Both A and B

59. Forest soils rich in organic matter contain higher percentage of ________ in soil air
A. O_2
B. CO_2
C. CH_4
D. N_2

60. જંગલની જમીનનું તાપમાન શેનાથી પ્રભાવિત થતું નથી.

A. સાઇટ ટોપોગ્રાફી
B. હવામાન પરિસ્થિતિઓ
C. છોડની કેનોપી દ્વારા આવરી લેવામાં આવતો વિસ્તાર
D. મૂળનું શ્વસન

61. ફોરેસ્ટ ફ્લોર પર ના પાંદડા __________ તરીકે કામ કરે છે, જે ગરમીને સાચવીને શિયાળામાં જમીનને ગરમ રાખે છે.

A. શોષક B. પરાવર્તક
C. અવાહક D. ઉપર્યુક્તમાંથી કોઈ નહિ

62. જમીનને પકડી શકે તેવા વિનિમયક્ષમ કેટાયન્સ ને શું કહેવાય છે?

A. ECEC B. EVEC
C. ECEV D. EVEV

63. FIA એટલે

A. Forest Industry and Analysis
B. Forest Inventory and Agriculture
C. Forest Inventory and Analysis
D. Forest Industry and Agriculture

64. છોડની જાતિના મહત્તમ વિકાસ માટે માટીની pH કેટલી હોવી જોઈએ?

A. 4.0-5.5 B. 5.5-7.2
C. 7.2-8.5 D. > 8.5

65. ISSS એટલે

A. International Soil Science Society
B. International Society of Soil Science
C. Indian Society of Soil Science
D. Indian Society of Soil Science

66. __________ ને જંગલની જમીનના મુખ્ય વિઘટનકર્તા માનવામાં આવે છે.

A. બેક્ટેરિયા B. મોલ્ડ્સ
C. યીસ્ટ D. આલ્ગી

60. Temperature of forest soil is not influenced by

A. Site topography
B. Meteorological conditions
C. Surface area covered by canopies of plants
D. Root respiration

61. Leaf litters on the forest floor act as __________, keeping the soils warmer in the winter by trapping radiant heat

A. Absorbers B. Reflectors
C. Insulators D. None of the above

62. The total amount of exchangeable cations that a soil can hold is referred to as

A. ECEC B. EVEC
C. ECEV D. EVEV

63. FIA stands for

A. Forest Industry and Analysis
B. Forest Inventory and Agriculture
C. Forest Inventory and Analysis
D. Forest Industry and Agriculture

64. Soil pH optimum for growth of maximum plant species

A. 4.0-5.5 B. 5.5-7.2
C. 7.2-8.5 D. >8.5

65. ISSS stands for

A. International Soil Science Society
B. International Society of Soil Science
C. Indian Society of Soil Science
D. Indian Society of Soil Science

66. __________ are considered the main decomposers in forest soils

A. Bacteria B. Molds
C. Yeast D. Algae

67. રાયસોસ્ફિયર શું છે?
A. છોડની નીચેનો માટીનો વિસ્તાર
B. છોડની કેનોપી હેઠળ જમીનનો વિસ્તાર
C. છોડના મૂળની આસપાસનો માટીનો વિસ્તાર
D. ઉપર્યુક્તમાંથી કોઈ નહિ

68. જંગલની જમીન ટેરેસ્ટીયલ કાર્બનનો ______ હિસ્સો સંગ્રહિત કરે છે .
A. 1/3 B. 1/2
C. 1/4 D. 2/3

69. ______ બેક્ટેરિયાની હાજરીને લીધે, જંગલની જમીન વાતાવરણના મીથેનનું સિંક કરે છે.
A. થર્મોફિલિક
B. મેથેનોટ્રોફિક
C. સાયકોટ્રોફિક
D. મેસોફિલિક

70. સમશીતોષ્ણ જંગલની જમીનને ______ N ડિપોઝીશન દ્વારા તીવ્ર અસર કરે છે
A. એન્થ્રોપોજેનિક B. નેચુરોજેનિક
C. માઇકોર્રિઝલ D. બેક્ટેરિયલ

71. જમીનમાં વિવિધ કદના કણો ભેગા થવાથી થતી ફેરનેશ કે કોરનેશ કયા નામ ઓળખાય છે?
A. સોઇલ પેકેજીંગ
B. સોઇલ ટેક્ષર
C. સોઇલ ડેન્સેટી
D. સોઇલ કોમ્પેક્શન

72. રેતાળ જંગલની જમીન પર કયું વૃક્ષ ઉગાડવા માટે યોગ્ય નથી?
A. *પોપ્લસ* B. *પાઈનશ*
C. *બાવળ* D. *ફેલોડેન્ડ્રોન*

67. Rhizosphere refers to
A. Soil area under the plant leaves
B. Soil area under plant canopy
C. Soil area around the plant roots
D. None of the above

68. Forest soils store ______ of the terrestrial C.
A. 1/3 B. 1/2
C. 1/4 D. 2/3

69. Due to the presence of ______ bacteria, forest soils represent a sink of atmospheric methane.
A. Thermophilic
B. Methanotrophic
C. Psychrotrophic
D. Mesophilic

70. Temperate forest soils are strongly affected by ______ N deposition
A. Anthropogenic B. Naturogenic
C. Mycorrhizal D. Bacterial

71. The degree of fineness or coarseness created by the close packing of variously sized particles together in a soil is refer as
A. Soil packaging
B. Soil texture
C. Soil density
D. Soil compaction

72. Which of the tree is not suitable to grow on sandy forest soil?
A. *Populus* B. *Pinus*
C. *Acacia* D. *Phellodendron*

73. ઉત્તરી ગોળાર્ધમાં જંગલની જમીન દક્ષિણ ગોળાર્ધની તુલનામાં __________ હોય છે
A. વધુ ઊંડા ઓર્ગેનિક મેટરના સ્તર અને ધીમું વિઘટન
B. ઓછા ઊંડા ઓર્ગેનિક મેટરના સ્તર અને ઝડપી વિઘટન
C. વધુ ઊંડા ઓર્ગેનિક મેટરના સ્તર અને ઝડપી વિઘટન
D. ઓછા ઊંડા ઓર્ગેનિક મેટરના સ્તર અને ધીમું વિઘટન

73. In northern hemisphere the forest soils have _________ as compare to southern hemisphere
A. Excessively deep organic matter layers and slow to decompose
B. Shallow organic matter layers and subject to rapid decomposition
C. Excessively deep organic matter layers with rapid decomposition
D. Shallow organic matter layers and slow to decompose

74. ભારતમાં ક્યાં જંગલની જમીન જોવા મળતી નથી?
A. હિમાચલ પ્રદેશ B. કર્ણાટક
C. રાજસ્થાન D. પંજાબ

74. In India forest soils does not found in
A. Himachal Pradesh B. Karnataka
C. Rajasthan D. Punjab

75. માટીની રચના એ વિવિધ કદ અને આકારના એકમોમાં જમીનના કણોની ગોઠવણી છે. આ એકમો શું કહેવામાં આવે છે?
A. પેડ્સ B. ક્લોડ્સ
C. ચંક D. કોન્ક્રેશન્સ

75. Soil structure is the arrangement of soil particles into units of different sizes and shapes. These units are called
A. Peds B. Clods
C. Chunks D. Concretions

76. ક્લાઈમેટ ચેન્જના લગભગ તમામ મોડેલો ગ્લોબલ વોર્મિન્ગના પરિણામે માટીમાંથી _____ ની આગાહી કરે છે
A. કાર્બનનો લાભ
B. કાર્બનનું નુકસાન
C. નાઇટ્રોજનનો લાભ
D. A અને C બન્ને

76. Nearly all models of global climate change predict a _________ from soils as a result of global warming
A. Gain of carbon
B. Loss of carbon
C. Gain of nitrogen
D. Both A and C

77. નીચેનામાંથી શું માટીમાંની હવાની માત્રા અને રચનાને અસર કરે છે?
A. સોઇલ ટેક્ષર
B. સોઇલ સ્ટ્રક્ચર
C. સોઇલ પોરોસિટી
D. ઉપરોક્ત તમામ

77. Which of the following affect the amount and composition of soil air and its renewal?
A. Soil texture
B. Soil structure
C. Soil porosity
D. All of the above

78. જંગલની જમીનમાં વાયુમિશ્રણ બે ભૌતિક પ્રક્રિયાઓ દ્વારા થાય છે:
A. એર ફ્લો અને ડીફ્યુઝન
B. માસ ફ્લો અને ડીફ્યુઝન
C. એર ફ્લો અને ડિસ્પર્શન
D. માસ ફ્લો અને ડિસ્પર્શન

79. ઊંચાઈ વધવાની સાથે જંગલની જમીનમાં જૈવિક ઘટકો
A. વધે છે
B. ઘટે છે
C. સતત રહે છે
D. ક્યારેક વધે છે ક્યારેક ઘટે છે

80. ભારતમાં ____________ જંગલની જમીનના કુલ ઓર્ગેનિક કાર્બન પૂલમાં મહત્તમ ફાળો આપે છે.
A. ઉષ્ણકટિબંધીય સદાબહાર જંગલો
B. મોન્ટેન સમશીતોષ્ણ જંગલો
C. ઉષ્ણકટિબંધીય શુષ્ક પાનખર જંગલો
D. ઉષ્ણકટિબંધીય ભેજવાળા પાનખર જંગલો

81. ઉપલા સ્તરની માટીનું તાપમાન તેનાથી નીચલા સ્તરની માટી કરતા સામાન્ય રીતે ____________ હોય છે.
A. નીચું
B. ઊંચું
C. સમાન
D. ક્યારેક નીચું તો ક્યારેક ઊંચું

82. વધુ ક્લે અને ઓછું ઓર્ગેનિક મેટર કઈ માટીમાં જોવા મળે છે ?
A. ઉષ્ણકટિબંધીય જંગલો
B. સમશીતોષ્ણ જંગલો
C. A અને B બન્ને
D. પર્વતનાં જંગલો

78. Aeration in forest soils takes place by two physical processes:
A. Air flow and diffusion
B. Mass flow and diffusion
C. Air flow and dispersion
D. Mass flow and dispersion

79. With increase in altitude the forest soil organic matter
A. Increase
B. Decrease
C. Remain constant
D. Sometimes increase sometimes decrease

80. In India ____________ forests contribute maximum to the total forest soil organic C pool
A. Tropical evergreen forests
B. Montane temperate forests
C. Tropical dry deciduous forests
D. Tropical moist deciduous forests

81. Surface soil temperature is generally ____________ than the air in the immediate soil surface.
A. Lower
B. Higher
C. same
D. Sometimes lower sometimes higher

82. High clay content and lower soil organic matter is found in the soils of
A. Tropical forests
B. Temperate forests
C. Both A and B
D. Mountain forests

83. જમીનની સપાટી શેના માટે ઓળખવામાં આવે છે?
A. જમીનની સપાટી પર સ્લાઇમ મોલ્ડનો વિકાસ
B. જમીનની સપાટી પર અકાર્બનિક સ્તરનો વિકાસ
C. જમીનની સપાટી પર કાર્બનિક સ્તરનો વિકાસ
D. જમીનની સપાટી પર સુક્ષ્મસજીવોનો વિકાસ

84. ફૂગ અને ઝાડના મૂળના સહનિવાસ ક્યાં નામે ઓળખાય છે?
A. લાઇકેન
B. માઇકોરિઝા
C. ડાયઝોટ્રોફિક એસોસિએશન
D. પ્રોટોકોપરેશન

85. જમીનની પ્રોફાઇલ દ્વારા પાણીની નીચેની ગતિને ક્યાં નામે ઓળખવામાં આવે છે?
A. પર્કોલેશન B. રનઓફ
C. લીચિંગ D. ઇનફીલ્ટ્રેશન

86. સપાટીની હાઇડ્રોલોજી જમીનના ________ ગુણધર્મો હેઠળ આવે છે
A. રાસાયણિક
B. જૈવિક
C. ભૌતિક
D. A અને C બન્ને

87. જમીનના સંયોજનોના રાસાયણિક ઓક્સિડેશનને કારણે માટીમાંથી CO_2 મુક્ત થવાની પ્રક્રિયા ક્યાં નામે ઓળખાય છે?
A. C સિકવેસ્ટ્રેશન
B. C એમિશન
C. ગ્રીન હાઉસ ઇફેક્ટ
D. સોઇલ રેસ્પિરેશન

83. Forest floors are referred as
A. development of slime molds on the soil surface
B. development of an inorganic layer on the soil surface
C. development of an organic layer on the soil surface
D. development of microorganisms on the soil surface

84. Association of fungi and tree roots is known as
A. Lichens
B. Mycorrhiza
C. Diazotrophic association
D. Protocooperation

85. Downward movement of water through the soil profile is known as
A. Percolation B. Runoff
C. Leaching D. Infiltration

86. Surface hydrology comes under the ________ properties of soil
A. Chemical
B. Biological
C. Physical
D. Both A and C

87. The release of CO_2 from soils due to chemical oxidation of soil compounds is known as
A. C Sequestration
B. C Emission
C. Greenhouse effect
D. Soil respiration

88. માટી જેમાં જૈવિક ઘટકો ખનિજો વાળી જમીનમાં 2 ફુટ અથવા વધુ ઊંડાઈએ ભળી જાય તેને શું કહે છે ?

A. મોર સોઇલ B. મૂલ સોઇલ

C. પેડઝોલ સોઇલ D. લેસ્સીવ સોઇલ

89. રેતાળ જમીન કેવી હોય છે?

A. ભારે ટેક્સચરવાળી જમીન

B. મધ્યમ ટેક્ષ્ચર જમીન

C. આછા ટેક્ષ્ચર જમીન

D. ખૂબ ભારે રચનાવાળી જમીન

90. લાકડાની કાપણીમાં વપરાયેલી ભારે મશીનરીને લીધે શું થાય છે ?

A. જંગલની જમીનોનું સંકોચન

B. વધતી વન જમીન છિદ્રાળુતા

C. વન જમીનોની ઘનતામાં ઘટાડો

D. જંગલની જમીનોનું ઉચ્ચ ધોવાણ

91. રુટ રેસ્પિરેશનથી જંગલની જમીનનું તાપમાન

A. વધે છે B. ઘટાડો થાય છે

C. અસર કરતું નથી D. નિયમન કરે છે

92. સામાન્ય રીતે વન જમીનોની રાસાયણિક ફળદ્રુપતા ______ હોય છે.

A. ઊંચી B. ખૂબ જ ઊંચી

C. મર્યાદિત D. અનહદ

93. કૃષિ માટે જંગલની શ્રેષ્ઠ જમીન કઈ છે?

A. શંકુદ્રુપ જંગલો

B. પાનખર જંગલો

C. ભૂમધ્ય જંગલો

D. વરસાદી જંગલો

94. Gelisols (Cryosolsand Histosols) અને Spodosols લાક્ષણિક ક્યાંની જમીન છે?

A. બોરિયલ જંગલો

B. ઉષ્ણકટિબંધીય જંગલો

C. સમશીતોષ્ણ જંગલો

D. ઉપર્યુક્તમાંથી કોઈ નહિ

88. Soil in which the organic matter is intimately mixed with the mineral soil for a depth of 2 feet or more is known as

A. Mor soil B. Mull soil

C. Podzol soil D. Lessive soil

89. Sandy soils are

A. Heavy textured soils

B. Moderate textured soils

C. Light textured soils

D. Very heavy textured soils

90. Heavy machinery used during timber harvesting cause

A. Compaction of forest soils

B. Increased porosity of forest soils

C. Decrease in density of forest soils

D. High drainage of forest soils

91. Root respiration ______ the temperature of forest soils

A. Increases B. Decreases

C. Does not affect D. Regulates

92. Chemical fertility of forest soils is generally

A. High B. Very high

C. Limited D. Boundless

93. Best forest soils for agriculture lands are

A. Coniferous forests

B. Deciduous forests

C. Mediterranean forests

D. Rainforests

94. Gelisols (Cryosolsand Histosols) and Spodosols are the typical soils of

A. Boreal forests

B. Tropical forests

C. Temperate forests

D. None of the above

95. ઉષ્ણકટિબંધીય ચેરના જંગલોની જમીનમાં શું હોય છે?
A. રેતી અને માટી
B. કાદવ અને રેતી
C. માટી અને કાદવ
D. રેતી, માટી અને કાદવ

96. સ્પોડોસોલ્સ (Pedzol in WRB) એ ક્યાંની લાક્ષણિક જમીન છે?
A. ઉષ્ણકટિબંધીય શંકુદ્રુપ અને બોરિયલ જંગલ
B. સમશીતોષ્ણ શંકુદ્રુપ અને બોરિયલ જંગલ
C. ઉષ્ણકટિબંધીય પાનખર અને બોરિયલ જંગલ
D. સમશીતોષ્ણ પાનખર અને બોરિયલ જંગલ

97. વનની જમીન પાણી શાને લીધે જાળવી રાખે છે?
A. સંલગ્નતા બળ
B. સંઘાત બળ
C. પૃષ્ઠતાણ
D. બન્ને સંલગ્નતા અને સંઘાત બળો

98. ____________ એ જડ સામગ્રી નથી.
A. રેતી (Sand)
B. કાંપ (Silt)
C. માટી (Clay)
D. ઉપર્યુક્તમાંથી કોઈ નહિ

99. જો પેરન્ટ મટેરીયલ લાઇમસ્ટોન હોય, તો વનની જમીનમાં ____________ હોઈ શકે.
A. કલ્કેરિયાસ નું સ્તર
B. મીનરલનું સ્તર
C. ઓર્ગેનિક સ્તર
D. સલ્ફરનું સ્તર

95. The soil substrates of tropical mangrove forests consist of
A. Sand and clay
B. Mud and sand
C. Clay and Mud
D. Sand, Clay and Mud

96. Spodosols (Podzols in WRB) are the typical soils of
A. Tropical coniferous and boreal forest
B. Temperate coniferous and boreal forest
C. Temperate deciduous and boreal forest
D. Tropical deciduous and boreal forest

97. Water in forest soils is retained by
A. Adhesion force
B. Cohesion force
C. Surface tension
D. Both adhesion and cohesion forces

98. ____________ is not an inert material
A. Sand
B. Silt
C. Clay
D. None of the above

99. If the parent material is limestone, a ____________ may be present in the forest soil
A. Calcareous layer
B. Mineral layer
C. Organic layer
D. Sulphur layer

100. પીટ માટી ક્યાં મળી આવે છે?

A. બોરિયલ જંગલો
B. સ્વેમ્પ જંગલો
C. વરસાદી જંગલો
D. પાનખર જંગલો

101. ચેરના જંગલોની જમીન મુખ્યત્વે શાને આધારે વર્ગીકૃત કરવામાં આવી છે?

A. એન્ટિસોલ્સ B. ઓક્સિસોલ્સ
C. ગેલીસોલ્સ D. અલ્ટિસોલ્સ

102. બોરિયલ વન જમીનની લાક્ષણિકતા શું છે?

A. પાતળા પોડઝોલ B. જાડા પોડઝોલ
C. ઊંડા પોડઝોલ D. ચંકી પોડઝોલ

103. બોરિયલ જંગલની જમીન કેવી છે?

A. એસિડિક અને વધુ મીનરલ વાળી
B. બેઈઝીક અને વધુ મીનરલ વાળી
C. એસિડિક અને મિનરલની ઊણપ ધરાવતી
D. બેઈઝીક અને મિનરલની ઊણપ ધરાવતી

104. ઉષ્ણકટિબંધીય વરસાદી જંગલોની જમીન કેવી છે?

A. ઓક્સિસોલ્સ B. લાલ લેટોસોલ્સ
C. ફેરલ્સલ્સ D. ઉપરોક્ત તમામ

105. વિશ્વ ભૂમિ દિવસ ક્યારે ઉજવવામાં આવે છે?

A. 6 જૂન B. 5 ડિસેમ્બર
C. 19 ઓક્ટોબર D. 11 ફેબ્રુઆરી

106. ઉષ્ણકટિબંધીય કાંટાવાળા જંગલોની જમીન કેવી છે?

A. લાલ માટી B. પીળી માટી
C. કાળી માટી D. ભૂરી માટી

107. જંગલો માટીમાંથી ભેજને તેમના તાત્કાલિક વાતાવરણમાં કયા સ્વરૂપે રિસાયકલ કરે છે?

A. પ્રેસિપિટેશન B. ટ્રાન્સ્પિરેશન
C. ભેજ D. વરસાદ

100. Peat soil is found in

A. Boreal forests
B. Swamp forests
C. Rainforests
D. Deciduous forests

101. The soils of mangrove forests are classified mainly as

A. Entisols B. Oxisols
C. Gelisols D. Ultisols

102. Boreal forest soils are characterized by

A. thin podzols B. thick podzols
C. deep podzols D. chunky podzols

103. Boreal forest soils are

A. acidic and high in minerals
B. basic and high in minerals
C. acidic and mineral deficient
D. basic and mineral deficient

104. Soil of tropical rainforests are

A. oxisols B. red latosols
C. ferralsols D. all of the above

105. World soil science day is celebrated on

A. 6 June B. 5 December
C. 19 October D. 11 February

106. Tropical thorn forests have

A. Red soil B. Yellow soil
C. Black soil D. Brown soil

107. Forests recycle moisture from soil into their immediate atmosphere by

A. Precipitation B. Transpiration
C. Humidity D. Rainfall

108. USDA અનુસાર, જમીન _________ ઓર્ડરમાં વહેંચાયેલી છે

A. 12 B. 8
C. 14 D. 10

109. લાટોસોલ્સમાં શું વધુ હોય છે?

A. આયર્ન અને એલ્યુમિનિયમ ઓક્સાઇડ
B. આયર્ન અને સિલિકોન ઓક્સાઇડ
C. આયર્ન અને નાઇટ્રોજન ઓક્સાઇડ
D. આયર્ન અને મેગ્નેશિયમ ઓક્સાઇડ

110. લાલ જમીનમાં કયું ક્લે મિનરલ પ્રબળ હોય છે?

A. ઇલાઇટ B. કૌલિનાઇટ
C. મોન્ટમોરીલોનાઇટ D. નોન્ટ્રોનાઇટ

111. ક્લે મિનરલ મુખ્યત્વે કયું હાઇડ્રોસ હોય છે?

A. આયર્ન ફાયલોસિલીકેટ્સ
B. મેગ્નેશિયમ ફાયલોસિલીકેટ્સ
C. એલ્યુમિનિયમ ફાયલોસિલીકેટ્સ
D. સોડિયમ ફાયલોસિલીકેટ્સ

112. કુલ છિદ્રાળુતા શેમાં સૌથી વધુ હોય છે?

A. માટી B. રેતી
C. કાંપ D. B અને C બન્ને

113. હાઇડ્રોમેટ્રિક પદ્ધતિનો ઉપયોગ શેના માટે થાય છે?

A. સોઇલ સેપરેશનનું રાસાયણિક વિશ્લેપણ
B. સોઇલ સેપરેશનનું યાંત્રિક વિશ્લેષણ
C. સોઇલ સેગ્રીગેશન નું રાસાયણિક વિશ્લેષણ
D. સોઇલ સેગ્રીગેશન યાંત્રિક વિશ્લેષણ

114. જમીનનું સૌથી ઉપરનું હોરાઈઝન ક્યાં નામે ઓળખાય છે ?

A. M હોરાઈઝન B. C હોરાઈઝન
C. B હોરાઈઝન D. A હોરાઈઝન

108. According to USDA, soils are divided into _______ orders

A. 12 B. 8
C. 14 D. 10

109. Latosols have high content of

A. Iron and aluminium oxides
B. Iron and silicon oxides
C. Iron and nitrogen oxides
D. Iron and magnesium oxides

110. In red soils, the dominant clay mineral is

A. Illite B. Kaolinite
C. Montmorillonite D. Nontronite

111. Clay minerals are majorly hydrous

A. Iron phyllosilicates
B. Magnesium phyllosilicates
C. Aluminium phyllosilicates
D. Sodium phyllosilicates

112. Total pore space is highest in

A. Clay B. Sand
C. Silt D. Both B and C

113. Hydrometric method is used for

A. Chemical analysis of soil separation
B. Mechanical analysis of soil separation
C. Chemical analysis of soil segregation
D. Mechanical analysis of soil segregation

114. Top most mineral horizon is known as

A. M horizon B. C horizon
C. B horizon D. A horizon

115. ________ વર્ટીઝોલ વન-ઘાસના મેદાનના ક્ષેત્રમાં જોવા મળે છે

A. એકવર્ટ્સ B. ઝેરેટ્સ
C. ક્રાયર્ટ્સ D. ટોરેર્ટ્સ

116. Indian Institute of Soil Science ક્યાં આવેલ છે?

A. ભોપાલ B. મુંબઈ
C. અમદાવાદ D. દિલ્હી

117. કેરળમાં મોટાભાગે ________ જમીન જોવા મળે છે.

A. આલ્કલાઈન B. ન્યુટ્રલ
C. એસિડિક D. A અને C બન્ને

118. ટુરમાલાઇન એ શેનો મુખ્ય સ્રોત છે?

A. K B. B
C. Ca D. Si

119. જંગલની જમીનમાં મિનરલ્સની ભૂમિકા શેના પર ખૂબ આધાર રાખે છે?

A. મિનરલ બોન્ડ્સની કેમેસ્ટ્રી
B. મિનરલ્સની 3-ડી રચનાઓ
C. મિનરલના કણોના કદ
D. ઉપરોક્ત તમામ

120. મોટાભાગની જંગલની જમીન જે પેરન્ટ મટેરીયલ પર બને છે તે ખનીજ થી ________ હોય છે .

A. ઉણપવાળી
B. પુરતી
C. ભરપૂર
D. ખનીજ પર આધારિત નથી

121. ઉષ્ણકટિબંધીય જંગલની જમીન, લિમોનાઇટ અને અન્ય આયર્ન અને એલ્યુમિનિયમ ઓક્સાઇડ અને હાઇડ્રોક્સાઇડ્સમાં ક્યાં એકઠા થાય છે?

A. A હોરાઈઝન B. E હોરાઈઝન
C. B હોરાઈઝન D. C હોરાઈઝન

115. ________ vertisols are found in forest-grassland transitions zones

A. Aquerts B. Xererts
C. Cryerts D. Torrerts

116. Indian Institute of Soil Science is situated in

A. Bhopal B. Mumbai
C. Ahmedabad D. Delhi

117. In Kerala, the mostly ________ soils arc found

A. Alkaline B. Neutral
C. Acidic D. Both A and C

118. Tourmaline is the major source of

A. K B. B
C. Ca D. Si

119. The role of minerals in forest soils depends heavily on the

A. chemistry of the mineral bonds
B. 3-D structures of minerals
C. sizes of the mineral particles
D. All of the above

120. Most forest soils form on parent material ________ in minerals.

A. Poor
B. Sufficient
C. Rich
D. Doesn't depend on mineral content

121. In tropical forest soils, limonite and other iron and aluminum oxides and hydroxides accumulate in the

A. A horizon B. E horizon
C. B horizon D. C horizon

122. સલ્ફેટ્સ, સલ્ફાઇડ્સ અને ક્લોરાઇડ વન જંગલની જમીનમાં ________ હોય છે.

A. અસામાન્ય B. પુરતાં

C. વિપુલ D. સામાન્ય

123. પાયરાઇટમાં વધુ માટી અથવા સ્પોઈલ મટેરીયલ હોય તો તે ________ હોય છે.

A. તટસ્થ

B. ખૂબ એસિડિક

C. ખૂબ આલ્કલાઇન

D. સહેજ આલ્કલાઇન

124. કેટલીક જંગલની જમીનમાં ________ સામાન્ય છે અને અને તે હાર્ડપાનમાં એક મહત્વપૂર્ણ સિમેન્ટિંગ એજન્ટ છે.

A. સીડરાઇટ B. કેલસાઇટ

C. ડોલોમાઇટ D. બાયોટાઇટ

125. કાર્બોનેટ સામાન્ય રીતે માત્ર ભેજવાળા વન જમીનમાં ક્યાં હાજર હોય છે?

A. ઉપરના ક્ષિતિજ પર

B. સપાટી પર

C. નીચલા હોરાઈઝોનમાં

D. A અને B બન્ને

126. ક્લે મિનરલ્સ _____ એલ્યુમિનો- સિલિકેટ્સ છે જે જમીનમાં ખૂબ મહત્વ ધરાવે છે.

A. પ્રાથમિક B. દ્વિતીય

C. તૃતીય D. ક્વાર્ટરનરી

127. ફાઇન - ગ્રેનના ગ્રેનાઈટ વેરાયટી માટી બનાવવાની પદ્ધતિમાં કયા નામે ઓળખાય છે?

A. ટ્રેકીટ્સ B. પેરીડોટાઇટ

C. એપ્લાઈટ D. શેલ

128. હ્યુમસ પ્રકૃતિમાં ________ છે

A. એસિડિક B. બેઈઝીક

C. ન્યુટ્રલ D. સલાઈન

122. Sulfates, sulfides, and chlorides are ________ in forest soils

A. Uncommon B. Sufficient

C. Abundance D. Common

123. Soil or spoil material high in pyrite is

A. Neutral

B. Highly Acidic

C. Highly Alkaline

D. Slightly Alkaline

124. ________ is common in some forest soils and is an important cementing agent in hardpans.

A. Siderite B. Calcite

C. Dolomite D. Biotite

125. Carbonates are generally present only in the ________ of humid zone forest soils.

A. Upper horizons

B. Surface

C. Lower horizons

D. Both A and B

126. Clay minerals are ________ alumino-silicates that are of great importance in soils

A. Primary B. Secondary

C. Tertiary D. Quaternary

127. Fine-grained granite variety soil forming rock type is known as

A. Trachytes B. Peridotite

C. Aplite D. Shale

128. Humus is ________ in nature

A. Acidic B. Basic

C. Neutral D. Saline

129. ________ જમીનનો 70 - 90 % હિસ્સો ધરાવે છે અને ખનીજ બનાવવામાં ભાગ ભજવે છે.
A. સિલિકેટ્સ
B. એલ્યુમિનિયમ ઓક્સાઇડ
C. હાઇડ્રોસ ઓક્સાઇડ
D. મેગ્નેટાઇટ

130. માઇકોરાઈઝલ ફૂગ "માઇન" સિલિકેટ મિનરલ્સ મેળવવામાં કઈ રીતે ઉપયોગી છે.
A. ન્યૂટ્રેશન કેટાયન્સ અને સિલિકેટ્સ
B. ન્યૂટ્રેશન કેટાયન્સ અને ફોસ્ફેટ
C. ન્યૂટ્રેશન કેટાયન્સ અને કાર્બોનેટ્સ
D. ન્યૂટ્રેશન કેટાયન્સ અને મેગ્નાઈટસ

131. ક્વાર્ટઝાઇટ કયા પ્રકારની માટી રાચે છે?
A. અલ્ટ્રાબેસિક B. મધ્યમ
C. અતિશય એસિડિક D. કાર્બોનેસિયસ

132. કયા વિસ્તારમાં લેટેરાઈટ જમીન વધુ જોવા મળે છે ?
A. કર્ણાટક B. કેરળ
C. મહારાષ્ટ્ર D. A અને B બન્ને

133. સેકવિઓક્સાઇડ માંથી બહાર નીકળવાની પ્રક્રિયા કયા નામે ઓળખવામાં આવે છે?
A. લેટરલાઈઝેશન
B. એલ્યુવિએશન
C. ઇલ્યુવિએશન
D. પોડઝોલાઇઝેશન

134. કેંકર નોડ્યુલ્સ મોટે ભાગે ક્યાં જોવા મળે છે?
A. કાળી માટી B. લાલ જમીન
C. ભૂરી માટી D. પીળી જમીન

135. લીચિંગનો ઉપયોગ શાની સારવાર માટે થાય છે?
A. એસિડિક જમીન B. બેઈઝીક જમીન
C. ખારી જમીન D. ન્યુટ્રલ જમીન

129. ________ account for 70-90% of the soil-forming minerals
A. Silicates
B. Aluminium oxides
C. Hydrous oxides
D. Magnetite

130. Myccorhizal fungi directly involved in "mine" silicate minerals to obtain
A. nutrient cations and silicates
B. nutrient cations and phosphate
C. nutrient cations and carbonates
D. nutrient cations and magnetite

131. Quartzite is the following type of soil forming rock
A. Ultrabasic B. Intermediate
C. Extremely acidic D. Carbonaceous

132. Laterite soils are dominant in
A. Karnataka B. Kerala
C. Maharashtra D. Both A and B

133. Process of moving out of Sesquioxide is known as
A. Lateralization
B. Eluviation
C. Illuviation
D. Podzolization

134. Canker nodules are mostly found in
A. Black soils B. Red soils
C. Brown soils D. Yellow soils

135. Leaching is used for the treatment of
A. Acidic soils B. Basic soils
C. Saline soils D. Neutral soils

136. ક્લેમાં જોવા મળતું સામાન્ય મિનરલ કયું છે?

A. મીકા B. ક્વાર્ટઝ
C. સર્પેન્ટીન D. ફેલ્ડસ્પર

137. હ્યુમશનો C:N રેશિયો શું છે?

A. 8: 1 B. 5: 1
C. 7: 1 D. 10: 1

138 . ICAR એ જમીનને કેટલા જૂથમાં વહેંચી છે?

A. 7 જૂથ B. 10 જૂથ
C. 8 જૂથ D. 9 જૂથ

139. લાલ જમીન લાલ રંગની શાને લીધે હોય છે?

A. આયર્ન ઓક્સાઇડ
B. એલ્યુમિનિયમ ઓક્સાઇડ
C. સિલિકોન ઓક્સાઇડ
D. મેગ્નેશિયમ ઓક્સાઇડ

140. તાજેતરમાં શોધાયેલ જમીનનો ઓર્ડર કયો છે?

A. અલ્ફિસોલ્સ B. ઇનસેપ્ટિસોલ્સ
C. એન્ટિસોલ્સ D. એરિડીસોલ્સ

141. ઓર્ગેનીક મેટરનું અદ્રાવ્ય ભાગ શું છે?

A. હ્યુમિક એસિડ B. ફુલવિક એસિડ
C. હુમિન D. A અને B બન્ને

142. જમીનમાં સલ્ફરનો સ્ત્રોત શું છે?

A. ઓર્ગેનિક સલ્ફર B. અકાર્બનિક સલ્ફર
C. પિરાઇટ D. સલ્ફાઇડ

143. ઓર્ગેનિક માટીની પ્રકૃતિ કેવી છે?

A. બેઈઝીક B. ન્યુટ્રલ
C. ખારી D. એસીડીક

144. વરસાદ પછી જમીનની ગંધ શાને કારણે હોય છે?

A. ફૂગ
B. એક્ટિનોમાઇસેટ્સ
C. ગ્રામ-નેગેટિવ બેક્ટેરિયા
D. આલ્ગી

136. The common clay mineral found in silt is

A. Mica B. Quartz
C. Serpentine D. Feldspar

137. C:N ratio of humus is

A. 8:1 B. 5:1
C. 7:1 D. 10:1

138. ICAR classified soils in

A. 7 groups B. 10 groups
C. 8 groups D. 9 groups

139. Red soils are red in color due to the presence of

A. Iron oxides
B. Aluminium oxides
C. Silicon oxides
D. Magnesium oxides

140. Recently formed soil order is

A. Alfisols B. Inceptisols
C. Entisols D. Aridisols

141. Insoluble fraction of organic matter is

A. Humic acid B. Fulvic acid
C. Humin D. Both A and B

142. The main reservoir of available sulfur in soil is

A. Organic sulfur B. Inorganic sulfur
C. Pyrite D. Sulfide

143. Nature of organic soil is

A. Basic B. Neutral
C. Saline D. Acidic

144. The earthy smell of soil after rain is due to

A. Fungi
B. Actinomycetes
C. Gram-negative bacteria
D. Algae

145. જમીનમાં સુક્ષ્મ પોષકતત્ત્વોની ઉપલબ્ધતા ક્યારે વધુ હોય છે?
A. ઉંચી pH B. નીચી pH
C. ન્યુટ્રલ pH D. ખુબ ઉંચી pH

146. નક્કર કોલોઇડ્સ દ્વારા વાતાવરણમાંથી પાણી આકર્ષિત થવાની પ્રક્રિયાને શું કહે છે ?
A. એબ્સોર્પશન B. ઇવેપોરેશન
C. પ્રેસિપિટેશન D. એડસોર્બશન

147. વેધરીંગ જમીનની રચનાનો ______ તબક્કો છે.
A. પ્રથમ B. બીજો
C. ત્રીજો D. ચોથો

148. એગ્રિલિએશન એક પ્રક્રિયા છે જેમાં વિખરાયેલા માટીના કણો ક્યાંથી સ્થાનાંતરિત થાય છે?
A. નીચલાથી ઉપરના હોરાઈઝનમાં
B. મધ્યથી ઉપરની હોરાઈઝનમાં
C. ઉપરથી નીચલા હોરાઈઝનમાં
D. ઉપરથી મધ્ય હોરાઈઝનમાં

149. ઓર્ગેનિક માટીની પ્રકૃતિ કેવી હોય છે?
A. એસિડિક B. બેઈઝીક
C. ન્યુટ્રલ D. ખારી

150. ______ હોરાઈઝન એ ઇલ્લુવીયલ હોરાઈઝન છે.
A. A B. B
C. O D. E

151. ક્લે, સિલ્ટ અને સેન્ડનું પ્રમાણ શું નક્કે કરે છે?
A. સ્ટ્રક્ચર
B. ન્યુટ્રિયન્ટ પોટેન્શિયલ
C. ફર્ટિલિટી લેવલ
D. ટેક્ષ્ચર

145. Availability of micronutrients in soil is more at
A. High pH B. Low pH
C. Neutral pH D. Very high pH

146. Attraction of water from atmosphere by the solid colloids is known as
A. Absorption B. Evaporation
C. Precipitation D. Adsorption

147. Weathering is the ______ phase of soil formation
A. First B. Second
C. Third D. Fourth

148. Agrilliation is a process in which the dispersed clay particles are migrated from
A. Lower to upper horizon
B. Middle to upper horizon
C. Upper to lower horizon
D. Upper to middle horizon

149. Organic soils are ______ in nature
A. Acidic B. Basic
C. Neutral D. Saline

150. ____ horizon is illuviation horizon
A. A B. B
C. O D. E

151. The proportion of clay, silt and sand in soil determines
A. structure
B. nutrient potential
C. fertility level
D. texture

152. કઇ માટીમાં સુક્ષ્મસજીવો માટે પુષ્કળ પાણી અને પોષક તત્વો છે?

A. વન B. રણ
C. ઘસના મેદાનો D. ખુલ્લા ખેતર

153. ફૂગ શેમાં મહત્તમ પ્રવૃત્તિ દર્શાવે છે?

A. આલ્કલાઇન જમીન B. એસિડિક જમીન
C. ન્યુટ્રલ જમીન D. A અને B બન્ને

154. પાણી અને ખનિજોની ઉપલબ્ધતા જંગલની જમીનના ક્યાં હોરાઈઝનમાં હોય છે ?

A. A B. B
C. O D. E

155. લિથોસ્ફીયર શું છે?

A. એથેનોસ્ફિયરનો ઝોન
B. એટમોસ્ફીયરનો ઝોન
C. હાઇડ્રોસ્ફિયરનો ઝોન
D. એટમોસ્ફીયર અને એથેનોસ્ફિયર વચ્ચેનો ઝોન

156. ગ્રેવીમેટ્રિક પાણી ક્યાં હોય છે ?

A. મેસો પોરમાં B. મેક્રો પોરમાં
C. માઈક્રો પોરમાં D. B અને C બન્ને

157. સમાન કદના કેટાયન બદલવાની પ્રક્રિયા શા નામથી ઓળખવામાં આવે છે?

A. આઇસોમોર્ફસ સબસ્ટિટ્યુશન
B. આઇસોમોર્ફિક રિપ્લેસમેન્ટ
C. કેટાયનિક સબસ્ટિટ્યુશન
D. કેટાયનિક રિપ્લેસમેન્ટ

158. ભૂમિ સંરક્ષણ પગલાંનો મુખ્યત્વે લક્ષ્ય શું છે?

A. પેટા માટી
B. ઊંડી માટી
C. ટોચની માટી
D. વૃક્ષો અને છોડ

152. Which of the following soil has plenty of water and nutrients for microorganisms

A. Forest B. Desert
C. Grassland D. Open field

153. Fungi show maximum activity in

A. Alkaline soils B. Acidic soils
C. Neutral soils D. Both A and B

154. Availability of water and minerals in the forest soils for maximum absorption by tree roots is in __________ horizon

A. A B. B
C. O D. E

155. Lithosphere is referred as

A. Zone of asthenosphere
B. Zone of atmosphere
C. Zone of hydrosphere
D. Zone situated between atmosphere and asthenosphere

156. Gravimetric water is present in

A. Meso pores B. Macro pores
C. Micro pores D. Both B and C

157. The process of replacing one structural cation for another of similar size is referred to as

A. isomorphous substitution
B. isomorphic replacement
C. Cationic substitution
D. Cationic replacement

158. Soil conservation measures are mainly aimed at protecting

A. Sub soil
B. Deep soil
C. Top soil
D. Trees and plants

159. જમીનની રચનાની લાક્ષણિકતા કોણે સમજાવી?

A. બીજેન્ડરિંક B. વિનોગ્રાડસ્કી
C. વક્સમેન D. સિમોસન

160. ભૂમિ વિજ્ઞાનના પિતા તરીકે કોણ ઓળખાય છે?

A. ઝિકિંગ B. વાસિલી ડોકુચૈવ
C. વક્સમેન D. વિનોગ્રાડસ્કી

161. B માસ્ટર હોરાઈઝોન જેને ક્યારેક ______ તરીકે ઓળખવામાં આવે છે

A. એક્યુમ્યુલેશન ઝોન
B. વેધરિંગ ઝોન
C. ડીપોઝીશન ઝોન
D. પ્રેસિપિટેશન ઝોન

162. પેડોજેનેસિસ શાનો અભ્યાસ છે?

A. માટી વર્ગીકરણ
B. છોડ પર જમીનનો પ્રભાવ
C. જમીનની રચના
D. જમીનના રાસાયણિક ગુણધર્મો

163. માટીના સમૂહનું નિર્માણ જમીનના પ્રાથમિક કણો દ્વારા ક્રિયાપ્રતિક્રિયાને કારણે થાય છે જેને શું કહે છે

A. રીઅરેન્જમેન્ટ
B. ફ્લોક્યુલેશન
C. સિમેન્ટિશન
D. ઉપરોક્ત તમામ

164. એક સ્તરીય પથ્થર કે જે કોમ્પેક્શન, સિમેન્ટિશન, અને રિક્રિસ્ટલાઇઝેશનથી રચાય છે તે ક્યાં નામે ઓળખાય છે?

A. સેન્ડસ્ટોન
B. કલે રોક
C. સેડિમેન્ટરી રોક
D. ઉપર્યુક્તમાંથી કોઈ નહિ

159. Who characterized soil formation as set of process

A. Beijerinck B. Winogradsky
C. Waksman D. Simonson

160. The father of soil science is

A. Ziking B. Vasily Dokuchaev
C. Waksman D. Winogradsky

161. B master horizon which is sometimes referred to as the

A. Zone of accumulation
B. Zone of weathering
C. Zone of deposition
D. Zone of precipitation

162. Pedogenesis is the study of

A. Soil taxonomy
B. Influence of soil on plants
C. Soil formation
D. Chemical properties of soil

163. The formation of soil aggregates occurs due to interactions of primary soil particles through

A. Rearrangement
B. Flocculation
C. Cementation
D. All of the above

164. A layered rock that is formed from the compaction, cementation, and the recrystallization of sediments is known as

A. Sand stone
B. Clay rock
C. Sedimentary rock
D. None of the above

165. **એસિડિક જમીનનો pH શું ઉમેરીને વધારી શકાય?**

A. સલ્ફર B. લાઇમ
C. સેન્ડ D. સિલિકેટ

166. **તે કઈ પ્રક્રિયા છે જેના દ્વારા પોષક તત્વો ઓગળી જાય અને પાણી દ્વારા તેનું વાહન થાય અને જમીનના નીચલા સ્તરોમાં જમા થાય છે?**

A. લીચિંગ B. મલ્ચિંગ
C. પ્રેસિપિટેશન D. ડીફોરેસ્ટેશન

167. **માસ્ટર હોરાઈઝન માં ક્લે કદના કણો કે ખનીજો ક્રમબદ્ધ જમા થવાની પ્રક્રિયા ક્યા નામે ઓળખાય છે?**

A. ડીપોઝીશન B. એલ્લુવિએશન
C. ઇલ્લુવિએશન D. પ્રેસિપિટેશન

168. **ઉષ્ણકટિબંધીય ક્ષેત્રમાં વિકસિત માટી સમશીતોષ્ણકટિબંધીય માટી કરતા ________ ઘડાયેલ અને વિકસિત હોય છે.**

A. ઓછી
B. વધુ
C. સતત
D. કદાપિ નહી

169. **માટીની અંદર દેખાતા સ્તરોને ________ તરીકે ઓળખવામાં આવે છે**

A. એલ્લુવિએશન હોરાઈઝન
B. ઇલ્લુવિએશન હોરાઈઝન
C. માસ્ટર હોરાઈઝન
D. પ્રોફાઇલ હોરાઈઝન

170. **માટીના કયા હોરાઈઝનમાં ખનિજ અને પથ્થરના ટુકડાઓ હ્યુમશયુક્ત હોય છે?**

A. A B. O
C. E D. B

165. **The pH of the acidic soils can be raised by adding**

A. Sulphur B. Lime
C. Sand D. Silicate

166. **The process by which nutrients get dissolved and carried away by water and are moved into lower layers of soil is known as**

A. Leaching B. Mulching
C. Precipitation D. Deforestation

167. **Deposition of clay-sized particles or minerals, respectively, within a master horizon is referred to as**

A. Deposition B. Eluviation
C. Illuviation D. Precipitation

168. **Soils developed in the tropics are ________ weathered and developed than those developed in temperate**

A. Less
B. More
C. Constant
D. Never

169. **Layers that appear within a soil are referred to as ________**

A. Eluviation horizons
B. Illuviation horizons
C. Master horizons
D. Profile horizons

170. **Which soil horizon has humus mixed with mineral and rock fragments?**

A. A B. O
C. E D. B

171. ક્યાં બે ઓર્ડર માટીના 12 મોટા ઓર્ડરમાંથી હોરાઈઝનની ઉત્ક્રાંતિ દ્વારા વર્ગીકૃત થયેલ છે?
A. મોલિસોલ્સ અને એરિડીસોલ્સ
B. અલ્ફિસોલ્સ અને મોલિસોલ્સ
C. સ્પોડોસોલ્સ અને અલ્ફિસોલ્સ
D. એન્ટિસોલ્સ અને ઇનસેપ્ટિસોલ

172. __________ ઠંડા ભેજવાળા વાતાવરણમાં શંકુદ્રુપ જંગલોના લેન્ડસ્કેપ્સમાં સામાન્ય છે.
A. અલ્ટિસોલ્સ B. વર્ટિસોલ્સ
C. સ્પોડોસોલ્સ D. ઓક્સિસોલ્સ

173. R હોરાઈઝન કઈ રીતે C હોરાઈઝનથી ભિન્ન છે?
A. વેધર્ડ મટેરીયલ
B. A અને E હોરાઈઝનની અનકોન્સોલિડેટેડ વેધર્ડ મટેરીયલ
C. A, E અને B હોરાઈઝનની અનકોન્સોલિડેટેડ વેધર્ડ મટેરીયલ
D. અનવેધર્ડ મટેરીયલ

174. એન્ટિસોલ્સથી અલ્ફિસોલ્સ કેવી રીતે અલગ પડે છે?
A. સારી રીતે વિકસિત A હોરાઈઝન
B. સારી રીતે વિકસિત B હોરાઈઝન
C. સારી રીતે વિકસિત A અને B હોરાઈઝન
D. ઉપર્યુક્તમાંથી કોઈ નહિ

175. ક્લે પાર્ટિકલનો વ્યાસ કેટલો હોય છે?
A. 2.00–0.20 મીમી
B. 0.20–0.02 મીમી
C. 0.02–0.002 મીમી
D. <0.002 મીમી

176. વર્મિક્યુલાઇટ શું છે?
A. કાંપ B. રેતી
C. માટી D. લોમ

171. Of the 12 major soil orders, two are characterized by minor development of horizons, those are
A. Mollisols and Aridisols
B. Alfisols and Mollisols
C. Spodosols and Alfisols
D. Entisols and Inceptisols

172. __________ are common in coniferous forest landscapes in cool moist climates
A. Ultisols B. Vertisols
C. Spodosols D. Oxisols

173. R horizon differs from C horizon in having
A. Weathered material
B. Unconsolidated weathered material of A & E horizon
C. Unconsolidated weathered material of A, E & B horizon
D. Unweathered material

174. Alfisols are distinguished from Entisols by having
A. Well developed A horizon only
B. Well developed B horizon only
C. Well developed A and B horizons
D. None of the above

175. Diameter of clay particles are
A. 2.00–0.20 mm
B. 0.20–0.02 mm
C. 0.02–0.002 mm
D. < 0.002 mm

176. Vermiculite comes under
A. Silt B. Sand
C. Clay D. Loam

177. જંગલની માટીના નિર્માણમાં મુખ્ય ઓર્ગેનિક સિમેન્ટિંગ એજન્ટ કયો છે?
A. ઓર્ગેનિક એસિડ્સ
B. પોલિસકેરાઇડ્સ
C. લિપિડ્સ
D. પ્રોટીન

178. આયર્ન અને એલ્યુમિનિયમ ઓક્સાઇડના સંચયની પ્રક્રિયા _______ તરીકે ઓળખાય છે.
A. પોડોલાઇઝેશન B. લેટરલાઈઝેશન
C. ડીપોઝીશન D. એગ્રિગેશન

179. માટીના વર્ગીકરણની નીચલી શ્રેણીમાં _____ સામેલ છે.
A. સિરીઝ B. ઓર્ડર
C. ગ્રુપ D. ફેમિલી

180. હાઇડ્રોજન બોન્ડિંગ _____________ ક્લે ખનીજમાં જોવા મળે છે.
A. વર્મિક્યુલાઇટ
B. કઓલિનાઇટ
C. મોન્ટમોરીલોનાઇટ
D. નોન્ટ્રોનાઇટ

181. બેક્ટેરિયાના ઉપયોગ માટે ફૂગ દ્વારા જમીનમાં જટિલ પરમાણુઓનું અપઘટન શાનું ઉદાહરણ છે?
A. પરસ્પરવાદ B. વિરોધીતા
C. તટસ્થતા D. સહભોજિતા

182. મોટાભાગની જમીનમાં પ્રબળ મિનરલ પાર્ટિકલ ક્યાં છે?
A. આયર્ન B. મેગ્નેશિયમ
C. સોડિયમ D. એલ્યુમિનિયમ

183. ________ તાજી ખુલ્લી ખડકોની સપાટી પર વધે છે.
A. ફૂગ B. એક્ટિનોમિસેટ્સ
C. સાયનોબેક્ટેરિયા D. લીલો શેવાળ

177. The major organic cementing agent in forest soil aggregate formation is
A. Organic acids
B. Polysaccharides
C. Lipids
D. Proteins

178. Process of accumulation of iron and aluminium oxides is known as
A. Podolization B. Laterization
C. Deposition D. Aggregation

179. Lower category of soil classification includes
A. Series B. Order
C. Group D. Family

180. Hydrogen bonding is found in _________ clay mineral
A. Vermiculite
B. Kaolinite
C. Montmorillonite
D. Nontronite

181. The degradation of complex molecules in soil by fungi for utilization by bacteria is an example of
A. Mutualism B. Antagonism
C. Neutralism D. Commensalism

182. The dominant mineral particles in most soils are compounds of
A. Iron B. Magnesium
C. Sodium D. Aluminium

183. _________ grows on the surfaces of freshly exposed rocks
A. Fungi B. Actinomycetes
C. Cyanobacteria D. Green algae

184. જંગલની જમીનોમાં ________ કાર્બનિક પદાર્થ છે

A. ફેલ્ડસ્પર
B. હ્યુમસ
C. બેડરોક
D. મેગ્નેટાઇટ

185. આમાંથી શું વેધરિંગનું કારણ નથી?

A. ઝડપી પવન
B. ભારે વરસાદ
C. વહેતું પાણી
D. વરસાદનું પાણી થીજી જવું

186. ________ માટીમાં મહત્તમ હવાની જગ્યા હોય છે અને છૂટીછવાઈ હોય છે.

A. રેતાળ માટી
B. કાંપવાળી માટી
C. લોમી માટી
D. ક્લે માટી

187. જમીનમાં પાણીનો સંચાર થાય છે તે ________ હોરાઈઝનમાં એકત્રિત થાય છે.

A. A B. C
C. B D. O

188. ________ માટી સૂકી સ્થિતિમાં તિરાડો અને સંકોચો બનાવે છે

A. રેતાળ માટી
B. લાલ છિદ્રાળુ માટી
C. કાળી ક્લે માટી
D. લોમી માટી

189. ખડકમાં ખનિજોનું ઓક્સિડેશન ________ નું ઉદાહરણ છે.

A. રાસાયણિક વેધરિંગ
B. ભૌતિક વેધરિંગ
C. જૈવિક વેધરિંગ
D. કોઈ વેધરિંગ નથી

184. The organic material in forest soils is

A. Feldspar
B. Humus
C. Bedrock
D. Magnetite

185. Which of these do not cause weathering?

A. Fast wind
B. Heavy Rains
C. Flowing water
D. Freezing of rainwater

186. ________ soil has maximum air space and loosely packed

A. Sandy soil
B. Silt soil
C. Loamy soil
D. Clayey soil

187. Water percolating into the soil collects in ________ Horizon

A. A B. C
C. B D. O

188. ________ soil forms cracks and shrinks in dry condition

A. Sandy soil
B. Red porous soil
C. Black clay soil
D. Loamy soil

189. Oxidation of minerals in a rock is an example of

A. Chemical weathering
B. Physical weathering
C. Biological weathering
D. No weathering at all

190. સોઇલ ક્રિપ અને લેન્ડ સ્લાઇડ ઘટના શા કારણે થાય છે?

A. માસ અલાયન્સ

B. માસ મુવમેન્ટ

C. પ્રેશર અલાયન્સ

D. પ્રેશર મુવમેન્ટ

190. Soil creep and Land slide comes under the phenomenon of

A. Mass alliance

B. Mass movement

C. Pressure alliance

D. Pressure movement

191. લેટરાઇટ માટી મુખ્યત્વે _____________ ની પ્રક્રિયાને કારણે રચાય છે.

A. ઈરોઝન B. વેધરિંગ

C. ડીપોઝીશન D. લીચિંગ

191. Laterite soil is mainly formed due to the process of

A. Erosion B. Weathering

C. Deposition D. Leaching

192. જંગલની જમીનમાં સામાન્ય ખનિજ કયા છે?

A. ગોથાઇટ B. ગિબ્સાઇટ

C. બોહેમાઇટ D. ઉપરોક્ત તમામ

192. The common mineral/s in forest soils is/are

A. Goethite B. Gibbsite

C. Boehmite D. All of the above

193. પીટિ માટી ક્યાં મળી આવે છે?

A. ગુજરાત B. આસામ

C. કેરળ D. કાશ્મીર

193. Peaty soil is found in

A. Gujarat B. Assam

C. Kerala D. Kashmir

194. ડેક્કન પ્લેટુ શેમાંથી બનેલ છે ?

A. ઈગ્નીઅશ રોક

B. સેડિમેન્ટરી રોક

C. મેટામોર્ફિક રોક

D. સેન્ડસ્ટોન

194. Deccan Plateau is formed from

A. Igneous rocks

B. Sedimentary rocks

C. Metamorphic rocks

D. Sand stone

195. મહત્તમ પાણી _______________ માટી દ્વારા જળવાય છે.

A. લોમી માટી B. ક્લે માટી

C. લાલ માટી D. લેટરિટિક માટી

195. Maximum water is retained by

A. Loamy soil B. Clayey soil

C. Red soil D. Lateritic soil

196. જંગલની જમીનમાં ફોસ્ફરસનો મુખ્ય સ્રોત __________છે.

A. અપાટાઇટ

B. ઓલિવિન

C. ટૂરમાલાઇન

D. ક્વાર્ટ્ઝ

196. Major source of P in forest soil is

A. Apatite

B. Olivine

C. Tourmaline

D. Quartz

197. વિદારણ પ્રતિરોધક ભારે ખનીજ મેગ્નાઈટ જમીનમાં કઈ રીતે થાય છે?
A. મજબૂત મેંગેનેટિઝમ સાથે લાલ રેતી
B. મજબૂત મેંગેનેટિઝમ સાથે પીળી રેતી
C. મજબૂત મેંગેનેટિઝમ સાથે કાળી રેતી
D. મજબૂત મેંગેનેટિઝમ સાથે સફેદ રેતી

198. સમશીતોષ્ણ પાનખર વનમાં માટીના ક્યાં ઓર્ડર સામાન્ય છે?
A. એન્ટિસોલ્સ અને ઇનસેપ્ટીસલ્સ
B. અલ્ફિસોલ્સ અને અલ્ટિસોલ્સ
C. એન્ટિસોલ્સ અને અલ્ટિસોલ્સ
D. A અને B બન્ને

199. જંગલની જમીનના ભૌતિક ગુણધર્મો મુખ્યત્વે શું નિયંત્રિત કરે છે ?
A. વૃક્ષ ટર્નઓવર દર
B. વન વિકાસ દર
C. ઉપલબ્ધ નાઇટ્રોજન અને ફોસ્ફરસ
D. વેધરિંગની પ્રક્રિયા

200. ભૂમિ નિર્માણના કયા તબક્કામાં માત્ર સૌથી પ્રતિરોધક ખનિજો જ ટકી રહે છે?
A. જુવેનાઇલ તબક્કો
B. વેરાઇલ તબક્કો
C. સેનાઈલ તબક્કો
D. ફાઇનલ તબક્કો

197. Magnetite is a heavy mineral resistant to weathering and often occurs in soils as a
A. red sand with strong magnetism
B. yellow sand with strong magnetism
C. black sand with strong magnetism
D. white sand with strong magnetism

198. Dominant soil orders common to the temperature deciduous forest include
A. Entisols and Inceptisols
B. Alfisols and Ultisols
C. Entisols and Ultisols
D. Both A and B

199. The physical properties of forest soil majorly control the
A. tree turnover rates
B. forest growth rates
C. available nitrogen and phosphorus
D. Weathering process

200. In Which stage of soil formation, only most resistant minerals survive?
A. Juvenile stage
B. Verile stage
C. Senile stage
D. Final stage

Answers | જવાબો

Que.	Ans.	Que.	Ans.	Que.	Ans.	Que.	Ans.	Que.	Ans.	Que.	Ans.	Que.	Ans.
1.	A	**31.**	D	**61.**	C	**91.**	C	**121.**	C	**151.**	D	**181.**	D
2.	D	**32.**	D	**62.**	A	**92.**	C	**122.**	A	**152.**	A	**182.**	A
3.	D	**33.**	C	**63.**	C	**93.**	B	**123.**	B	**153.**	B	**183.**	C
4.	B	**34.**	D	**64.**	B	**94.**	A	**124.**	A	**154.**	A	**184.**	B
5.	C	**35.**	C	**65.**	A	**95.**	C	**125.**	C	**155.**	D	**185.**	D
6.	A	**36.**	B	**66.**	B	**96.**	B	**126.**	B	**156.**	B	**186.**	A
7.	A	**37.**	A	**67.**	C	**97.**	D	**127.**	C	**157.**	A	**187.**	B
8.	B	**38.**	C	**68.**	D	**98.**	C	**128.**	A	**158.**	C	**188.**	C
9.	C	**39.**	B	**69.**	B	**99.**	A	**129.**	A	**159.**	D	**189.**	A
10.	D	**40.**	D	**70.**	A	**100.**	B	**130.**	B	**160.**	B	**190.**	B
11.	D	**41.**	A	**71.**	B	**101.**	A	**131.**	C	**161.**	A	**191.**	D
12.	A	**42.**	D	**72.**	D	**102.**	A	**132.**	D	**162.**	C	**192.**	D
13.	B	**43.**	B	**73.**	A	**103.**	C	**133.**	D	**163.**	D	**193.**	C
14.	D	**44.**	D	**74.**	D	**104.**	D	**134.**	B	**164.**	C	**194.**	A
15.	A	**45.**	B	**75.**	A	**105.**	B	**135.**	C	**165.**	B	**195.**	B
16.	B	**46.**	C	**76.**	B	**106.**	C	**136.**	A	**166.**	A	**196.**	A
17.	C	**47.**	B	**77.**	D	**107.**	B	**137.**	D	**167.**	C	**197.**	C
18.	A	**48.**	A	**78.**	B	**108.**	D	**138.**	C	**168.**	B	**198.**	D
19.	D	**49.**	D	**79.**	A	**109.**	A	**139.**	A	**169.**	C	**199.**	A
20.	B	**50.**	D	**80.**	D	**110.**	B	**140.**	C	**170.**	A	**200.**	C
21.	A	**51.**	C	**81.**	B	**111.**	C	**141.**	C	**171.**	D		
22.	C	**52.**	A	**82.**	A	**112.**	A	**142.**	A	**172.**	C		
23.	D	**53.**	C	**83.**	C	**113.**	B	**143.**	D	**173.**	D		
24.	C	**54.**	B	**84.**	B	**114.**	D	**144.**	B	**174.**	C		
25.	C	**55.**	D	**85.**	A	**115.**	C	**145.**	B	**175.**	D		
26.	D	**56.**	A	**86.**	C	**116.**	A	**146.**	D	**176.**	C		
27.	C	**57.**	B	**87.**	D	**117.**	C	**147.**	A	**177.**	B		
28.	D	**58.**	D	**88.**	B	**118.**	B	**148.**	C	**178.**	B		
29.	B	**59.**	B	**89.**	C	**119.**	D	**149.**	A	**179.**	A		
30.	A	**60.**	D	**90.**	A	**120.**	C	**150.**	B	**180.**	B		

Chapter 9
પ્રકરણ ૯

Pollution and its impact on Environment
પ્રદુષણ અને તેની પર્યાવરણ પર અસરો

Central Pollution Control Board

Year of establishment	1974
HQ	New Delhi
Chairperson	S. P. S. Parihar
Regional Offices	6 (Bhopal, Bengaluru, Lucknow, Kolkata, Vadodara, Chennai, Shilong)

CPCB 2020

કેન્દ્રીય પ્રદૂષણ નિયંત્રણ બોર્ડ

સ્થાપના	1974
મુખ્ય મથક	નવી દિલ્હી
અધ્યક્ષ	એસ.પી.એસ. પરિહાર
પ્રાદેશિક કચેરીઓ	6 (ભોપાલ, બેંગાલુરુ, લખનૌ, કોલકાતા, વડોદરા, ચેન્નાઈ, શિલોંગ)

CPCB 2020

Pollution types

Type	Pollutants/ Source
Air Pollution	CO, CO_2, CFC, Ozone, SPM, Smog, Fly ash
Water Pollution	Fertilizers, Pesticides, Oil spill, Sewage, Garbage, Industrial water, Plastic
Soil Pollution	Fertilizers, Manures, Pesticides, Plastic, Discarded material
Noise Pollution	Machines, Industries, People

પ્રદૂષણના પ્રકારો

પ્રકાર	પ્રદૂષકો / સ્રોત
વાયુ પ્રદૂષણ	CO, CO_2, CFC, ઓઝોન, એસપીએમ, સ્મોગ, ફ્લાય એશ
જળ પ્રદૂષણ	ખાતરો, જંતુનાશક દવાઓ, તેલનું પ્રસરણ, ગટર, કચરો, ઔદ્યોગિક પાણી, પ્લાસ્ટિક
જમીન પ્રદૂષણ	ખાતરો, જંતુનાશકો, પ્લાસ્ટિક, કાઢી નાખેલી સામગ્રી
ધ્વનિ પ્રદૂષણ	મશીનો, ઉદ્યોગો, લોકો

Global Warming Potential and Lifetime of Green House Gases
ગ્રીન હાઉસ ગેસોની અવધિ અને ગ્લોબલ વોર્મિંગનું સંભાવ્ય

GAS	GWP (100 year)	Lifetime (years)
Carbon di oxide	1	100
Methane	21	12
Nitrous oxide	310	120
Hydro fluoro carbons (HFCs)	140-11,700	1-270
Perfluoro carbons (PFCs)	6,500-9,200	800-50,000
Sulfur hexafluoride (SF6)	23,900	3,200

Pollutant Effects on Human | માનવ પર પ્રદૂષકોની અસરો

Pollutant	Health hazard	Cause
Asbestos dust	Respiratory issues and cancer	Mine and manufacturing units
Cadmium	Heart functioning	Factories
Carbon Dioxide	Eye vision, heart strain and headache.	Fuel burning, Fossil fuels
Carbon Monoxide	Headache, irritation to mucous membrane, unconsciousness and death	Vehicles, Fossil fuels
Chlorofluorocarbon	Global warming impacts on human health	Spray, Refrigerator
Coal particles	Respiratory issues, Pulmonary fibrosis and cancer	Coal mine
Cotton dust	Lung problems, chronic cough and bronchitis	Textile industries using cotton
Hydrocarbons	Kidney problems hypertension, respiratory problems, asthma, bronchitis and lung cancer.	Fossil fuel burning
Lead	Nervous system, kidneys and brains, Effect on RBCs	Petrol emissions containing lead
Mercury	Insomnia, Nervous disorders, memory issues, and gingivitis.	Industries using mercury
Nitrogen Oxides	Respiratory system problems, lung issues, bronchitis and asthma.	Vehicles, Industries and Thermal power station
Ozone	Respiratory issues, asthma, breathlessness, chest pain, emphysema and bronchitis.	Automobile industries
Radioactive rays	Tissues and blood cell destruction, enzyme functional issues, leukemia and permanent genetic modifications.	X-rays and radium
Silica particles/ dust	Respiratory issues and effect on lungs	Quarries of Silicon
Smog	Respiratory issues and irritation to skin and eyes.	Vehicles and Industries
Sulphur Oxides	Respiratory issues, Allergy and enzyme functional issues	Thermal power stations

Pollutant	Health hazard	Cause
Suspended Particulate Matter (SOM)	Respiratory issues, Pulmonary malfunction and Reduction of Red Blood Cells	Vehicles and Fossil fuel burning

National Ambient Air Quality Standard (NAAQS)

રાષ્ટ્રીય એમ્બિયન્ટ હવા ગુણવત્તા ધોરણ (NAAQS)

Pollutant	Time Weighted Average	Concentration in Ambient Air	
		Industrial, Residential, Rural and Other Areas	Ecologically Sensitive Area (notified by Central Government)
Sulphur Dioxide (SO2) μg/ m^3	Annual	50	20
	24 hours	80	80
Nitrogen Dioxide (NO2) μg/ m^3	Annual	40	30
	24 hours	80	80
Particulate Matter (size less than 10 μm) or PM10 μg/m^3	Annual	60	60
	24 hours	100	100
Particulate Matter (size less than 2.5 μm) or PM2.5μg/ m^3	Annual	40	40
	24 hours	60	60
Ozone (O3) μg/ m^3	8 hours	100	100
	1 hour	180	180
Lead (Pb) μg/ m^3	Annual	0.50	0.50
	24 hours	1.0	1.0
Carbon Monoxide (CO) mg/ m^3	8 hours	02	02
	1 hour	04	04
Ammonia (NH3) μg/ m^3	Annual	100	100
	24 hours	400	400
Benzene (C6H6) μg/ m^3	Annual	5	5
Benzo(a) Pyrene (BaP)-particulate phase only, ng/m^3	Annual	1	1
Arsenic(As) ng/m^3	Annual	6	60
Nickel (Ni), ng/m^3	Annual	20	20

Note:

Annual are arithmetic mean of minimum 104 measurements in a year at a particular site taken twice a week 24 hourly at uniform intervals.

4 hourly or 8 hourly or 1 hourly are monitored value, as applicable, shall be complied with 98% of the time, they may exceed the limits but not on two consecutive days of monitoring.

CPCB 2020

Operating Stations in Gujarat under National Air Quality Monitoring Programme (NAMP)

રાષ્ટ્રીય હવા ગુણવત્તા મોનીટરીંગ પ્રોગ્રામ (NAMP) હેઠળ ગુજરાતમાં ઓપરેટિંગ સ્ટેશનો

City	No. of Stations
Ahmedabad	9
Ankaleshwar	2
Jamnagar	1
Rajkot	2
Surat	3
Vadodara	5
Vapi	2

CPCB 2020

Ambient Noise level Standards | એમ્બિયન્ટ અવાજ સ્તરનાં ધોરણો

Area code	Category of Area / Zone	Limits in dB(A) Leq	
		Day Time 6 am to 10 pm	Night Time 10 pm to 6 am
A	Industrial area	75	70
B	Commercial area	65	55
C	Residential area	55	45
D	Silence Zone	50	40

CPCB 2020

Categorization of Industries | ઉદ્યોગનું વર્ગીકરણ

Industry's Pollution Index (PI) score	Category	Remarks
60 and above	Red	Thermal power plants, paints and other highly polluting industries
41 to 59	Orange	Coal, Washeries and Automobile
21 to 40	Green	Low polluting industries
upto 20	White	Non-polluting industries and exempted from environmental clearance.

MoEFCC 2020

Pollution Legislation in India | ભારતમાં પ્રદૂષણ કાયદાઓ

Year	Act/ Rules
1974	The Water (Prevention and Control of Pollution) Act
1986	Environment (Protection) Act
2000	The Noise Pollution (Regulation and Control) Rules,
2016	Hazardous Waste (Management, Handling & Transboundary Movement) Rules
2016	Bio-medical Waste Management Rules
2016	Plastic Waste Management Rules
2016	E-Waste Rules

CPCB 2020

Trees for Pollution | પ્રદૂષણ નિવરવા માટેના વૃક્ષો

Type	Suitable Species
Dust pollution	*All broad leaved species Teak, Ficus, Mango, Peltoforum, Terminalia*
Lead pollution	*Cassis siamea. Zisyphus mauritiana*
Noise pollution	*Alstonia scholaris, Acer negundo, Erythrina spp. Butea monosperma, Alnus indica, Grevellia, Juniperus chinensis, Terminalia arjuna*
Oxides of nitrogen	*Quercus rubra, Robinia pseudocacia, Sambucus nigra, Taxus buccata, Alnus montana, Pinus* spp.
Ozone	*Acer campestral, A. platanoides, A negundo, Quercus rubra*
Peroxy Acetyl Niritie (PAN)	*Acer platanoides, Juniperus, Quercus* spp. *Eunonymus europalaus*
SO_2	*Albizzia lebbeck, Ailanthus Alstonia, Neem, Ficus, Polyalthia, Terminalia arjuna, Quercus palustrie*

Multiple Choice Questions | હેતુલક્ષી પ્રશ્નો

1. પ્રદૂષણ નિવારણ અને નિયંત્રણનો કાયદો ક્યારે ઘડાયો હતો?
A. 1979 B. 1980
C. 1982 D. 1981

2. દિલ્હી એનસીઆરમાં પ્રદૂષણ માટે કયા કણ વધારે જવાબદાર છે?
A. PM 11 અને PM 3.0
B. PM 9 અને PM 2.0
C. PM 10 અને PM 2.5
D. PM 8 અને PM 1.5

3. એર એક્ટમાં કેટલા સેક્શન છે?
A. 56 સેક્શન B. 54 સેક્શન
C. 58 સેક્શન D. 62 સેક્શન

4. માનવ પ્રવૃત્તિઓને લીધે વાતાવરણના બદલાવને ________ તરીકે ઓળખવામાં આવે છે.
A. ઇકોલોજીકલ ડિગ્રેડેશન
B. પ્રદૂષણ
C. ઇકોલોજીકલ ડિસ્ટર્બન્સ
D. આપત્તિ

5. લાયકેન એ શેનું શ્રેષ્ઠ સૂચક છે?
A. NO પ્રદૂષણ
B. CO_2 પ્રદૂષણ
C. SO_2 પ્રદૂષણ
D. ઉપર્યુક્તમાંથી એક પણ નહિ

6. સિમેન્ટ ફેક્ટરી માં કામ કરતા મજૂરોમાં શેનું જોખમ હોય વધુ છે?
A. અસ્થિ મજ્જાના રોગો
B. લ્યુકેમિયા
C. એસ્બેસ્ટોસિસ
D. સાયટોસિલીકોસિસ

1. **Prevention and Control of Pollution Act was enacted in**
A. 1979 B. 1980
C. 1982 D. 1981

2 **Which Particulate matter is of Major concerns for pollution in Delhi NCR**
A. PM 11 and PM 3.0
B. PM 9 and PM 2.0
C. PM 10 and PM 2.5
D. PM 8 and PM 1.5

3. **The air Act has**
A. 56 sections B. 54 sections
C. 58 sections D. 62 sections

4. **The unfavourable alteration of environment due to human activities is termed as**
A. Ecological degradation
B. Pollution
C. Ecological disturbance
D. Catastrophe

5. **Lichen is the best indicator for**
A. NO pollution
B. CO_2 pollution
C. SO_2 pollution
D. None of the above

6. **Cement factory labours are more prone to**
A. Bone marrow diseases
B. Leukemia
C. Asbestosis
D. Cytosillicosis

7. **BOD નુ પૂરું નામ શું છે?**
A. બાયોલોજિકલ ઓક્સિજન ડિમાન્ડ
B. બાયોકેમિકલ ઓક્સિજન ડિમાન્ડ
C. બાયોલોજિકલ ઓક્સિડેશન ડિમાન્ડ
D. બાયોટિક ઓક્સિજન ડિમાન્ડ

8. **ભારે ધૂળના લીધે શું થાય છે?**
A. પાંદડાનું બ્રાઉનીંગ
B. સ્ટોમાટા ખુલવું
C. સ્ટોમાટા બંધ થવું
D. લિફબ્લાઈટ

9. **નીચેના માંથી કયું નિવેદન "ગ્રીન હાઉસ ઇફેક્ટ" વિશે સાચું છે?**
A. ફક્ત CO દ્વારા થાય છે
B. ઘણા વાયુઓના સંયોજન દ્વારા થાય છે
C. CO_2, CFC, CH અને NO_2 દ્વારા થાય છે
D. ઉપર્યુક્તમાંથી એક પણ નહિ

10. **હવામાં સ્થગિત ફાઇન ઓર્ગેનિક અને ઈનઓર્ગેનિક કણોને શું કહેવામાં આવે છે?**
A. એરોસોલ
B. વાયુયુક્ત પ્રદૂષક
C. કણો પ્રદૂષકો
D. ઉપર્યુક્તમાંથી એક પણ નહિ

11. **નીચેનામાંથી સેકન્ડરી પ્રદૂષક કયુ છે?**
A. CO B. SO_2
C. CO_2 D. O_3

12. **પર્યાવરણીય વિરામ દર કયો છે?**
A. -6.5°C / કિમી B. 6.5°C / કિમી
C. 8.5°C / કિમી D. 5.6°C / કિમી

13. **નીચેના કયા છોડના જૂથનો હવામાંના SO પ્રદૂષણના સૂચક તરીકે ઉપયોગ કરી શકાય છે?**
A. ફર્ન્સ B. એપિફેટિક લાઇકેન્સ
C. જીવંત વોર્ટ્સ D. હોર્ન વોર્ટ્સ

7. **BOD stands for**
A. Biological oxygen demand
B. Biochemical oxygen demand
C. Biological oxidation demand
D. Biotic oxygen demand

8. **Heavy dust can cause**
A. Browning of leaves
B. Opening of stomata
C. Closing of stomata
D. leaf blights

9. **True statement about "Green house effect" is that**
A. caused only by CO
B. caused by combination of many gases
C. caused by CO_2, CFC, CH and NO_2
D. None of the above

10. **Fine organic and inorganic particles suspended in air is called**
A. Aerosol
B. Gaseous pollutants
C. particulate pollutants
D. None

11. **Which of the following is a secondary pollutant**
A. CO B. SO_2
C. CO_2 D. O_3

12. **Environmental lapse rate is found to be**
A. -6.5°C/km B. 6.5°C/km
C. 8.5°C/km D. 5.6°C/km

13. **Which of the following group of plants can be used as an indicators of SO pollution in air?**
A. Ferns B. Epiphytic lichens
C. Live worts D. Horn worts

14. મુખ્ય ફોટોકેમિકલ ઓક્સિડેન્ટ કયો છે?

A. હાઇડ્રોજન પેરોક્સાઇડ

B. નાઇટ્રોજન ઓક્સાઇડ

C. ઓઝોન

D. પેરોક્સિલ એસિટિલ નાઇટ્રેટ

15. મોબાઇલ કમ્બશન આશરે કેટલા % પ્રદુષણ કરે છે?

A. વાયુ પ્રદૂષણના 10-20%

B. વાયુ પ્રદૂષણના 50-60%

C. વાયુ પ્રદૂષણના 5-10%

D. વાયુ પ્રદૂષણના 40-50%

16. બાયોકેમિકલ ઓક્સિજન માંગ શું માપે છે?

A. જૈવિક કચરાના વિઘટન માટે સૂક્ષ્મજીવાણુઓ દ્વારા ઓગળેલ ઓક્સિજનની જરૂરિયાત

B. અસરોની પ્રદૂષણ ક્ષમતા

C. વાયુ પ્રદૂષણ

D. ઓદ્યોગિક પ્રદૂષણ

17. વાહનોમાં શાનાથી હવાના પ્રદૂષણને નિયંત્રિત કરી શકાય છે?

A. વેટ સ્ક્બર

B. ફેબ્રિક ફિલ્ટર

C. સાઈક્લોન સેપરેટર

D. કેટાલિટીક કન્વરટર

18. BOD / COD ગુણોત્તર હંમેશા કેટલો હોવો જોઈએ?

A. 1 કરતા વધારે

B. 1 કરતા ઓછો

C. 1

B. ઉપર્યુક્તમાંથી એક પણ નહિ

19. હવામાં પીએમ 10 નું કેટલું કોન્સન્ટ્રેશન માન્ય છે?

A. 20 μg / m³ B. 50 μg / m³

C. 60 μg / m³ D. 30 μg / m³

14. The major photochemical oxidant is

A. Hydrogen peroxide

B. Nitrogen oxides

C. Ozone

D. Peroxyl Acetyl Nitrate (PAN)

15. Mobile combustion accounts

A. 10-20% of air pollution

B. 50-60% of air pollution

C. 5-10% of air pollution

D. 40-50% of air pollution

16. Biochemical oxygen demand measures

A. Dissolved O_2 need by microbes to decompose organic waste

B. Pollution capacity of effects

C. Air Pollution

D. Industrial pollution

17. Air pollution from automobiles can be controlled by fitting

A. Wet Scrubber

B. Fabric filter

C. Cyclone separator

D. Catalytic converter

18. BOD/COD ration should always be

A. More than 1

B. Less than 1

C. Equal to 1

D. None of the above

19. The permitted concentration of PM 10 in the air is

A. 20 $\mu g/m^3$ B. 50 $\mu g/m^3$

C. 60 $\mu g/m^3$ D. 30 $\mu g/m^3$

20. ધુમાડાના કણનું ન્યુનત્તમ કદ કેટલું હોય છે?

A. 0.5 um B. 0.2 um
C. 1.0 um D. 0.8 um

21. નીચેનામાંથી કયો ગેસ કાર્બોક્સી હિમોગ્લોબિન માટે જવાબદાર છે?

A. SO_2 B. CO
C. CO_2 D. NO_2

22. નીચેનામાંથી મુખ્ય ફોટોકેમિકલ સ્મોગ કયો છે?

A. પેરોક્સિલ એસિટિલ નાઇટ્રેટ
B. હાઇડ્રોજન પેરોઝાઇડ
C. નાઇટ્રિક ઓકસાઈડ
D. ક્લોરોફ્યુરોકાર્બન

23. ___________ એ ધુમ્મસ શબ્દ આપ્યો હતો?

A. નિકોલસ કોપરનીકસ
B. જો રોમ
C. હેનરી એન્ટોન
D. જેમ્સ હેન્સન

24. Air Quality Index શું છે?

A. હવાના પ્રદૂષણની માત્રાને માપે છે
B. હવામાં સલ્ફરનું પ્રમાણ માપે છે
C. હવામાં ઓઝોન સ્તરની માત્રા માપે છે
D. હવામાં ઓક્સિજન સ્તરની માત્રા માપે છે

25. કયા વાયુ પ્રદૂષક આંખો અને ગળામાં બળતરા ઉત્પન્ન કરે છે?

A. આર્સેનિક
B. સલ્ફર ડાયોક્સાઇડ
C. હાઇડ્રોજન સાયનાઇડ
D. હાઇડ્રોજન સલ્ફાઇડ

26. અપૂર્ણ દહન પ્રતિક્રિયાઓ દરમિયાન નીચેનામાંથી ક્યા શ્વાસાવરોધક વાયુ ઉદ્દભવે છે?

A. પારો B. કાર્બન મોનોક્સાઇડ
C. ઓઝોન D. કાર્બન ડાયોક્સાઇડ

20. The minimum size of smoke particle is

A. 0.5 um B. 0.2 um
C. 1.0 um D. 0.8 um

21. Which of the following gas is responsible for carboxy haemoglobin

A. SO_2 B. CO
C. CO_2 D. NO_2

22. Which of the following is major photochemical smog

A. Peroxyl Acetyl Nitrate
B. Hydrogen perozide
C. Nitric oxide
D. Chlorofurocarbon

23. Who had coined the term SMOG

A. Nicolaus Copernicus
B. Joe Romm
C. Henry Antoine
D. James Hanson

24. What is air quality index?

A. It measures the amount of air pollution
B. It measures the amount of sulphur in air
C. It measures the amount of ozone level in air
D. It measures the amount of Oxygen in the air

25. Which air pollutant produce burning sensation of eyes and throat

A. Arsenic
B. Sulphur dioxide
C. Hydrogen cyanide
D. Hydrogen sulphide

26. Which of the following is an asphyxiant during incomplete combustion reactions?

A. Mercury B. Carbon monoxide
C. Ozone D. Carbon dioxide

27. ઇકોસિસ્ટમને લાંબા ગાળાના પર્યાવરણીય નુકસાનમાં નીચેનામાંથી ઓછામાં ઓછું કારણભૂત કયુ છે?

A. કાર્બન મોનોક્સાઇડ

B. ઓઝોન

C. કાર્બન ડાયોક્સાઇડ

D. સલ્ફર

28. હવાના પ્રદુષક ક્લોરોફ્લોરોકાર્બન નો સ્ત્રોત શું છે?

A. મોટર વાહનોના એક્ઝોસ્ટ

B. લાકડાનું અપૂર્ણ દહન

C. એરોસોલ સ્પ્રે

D. ઉપર્યુક્તમાંથી એક પણ નહિ

29. ઓઝોન સ્તરની જાડાઈ માપવાનો એકમ કયો છે?

A. Psi B. dB

C. AMU D. D.U.

30. CPCB મુજબ, વાયુ પ્રદૂષણમાં કયા કદનાં ધૂળનાં કણો માનવ સ્વાસ્થ્ય માટે હાનિકારક છે?

A. 0.25 μm B. 2.5 μm

C. 2.5 μm કરતાં ઓછા D. બંને B અને C

31. પૃથ્વીના ધ્રુવમાં ઓઝોન છિદ્રની રચના ચિંતાનું કારણ છે. આ છિદ્રની રચનાનું કારણ શું હોઈ શકે?

A. પ્રોમીનન્ટ ટ્રોપોસ્ફેરીક ટર્બ્યુલન્સની રચના અને ક્લોરોફ્લોરોકાર્બન્સનો પ્રવાહ

B. પ્રોમીનન્ટ ધ્રુવીય અને સ્ટ્રેટોસ્ફેરીક વાદળોની હાજરી અને ક્લોરોફ્લોરોકાર્બનનો પ્રવાહ

C. ધ્રુવીય ફ્રન્ટ અને સ્ટ્રેટોસ્ફેરીક વાદળોની ગેરહાજરી અને મિથેન અને ક્લોરોફ્લોરોકાર્બન્સનો પ્રવાહ

D. તાપમાનમાં વધારો

27. Which of the following is least likely result in long-term environmental damage to an ecosystem?

A. Carbon monoxide

B. Ozone

C. Carbon dioxide

D. Sulpher

28. What is the source of chlorofluoro-carbon pollutant of air?

A. Exhaust fumes of motor vehicles

B. Incomplete burning of woods

C. aerosol sprays foams

D. None of the above

29. The unit by which the thickness of ozone layer is measured

A. Psi B. dB

C. AMU D. D.U.

30. As per CPCB, which size of dust particles in air pollution are harmful to human health?

A. 0.25 μm B. 2.5 μm

C. Less than 2.5 μm D. Both B and C

31. The formation of ozone hole in the earth's pole has been a cause of concerns. What could be the reason for the formation of this hole?

A. Formation of prominent tropospheric turbulence and inflow of chloro-fluorocarbons

B. Presence of prominent polar front and stratospheric clouds and inflow of chlorofluorocarbons

C. Absence of polar front and stratospheric clouds and inflow of methane and chlorofluorocarbons

D. Increased temperature

32. નીચેનામાંથી કયું વિધાન "કાર્બન ક્રેડિટ" સંબંધિત ખોટું છે?

A. ક્યોટો પ્રોટોકોલ સાથે જોડાણમાં કાર્બન ક્રેડિટ સિસ્ટમને બહાલી આપવામાં આવી હતી

B. કાર્બન ક્રેડિટ એવા દેશો અથવા જૂથને આપવામાં આવે છે કે જેમણે તેમના ઉત્સર્જન ક્વોટાથી નીચે ગ્રીન હાઉસ ગેસ ઘટાડ્યો છે.

C. કાર્બન ક્રેડિટ સિસ્ટમનું લક્ષ્ય કાર્બન ડાયોક્સાઇડ ઉત્સર્જનના વધારાને મર્યાદિત કરવાનું છે

D. યુએન પર્યાવરણ પ્રોગ્રામ દ્વારા કાર્બન ક્રેડિટ સમય-સમય પર નિયત ભાવે વેચાય છે.

33. વાતાવરણ માં CO_2 નું પ્રમાણ વધી રહ્યું છે પરંતુ તેમાંથી મોટા ભાગ ના કણો નીચલા વાતાવરણ માં રહેતા નથી જેનું મુખ્ય કારણ શું છે?

A. બાહ્ય અવશેષોમાંથી બહાર નીકળવું

B. ધ્રુવીય બરફની ટોચમાં હવા ફસાઈ જવાથી

C. પ્રકાશસંશ્લેષણ માટે ફાયટો-પ્લાન્કટોન દ્વારા ઉપયોગમાં લેવામાં આવે છે

D. બંને B અને C

34. 2/3rd SO_2 અને 1/4th NO વાતાવરણમાં કયા સ્રોતમાંથી આવે છે?

A. જ્વાળામુખી

B. ઉત્પાદન એકમો

C. ઓઇલ રિફાઇનરીઓ

D. ઇલેક્ટ્રિક પાવર જનરેટર

35. એક ક્લોરિન અણુ કેટલા ઓઝોન પરમાણુઓને તોડી શકે છે?

A. 100000 B. 1000

C. 100 D. 10,000

32. Which of the following statement is incorrect regarding "Carbon Credit"?

A. The carbon credit system was ratified in conjunction with the Kyoto protocol

B. Carbon credit are awarded to countries or group that have reduced green house gas below their emission quota

C. The goal of carbon credit system is to limit the increase of carbon dioxide emission

D. Carbon credits are traded at a price fixed from time to time by the UN environment programme

33. There is an increased conc. of CO2 in the atmosphere but lot of it does not remain in the lower atmosphere because of the lead works of

A. Escapes into the outer stratosphere

B. Trapping of the air in the polar ice caps

C. Used by phyto-planktons for photosynthesis

D. Both B and C

34. 2/3rd of the SO2 and 1/4th of the NO in the atmosphere comes from which source?

A. Volcanoes

B. Manufacturing units

C. Oil refineries

D. Electric power generators

35. One chlorine atom can break apart how many ozone molecules

A. 100000 B. 1000

C. 100 D. 10,000

36. નીચેનામાંથી કયું વિશાળ પ્રાકૃતિક ગ્રીન હાઉસ તરીકે વર્તે છે?
A. પૃથ્વીનો સ્ટ્રેટોસ્ફીયર
B. પૃથ્વીનું ટ્રોપોસ્ફીયર
C. પૃથ્વીનું એટમોસ્ફીયર
D. પૃથ્વીનું લિથોસ્ફીયર

37. એસિડ કણો અને વાયુઓ ભેજની ગેરહાજરીમાં કયા વાતાવરણમાં પણ જમા થઈ શકે છે?
A. ભીના B. ભેજવાળા
C. સુકા D. ઉપર્યુક્તમાંથી એક પણ નહિ

38. સૂર્યપ્રકાશની હાજરીમાં નીચેનામાંથી હાઇડ્રોકાર્બન અને નાઇટ્રોજન ઓકસાઈડની પ્રતિક્રિયા દ્વારા શું રચાય છે?
A. સ્મોગ
B. ગ્રાઉન્ડ લેવલ ઓઝોન
C. વાતાવરણીય રજકણ પદાર્થ
D. ફાઇન કણો

39. કયા ઉપકરણમાં 3 માઇક્રોનથી ઓછા વ્યાસવાળા પદાર્થો કબજે કરી શકાય છે?
A. બેડ પેક સ્ક્રબર
B. સ્પ્રે ચેમ્બર
C. ઇલેક્ટ્રોસ્ટેટિક પ્રેસિપિટેટર્સ
D. હાઈ એનર્જી વેન્ટુરી સ્ક્રબર

40. વાતાવરણીય વિક્ષેપને અસર કરતા કયા પરિબળમાં "માર્ગ ટ્રાફિક" સંબંધિત છે?
A. હવામાન પરિબળ
B. સ્રોત પરિબળ
C. ભૌગોલિક પરિબળ
D. ઉત્સર્જનની ઉંચાઈ

41. કયું પરિબળ વાયુ પ્રદૂષણને વાતાવરણમાં વિખેરી નાખવાના અસર કરે છે?
A. ઉંચાઇ
B. વાતાવરણીય સ્થિરતા
C. અસ્થિરતા
D. પવનની ગતિ

36. Which of the following is behave as a nature's gigantic green house?
A. Earth's Stratosphere
B. Earth's troposphere
C. Earth's atmosphere
D. Earth's lithosphere

37. Acid particles and gases can also deposit in atmosphere in the absence of moisture as
A. Wet deposition B. Damp deposition
C. Dry Deposition D. None of these

38. Which of the following is formed by reaction of hydrocarbons and nitrogen oxide in the presence of sunlight?
A. Smog
B. Ground-level ozone
C. Atmospheric particulate matter
D. Fine particles

39. Which device can capture particulate matters (PM) less than 3 microns in diameter?
A. Bed packed scrubber
B. Spray chamber
C. Electrostatic Precipitators
D. High energy venturi scrubber

40. In which factor affecting atmospheric dispersion does "road traffic" belong?
A. Meteorological factor
B. Source factor
C. geographical factor
D. Emission height

41. Which factors affects majorly he dispersion of air pollution in the atmosphere?
A. Elevation
B. Atmospheric stability
C. Turbulence
D. Wind speed

42. મોટર વાહન પ્રદુષકો અને સૂર્યપ્રકાશની વચ્ચે થતી રાસાયણિક પ્રતિક્રિયા ______ તરીકે ઓળખાય છે.
A. એર ટોક્સિક્સ
B. ફોટોકેમિકલ સ્મોગ
C. સલ્ફયુરેશન
D. કાર્બોનેશન

43. નીચેના માંથી કયા ધાતુ ઝેરી નથી?
A. પારો B. કેડમિયમ
C. ડાયોક્સિન D. ક્રોમિયમ

44. કયા ચોક્કસ જગ્યાના વાયુ પ્રદુષણ વધારે હાનિકારક હોય છે?
A. આઉટડોર વાયુ પ્રદૂષણ
B. ઇન્ડોર વાયુ પ્રદૂષણ
C. આઉટર વાયુ પ્રદૂષણ
D. ઇન્ટર્નલ વાયુ પ્રદૂષણ

45. મોન્ટ્રીયલ પ્રોટોકોલ શેને સંબંધિત છે?
A. ગ્લોબલ વોર્મિંગ
B. ક્ષમતાલક્ષી વિકાસ
C. ઓઝોન સ્તરનું અવક્ષયન
D. ખાદ્ય સુરક્ષા

46. નીચેના માંથી કયું ભારતના નાગરિકો ના આરોગ્ય માટે સૌથી મોટો ખતરો છે?
A. જળ પ્રદૂષણ
B. હવાનું પ્રદૂષણ
C. ખાદ્ય સુરક્ષા
D. જમીન પ્રદૂષણ

47. EPA-USA ના જણાવ્યા મુજબ નીચેનામાંથી કયા છ મોટા પ્રદૂષકોમાંથી એક નથી?
A. ઓઝોન B. CO_2
C. NO D. CO

48. એસિડ વરસાદ કયા વાયુ પ્રદૂષકને કારણે થાય છે?
A. CO_2 B. CO
C. મિથેન D. NO અને SO_2

42. The chemical reaction occurs between motor vehicle pollutants and sunlight are known as
A. Air toxics
B. Photochemical smog
C. Sulphuration
D. Carbonation

43. Which is not a metal air toxics
A. Mercury B. Cadmium
C. Dioxin D. Chromium

44. Which air pollution is more harmful due to it is an isolated space pollution
A. Outdoor air pollution
B. Indoor air pollution
C. Outer air pollution
D. Internal air pollution

45. Montreal Protocol is related to the
A. Global warming
B. Sustainable development
C. Ozone layer depletion
D. Food security

46. Which is the biggest threat to health in India
A. Water pollution
B. Air pollution
C. Food security
D. Soil pollution

47. According to EPA USA which of the following is not one of six major pollutants?
A. Ozone B. CO_2
C. NO D. CO

48. Acid rain is due to the air pollution of
A. CO_2 B. CO
C. Methane D. NO and SO_2

49. ફ્યુઝીટીવ એમિશનમાં શું શામેલ છે?
A. શેરીની ધૂળ
B. બાંધકામની પ્રવૃત્તિઓમાંથી ધૂળ
C. ખેતીની ધૂળ
D. ઉપર્યુક્ત તમામ

50. ક્રિપ્ટો પ્રોટોકોલ શેને સંબંધિત છે?
A. ક્ષમતાલક્ષી વિકાસ
B. ગ્રીનહાઉસ ગેસ ઉત્સર્જનમાં ઘટાડો
C. માનવાધિકાર
D. પર્યાવરણનું રક્ષણ

51. નીચેનામાંથી કયુ વાયુ પ્રદુષણ નથી?
A. ધુમ્મસ
B. એસિડ વરસાદ
C. યુટ્રોફિકેશન
D. એસ્બેસ્ટોસિસ

52. "ભોપાલ ગેસ દુર્ઘટના" ક્યારે બની હતી?
A. ડિસેમ્બર 4, 1986
B. ડિસેમ્બર 3, 1985
C. ડિસેમ્બર 3, 1984
D. ડિસેમ્બર 2, 1982

53. EPA એ કેટલા વ્યક્તિગત સંયોજનો ઉપર કેટલા વાયુ પ્રદુષક સૂચિબદ્ધ કર્યા છે?
A. 1000 B. 1100
C. 2000 D. 2500

54. ભારતમાં સૌથી વધુ મૃત્યુ માટે કયુ વાયુ પ્રદૂષક જવાબદાર છે?
A. ઓઝોન B. CO
C. SO_2 D. ઉપર્યુક્તમાંથી એક પણ

55. ભારતમાં શિયાળા દરમિયાન પ્રદૂષણનો મુખ્ય સ્રોત કયો છે?
A. ગૌણ કણો
B. બાયોમાસ બર્નિંગ
C. વાહનનો ધુમાડો
D. કોલસો અને ફ્લાય એશ

49. Fugitive emissions consists of
A. Street dust
B. Dust from construction activities
C. Dust from farm cultivations
D. All of the above

50. Krypto protocol is related to
A. Sustainable development
B. Reduction in greenhouse gas emission
C. Human rights
D. Protection of Environment

51. Which of the following is not an air pollution
A. Smog
B. Acid rain
C. Eutrophication
D. Asbestosis

52. "Bhopal Gas Tragedy" happened on
A. December 4, 1986
B. December 3, 1985
C. December 3, 1984
D. December 2, 1982

53. EPA have listed over how many individual compounds which are classified as air toxics
A. 1000 B. 1100
C. 2000 D. 2500

54. Which air pollutant is responsible for the highest number of deaths in India
A. Ozone B. CO
C. SO_2 D. None of these

55. Major source of pollution during winter in India
A. Secondary particulates
B. Biomass burning
C. Vehicle smoke
D. Coal and fly ash

56. ભારતમાં કેટલા ઓપરેટિંગ સ્ટેશન NAMP પ્રોગ્રામ હેઠળ કાર્ય કરે છે?
A. 550 B. 691
C. 650 D. 791

57. ભારતમાં National Air Quality Index ની શરૂઆત ક્યારે કરવામાં આવી હતી?
A. 2010 B. 2012
C. 2014 D. 2015

58. ઔદ્યોગિક પ્લાન્ટ જેવા કે તેલ રિફાઇનરીઓ માંથી સામાન્ય રીતે કયા પ્રકારનાં પ્રદૂષકોને દૂર કરવામાં આવે છે?
A. VOCs અને PM
B. HAPs અને હાઇડ્રોજન સલ્ફાઇડ્સ
C. VOCs અને Large PM
D. VOCs અને HAPs

59. કયું હવાનું ઝેરી પ્રદૂષક ગેસોલીનમાં પણ જોવા મળે છે?
A. લીડ B. બેન્ઝિન
C. ટોલુએન D. પારો

60. SO_2 નું થ્રેસોલ્ડ કોન્સનટ્રેશન ઉદ્યોગો માં મહત્તમ કેટલું હોય તો પરવાનગી આપવામાં આવતી નથી?
A. 2 પીપીએમ B. 3 પીપીએમ
C. 4 પીપીએમ D. 5 પીપીએમ

61. અસ્થિર કાર્બનિક સંયોજનો નો મુખ્ય સ્રોત શું છે?
A. પરિવહન
B. ઔદ્યોગિક પ્રક્રિયાઓ
C. સ્થિર બળતણ દહન
D. જ્વાળામુખી

62. વિશાખાપટ્ટનમ ગેસ લિક કયા વર્ષ ની ઘટના છે?
A. 2020 B. 2019
C. 2018 D. 2017

56. **How many operating stations works in India under NAMP programme**
A. 550 B. 691
C. 650 D. 791

57. **National Air quality index was launched in India in**
A. 2010 B. 2012
C. 2014 D. 2015

58. **What kind of pollutants are usually removed from an industrial plant such as oil refineries?**
A. VOCs & PM
B. HAPs & Hydrogen sulphides
C. VOCs and large PM
D. VOCs and HAPs

59. **The toxic air pollutant is also found in gasoline**
A. Lead B. Benzene
C. Toluene D. Mercury

60. **The threshold concentration of SO_2 in any industrial activity should not be permitted beyond?**
A. 2 ppm B. 3 ppm
C. 4 ppm D. 5 ppm

61. **The principal source of volatile organic compounds (VOCs) is**
A. Transportation
B. Industrial processes
C. Stationary fuel combustion
D. Volcanoes

62. **Vishakhapatnam gas leak happened in**
A. 2020 B. 2019
C. 2018 D. 2017

63. કયો વાયુ વાતાવરણમાં ગ્રીનહાઉસ વાયુઓનું મહત્તમ પ્રમાણ ધરાવે છે?
A. CH_4 B. SO_2
C. CO_2 D. CO

64. સૌથી વધુ મિથેન ઉત્પન્ન કરનારા પ્રાણીઓ કયા છે?
A. મરઘાં B. રુમીનન્ટસ
C. સમુદ્રના પ્રાણીઓ D. સરિસૃપ

65. વાતાવરણમાંથી CO_2 ને દૂર કરવા અને તેને જળાશયમાં સંગ્રહિત કરવા ની પ્રક્રિયા ને શું તરીકે ઓળખાય છે?
A. કાર્બન કવીન્ચિંગ
B. કાર્બન સ્ટોરેજ
C. કાર્બન રિમૂવલ
D. કાર્બન સિક્વેસ્ટરેશન

66. ______________ એ એક કુદરતી ઘટના છે જે સપાટી પર તાપમાન ગરમ અને સંતુલિત રાખવામાં મદદ કરે છે, જે બધા જીવો માટે યોગ્ય છે.
A. ગ્રીનહાઉસ ઇફેક્ટ
B. ન્યુટન ઇફેક્ટ
C. થોમસન ઇફેક્ટ
D. મેગ્નેટિક ઇફેક્ટ

67. જંગલોના કાપવાથી શું અસર થાય છે?
A. વરસાદમાં વધારો
B. ગ્રીનહાઉસ ઇફેક્ટમાં વધારો
C. તાજી હવામાં વધારો
D. ગ્રીનહાઉસ ઇફેક્ટમાં ઘટાડો

68. ડાંગરના ભૂસિયા બળી જવાથી શેમાં વધારો થયો છે?
A. ભૂમિ પ્રદૂષણ
B. જળ પ્રદૂષણ
C. વાયુ પ્રદૂષણ
D. બંને B અને C

63. Which gas shares the maximum proportion of greenhouse gases in atmosphere
A. CH_4 B. SO_2
C. CO_2 D. CO

64. Highest methane producing animals are
A. Poultry B. Ruminants
C. Sea animals D. Reptiles

65. Removal of CO_2 from atmosphere and storing it in reservoir is known as
A. Carbon quenching
B. Carbon storage
C. Carbon removal
D. Carbon sequestration

66. ______________ is a natural phenomenon that helps in keeping the temperature at the surface warm and balanced, which is suitable for all organisms.
A. Greenhouse effect
B. Newton effect
C. Thomson effect
D. Magnetic effect

67. Deforestation results in
A. Increase in rainfall
B. Increase in Greenhouse effect
C. Increase in fresh air
D. Decrease in Greenhouse effect

68. Paddy straw burning results in increase in
A. Land pollution
B. Water Pollution
C. Air Pollution
D. Both B and C

69. અયોગ્ય માનવસર્જિત પદાર્થો ને ફેંકી દેવું ____ તરીકે ઓળખાય છે.
A. વિઝ્યુઅલ પ્રદૂષણ
B. માટી દૂષણ
C. પ્લાસ્ટિક પ્રદૂષણ
D. લિટરિંગ

70. CWAનું પુર્ણ નામ શું છે ?
A. ક્લીન વેસ્ટ એક્ટ
B. ક્લીન વોટર એક્ટ
C. ક્લિયરન્સ વેસ્ટ એક્ટ
D. ક્લિયરન્સ વોટર એક્ટ

71. કેટલાક સાયનોબેક્ટેરિયા, જળ સંસ્થાઓમાં ન્યુરોટોક્સિનનું ઉત્પાદન કરે છે, વર્ષ 2020 માં જેનાથી ક્યાં 300 થી વધુ હાથીઓને માર્યા ગયા?
A. બોત્સ્વાના B. ભારત
C. નમિબીઆ D. ઝિમ્બાબ્વે

72. વિશ્વ પર્યાવરણીય આરોગ્ય દિવસ ક્યારે ઉજવવામાં આવે છે?
A. 29 ડિસેમ્બર B. 24 નવેમ્બર
C. 26 સપ્ટેમ્બર D. 19 ઓક્ટોબર

73. બ્લેક કાર્બન એ એરોસોલ છે જે હવામાં નક્કર કણો અથવા પ્રવાહી ટીપાં તરીકે સસ્પેન્ડ કરવામાં આવે છે. હવામાન પલટામાં તેને બીજા ક્રમનું સૌથી મહત્વપૂર્ણ ____ માનવામાં આવે છે
A. કુદરતી એજન્ટ
B. પ્રદૂષક
C. ગ્રીનહાઉસમાં ફાળો આપનાર
D. એન્થ્રોપોજેનિક એજન્ટ

74. યુનાઇટેડ નેશન્સના અહેવાલ મુજબ, લોકડાઉન દરમિયાન નવી દિલ્હીમાં નાઇટ્રોજન ડાયોક્સાઇડનું સ્તર લગભગ ____ ઘટી ગયું હતું.
A. 90% B. 80%
C. 85% D. 70%

69. The criminal throwing of inappropriate man-made objects is known as
A. Visual Pollution
B. Soil Contamination
C. Plastic pollution
D. Littering

70. CWA stands for
A. Clean Waste Act
B. Clean Water Act
C. Clearance Waste Act
D. Clearance Water Act

71. Some Cyanobacteria produce neuro-toxins in water bodies, In the year 2020 that killed more than 300 elephants in
A. Botswana B. India
C. Namibia D. Zimbabwe

72. World Environmental Health Day is celebrated on
A. 29th December B. 24th November
C. 26th September D. 19th October

73. Black carbon is an aerosol suspended as fine solid particle or liquid droplet in air. It is considered as the second most important ____ in climate change
A. Natural agent
B. Pollutant
C. Greenhouse contributor
D. Anthropogenic agent

74. According to the report of United Nations, the nitrogen dioxide levels in New Delhi had fell by about ____ during the lock down.
A. 90 % B. 80 %
C. 85 % D. 70 %

75. મોટાભાગના વાતાવરણીય ઓઝોન વાતાવરણના ______ સ્તરમાં કેન્દ્રિત હોય છે

A. સ્ટ્રેટોસ્ફિયર B. આયનોસ્ફિયર

C. મેસોસ્ફિયર D. ટ્રોપોસ્ફિયર

76. O_3 શું રજૂ કરે છે?

A. ઓક્સિજન પરમાણુ

B. ટ્રાયઓક્સિજન

C. ઓઝોન

D. બંને B અને C

77. ડેસિબલ એ કયું પરિબળ માપવા માટેનું એકમ છે?

A. ધ્વનિ પ્રદૂષણ B. વાયુ પ્રદૂષણ

C. જળ પ્રદૂષણ D. પરમાણુ પ્રદૂષણ

78. ઓઝોન શેના દ્વારા રચાય છે?

A. ઓક્સિજન સાથે યુવી કિરણોત્સર્ગની ક્રિયાપ્રતિક્રિયા

B. ઓક્સિજન પર ડેલાઇટની ક્રિયા

C. ઓક્સિજન પરમાણુ અને મુક્ત ઓક્સિજન અણુના સંયોજન દ્વારા

D. ઉપરના બધા

79. ______ એક્ટ 1974 ની સાલમાં અમલમાં આવ્યો હતો

A. ભારતીય વન અધિનિયમ

B. જળ (પ્રદૂષણ નિવારણ અને નિયંત્રણ) અધિનિયમ

C. વન સંરક્ષણ અધિનિયમ

D. હવા (પ્રદૂષણ નિવારણ અને નિયંત્રણ) અધિનિયમ

80. *Pitheolobium dulce* નું વાવેતર શેમાં ઘટાડો કરે છે?

A. કિરણોત્સર્ગી પ્રદૂષણ

B. ધ્વનિ પ્રદૂષણ

C. કોલસાનુ પ્રદૂષણ

D. થર્મલ પ્રદૂષણ

75. Most atmospheric ozone is concentrated in ______ layer of the atmosphere

A. Stratosphere B. Ionosphere

C. mesosphere D. troposphere

76. O_3 represents

A. Oxygen molecule

B. Trioxygen

C. Ozone

D. Both B and C

77. Decibel is a unit to measure

A. Noise pollution B. Air pollution

C. Water pollution D. Nuclear pollution

78. Ozone is formed by

A. interaction of UV radiation with oxygen

B. action of daylight on oxygen

C. By combination of oxygen molecule and free oxygen atom

D. All of the above

79. ______ Act was enacted in the year 1974

A. Indian Forest Act

B. Water (Prevention & control of pollution) Act

C. Forest Conservation Act

D. Air (Prevention & control of pollution) Act

80. *Pitheolobium dulce* dense plantation reduces

A. Radioactive pollution

B. Noise pollution

C. Burning coal pollution

D. Thermal pollution

81. વાયુ પ્રદૂષણ ઘટાડવા માટે ઔદ્યોગિક એકમોને પ્રોત્સાહિત કરવા અને ઇટીએસ ___ રાજ્યમાં શરૂ કરવામાં આવી હતી

A. દિલ્હી B. મહારાષ્ટ્ર
C. હરિયાણા D. ગુજરાત

82. એનસીએપી નું આખું નામ ___

A. નેશનલ ક્લીન એર પ્રોગ્રામ
B. નેશનલ સેન્ટ્રલ એર પ્રોજેક્ટ
B. નેશનલ ક્લીન એર પ્રોજેક્ટ
D. નેશનલ સેન્ટ્રલ એર પ્રોગ્રામ

83. NCAPની શરૂઆત કેન્દ્ર સરકાર દ્વારા ક્યારે કરવામાં આવી હતી?

A. 2016 B. 2017
C. 2018 D. 2019

84. હવામાં ગુણવત્તાની ઇન્ડેક્સ રેંજમાં 201 થી 300 ની વચ્ચે હોય તો, હવાની ગુણવત્તા કેવી ગણવામાં આવે છે?

A. સંતોષકારક B. મધ્યમ
C. નબળી D. ગંભીર

85. સ્ટબલ બર્નિંગ હવાના પ્રદૂષણના મુખ્ય કારણો નીચેના માંથી કયા રાજ્યોમાં મુખ્ય જોવા મળે છે?

A. ભારતના ઉત્તરી રાજ્યો
B. ભારતના દક્ષિણ રાજ્યો
C. ભારતના પૂર્વી રાજ્યો
D. ભારતના પશ્ચિમી રાજ્યો

86. ભારતમાં, ભૂગર્ભ જળનો મોટો જથ્થો કોના દ્વારા કાઢવામાં આવે છે?

A. ઘરો B. ઉદ્યોગો
C. ખેડુતો D. મજૂરો

87. 2020 ની વર્લ્ડ વોટર ડે થીમ શું હતી?

A. જળ અને હવામાન પરિવર્તન
B. પાણી અને પ્રદૂષણ
C. બધા માટે પાણી
D. પાણી માટે કુદરત

81. ETS was launched in ___ state to encourage and incentivize the industrial units to cut air pollution

A. Delhi B. Maharashtra
C. Haryana D. Gujarat

82. NCAP stands for

A. National Clean Air Programme
B. National Central Air Project
C. National Clean Air Project
D. National Central Air Programme

83. NCAP was launched by Centre government in

A. 2016 B. 2017
C. 2018 D. 2019

84. In air quality index ranges between 201-300, the quality of air is

A. Satisfactory B. Moderate
C. Poor D. Severe

85. Stubble burning is one of the major causes of air pollution in

A. Northern states of India
B. Southern states of India
C. Eastern states of India
D. Western states of India

86. In India, major amount of ground water is extracted by

A. Households B. Industries
C. Farmers D. Laborers

87. World Water Day Theme- 2020 was

A. Water and Climate Change
B. Water and Pollution
C. Water for all
D. Nature for Water

88. સતત ત્રણ વર્ષથી વિશ્વની સૌથી મોટી પ્લાસ્ટિક પ્રદૂષણ કરતી બ્રાંડ કઈ છે?

A. પેપ્સિકો
B. કોકા કોલા અને નેસ્લે
C. પેપ્સિકો અને નેસ્લે
D. કોકા-કોલા, પેપ્સિકો અને નેસ્લે

89. ચેર્નોબિલ પરમાણુ દુર્ઘટના ક્યારે સર્જાઈ હતી?

A. 1983 B. 1986
C. 1988 D. 1991

90. એસિડ વરસાદ દ્વારા કૃષિ ઉત્પાદનમાં ક્યા કારણે ઘટાડો થઈ શકે છે?

A. જમીનના રાસાયણિક ગુણધર્મોને બદલવું
B. માઇક્રોબાયલ પ્રક્રિયાઓને ધીમું કરવું
C. જમીનના પોષક તત્વોમાં ઘટાડો
D. ઉપર્યુક્ત તમામ

91. એસિડ વરસાદથી જમીનની pH માં ______ થાય છે

A. ઘટાડો B. વધારો
C. તીવ્ર વધારો D. કોઈ અસર નહીં

92. WCED નું આખું નામ શું છે?

A. ઊર્જા અને વિકાસ પર વિશ્વ પંચ
B. પર્યાવરણ અને વિકાસ પર વિશ્વ પંચ
C. વિશ્વ સ્વચ્છ પર્યાવરણ દિવસ
D. વિશ્વ સ્વચ્છ ઊર્જા દિવસ

93. સસ્ટેનેબલ ડેવલોપમેન્ટની આધુનિક વિભાવના ક્યાંથી ઉદ્દભવી છે?

A. 1984 બ્રુંડટલેન્ડ અહેવાલ
B. 1992 અર્થ સબમિટ
C. 1987 બ્રુંડટલેન્ડ રિપોર્ટ
D. 2002 અર્થ સબમિટ

88. The world's biggest plastic pollution brand from three consecutive years is/are

A. Pepsico
B. Coca-cola and Nestle
C. Pepsico and Nestle
D. Coca-cola, Pepsico and Nestle

89. The Chernobyl disaster was caused by a nuclear accident occurred in

A. 1983 B. 1986
C. 1988 D. 1991

90. Acid rain can reduce agricultural production by

A. changing the chemical properties of soil
B. slowing down microbial processes
C. reducing soil nutrients
D. all of the above

91. Acid rain causes ____________ in pH of the soil

A. Decrease B. Increase
C. Sharp increase D. No effect

92. WCED stands for

A. World Commission on Energy and Development
B. World Commission on Environment and Development
C. World Clean Environment Day
D. World Clean Energy Day

93. The modern concept of sustainable development is mostly derived from the

A. 1984 Brundtland Report
B. 1992 Earth Submit
C. 1987 Brundtland Report
D. 2002 Earth Submiit

94. તમામ પ્રકારના જંગલોના વ્યવસ્થાપન, સંરક્ષણ અને ટકાઉ વિકાસ પર વૈશ્વિક સંમતિ માટે સિદ્ધાંતોનું કાયદેસર રીતે બંધનકર્તા અધિકૃત નિવેદન કયા નામે ઓળખાય છે?

A. રિયો વન સિદ્ધાંતો
B. વન સંરક્ષણના સિદ્ધાંતો
C. વન સિદ્ધાંતો
D. બંને A અને C

95. ત્રણ "R" જેનો ઉપયોગ પર્યાવરણને બચાવવા માટે કરવામાં આવે છે. Reduce, Reuse Recycling અને નીચેનામાંથી શેમાં શામેલ છે

A. જમીન B. હવા
C. જંગલો D. આબોહવા

96. સસ્ટેનેબલ વન વ્યવસ્થાપન હંમેશા શેમાં સંતુલન જાળવવાનો પ્રયાસ કરે છે?

A. વનના પ્રાકૃતિક સંશોધનોની માંગ અને વન ઇકોસિસ્ટમની વચ્ચે
B. વન ઉત્પાદનો માટે સમાજની માંગને પહોંચી વળવા
C. વન ઉત્પાદનો માટે ઉદ્યોગોની માંગને પહોંચી વળવા
D. વન આરોગ્ય સંરક્ષણ માટે

97. _______________ પાણીની કાયમી કઠિનતાનું કારણ બને છે.

A. મેગ્નેશિયમ ક્લોરાઇડ
B. કેલ્શિયમ ક્લોરાઇડ
C. મેગ્નેશિયમ સલ્ફેટ
D. મેગ્નેશિયમ કાર્બોનેટ

98. તાજા પાણીના તળાવોમાં, નીચેનામાંથી કયા અકાર્બનિક આયનોની ઉપલબ્ધતાને લીધે અલ્ગાલ બ્લૂમ થાય છે.

A. નાઇટ્રોજન B. ફોસ્ફરસ
C. સલ્ફર D. પોટેશિયમ

94. Non-legally binding authoritative statement of principles for a global consensus on the management, conservation and sustainable development of all types of forests is known as

A. Rio Forest Principles
B. Forest Protection Principles
C. Forest Principles
D. Both A and C

95. The three R's which are used to save the environment; Reduce, Reuse and Recycle include which of the following

A. Soil B. Air
C. Forests D. Climate

96. Sustainable forest management always tries to maintain a balance

A. between the demand for the forest's natural resources and the vitality of the forest ecosystem
B. for meeting the demand of society for forest products
C. for meeting the demand of industries for forest products
D. for conserving forest health

97. _______________ salt causes permanent hardness of water

A. Magnesium chloride
B. Calcium chloride
C. Magnesium sulphate
D. Magnesium carbonate

98. In freshwater lakes, algal blooms are caused by availability of which of the following inorganic ions

A. Nitrogen B. Phosphorus
C. Sulphur D. Potassium

99. કોલિફોર્મ્સ પાણીમાં __________ દૂષણના સૂચક સજીવ છે

A. ઔદ્યોગિક B. પ્લાસ્ટિક
C. ભારે ધાતુ D. ફેકલ

100. ગ્રીનહાઉસ વાયુઓના પ્રાકૃતિક સ્રોતોને અંકુશમાં લેવાનો કોનો પ્રાથમિક એજન્ડા છે?

A. ક્યોટો પ્રોટોકોલ
B. મોન્ટ્રીયલ પ્રોટોકોલ
C. દોહા સુધારો
D. પેરિસ ક્લાઈમેટ કરાર

101. બાયોકેમિકલ ઓક્સિજન માંગ મુખ્યત્વે શેના માટે વપરાય છે?

A. કાર્બનનો કુલ અંદાજ
B. વિગલિત ઓક્સિજનનું સ્તર નક્કી કરવું
C. ગટરના પાણીમાં કાર્બનિક પદાર્થોની માત્રાનો અંદાજ
D. ઉપર્યુક્તમાંથી એક પણ નહિ

102. __________ને લાયકેન્સ નો સૂચક તરીકે માનવામાં આવે છે.

A. વાયરલ રોગો
B. પ્રદૂષણ
C. ઉત્ક્રાંતિ
D. પર્યાવરણ

103. __________ ધાતુઓ અને કિરણોત્સર્ગી પદાર્થો દ્વારા દરિયાકાંઠાના અને એસ્ચ્યુઅરીઝ વાતાવરણના પ્રદૂષણ માટે બાયોઇન્ડિકેટર તરીકે ગણવામાં આવે છે

A. મ્યુસેલ્સ B. કાચબા
C. વ્હેલ D. ફાયટોપ્લાંકટોન્સ

104. વિશ્વનું પ્રથમ ગ્રીનહાઉસ ગેસ નિરીક્ષણ સેટેલાઇટ કોના દ્વારા લોંચ કરવામાં આવ્યું હતું?

A. 2007 માં ભારત દ્વારા
B. 2005 માં જર્મની દ્વારા
C. 2011 માં રશિયા દ્વારા
D. 2009 માં જાપાન દ્વારા

99. Coliforms are the indicator organisms of __________ contamination in water

A. Industrial B. Plastic
C. Heavy metal D. Fecal

100. Controlling anthropogenic sources of greenhouse gases is a primary agenda of

A. Kyoto protocol
B. Montreal Protocol
C. Doha Amendment
D. Paris Climate Agreement

101. Biochemical oxygen demand is primarily used for

A. Estimating the total amount of carbon
B. Determining the level of dissolved oxygen
C. Estimating the quantity of organic matter in sewage water
D. None of the above

102. Lichens are considered as indicators of

A. Viral disease
B. Pollution
C. Evolution
D. Environemnt

103. __________ are considered as bioindicators for pollution of coastal and estuarine environments by metals and radioactive substances

A. Mussels B. Turtles
C. Whales D. Phytoplanktons

104. World's first Greenhouse Gases Observing Satellite was launched by

A. India in 2007
B. Germany in 2005
C. Russia in 2011
D. Japan in 2009

105. સ્વચ્છ ઊર્જા રોકાણો દ્રષ્ટિએ વિશ્વનો ટોચનો દેશ કયો છે?

A. જાપાન B. યુએસએ
C. ચાઇના D. જર્મની

106. ભારતમાં પહેલો મોટો પર્યાવરણીય સંરક્ષણ અધિનિયમ કયો આવ્યો?

A. જળ અધિનિયમ
B. વાયુ અધિનિયમ
C. પર્યાવરણીય અધિનિયમ
D. અવાજ પ્રદૂષણ અધિનિયમ

107. કયા વર્ષમાં જળ અધિનિયમ લાગુ કરાયો હતો?

A. 1986 B. 1994
C. 1974 D. 1975

108. યુટ્રોફિકેશનના સંદર્ભમાં, નીચેનામાંથી કયું સાચું છે?

A. તે ઓગળેલા ઓક્સિજનનું પાણી ઘટાડે છે
B. માનવ પ્રવૃત્તિઓ મુખ્યત્વે જવાબદાર છે
C. તે જળ પદાર્થોના પોષક સંવર્ધનની ઘટના છે
D. ઉપર્યુક્ત તમામ

109. કઈ નદી સૌથી પ્રદૂષિત નદી તરીકે ગણવામાં આવે છે?

A. ગંગા B. યમુના
C. ચેનાબ D. કાવેરી

110. પાણીમાં ઓગળેલા સોલિડ્સ (TDS) ઘટાડવા માટે કઈ તકનીકોનો ઉપયોગ થાય છે?

A. આયન એક્સચેંજ
B. નિસ્યંદન
C. બંને A અને B
D. ઉપર્યુક્તમાંથી એક પણ નહિ

105. The world's top country in terms of clean energy investment is

A. Japan B. USA
C. China D. Germany

106. The first major environmental protection act to be laid in India

A. Water act
B. Air act
C. Environmental act
D. Noise pollution act

107. The water act was enacted in the year

A. 1986 B. 1994
C. 1974 D. 1975

108. With reference to eutrophication, which of the following is/are true?

A. It depletes the water of dissolved oxygen
B. Human activities are mainly responsible
C. It is a phenomenon of nutrient enrichment of water bodies
D. All of the above

109. Which river is consider as the most polluted river?

A. Ganga B. Yamuna
C. Chenab D. Cauvery

110. Which techniques is/are used to reduce the total dissolved solids (TDS) in the water?

A. Ion Exchange
B. Distillation
C. Both A & B
D. None of the above

111. કયો ક્ષાર પાણીની કાયમી કઠિનતાનું મુખ્ય કારણ છે?

A. મેગ્નેશિયમ બાયકાર્બોનેટ
B. મેગ્નેશિયમ કાર્બોનેટ
C. મેગ્નેશિયમ સલ્ફેટ
D. ઉપર્યુક્ત તમામ

112. પાણીના નમૂનામાં ગંધ સંયોજનોનું સાંદ્રતા નક્કી કરવા માટે કઈ તકનીકનો ઉપયોગ કરવામાં આવે છે?

A. ફ્લશિંગ B. સ્ટ્રિપિંગ
C. ક્લોરીનેશન D. સેટલીંગ

113. પાણીમાં આર્સેનિક પ્રદુષકનો મુખ્ય સ્રોત કયો છે?

A. પૂર B. ખાતરો
C. ઔદ્યોગિક કચરો D. બંને B અને C

114. પાણીની ટર્બીડિટી માપવા માટેનું પરિમાણ કયું છે?

A. ઇએયુ ફોર્માઝિન એટેન્યુએશન એકમો
B. નેફેલિમેટ્રિક ટર્બિડિટી એકમો
C. ફોર્માઝિન નેફેલોમેટ્રિક એકમ
D. ઉપર્યુક્ત તમામ

115. માછલી કઈ ભારે ધાતુ ને વધુ માત્રામાં સંગ્રહ કરી શકે છે?

A. બિસ્મથ B. પારો
C. પેલેડિયમ D. ક્લોરિન

116. પાણીમાં હેલમિન્થના કયા કારણે થાય છે?

A. એમોબિક મરડો
B. કોલેરા
C. હૂક વૉર્મ
D. ટાઇફોઇડ

117. કુદરતી પાણીમાં ઓગળેલા ઓક્સિજનનું મહત્તમ મૂલ્ય કેટલું હોય છે?

A. 4-6 પીપીએમ B. 2-4 પીપીએમ
C. 4-7 પીપીએમ D. 5-7 પી.પી.એમ

111. Which salt is the main cause of permanent hardness of water?

A. Magnesium bicarbonate
B. Magnesium carbonate
C. Magnesium sulphate
D. All of the above

112. Which technique is use to determine the concentration of odour compounds in the water sample?

A. Flushing B. Stripping
C. Chlorination D. Settling

113. The main source of arsenic contamination in water

A. Floods B. Fertilizers
C. Industrial wastes D. Both B & C

114. The parameter to measure the turbidity of water is

A. EAU Formazin attenuation units
B. Nephelometric turbidity units
C. Formazin Nephelometric unit
D. All of the above

115. Which heavy metal fish can stored in more quantity?

A. Bismuth B. Mercury
C. Palladium D. Chlorine

116. Helminth in water causes due to ______

A. Amoebic dysentery
B. Cholera
C. Hook worm
D. Typhoid

117. What is the optimum value of dissolved oxygen in natural water?

A. 4-6 ppm B. 2-4 ppm
C. 4-7 ppm D. 5-7 ppm

118. કેન્દ્રીય પ્રદૂષણ નિયંત્રણ બોર્ડ અને દરિયાઇ અને પર્યાવરણ વિભાગે દરિયાકાંઠા પર કેટલા સ્ટેશનો સ્થાપિત કર્યા છે?

A. 171 B. 172
C. 173 D. 174

119. કાંઠાના પાણીમાં શેના મોટા તફાવતો આવે છે?

A. વાહકતા B. પ્રદૂષણ
C. ખારાશ D. સીવેજ

120. નીચેનામાંથી શું તળાવોના રાસાયણિક દેખરેખ હેઠળ આવે છે?

A. અસ્થિરતા B. પેથોજેન્સ
C. ડીટરજન્ટ D. વાહકતા

121. કાચા પાણીની ગુણવત્તા સુધારવાનું પ્રથમ પગલું નીચેનામાંથી કયું છે?

A. ક્લોરીનેશન
B. ફિલ્ટરેશન
C. કોગ્યુલેશન
D. સેડિમેંટેશન

122. પલાર પાની શું છે?

A. નદીનું પાણી
B. મહાસાગરનું પાણી
C. વરસાદનું પાણી
D. ઉપર્યુક્તમાંથી એક પણ નહિ

123. જંતુનાશક દવાઓનો વધુ પડતો ઉપયોગથી શું થાય છે?

A. બાયોમેગ્નિફિકેશન
B. બાયોએક્યુમ્યુલેશન
C. બંને A અને B
D. ઉપર્યુક્તમાંથી એક પણ નહિ

124. ગંગા એક્શન પ્લાન કયા વર્ષે શરૂ કરવામાં આવી હતી?

A. 1975-76 B. 1985-86
C. 1995-96 D. 1997-98

118. How many stations have Central pollution control board and the department of ocean and environment have established over the coastal line?

A. 171 B. 172
C. 173 D. 174

119. Coastal water shows major differences in

A. Conductivity B. Pollution
C. Salinity D. Sewage

120. Which of the following is comes under the chemical monitoring of lakes?

A. Turbidity B. Pathogens
C. Detergent D. Conductivity

121. Which of the following is first step of raw water treatment?

A. Chlorination
B. Filtration
C. Coagulation
D. Sedimentation

122. What is Palar Pani?

A. River water
B. Ocean water
C. Rain water
D. None of the above

123. Excess use of pesticides causes

A. Biomagnification
B. Bioaccumulation
C. Both A & B
D. None of the above

124. Ganga action plan was launched in which year?

A. 1975-76 B. 1985-86
C. 1995-96 D. 1997-98

125. ડીડીટી કયા પ્રકારનાં પ્રદૂષક છે?
A. નોન બાયોડિગ્રેડેબલ
B. બાયોડિગ્રેડેબલ
C. કુદરતી રીતે ડિગ્રેડેબલ
D. ઉપર્યુક્તમાંથી એક પણ નહિ

126. નીચેનામાંથી કયું શુદ્ધિકરણ પ્લાન્ટનો ભાગ નથી?
A. હાઇડ્રોજન ટેન્ક B. ક્લોરીનેટર
C. સક્રિય કાર્બન D. રેતી ગાળક

127. પીવાના પાણીમાં વધુ પડતો પારો અને લીડ શું અસર કરે છે?
A. ચેતાતંત્રને અસર કરે છે
B. યકૃતમાં ખામી
C. હિમોગ્લોબીનમાં ઘટાડો
D. ઉપર્યુક્ત તમામ

128. પાણીમાં ભંગાણના કારણો શું છે?
A. ઓક્સિજનની માંગમાં ઘટાડો
B. ઓગળેલ ઓક્સિજનમાં ઘટાડો.
C. ઓક્સિજનની માંગમાં વધારો
D. બંને B અને C

129. જો નદીના પાણીનો બીઓડી ખૂબ ઊંચો હોય. તે જણાવે છે કે,
A. પાણી ખૂબ જ ચોખ્ખું છે
B. પાણીમાં શેવાળ હોય છે
C. પાણી ખૂબ પ્રદૂષિત છે
D. પાણીમાં ઓગળેલા ખનીજ છે

130. નીચેનામાંથી કયો જળ સંસ્થાઓમાં થર્મલ પ્રદૂષણનો મોટો સોત છે?
A. થર્મલ પાવર પ્લાન્ટ
B. નક્કર કચરો નિકાલ કરવાની સાઇટ્સ
C. સીવેજ ટ્રીટમેન્ટ પ્લાન્ટ
D. ઉપર્યુક્ત તમામ

125. DDT is which type of pollutant?
A. Non-biodegradable
B. Biodegradable
C. Naturally degradable
D. None of the above

126. Which of the following is not a part of water treatment plant?
A. Hydrogen Tank B. Chlorinator
C. Activated carbon D. Sand filter

127. Excess amount of mercury and lead in drinking water leads to .....
A. Affecting nervous system
B. Liver malfunction
C. Reduction in Hb content
D. All of the above

128. What causes in water by hear dissipation in water?
A. Decrease oxygen demand
B. Decrease dissolved oxygen conc.
C. Increase oxygen demand
D. Both B & C

129. If BOD of a river water is very high. It states that,
A. water is very clear
B. water contains algae
C. water is highly polluted
D. Water contain dissolved minerals

130. Which of the following is a major source of thermal pollution in water bodies?
A. Thermal power plant
B. Solid waste disposal sites
C. Sewage treatment plant
D. All of the above

131. ઇ.આઇ.એ. માટેનું મૂળભૂત કારણ શું છે?
A. સૂચિત વિકાસનું વર્ણન
B. વિકાસની અસરોના કદની આગાહી
C. વિકાસના હાલના પર્યાવરણીય પ્રભાવોને ઘટાડવા
D. વિકાસના પર્યાવરણીય પરિણામોને અગાઉથી ઓળખવા

131. **The principle reason for EIA is to**
A. Describe proposed developments
B. Predict the size of impacts of developments
C. mitigate existing environmental impacts of development
D. identifies the environmental consequences of development in advance

132. કઈ તકનીક પાણીના સુક્ષ્મજીવાણુના દૂષણને ઓળખવા માટે ઝડપી, અસરકારક અને માત્રાત્મક અર્થ પ્રદાન કરે છે?
A. પ્યોર ક્લચર
B. મોલેક્યુલર
C. સેરોલોજિકલ પદ્ધતિ
D. ઉપર્યુક્તમાંથી એક પણ નહિ

132. **Which technique provides a quick, effective and quantitative means of identifying microbial contamination of water?**
A. Pure culture
B. Molecular
C. Serological method
D. None of the above

133. બાયોમ નીચેના કયા પરિમાણો પર આધારિત છે?
A. ભેજ અને પવન
B. તાપમાન અને પવન
C. તાપમાન અને વરસાદ
D. ઉપર્યુક્ત તમામ

133. **Biome distribution is based on which of the following parameters**
A. Humidity and wind
B. Temperature and wind
C. Temperature and Precipitation
D. All of the above

134. બાયોસ્ફિયરનો જમીની ભાગ કેટલા બાયોમમાં વહેંચાયેલો છે?
A. 4 B. 7
C. 9 D. 5

134. **The terrestrial part of biosphere are divided in to how many biomes?**
A. 4 B. 7
C. 9 D. 5

135. ENVIS કયા વર્ષમાં લાગુ કરવામાં આવ્યું હતું?
A. 1980 B. 1981
C. 1982 D. 1985

135. **ENVIS is implemented in which year?**
A. 1980 B. 1981
C. 1982 D. 1985

136. પાણીના સમૂહના તળિયે કયા પ્રકારના જીવંત જીવો જોવા મળે છે?
A. પ્લેન્કટોન B. બેન્થોસ
C. નેક્ટોન D. ન્યુસ્ટન

136. **The living organisms which are found living in the bottom of the water mass is** ________
A. Plankton B. Benthos
C. Nekton D. Neuston

137. ભારતનું સૌથી જૂનું તળાવ "સુદર્શન" ક્યાં આવેલું છે?

A. ગિરનાર, ગુજરાત

B. હિમાલય

C. પશ્ચિમ ઘાટ

D. ઉપર્યુક્તમાંથી એક પણ નહિ

138. હાર્મફુલ એલ્ગલ બ્લૂમ્સ (HASBs) ________ સાથે પ્રતિક્રિયા કરે છે જે માછલીઓ અને માનવો માટે જોખમી છે.

A. પાણીમાં ઓક્સિજન ઘટાડે છે

B. પાણીમાં હાનિકારક ઉમેરે છે

C. પાણીમાં ઝેર ઉમેરે છે

D. પાણીની ક્ષારતા ઘટાડે છે

139. સોસાયટી ઓફ ઇન્ટીગ્રેટેડ કોસ્ટલ મેનેજમેન્ટ (SICOM) કયા વર્ષમાં શરૂ કરવામાં આવ્યું છે?

A. 2008 B. 2010

C. 2009 D. 2012

140. રાષ્ટ્રીય એમ્બિયન્ટ એર ક્વોલિટી સ્ટાન્ડર્ડ્સ (NAAQS) ને કયા વર્ષમાં સૂચિત કરાયું હતું?

A. 1980 B. 1985

C. 1982 D. 1984

141. અતિશય દૂષિત પાણીમાં ઓગળેલા ઓક્સિજન નું પ્રમાણ કેટલું હોય છે?

A. 4 mg/l થી નીચે

B. 8 mg/l થી નીચે

C. 10 mg/l થી નીચે

D. 6 mg/l થી નીચે

142. જૈવિક કચરા દ્વારા જળ પ્રદૂષણ શેમાં માપવામાં આવે છે?

A. કેમિકલ ઓક્સિજન ડિમાન્ડ

B. બાયોકેમિકલ ઓક્સિજન ડિમાન્ડ

C. બંને A અને B

D. ઉપર્યુક્તમાંથી એક પણ નહિ

137. The Lake "Sudarshan" the oldest lake in India is located at

A. Girnar, Gujarat

B. Himalayan

C. Western Ghats

D. None of the above

138. What does Harmful Algal Blooms (HABs) reacts to which is dangerous to fish and humans ?

A. Deplete oxygen in water

B. Add harmful chemicals to water

C. Add toxicity to water

D. Reduce alkalinity of water

139. Society of Integrated coastal management (SICOM) is launched during which year?

A. 2008 B. 2010

C. 2009 D. 2012

140. National Ambient Air Quality Standards (NAAQS) were notified in which year?

A. 1980 B. 1985

C. 1982 D. 1984

141. What is the content of dissolved oxygen in highly contaminated water?

A. Below 4 mg/L

B. Below 8 mg/L

C. Below 10 mg/L

D. Below 6 mg/L

142. Water pollution by organic wastes is measure in which term?

A. Chemical oxygen demand

B. Biochemical oxygen demand

C. Both A & B

D. None of the above

143. માનવ શરીર ફ્લોરાઇડનું કેટલું સ્તર સહન કરી શકે છે?

A. 2.0 પીપીએમ B. 4 પીપીએમ
C. 1.5 પીપીએમ D. 2.5 પીપીએમ

144. આર્સેનિક દૂષણ એ એક ગંભીર સમસ્યા છે કે કયા પ્રકારના ક્ષેત્રમાં છે?

A. નદી બાજુના વિસ્તારો
B. દરિયા કિનારા
C. ટ્યુબ-વેલ ખોદાયેલા વિસ્તારો
D. પહાડી વિસ્તારો

145. બ્રેગોલી એ કયા ઉદ્યોગનુ પેટા-ઉત્પાદન છે કે જેનાથી પાણીમાંઓઇલ સ્પીલ સાફ કરવા માટે ઉપયોગ કરવામાં આવે છે?

A. પેટ્રોકેમિકલ ઉદ્યોગો
B. કાગળ ઉદ્યોગો
C. પ્લાસ્ટિક ઉદ્યોગો
D. એગ્રોકેમિકલ ઉદ્યોગો

146. અવાજના કયા સ્તર પર વ્યક્તિની સંભાળવાની ક્ષમતાને નુકસાન થઈ શકે છે?

A. 50 ડીબી B. 35 ડીબી
C. 75 ડીબી D. 90 ડીબી

147. ઘોંઘાટ પ્રદૂષણ નિયંત્રણ અને નિયમન નિયમો કયા વર્ષમાં અમલમાં આવ્યા?

A. 2004 B. 2002
C. 2006 D. 2000

148. માછલી દૂષિત પાણીમાંથી કયા કિરણોત્સર્ગી પદાર્થો એકઠા કરે છે?

A. 65 Zn
B. 55 Fe
C. 90 Sr
D. ઉપર્યુક્તમાંથી એક પણ નહિ

149. મહત્તમ ઇ-કચરો પેદા કરતુ અગ્રણી શહેર કયું છે?

A. દિલ્હી B. બેંગલોર
C. મુંબઈ D. ચેન્નાઈ

143. How much level of fluoride a human body can tolerate?

A. 2.0 ppm B. 4 ppm
C. 1.5 ppm D. 2.5 ppm

144. Arsenic contamination is a serious problem in which types of area

A. River-side areas
B. Seashores
C. Tube-well dug areas
D. Hilly areas

145. Bregoli is a by-product of which industry used to clean oil spills from water

A. Petrochemical industries
B. Paper industries
C. Plastic industries
D. Agrochemical industries

146. Over which level of noise a person's hearing can be damaged?

A. 50 dB B. 35 dB
C. 75 dB D. 90 dB

147. Noise pollution control and regulation rules implemented in which year?

A. 2004 B. 2002
C. 2006 D. 2000

148. Which radioactive material a fish accumulated from contaminated water?

A. 65Z n
B. 55 Fe
C. 90 Sr
D. None of the above

149. Which is the leading city for generating maximum E-waste?

A. Delhi B. Banglore
C. Mumbai D. Chennai

150. પીવીસી પ્લાસ્ટિકના દહન થી કયું કાર્સિનોજેનિક કમ્પાઉન્ડ પેદા થાય છે?

A. બેન્ઝિન B. SO_2
C. સાયનાઇડ D. ડાયોક્સિન

151. બાયોમેડિકલ વેસ્ટ હેન્ડલિંગ અને મેનેજમેન્ટનાં નિયમો કયા વર્ષથી અમલમાં આવ્યા?

A. 1998 B. 2000
C. 1995 D. 2002

152. પાયરોલિસીસની પ્રક્રિયા માટે નીચેનામાંથી કયું વિધાન યોગ્ય છે?

A. ઓક્સિજન ની હાજરીમાં દહન
B. ઓઝોનની હાજરીમાં દહન
C. ઓક્સિજન ની ગેરહાજરીમાં દહન
D. ઉપર્યુક્તમાંથી એક પણ નહિ

153. થર્મલ પ્રદૂષણ દરમિયાન, પાણીના ગરમ થવાથી_____થાય છે.

A. ડીઓ (Dissolved Oxygen) ના સ્તરમાં ઘટાડો
B. ડીઓ ના સ્તરમાં વધારો
C. CO_2 ના સ્તરમાં ઘટાડો
D. ઉપર્યુક્ત તમામ

154. *In-situ* બાયોરેમેડીસેશન વ્યૂહરચનામાં કઈ તકનીક શામેલ છે?

A. બાયવન્ટિંગ B. બાયોઓગમેન્ટેશન
C. બાયોસ્પાર્જિંગ D. ઉપર્યુક્ત તમામ

155. "ઓઇલઝેપર" બેક્ટેરિયા કઈ સંસ્થા દ્વારા વિકસિત થાય છે?

A. એનબીઆરઆઈ B. ટેરી
C. ઇપીએ D. આઈસીએઆર

156. કયા બેક્ટેરિયમનો ઉપયોગ અણુ કચરામાંથી ટોલ્યુએન અને આયનીય પારોને ડિટોક્સિફાઇ કરવા માટે થાય છે?

A. ડીનોકોકસુરાદોડ્યુરાન્સ
B. આઇડિઓનેલાસ્કેઇનેન્સીસ
C. *મોરેક્સેલા* સ્પી.
D. ઉપર્યુક્ત તમામ

150. The carcinogenic compound releases during burning PVC plastics

A. Benzene B. SO_2
C. Cyanide D. Dioxin

151. Biomedical waste handling and management rules comes in act in which year?

A. 1998 B. 2000
C. 1995 D. 2002

152. Which of the following statement is correct for the process of pyrolysis?

A. Combustion in the presence of O_2
B. Combustion in the presence of Ozone
C. Combustion in the absence of O_2
D. None of the above

153. During thermal pollution, warming of water leads_________.

A. Decrease in level of the DO
B. Increase in level of DO
C. Decrease in level of CO_2
D. All of the above

154. *In-situ* bioremediation strategies includes which technique(s)?

A. Bioventing B. Bioaugmentation
C. Biosparging D. All of the above

155. "Oilzapper" bacteria are developed by which organisation?

A. NBRI B. TERI
C. EPA D. ICAR

156. Which bacterium is used to detoxify toluene and ionic mercury from nuclear waste?

A. Deinococcusradiodurans
B. Ideonellasakainensis
C. *Moraxella* sp.
D. All of the above

157. માયકોરેમેડિએશન માં કયા સુક્ષ્મસજીવોનો ઉપયોગ થાય છે?
A. બેક્ટેરિયા
B. આલ્ગી
C. ફૂગ
D. ઉપર્યુક્તમાંથી એક પણ નહિ

158. ફોર્મીક એસિડ (HCOOH) નું પ્રદુષણ કઈ કુદરતી આફત દરમિયાન વધારે થાય છે?
A. ઔદ્યોગિક આગ
B. જ્વાળામુખી
C. દાવાનળ
D. ઉપર્યુક્ત તમામ

159. ભારતના કયા શહેરમાં પ્રથમ એસિડ વરસાદ થયો હતો?
A. નવી દિલ્હી B. મુંબઈ
C. ચેન્નાઈ D. કોલકાતા

160. પહાડી જંગલોને ટકાવી રાખવા માટે, આસામ હિલની જમીન અને ઇકોલોજીકલ સાઇટ્સ અધિનિયમ કયા વર્ષ દરમિયાન અમલમાં આવ્યા છે?
A. 2004 B. 2001
C. 2006 D. 2008

161. ભારત સરકાર દ્વારા "હરિત દિવાળી અને સ્વસ્થ દિવાળી" નામનું અભિયાન કયા વર્ષ દરમિયાન શરૂ કરવમાં આવ્યું હતું?
A. 2015 B. 2017
C. 2016 D. 2019

162. કયા વર્ષમાં પ્રદૂષણ ઘટાડવા માટેની સહાય યોજના અમલમાં આવી?
A. 1992 B. 1991
C. 1990 D. 1995

163. ગટરના ઉપદ્રિત પ્રવાહી ની pH રેન્જની પરવાનગી શું હોવી જોઈએ?
A. 5.0 થી 7.0 B. 6.0 થી 7.0
C. 4.0 થી 7.0 D. 6.5 થી 9.0

157. Which microorganism is used in mycoremediation?
A. Bacteria
B. Alage
C. Fungi
D. None of the above

158. Formic acid (HCOOH) pollution is high during which calamity?
A. Industrial fires
B. Volcanic eruption
C. Forest fires
D. All of the above

159. In which city of India the first acid rain was reported?
A. New Delhi B. Mumbai
C. Chennai D. Kolkata

160. To sustain hill forests, Assam Hill land and ecological sites act implemented during which year?
A. 2004 B. 2001
C. 2006 D. 2008

161. Govt of India's campaign called "Harit Diwali and Swasth Diwali" was launched during which year?
A. 2015 B. 2017
C. 2016 D. 2019

162. In which year the Scheme of Assistance for Abatement of Pollution came into act?
A. 1992 B. 1991
C. 1990 D. 1995

163. What should be the permitted pH range of treated effluent of sewage?
A. 5.0 to 7.0 B. 6.0 to 7.0
C. 4.0 to 7.0 D. 6.5 to 9.0

164. ભારતમાં હાલમાં કેટલા CAAQMS સ્ટેશનો કાર્યરત છે?

A. 85 B. 110

C. 90 D. 120

165. ખાતર ઉદ્યોગો માટે સુધારેલા પર્યાવરણીય ધોરણો કયા વર્ષ દરમિયાન આવ્યા?

A. 2017 B. 2015

C. 2019 D. 2014

166. હવાની ગુણવત્તાની માહિતી બુલેટિનમાં દરરોજ કયા સમયે પ્રકાશિત થાય છે?

A. બપોરે 2.00 વાગ્યે B. સાંજે 5.00 વાગ્યે

C. સાંજે 4.00 વાગ્યે D. સવારે 9.00 વાગ્યે

167. ગંગા નદી પર કેટલા વાસ્તવિક સમયના પાણીની ગુણવત્તાના નિરીક્ષણ કેન્દ્રો સ્થપાયેલા છે?

A. 40 B. 44

C. 50 D. 55

168. ટ્રોપોસ્ફિયરમાં 6.5°C /1000m ના દરે હવાના તાપમાનમાં ઘટાડો ક્યા નામે ઓળખાય છે?

A. પર્યાવરણીય અધોગતિ

B. પર્યાવરણીય વિરામ દર

C. ગ્રીનહાઉસ અસર

D. ગ્લોબલ વોર્મિંગ

169. જે પ્રદેશમાં તાજા નદીનું પાણી ખારા સમુદ્રના પાણીને મળે છે તેને શું કરે છે?

A. ક્ષિતિજ

B. એસચુરી

C. ડેલ્ટા

D. ઉપર્યુક્તમાંથી એક પણ નહિ

170. આગ્રા ખાતે તાજમહેલને મુખ્યત્વે કયા કમ્પાઉન્ડથી નુકસાન થાય છે?

A. હાઇડ્રોજન પેરોક્સાઇડ

B. સલ્ફર ડાયોક્સાઇડ

C. ક્લોરિન

D. ઉપર્યુક્ત તમામ

164. How many CAAQMS stations are currently operating in India?

A. 85 B. 110

C. 90 D. 120

165. The revised environmental standards for Fertilizer industries came during which year?

A. 2017 B. 2015

C. 2019 D. 2014

166. The air quality data in bulletin is published every day at what time?

A. 2.00 PM B. 5.00 PM

C. 4.00 PM D. 9.00 AM

167. How many real time water quality monitoring stations are established on river ganga?

A. 40 B. 44

C. 50 D. 55

168. The drop in air temperature at a rate of 6.5°C/1000m in troposphere is known as

A. Environmental degradation

B. Environmental lapse rate

C. Green-house effect

D. Global warming

169. The region where fresh river water meets salty sea water is called as?

A. Horizon

B. Estuarine

C. Delta

D. None of the above

170. The Taj Mahal at Agra is damaged mainly by which compound?

A. Hydrogen Peroxide

B. Sulphur dioxide

C. Chlorine

D. All of the above

171. રાષ્ટ્રીય વન નીતિ દ્વારા ભલામણ કરેલ વન વિસ્તારની ટકાવારી કેટલી છે?

A. 33% B. 43%
C. 55% D. 13%

172. કયું ચક્ર સમુદ્ર એસિડિફિકેશનનું પ્રતિનિધિત્વ કરે છે?

A. ફોસ્ફરસ ચક્ર B. સલ્ફર ચક્ર
C. નાઇટ્રોજન ચક્ર D. કાર્બન ચક્ર

173. ગ્રે ધુમ્મસ થવાનું મુખ્ય કારણ શું છે?

A. કાર્બનિક ધુમ્મસ
B. ફોટોકેમિકલ ધુમ્મસ
C. શહેરી ધુમ્મસ
D. ઔદ્યોગિક ધુમ્મસ

174. નીચેની કઈ ઘટના દરમિયાન, હવાના પ્રદૂષકો ગરમ હવાના સ્તરની નીચે હવાના એક ઠંડા સ્તર માં ફસાઈ જાય છે?

A. તાપમાન માં ઉલટસુલટ
B. અપવેલિંગ
C. ધુમ્મસ
D. ઉપર્યુક્તમાંથી એક પણ નહિ

175. કયા અકાર્બનિક પદાર્થ ને કારણે અલ્ગાલ બ્લૂમ મરી જાય છે અને ઓક્સિજનની ઊણપ મહાસાગરોમાં માછલીઓના મૃત્યુ તરફ દોરી જાય છે?

A. ફોસ્ફેટ્સ
B. નાઈટ્રેટ્સ
C. કેલટ્રેટ્સ
D. ઉપર્યુક્ત તમામ

176. નીચેનામાંથી કયાને પ્રદૂષણના નોન-પોઇન્ટ સ્રોત તરીકે ગણવામાં આવે છે?

A. ફેક્ટરી ધુમાડો સ્ટેક
B. ખાતરો અને જંતુનાશક રન ઓફ
C. ઓટોમોબાઈલ એક્ઝોસ્ટ પાઇપ
D. ઉપર્યુક્તમાંથી એક પણ નહિ

171. Percentage of forest area recommended by the national forest policy of plain is?

A. 33% B. 43%
C. 55% D. 13%

172. Which cycle represents the ocean acidification?

A. Phosphorus cycle B. Sulfur cycle
C. Nitrogen cycle D. Carbon cycle

173. The grey smog is mainly due to

A. Organic smog
B. Photochemical smog
C. Urban smog
D. Industrial smog

174. During which of the following phenomenon, air pollutants build up and trapped in a dense, cool layer of air beneath a layer of warm air?

A. Temperature inversion
B. Upwelling
C. Smog
D. None of the above

175. Which inorganic material cause algal blooms to die and create oxygen deficiencies leads to fish death in oceans?

A. Phosphates
B. Nitrates
C. Caltrates
D. All of the above

176. Which of the following is considered as a non-point source of pollution?

A. Factory smoke stack
B. Fertilizers and pesticides runoff
C. Automobile exhaust pipe
D. None of above

177. નીચેનામાંથી કયો ઝેરી પ્રદૂષક પદાર્થ બેટરી અને ગંધના પ્લાન્ટની નજીકમાં જોવા મળે છે જે આરોગ્ય માટે જોખમો ધરાવે છે.

A. ઓઝોન B. કાર્બન મોનોક્સાઈડ

C. લીડ D. આયર્ન

178. વિશ્વના ટોચના 20 પ્રદૂષિત શહેરોમાંથી, ભારતમાં કેટલા છે?

A. 15 B. 13

C. 10 D. 5

179. ગ્લોબલ એર ક્વોલિટી ઇન્ડેક્સ હેઠળ ભારતને કયો ક્રમ મળ્યો છે?

A. 140 B. 125

C. 155 D. 135

180. "મીનામાતા રોગ" કયા પ્રદૂષકથી દૂષિત માછલી ખાવાથી થયો હતો?

A. પારો

B. લીડ

C. ઝીંક

D. ઉપર્યુક્તમાંથી એક પણ નહિ

181. ઓઇલ સ્પીલથી થતા પાણી ના ઓક્સિજીનેશન ને કયા પ્રકારનું પ્રદૂષણ કહેવામાં આવે છે?

A. જૈવિક પ્રદૂષક

B. ભૌતિક પ્રદૂષક

C. રાસાયણિક પ્રદૂષક

D. ઉપર્યુક્ત તમામ

182. એસિડ વરસાદને કારણે માટીના કાર્બનિક પદાર્થોના વિઘટનમાં ઘટાડો કરી કઈ પ્રક્રિયામાં વધારો કરે છે?

A. જળચર જીવન અને જંગલોમાં ફૂગ રોગ

B. જળચર જીવન અને જંગલોમાં બેક્ટેરિયલ રોગ

C. જળચર જીવનમાં ફૂગ રોગ

D. ફક્ત જંગલોમાં બેક્ટેરિયલ રોગ

177. Which of the following toxic pollutant is largely found near battery and smelter plants having numerous health hazards?

A. Ozone B. Carbon monoxide

C. Lead D. Iron

178. Out of world's top 20 polluted cities, how many are from India?

A. 15 B. 13

C. 10 D. 5

179. Under Global Air quality index, what rank has India got?

A. 140 B. 125

C. 155 D. 135

180. Minamata disease was caused by eating fish contaminated with which pollutant?

A. Mercury

B. Lead

C. Zinc

D. None of the above

181. Oil spills prevent oxygenation of water, which type of pollutant is it?

A. Biological pollutant

B. Physical pollutant

C. Chemical pollutant

D. All of the above

182. Acid rain causes delay in the decomposition of soil organic material, and an increase in

A. fungal disease in aquatic life and forests

B. bacterial disease in aquatic life and forests

C. fungal disease only in aquatic life

D. bacterial disease only in forests

183. નીચેનામાંથી કયા બાયોપેસ્ટીસાઈડ નથી?
A. બાયોનિયમ B. કાર્બેરિલ
C. બાયોલેપ D. ડાઇપલ

184. ગુજરાતમાં, કેન્દ્રિય પ્રદૂષણ નિયંત્રણ બોર્ડની પ્રાદેશિક કચેરી ક્યાં સ્થિત છે?
A. અમદાવાદ B. ગાંધીનગર
C. વડોદરા D. અંકલેશ્વર

185. સીપીસીબી કયા મંત્રાલય હેઠળ કામ કરે છે?
A. વિજ્ઞાન અને ટેકનોલોજી મંત્રાલય
B. પેટ્રોલિયમ અને કુદરતી ગેસ મંત્રાલય
C. નવા અને નવીનીકરણીય ઉર્જા મંત્રાલય
D. પર્યાવરણ, વન અને જલવાયુ પરિવર્તન મંત્રાલય

186. પ્લાસ્ટિકનો મોટાભાગનો કચરો __________ માં ભળે છે?
A. મહાસાગરો
B. લેન્ડફિલ્સ
C. રિસાયકલ થાય છે
D. સળગાવી દેવાય છે

187. નિકાલની પ્રકૃતિ અનુસાર પ્રદૂષકોને શેમાં વિભાજિત કરવામાં આવ્યા છે?
A. કુદરતી અને માનવકેન્દ્રીય
B. બાયોડિગ્રેડેબલ અને નોન બાયો-ડિગ્રેડેબલ
C. ગુણાત્મક અને માત્રાત્મક
D. પ્રાથમિક અને ગૌણ

188. વાયુના પ્રદૂષણમાં વધારો કયા કારણે થાય છે?
A. ટ્રાફિકમાં વધારો
B. વિકસતા શહેરો અને ઝડપી આર્થિક વિકાસ
C. ઔદ્યોગિકરણ
D. ઉપર્યુક્ત તમામ

183. Which of the following is not a bio pesticide?
A. Bioneem B. Carbaryl
C. Biolap D. Dipel

184. Where in Gujarat, Central Pollution Control Board's regional Office is located?
A. Ahmedabad B. Gandhinagar
C. Vadodara D. Ankleshwar

185. CPCB is a board under
A. Ministry of Science and Technology
B. Ministry of Petroleum and Natural Gas
C. Ministry of New and Renewable Energy
D. Ministry of Environment, Forest and Climate Change

186. The majority of plastic waste ends up in __________
A. Oceans
B. Landfills
C. Recycled
D. Burned

187. According to the nature of disposal pollutants are divided into
A. Natural and anthropogenic
B. Biodegradable and non-biodegradable
C. Qualitative and quantitative
D. Primary and secondary

188. Air pollution is aggravated because of
A. increasing traffic
B. growing cities and rapid economic development
C. industrialization
D. All of the above

189. વાયુ પ્રદૂષણને વર્ણવવા માટે વપરાયેલ શબ્દ ____________, જે વાતાવરણમાં અમુક રસાયણો સાથે સૂર્યપ્રકાશની ક્રિયાપ્રતિક્રિયાના પરિણામ રૂપે ઉદ્દભવે છે.

A. ફોટોકેમિકલ ધુમ્મસ
B. ફોટો-રિએક્ટિવ ધુમ્મસ
C. ફોટોકેમિકલ સ્મોગ
D. ફોટો-રિએક્ટિવ સ્મોગ

190. ____________ સ્ટ્રેટોસ્ફેરીક ઓઝોન સ્તરના ઘટાડા માટે જવાબદાર છે.

A. એસ્બેસ્ટોસ ધૂળ
B. કિરણોત્સર્ગી પ્રદૂષકો
C. ક્લોરોફ્લોરોકાર્બન્સ
D. સસ્પેન્ડેડ પાર્ટિક્યુલેટ મેટર

191. રાષ્ટ્રીય એમ્બિયન્ટ એર ક્વોલિટી સ્ટાન્ડર્ડ્સ ને વર્ષ 1982 માં ઘડી અને નવેમ્બર 2009 માં કેટલા પ્રદૂષકો ઉમેરવામાં આવ્યા?

A. 10 પ્રદૂષકો B. 12 પ્રદૂષકો
C. 14 પ્રદૂષકો D. 16 પ્રદૂષકો

192. BOD અને COD એ શેના માપદંડ છે?

A. જળચર ઇકોસિસ્ટમ્સમાં પ્રદૂષણનું પૃથક્કરણ
B. લોહીમાં ઓક્સિજનનું સ્તર માપવા
C. વન ઇકોસિસ્ટમ્સમાં ઓક્સિજન સ્તરની ગણતરી
D. ઉંચાઇવાળા પ્રદેશોમાં ઓક્સિજનના સ્તરોનું મૂલ્યાંકન

193. ફૂગના ઉપયોગ દ્વારા કરવામાં આવતા ડેકોન્ટામીનેશન ને ____________ તરીકે ઓળખાય છે.

A. રાઇઝોફિલ્ટરેશન
B. માઈકોરેમેડિએશન
C. માઇકોફિલ્ટરેશન
D. ફાયટોરેમેડિએશન

189. ____________ term used to describe air pollution that is a result of the interaction of sunlight with certain chemicals in the atmosphere

A. Photochemical fog
B. Photo-reactive fog
C. Photochemical smog
D. Photo-reactive smog

190. ____________ results in depletion of stratospheric ozone layer

A. Asbestos dust
B. Radioactive pollutants
C. Chlorofluorocarbons
D. Suspended Particulate Matter

191. National Ambient Air Quality Standards (NAAQS) were notified in the year 1982, and revised in November 2009 for

A. 10 Pollutants B. 12 Pollutants
C. 14 Pollutants D. 16 Pollutants

192. BOD and COD are standard criterion for

A. Pollution assay in aquatic ecosystems
B. Measuring oxygen levels in blood
C. Computing oxygen levels in forest ecosystems
D. Assessing oxygen levels in high altitude regions

193. When fungi are used to decontaminate the area, it is known as

A. Rhizofiltration
B. Mycoremediation
C. Mycofilteration
D. Phytoremediation

194. થર્મલ પ્રદૂષણના મુખ્ય સ્રોત કયા છે?
A. થર્મલ પ્લાન્ટ B. પરમાણુ પ્લાન્ટ
C. ખાદ્ય ઉદ્યોગો D. બંને A અને B

195. બધા કુદરતી અને માનવસર્જિત પ્રદૂષકોનો અંતે ક્યાં ઠલવાય છે?
A. જમીનમાં B. મહાસાગરોમાં
C. નદીઓમાં D. જંગલોમાં

196. સોલિડ કચરાને શેના આધારે વર્ગીકૃત કરવામાં આવે છે?
A. મ્યુનિસિપલ અને જોખમી કચરો
B. જોખમી અને બાયોમેડિકલ કચરો
C. મ્યુનિસિપલ અને બાયોમેડિકલ કચરો
D. મ્યુનિસિપલ, જોખમી અને બાયોમેડિકલ કચરો

197. જૈવિક પ્રદૂષકો જેવાકે તેલ, ફેનોલ અને નેપ્થા નીચેનામાંથી કયા ઉદ્યોગ દ્વારા પેદા થાય છે?
A. સાબુ અને સફાઈકારક
B. ફાર્માસ્યુટિકલ
C. ફૂડ પ્રોસેસિંગ
D. આયર્ન અને સ્ટીલ

198. પર્યાવરણીય પ્રદૂષકોને ઓછા ઝેરી સ્વરૂપોમાં રૂપાંતર કરવા માટે સુક્ષ્મસજીવોનો ઉપયોગની પ્રક્રિયા ને શું કહેવામાં આવે છે?
A. ફાયટોરેમેડીએશન
B. માઇક્રોરેમેડીએશન
C. બાયોરેમિડિએશન
D. બાયોડિગ્રેડેશન

199. વડા પ્રધાન દ્વારા એપ્રિલ, 2015 માં હવામાન ગુણવત્તાની માહિતીના પ્રસાર માટે 14 શહેરોથી _______________ શરૂ કરવામાં આવી હતી.
A. રાષ્ટ્રીય હવા ગુણવત્તા સૂચકાંક
B. રાષ્ટ્રીય જળ ગુણવત્તા સૂચકાંક
C. રાષ્ટ્રીય માટી ગુણવત્તા સૂચકાંક
D. રાષ્ટ્રીય પર્યાવરણ ગુણવત્તા સૂચકાંક

194. The main sources of thermal pollution are
A. Thermal Plants B. Nuclear Plants
C. Food industries D. Both A and B

195. The ultimate sink of all natural and manmade pollutants is
A. Soil B. Oceans
C. Rivers D. Forests

196. Solid wastes are classified depending on their source into
A. Municipal and hazardous waste
B. Hazardous and biomedical waste
C. Municipal and biomedical waste
D. Municipal, hazardous and biomedical waste

197. Organic pollutants viz. oil, phenol and neptha are produced by which of the following industry?
A. Soap and detergent
B. Pharmaceutical
C. Food processing
D. Iron and Steel

198. _______________ is the use of microorganisms to degrade the environmental contaminants into less toxic forms
A. Phytoremediation
B. Microremediation
C. Bioremediation
D. Biodegradation

199. _______________ was launched by the Prime Minister in April, 2015 starting with 14 cities to disseminate air quality information
A. National Air Quality Index
B. National Water Quality Index
C. National Soil Quality Index
D. National Environment Quality Index

200. વૈશ્વિક પ્રદૂષણમાં નીચેનામાંથી સૌથી વધુ યોગદાન આપનાર ________ છે.

A. જમીન પ્રદૂષણ

B. વાયુ પ્રદૂષણ

C. ઔદ્યોગિક પ્રદૂષણ

D. જળ પ્રદૂષણ

200. Which of the following are the largest contributors to the global pollution?

A. Soil pollution

B. Air pollution

C. Industrial pollution

D. Water pollution

Answers | જવાબો

Que.	Ans.	Que.	Ans.	Que.	Ans.	Que.	Ans.	Que.	Ans.	Que.	Ans.	Que.	Ans.
1.	D	**31.**	B	**61.**	B	**91.**	A	**121.**	D	**151.**	A	**181.**	B
2.	C	**32.**	B	**62.**	A	**92.**	B	**122.**	C	**152.**	C	**182.**	A
3.	B	**33.**	D	**63.**	C	**93.**	C	**123.**	C	**153.**	A	**183.**	B
4.	B	**34.**	D	**64.**	B	**94.**	D	**124.**	B	**154.**	D	**184.**	C
5.	C	**35.**	A	**65.**	D	**95.**	C	**125.**	A	**155.**	B	**185.**	D
6.	D	**36.**	C	**66.**	A	**96.**	A	**126.**	A	**156.**	A	**186.**	A
7.	A	**37.**	C	**67.**	B	**97.**	C	**127.**	D	**157.**	C	**187.**	B
8.	C	**38.**	B	**68.**	C	**98.**	B	**128.**	D	**158.**	C	**188.**	D
9.	C	**39.**	D	**69.**	D	**99.**	D	**129.**	C	**159.**	B	**189.**	C
10.	A	**40.**	D	**70.**	B	**100.**	A	**130.**	A	**160.**	C	**190.**	C
11.	D	**41.**	C	**71.**	A	**101.**	C	**131.**	D	**161.**	B	**191.**	B
12.	A	**42.**	B	**72.**	C	**102.**	B	**132.**	A	**162.**	A	**192.**	A
13.	B	**43.**	C	**73.**	D	**103.**	A	**133.**	C	**163.**	D	**193.**	B
14.	B	**44.**	B	**74.**	D	**104.**	D	**134.**	B	**164.**	C	**194.**	D
15.	B	**45.**	C	**75.**	A	**105.**	C	**135.**	C	**165.**	A	**195.**	B
16.	A	**46.**	B	**76.**	D	**106.**	A	**136.**	B	**166.**	C	**196.**	D
17.	D	**47.**	B	**77.**	A	**107.**	C	**137.**	A	**167.**	B	**197.**	D
18.	B	**48.**	D	**78.**	D	**108.**	D	**138.**	A	**168.**	A	**198.**	C
19.	C	**49.**	D	**79.**	B	**109.**	A	**139.**	B	**169.**	B	**199.**	A
20.	A	**50.**	B	**80.**	C	**110.**	C	**140.**	C	**170.**	B	**200.**	C
21.	B	**51.**	C	**81.**	D	**111.**	C	**141.**	B	**171.**	A		
22.	D	**52.**	C	**82.**	A	**112.**	B	**142.**	B	**172.**	D		
23.	C	**53.**	B	**83.**	D	**113.**	D	**143.**	C	**173.**	D		
24.	C	**54.**	A	**84.**	C	**114.**	D	**144.**	C	**174.**	A		
25.	D	**55.**	A	**85.**	A	**115.**	B	**145.**	B	**175.**	D		
26.	B	**56.**	B	**86.**	C	**116.**	C	**146.**	C	**176.**	B		
27.	A	**57.**	D	**87.**	A	**117.**	A	**147.**	D	**177.**	C		
28.	C	**58.**	A	**88.**	D	**118.**	D	**148.**	B	**178.**	B		
29.	D	**59.**	B	**89.**	B	**119.**	A	**149.**	C	**179.**	C		
30.	D	**60.**	D	**90.**	D	**120.**	C	**150.**	D	**180.**	A		

Chapter 10

પ્રકરણ ૧૦

Silviculture

વનવૃક્ષ વિજ્ઞાન

Classification of Trees

Class symbol	Type of tree	Description	Sub class			
			A	B	C	D
I	Dominant trees	Trees of upper most canopy	Trees with normal crown development and good stem form	Trees with defective stem or crown	Tees with very defective stem or crown	Trees with very thin bole and very constricted crown
II	Dominated trees	Trees with a height of 3/4th of the tallest trees	Trees with normal crown development with good stem form	Trees with defective crown or stem with good stem form	—	—
III	Suppressed trees	Trees with 1/2 to 5/8 of the height of the tallest trees	—	—	—	—
IV	Dead and moribund tree	Bent over or badly leaning trees	—	—	—	—
V	Diseased trees	Trees that are infected with parasites	Dominant trees	Dominated and suppressed trees	—	—

વૃક્ષોનું વર્ગીકરણ

વર્ગ પ્રતીક	વૃક્ષનો પ્રકાર	વર્ણન	પેટા વર્ગ			
			A	B	C	D
I	ડોમિનન્ટ વૃક્ષો	ટોચની કેનોપીના વૃક્ષો	સામાન્ય ક્રાઉન ડેવલોપમેન્ટ અને સારા સ્ટેમ ફોર્મવાળા વૃક્ષો	ખામીયુક્ત સ્ટેમ અથવા ક્રાઉનવાળા વૃક્ષો	ખૂબ ખામીયુક્ત સ્ટેમ અથવા ક્રાઉનવાળા વૃક્ષો	ખૂબ પાતળા થડ અને ખૂબ જ સંકુચિત ક્રાઉનવાળા વૃક્ષો
II	ડોમિનેટેડ વૃક્ષો	સૌથી ઊંચા વૃક્ષ 3/4 ની ઊંચાઇવાળા વૃક્ષો	સામાન્ય ક્રાઉન ડેવલોપમેન્ટ અને સારા સ્ટેમ ફોર્મવાળા વૃક્ષો	ખામીયુક્ત સ્ટેમ અથવા ક્રાઉનવાળા સારા સ્ટેમ ફોર્મવાળા વૃક્ષો	—	—
III	સપ્રેસ્ડ વૃક્ષો	સૌથી ઊંચા વૃક્ષ 1/2 થી 2/5 ની ઊંચાઇવાળા વૃક્ષો	—	—	—	—
IV	મૃત અને મોરીબુંડ વૃક્ષ	ખરાબ રીતે વળેલા વૃક્ષો		—	—	—
V	રોગોવાળા વૃક્ષો	પરોપજીવી દ્વારા ચેપ લાગેલ વૃક્ષો	પ્રભાવશાળી વૃક્ષો	ડોમિનેટેડ અને સપ્રેસ્ડ વૃક્ષો	—	

Thinning Grades | થિનિંગ ગ્રેડસ

Grades	Thinning	Tree class	Remarks
A	Light thinning	V, IV and III	Removal of dead, dying, diseased and suppressed trees mostly
B	Moderate thinning	V, IV, III, IIb and Ia	Removal of dead, dying, diseased, suppressed and some defective trees in dominated class
C	Heavy thinning	V, IV. III, II, Ib, Ic and Id	Defective dominated as well as co-dominant trees
D	Very heavy thinning	V, IV, III, II, Id, Ic, Ib, and Ia	Some good dominants trees also be removed to create a larger gap
E	Very heavy thinning	—	This is carried mainly for research purposes

Rotation Types | રોટેશનના પ્રકારો

Type	Description	Examples
Physical rotation	Rotation which coincides with the natural lease of life of a species on a given site.	Amenity forest, park lands and road side avenues
Silvicultural rotation	Rotation through which a species is retained until satisfactory vigour of growth and reproduction on a given site.	Aesthetic and recreational forestry
Rotation of maximum volume production	Rotation that yields the maximum annual growth of material.	The age at which Mean Annual Increment (MAI) culminates
Rotation of highest income	Rotation which yields the highest average annual gross or net revenue irrespective of the value of the forest. Calculated without interest and irrespective of time.	—
Technical rotation	Rotation in which a species yields maximum material of a desired size for economic use. Applicable to industrial forestry.	Railway sleepers, pulp wood
Economic/ Financial rotation	Rotation which yields the highest net return on the invested capital. It gives the highest discounted benefit.	—

Root Types | મૂળના પ્રકારો

Type	Descreption	Example
Adventitious roots	Roots arise from the stem, branches, leaves, or old woody roots	Willow (*Salix*)
Aerating roots	Roots raised above the ground	*Avicennia, Sonneratia*
Aerial roots	Roots found entirely above the ground	*Ficus glomerata, Epiphytic orchids*
Haustorial roots	Roots of parasitic plants	Mistletoe (*Viscum album*) and Dodder
Stilt roots	Roots that grow down from lateral branches	Mangrove species
Storage roots	Roots for storage of food or water	Carrots and Beets

Classification of Silvicultural Systems | સિલ્વીકલ્ચરલ સિસ્ટમ્સનું વર્ગીકરણ

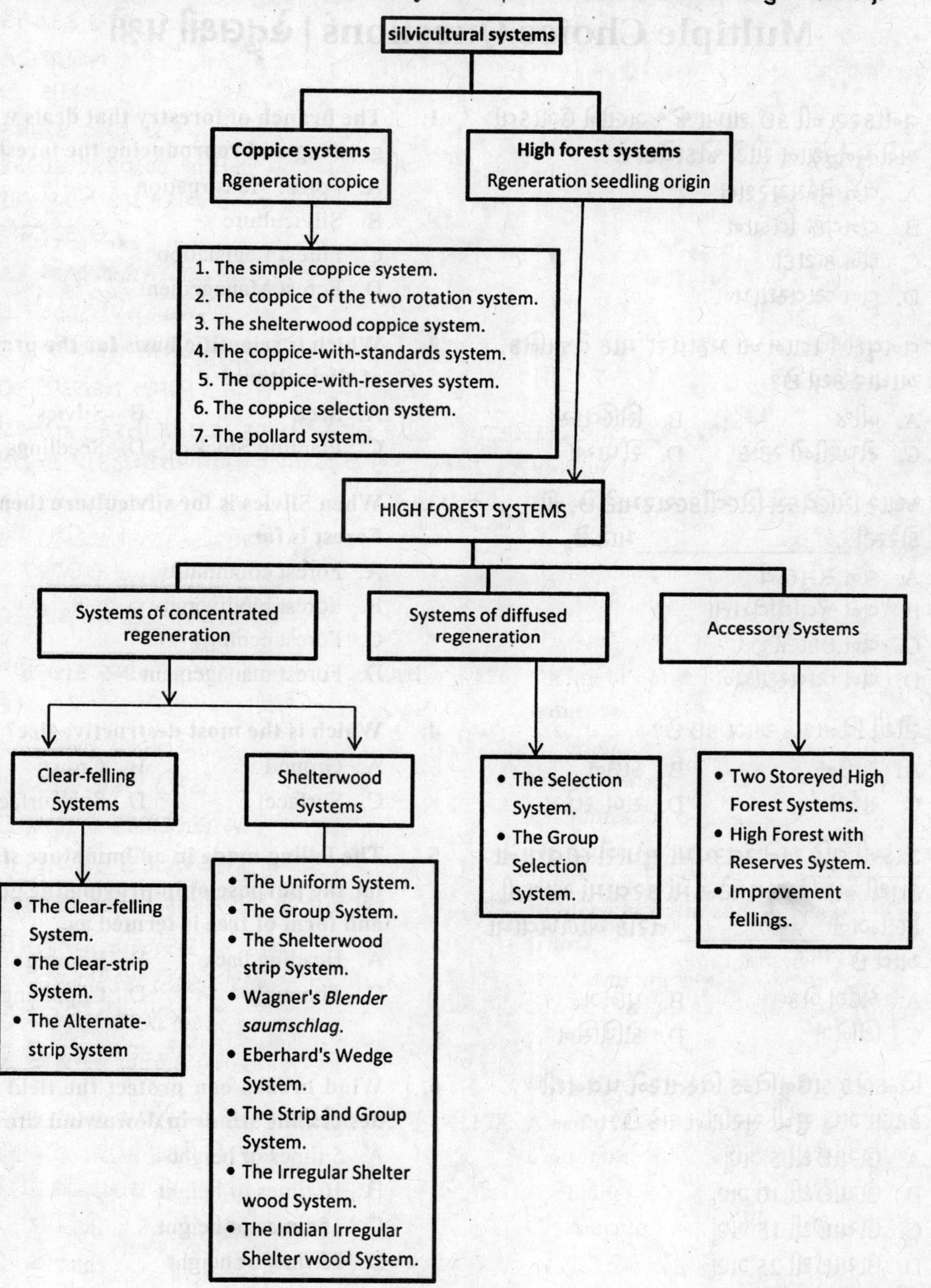

Prakash and Khanna 1983

Multiple Choice Questions | હેતુલક્ષી પ્રશ્નો

1. **વનીકરણ ની કઈ શાખા જે જંગલોને ઉગાડવા અને પુર્નરૂથાન સાથે જોડાયેલ છે?**
   A. વન મેન્સયુરેશન
   B. વનવૃક્ષ વિજ્ઞાન
   C. વન કાયદો
   D. વન વ્યવસ્થાપન

2. **વનવૃક્ષ વિજ્ઞાન ના મહાવરા માટે વૈજ્ઞાનિક આધાર કયો છે?**
   A. બીજ B. સિલ્વિક્સ
   C. રોપણીનો સ્ટોક D. રોપાઓ

3. **જ્યારે સિલ્વિક્સ સિલ્વીકલ્ચરમાટે છે, તો ફોરેસ્ટ્રી ________________ માટે છે.**
   A. વન સમુદાય
   B. વન જૈવવિવિધતા
   C. વન ઇકોલોજી
   D. વન વ્યવસ્થાપન

4. **સૌથી વિનાશક આગ કઇ છે?**
   A. ગ્રાઉન્ડ B. ક્રાઉન
   C. સર્ફેસ D. સબ સર્ફેસ

5. **ઝાડની વૃદ્ધિ અને સ્વરૂપમાં સુધારો લાવવાના હેતુથી અપરિપક્વ સ્ટેન્ડમાં કરવામાં આવતી ફેલીન્ગને ____________ તરીકે ઓળખવામાં આવે છે**
   A. હેડિંગ બેક B. પ્રુનિંગ
   C. થિનિંગ D. કોપિસિન

6. **વિન્ડબ્રેક ડાઉનવિન્ડ વિસ્તારને પવનથી કેટલા ગણું સુધી સુરક્ષિત શકે છે?**
   A. ઊંચાઈથી 5 ગણું
   B. ઊંચાઈથી 10 ગણું
   C. ઊંચાઈથી 15 ગણું
   D. ઊંચાઈથી 25 ગણું

1. **The branch of forestry that deals with growing and reproducing the forests**
   A. Forest Menstruation
   B. Silviculture
   C. Forest Legislation
   D. Forest Management

2. **Which is scientific basis for the practice of silviculture?**
   A. Seed B. Silvics
   C. Planting Stock D. Seedlings

3. **When Silvics is for silviculture then Forest is for**
   A. Forest community
   B. Forest biodiversity
   C. Forest ecology
   D. Forest management

4. **Which is the most destructive fire?**
   A. Ground B. Crown
   C. Surface D. Sub surface

5. **The felling made in an immature stand for the purpose of improving the growth and form of tree is termed as**
   A. Heading back B. Pruning
   C. Thinning D. Coppicing

6. **Wind breaks can protect the field from desiccating winds in downwind site up to**
   A. 5 times of height
   B. 10 times of height
   C. 15 times of height
   D. 25 time of height

7. શિફ્ટિંગ કલ્ટીવેશનની સૌથી વધુ નુકસાનકારક અસર ________ છે
A. પશુઓને ખલેલ
B. કુટુંબની નબળી આવક
C. જમીનને નુકશાન
D. નીંદણની વૃદ્ધિ

8. અગ્નિરોધક વાંસ પ્રજાતિઓ કઈ છે?
A. *Dendrocalamus hamiltonii*
B. *Melocanna bamboosoides*
C. બંને A અને B
D. ઉપર્યુક્તમાંથી એક પણ નહિ

9. ઝાડમાંથી ઘાસચારો સામાન્ય રીતે ________ માનવામાં આવે છે.
A. વરસાદી ઘાસચારો
B. શ્રેષ્ઠ ચારો
C. ઉનાળું પાવડર
D. અછતનો ઘાસચારો

10. DBH એ સ્થાયી ઝાડના વ્યાસને માપવાની એક માનક પદ્ધતિ છે જે સામાન્ય રીતે કેટલી ઊંચાઈએ લેવામાં આવે છે?
A. 2 મી B. 2.37 મી
C. 1 મી D. 1.37 મી

11. જંગલની આગને રોકવા માટેનો શ્રેષ્ઠ માર્ગ કયો છે?
A. આગ પર ફટકા મારવા
B. સપ્રેસન્ટ ફીણનો ઉપયોગ કરવો
C. ફાયર લાઇન બનાવવી
D. આગને બુઝાવવા માટે પાણીનો ઉપયોગ કરવો

12. ________ વન વિભાગની એક મહત્વપૂર્ણ શાખા છે
A. પ્રતિબંધિત વનીકરણ
B. પરંપરાગત વનીકરણ
C. વન વ્યવસ્થાપન
D. બીજ વ્યવસ્થાપન

7. The most deleterious effect of shifting cultivation is on
A. Animal disturbance
B. Poor family income
C. Soil loss
D. Weed growth

8. The fire resistant bamboo species
A. *Dendrocalamus hamiltonii*
B. *Melocanna bamboosoides*
C. Both
D. None of these

9. Fodder from tree is usually considered as
A. Rainy fodder
B. Best fodder
C. Summer powder
D. Scarcity fodder

10. Diameter at breast height- DBH is a standard method of measuring the diameter of the trunk a standing tree, it is generally taken at
A. 2 m B. 2.37 m
C. 1 m D. 1.37 m

11. Which is the best way to prevent the forest fire?
A. Beat the fire
B. Use suppressant foam
C. Make a fire line
D. Use water to suppress the fire

12. One of the important branch of forestry is
A. Restricted Forestry
B. Traditional Forestry
C. Forest Management
D. Seed Management

13. છાલ સહિતના ઝાડના ક્રોસ સેક્શનનું ક્ષેત્રફળ સામાન્ય રીતે DBH પર કયા ભાગનું હોય છે?
A. બેસલ એરિયા B. બ્રેસ્ટ એરિયા
C. બટ્રેસ D. કોપાઇસ

14. નીચેનામાંથી કયી પ્રજાતિને સાલના સૂચક તરીકે ગણવામાં આવે છે ?
A. *Clerodendron viscosum*
B. *Flamengia* sp
C. *Mogharia chappar*
D. ઉપર્યુક્ત તમામ

15. નીચેનામાંથી કઈ પ્રજાતિ પાણીના ભરાવા ની સ્થિતિને સહન કરી શકે છે?
A. *Terminalia arjuna*
B. *Dalbergia sissoo*
C. *Tectona grandis*
D. *Shorea robusta*

16. નીચેની કઇ પ્રજાતિ નું ખારા-આલ્કલાઇન જમીનમાં વાવેતર કરી શકાય છે?
A. બાંવળ B. બોમ્બેક્સ
C. સાગ D. સાલ

17. ઉષ્ણકટિબંધીય ભેજવાળા પાનખર જંગલમાં શેડ ડિમાન્ડર પ્રજાતિ કઈ છે?
A. સાલ B. *Depterocarpus* sp
C. *Pinus roxburghii* D. જાંબુ

18. એરિયલ સીડિંગ ______માં કરવામાં આવે છે.
A. મેદાનો
B. દરિયાકાંઠે
C. દુર્ગમ પર્વતીય વિસ્તારો
D. ઉપર્યુક્તમાંથી એક પણ નહિ

19. બાહ્ય વાતાવરણ દ્વારા પ્રેરિત નિષ્ક્રિયતા (Dormancy) માટે વપરાતો શબ્દ કયો છે?
A. કવીશેન્સ
B. રેસ્ટ
C. કરેક્ટેડ ઈન્હિબિશન
D. કોઈ ચોક્કસ શબ્દ નથી

13. What is the area of the cross section of a tree stem including the bark, generally at breast height?
A. Basal area B. Breast area
C. Buttress D. Coppice

14. Which of the following is considerd as Indicator plant of Sal?
A. *Clerodendron viscosum*
B. *Flamengia* sp
C. *Mogharia chappar*
D. All of the above

15. Which of the following species can tolerates water logging conditions?
A. *Terminalia arjuna*
B. *Dalbergia sissoo*
C. *Tectona grandis*
D. *Shorea robusta*

16. Which of the following species can be planted in Saline-alkaline soils?
A. Babul B. Bombax
C. Teak D. Sal

17. Shade demander species in tropical moist deciduous forest:
A. Sal B. *Depterocarpus* sp
C. *Pinus roxburghii* D. Jamun

18. Aerial seeding is done in
A. Plains
B. Along the coasts
C. Inaccessible hilly areas
D. None of the above

19. The term used as dormancy induced by external environment is:
A. Quiescence
B. Rest
C. Correlated inhibition
D. No specific word

20. થિનિંગ શેના માટે કરવામાં આવે છે?
A. વૃક્ષો વચ્ચેની સ્પર્ધામાં ઘટાડો કરવા
B. સ્ટેન્ડની સ્વચ્છતા માટે
C. નવી ડાળીઓ ના વિકાસ માટે
D. રસ્તા બનાવવા માટે

21. વિસ્તારો જ્યાં સમાન પ્રકારનો વરસાદ થાય છે એને શું કહેવામાં આવે છે?
A. કંટુર
B. ઇસોફાઇટસ
C. ઇસોક્યુઅન્ટસ
D. આઇસોબાર્સ

22. જે રીતે ફોરેસ્ટ્રી સિલ્વીકલ્ચર માટે છે તેમ એગ્રોનોમી શેના માટે છે?
A. ઓલેરીકલ્ચર
B. એન્ટોમોલોજી
C. કૃષિ
D. બાગાયત

23. એક વૃક્ષ કે જેમાં એક કરતા વધુ મુખ્ય દાંડી હોય છે તેને શું કહેવામાં આવે છે?
A. ફોર્કડ
B. બટ્રેસ્ડ બ્રાન્ચડ
C. એપીકલી બ્રાન્ચડ
D. ફ્લુટેડ

24. ન્યુમેટોફોર્સ સામાન્ય રીતે કઈ પ્રજાતિ માં જોવા મળે છે?
A. સુકા પાનખર જંગલ
B. મેંગ્રોવ વન
C. ખુલ્લા વન
D. સબ-ઉષ્ણકટિબંધીય વન

25. જે તે પ્રદેશમાં વનસ્પતિને પ્રભાવિત કરતા મુખ્ય સ્થાનિક પરિબળો કયા છે?
A. ક્લાઇમેટ અને ટોપોગ્રાફિક
B. ટોપોગ્રાફી અને એડિફિક
C. એડિફિક અને બાયોટિક
D. ઉપર્યુક્ત તમામ

20. Thinning are done for
A. Decreasing the competition between the trees
B. Sanitation of the stand
C. For inducing regeneration
D. Making roads

21. Places which receive equal rainfall are called
A. Contours
B. Isohyets
C. Isoquants
D. Isobars

22. Forestry is to Silviculture as agronomy is to
A. Olericulture
B. Entomology
C. Agriculture
D. Horticulture

23. A tree which has more than one leader stem is called
A. Forked
B. Buttressed
C. Epicormically branched
D. Fluted

24. Pneumatophores are commonly seen in
A. Dry deciduous forest
B. Mangrove forest
C. Open forest
D. Sub-tropical forest

25. Major locality factors influencing vegetation in region are
A. Climatic & topographic
B. Topography & edaphic
C. Edaphic & biotic
D. All of the above

26. ચેમ્પિયન અને શેઠ (1968) એ ભારતીય જંગલોને કેટલા જૂથોમાં વર્ગીકૃત કર્યા હતા?
A. 6 B. 16
C. 60 D. 66

27. એન્ડેમિક છોડની પ્રજાતિઓ એટલે શું
A. છૂટાછવાયા વિતરણ
B. તમામ સ્થળોએ વિતરણ
C. વિતરણ ચોક્કસ ક્ષેત્ર સુધી મર્યાદિત છે
D. ઉપર્યુક્તમાંથી એક પણ નહિ

28. સાલ કુદરતી રીતે કયા વિસ્તારમાં મળી આવે છે?
A. લિટોરલ અને સ્વેમ્પ જંગલો
B. ઉષ્ણકટિબંધીય ભેજવાળા પાનખર જંગલો
C. અલ્પાઇન જંગલો
D. સમશીતોષ્ણ જંગલો

29. ઉષ્ણકટિબંધીય શુષ્ક સદાબહાર જંગલો કયા વિસ્તારમાં જોવા મળે છે?
A. હિમાલય B. ઉત્તરપ્રદેશ
C. તમિલનાડુ D. પશ્ચિમ બંગાળ

30. વન પાકોના ફાયદા માટે બીજ અને પરિપક્વતાના તબ્બકા વચ્ચે જે કર્યો કરવામાં આવે છે તેને ________ તરીકે ઓળખવામાં આવે છે.
A. થિનિંગ B. પ્રુનિંગ
C. ટેન્ડિંગ D. પોલારર્ડિંગ

31. સિલ્વીકલ્ચરલ સિસ્ટમ્સમાં, ઝાડ ની કૅનોપી ને શા માટે ખુલ્લી રાખવામાં આવે છે?
A. રિજનરેશન
B. એકસરખી ગુણવત્તાવાળી લાકડા માટે
C. ઉપજ મેળવવા માટે
D. એકસરખા વાવેતર માટે

32. સમાન મૂળમાંથી ઉદ્દભવતી ડાળીઓ ને શું કહેવાય છે?
A. ક્લમ B. સ્ટેન્ડ
C. કલમ્પ D. બોલ

26. Champion & Seth (1968) classified the Indian forests into
A. 6 groups B. 16 groups
C. 60 groups D. 66 six groups

27. Endemic means plant species which has:
A. Scattered distribution
B. Distribution all over places
C. Distribution is limited to a specific area
D. None

28. Sal is naturally found in
A. Littoral & swamp forests
B. Tropical moist deciduous forests
C. Alpine forests
D. Temperate forests

29. Tropical dry evergreen forests are found in:
A. Himalayas B. Uttarpradesh
C. Tamilnadu D. West Bengal

30. An operation carried out for the benefit of forest crop at any stage of its life cycle between seedling and mature stage are
A. Thinning B. Pruning
C. Tending D. Pollarding

31. In Silvicultural systems, canopy is opened to get
A. Regeneration
B. Uniform quality timber
C. To get yield
D. Uniform plantation

32. The aggregate of stems originating from the same root is
A. Clum B. Stand
C. Culmp D. Bole

33. ચંદન એના વિકાસ માટે પોષક તત્વો પોતાના પરોપજીવી __________ની મદદથી મેળવે છે.

A. ન્યુમેટોફોર્સ B. સ્ટિલ્ટ મૂળ
C. હાઉસ્ટરીઆ D. લેગાયેમોગ્લોબિન

34. *વાયોલા કેન્સિસન્સ* શેની સૂચક પ્રજાતિ છે?

A. સાગ અને સેન્ડલ B. સાગ અને કૈલ
C. સાલ અને લીમડો D. દેવદાર અને કૈલ

35. સેલ્ફ થિનિંગ નો સિદ્ધાંત કોણે આપ્યો હતો?

A. યોદા B. ક્લેમેન્ટ
C. ડાર્વિન D. ડોસન

36. કોન્સેપ્ટ ઓફ ટોલરન્સ કોના દ્વારા આપવામાં આપ્યું?

A. યોદા B. ક્લેમેન્ટ
C. ડાર્વિન D. શેલ્ફોર્ડ

37. સાલનું શ્રેષ્ઠ કુદરતી પુનર્જીવન શેના દ્વારા પ્રાપ્ત કરી શકાય છે?

A. ક્લીઅર ફેલિન્ગ B. કોપીસ
C. લાકડું D. ઉપર્યુક્ત તમામ

38. સાલમાં નબળું રીજનરેશન શેને કારણે હોય છે?

A. બીજ નિષ્ક્રિયતા
B. બીજ સૂકવણી
C. ઓર્થોડોક્સ બીજ
D. બીજ સંકુરણ

39. ઝાડના આકાર પર પ્રકાશની શું અસર ને શું કહે છે?

A. ફોટોટ્રોપિઝમ B. ફોટો ડેવલપમેન્ટ
C. બંને A અને B D. ફોટોપેરિઓડિઝમ

40. લાઇઝિમીટર શું માપવા માટે વપરાય છે?

A. ઇવાપોરેશન
B. ઇવાપોટ્રાન્સ્પિરેશ
C. ટ્રાન્સ્પિરેશન
D. ઉપર્યુક્તમાંથી એક પણ નહિ

33. Sandal derives its nutrient requirement through its parasitic nature with the help of

A. Pneumatophores B. Stilt roots
C. Haustoria D. Leghaemoglobin

34. *Viola canescens* is an indicator species for:

A. Teak & Sandal B. Teak & Kail
C. Sal & Neem D. Deodar & Kail

35. The law of self-thinning was given by?

A. Yoda B. Clement
C. Darwin D. Dowson

36. Concept of tolerance was given by?

A. Yoda B. Clement
C. Darwin D. Shelford

37. Natural regeneration of Sal is best achieved under

A. Clear felling B. Coppice
C. Shelter wood D. All of above

38. Poor natural regeneration in Sal can be attributed to

A. Seed dormancy
B. Drying back of seeding
C. Orthodox nature of seed
D. Seed germination

39. Effect of light on the form of trees is

A. Phototropism
B. Photo development
C. Both
D. Photoperiodism

40. Lysimeter is used to measure

A. Evaporation
B. Evapotranspiration
C. Transpiration
D. None of the above

41. "ફોરેસ્ટ" શબ્દ માં "ફોરિસ" શબ્દ ક્યાંથી આવ્યો છે?
A. લેટિન B. ગ્રીક
C. સંસ્કૃત D. રશિયન

42. રોડોડેન્ડ્રોન પ્રજાતિ કયા વિસ્તારમાં મળી આવે છે?
A. સમશીતોષ્ણ વન
B. ઉષ્ણકટિબંધીય વન
C. ઉષ્ણકટિબંધીય વન
D. ઉપર્યુક્તમાંથી એક પણ નહિ

43. વન પાકોની વૈજ્ઞાનિક રીતે ખેતી કરવાની પ્રક્રિયા ને શું કહેવાય છે?
A. ફોરેસ્ટ્રી B. વાવેતર વનીકરણ
C. સિલ્વીકલ્ચર D. ઉપર્યુક્ત તમામ

44. ઝાડના થડ પર ફૂલ અને ફળ આવવાને વૈજ્ઞાનિક રીતે શું કહેવામાં આવે છે
A. પાર્થેનોકાર્પી
B. વિવિપરી
C. કોલિફ્લોરી
D. ઉપર્યુક્તમાંથી એક પણ નહિ

45. દુષ્કાળ સંવેદનશીલ પ્રજાતિ કઈ છે?
A. *Madhuca indica*
B. *Toona ciliate*
C. *Pterocarpus marsupium*
D. ઉપરોક્ત તમામ

46. કોપિસીંગ મુખ્યત્વે શેના માટે વપરાય છે?
A. બળતણ લાકડું
B. ઘાસચારો
C. શોર્ટ રોટેશન
D. ઉપરોક્ત તમામ

47. કયા વિસ્તારોમાં શંકન બેડ ની જરૂર છે?
A. ભીના
B. સુકા
C. શેલો વિસ્તાર
D. ઢોળાવો પર

41. **"Forest" term is derived from the word "Foris" in**
A. Latin B. Greek
C. Sanskrit D. Russian

42. **Rhododendron is a species of**
A. Temperate forest
B. Tropical forest
C. Sub-tropical forest
D. None of the above

43. **The art and science of cultivating forest crops is known as**
A. Forestry B. Plantation forestry
C. Silviculture D. All of the above

44. **Stem of trees bear flowers and fruits habit called**
A. Parthenocarpy
B. Vivipary
C. Cauliflory
D. None of the above

45. **Draught sensitive spp is**
A. *Madhuca indica*
B. *Toona ciliate*
C. *Pterocarpus marsupium*
D. All of the above

46. **Coppicing character is mainly exploited for**
A. Fuel wood
B. Fodder
C. Short rotation
D. All of the above

47. **Which areas requires Sunken bed?**
A. Wet
B. Dry
C. Shallow sites
D. Uphills

48. ખારી જમીનની pH ________ હોવી જોઈએ.

A. 7.5 કરતા ઓછા

B. 7.5 કરતા વધારે

C. 7.0 ની આસપાસ

D. 8.5 થી ઓછી

49. કયા પ્રકારની વનસ્પતિનું થડ હંમેશા લીલું રહે છે અને ઊંચાઈ 1 મી થી વધુ હોતી નથી.

A. વૃક્ષ B. હર્બ

C. શ્રબ D. ક્લાઈમ્બર

50. જંગલને ઉગાડવા અને મલ્ટિપ્લિકેશન કરવાની કળાને શું કહેવામાં આવે છે?

A. સિલ્વીકલ્ચર

B. મેન્સ્યુરેશન

C. ઇકોલોજી

D. ફોરેસ્ટ્રી

51. ભારતમાં કેટલા પ્રકારના જંગલો જોવા મળે છે?

A. 16 B. 17

C. 18 D. 19

52. વન ઉત્પાદનોને વ્યાપક રૂપે કેટલા વિભાગોમાં વર્ગીકૃત કરી શકાય છે?

A. મેજર ટિમ્બર પ્રોડક્ટ્સ

B. NTFPs

C. સબ્સિડરી ફોરેસ્ટ પ્રોડક્ટ્સ

D. બંને A અને B

53. સાઈટ ઈન્ડેક્સ શબ્દ એ શેના સંબંધને સમજાવવા માટે વપરાય છે?

A. ઊંચાઈ થી ઉંમર

B. ડાયામીટર થી ઉંમર

C. ડાયામીટર થી ઊંચાઈ

D. વોલ્યુમ થી ઉંમર

54. શેડનો સામનો કરવા માટે વૃક્ષની ક્ષમતાને શું કહેવામાં આવે છે?

A. ટોલરન્સ B. ઈનટોલરન્સ

C. હાર્ડિનેસ D. ટફનેશ

48. pH of saline soil should be

A. Less than 7.5

B. Greater than 7.5

C. Around 7.0

D. Less than 8.5

49. A plant whose stem is always green and not more than 1m in height is known as

A. Tree B. Herb

C. Shrub D. Climber

50. The art of growing and reproducing a forest is called

A. Silviculture

B. Mensuration

C. Ecology

D. Forestry

51. How many numbers of major forest types found in India?

A. 16 B. 17

C. 18 D. 19

52. The forest products can be broadly classified into

A. Major Timber Products

B. NTFPs

C. Subsidiary Forest Products

D. Both A and B

53. Site index is a term used by foresters to explain the relationship of ________ in forest stands.

A. Height to Age

B. Diameter to Age

C. Diameter to Height

D. Volume to Age

54. The ability of a tree to withstand shade is called

A. Intolerance B. Tolerance

C. Hardiness D. Toughness

55. વન કે જે સ્પ્રાઉટીંગ દ્વારા ફરી ઉગે છે તેને _______ વન કહેવામાં આવે છે.

A. એપીકોર્મિક B. સ્ટોક
C. કોપાઇસ D. એન્ડેમિક

56. ક્લીઅર કાપ બાદ કેવા પ્રકારના પ્રકાંડ ઉગે છે?

A. અનઇવન એજડ
B. ઇવન એજડ
C. ઓલ એજડ
D. ઉપર્યુક્ત તમામ

57. સારી ઉપજ લાવવા માટે ઝાડની સમયાંતરે કાપણી કરવાની પ્રક્રિયા ને શું કહેવામાં આવે છે?

A. ક્લિયરકટિંગ
B. સીડ-ટ્રી કટિંગ
C. પસંદગીયુક્ત કટીંગ
D. શેલ્ટરવુડ કટીંગ

58. નીચેની કઇ પદ્ધતિઓ દ્વારા વૃક્ષના બીજ વિખેરી શકાય છે?

A. પવન
B. પક્ષીઓ અને પ્રાણીઓ
C. પાણી
D. ઉપરોક્ત તમામ

59. પ્રજાતિઓ અનુસાર વૃક્ષોની ઓળખને _______ કહેવામાં આવે છે.

A. ફોરેસ્ટ્રી B. સિલ્વિક્સ
C. ઇકોલોજી D. ડેન્ડ્રોલોજી

60. પાનખર વૃક્ષો _______ તરીકે ઓળખવામાં આવતા નથી.

A. બ્રોડલીફ B. સોફ્ટવુડ
C. હાર્ડવુડ D. હાર્ટવુડ

61. ઝાડના થડનો બીજો પર્યાય_______ છે

A. બોલ B. ક્લબ
C. સ્કીડ D. પાઇક

55. A forest which has been reproduced through sprouting is called a _______ forest.

A. Epicormic B. Stocked
C. Coppice D. Endemic

56. A Stand of trees produced after a clear cut is most likely to be

A. Uneven aged
B. Even aged
C. All aged
D. All of the above

57. The periodic removal of individual trees in order to produce better yields is referred as

A. Clearcutting
B. Seed-tree cutting
C. Selective cutting
D. Shelterwood cutting

58. Tree seeds may be dispersed by which of the following methods?

A. Wind
B. Birds and animals
C. Water
D. All of the above

59. Identification of trees according to species is called _______.

A. Forestry B. Silvics
C. Ecology D. Dendrology

60. Deciduous trees are not referred to as _______.

A. Broadleaf B. Softwood
C. Hardwood D. Heartwood

61. Another term for the trunk of a tree is_______

A. Bole B. Club
C. Skid D. Pike

62. ઝાડના પાંદડા અને શાખાઓ ઝાડના ______ ના આધારે સૂચવવામાં આવે છે.

A. ડોમિનન્ટ B. બોલ
C. ક્રાઉન D. ડાયામીટર

63. વૃક્ષની વધારે ગીચતા ________ ઘટાડે છે.

A. સ્ટેન્ડ ડેન્સિટી B. ઉંચાઈ
C. નીંદણ D. વ્યાસ

64. છોડના રોપાઓ રોપવાનો વર્ષનો શ્રેષ્ઠ સમય ______ છે.

A. પ્રારંભિક વસંત B. પ્રારંભિક ઉનાળો
C. પ્રારંભિક વરસાદ D. પ્રારંભિક શિયાળો

65. ઝાડની નીચલી શાખાઓ દૂર કરવાને ________ કહેવામાં આવે છે.

A. થિનિંગ B. કટિંગ બેક
C. પિંચિંગ D. પ્રુનિંગ

66. સ્ટેન્ડ ડેન્સિટીમાં નીચેનામાંથી શાને ધ્યાનમાં લેવામાં આવતા નથી?

A. બેઝલ વિસ્તાર
B. વૃક્ષની ઉંચાઈ
C. ક્રાઉન ધનતા
D. એકર દીઠ છોડની સંખ્યા

67. સ્ટેન્ડિંગ ટિમ્બર વેચવાનું સ્કેલ ________ કહેવાય છે.

A. લાઇવ
B. ગ્રીન
C. સ્ટમ્પેજ
D. સ્ટેન્ડિંગ

68. સ્પ્રાઉટ્સથી લગભગ સંપૂર્ણ રીતે ટ્રી સ્ટેન્ડ પુનર્જીવનને ________ કહેવામાં આવે છે.

A. સ્પ્રાઉટ વન
B. સીડલેસ વન
C. સ્ટમ્પ વન
D. કોપીસ વન

62. The leaves and branches of a tree are referred based on ______ of the tree.

A. Dominant B. Bole
C. Crown D. Diameter

63. Overcrowding reduces ______ of the tree.

A. Stand density B. Height
C. Weeding D. Diameter

64. The best time of the year to plant seedlings is ______.

A. Early spring B. Early Summer
C. Early rains D. Early winter

65. Removal of the lower branches of a tree is called ________.

A. Thinning B. Cutting back
C. Pinching D. Pruning

66. Which of the followings is not considered in stand density?

A. Basal area
B. Tree height
C. Crown density
D. Number of stems per acre

67. Selling standing timber is called ________ scale.

A. Live
B. Green
C. Stumpage
D. Standing

68. A tree stand regeneration almost totally from sprouts is called as ________.

A. Sprout forest
B. Seedless forest
C. Stump forest
D. Coppice forest

69. ગર્ડલિંગ પદ્ધતિનો ઉપયોગ શેના માટે કરવામાં આવે?

A. અનિચ્છનીય ઝાડને કાઢી નાખવા માટે
B. પરિપક્વ ઝાડો ને કાપી નાખવા માટે
C. સારી ગુણવત્તાવાળા યોગ્ય અંતરે વાવેલા વિનીર વૃક્ષો માટે
D. ગાંઠ વિનાના લાકડાં માટે

70. ઘાસ જેવા નહિ પણ હર્બિસિયસ છોડને ____ કહેવામાં આવે છે.

A. ફોર્બ B. ફોરેજ
C. વન D. વૃક્ષ

71. ફૂલનો મુખ્ય હેતુ ____________ છે.

A. પ્રજનન
B. શ્વસન
C. લાભકારક કીટકો આકર્ષવા માટે
D. ટ્રાન્સપિરેશન

72. સાઇટ ઈન્ડેક્સ નક્કી કરવા માટે કયા બે માપનની જરૂર છે?

A. ઉંમર અને વ્યાસ
B. ઉંમર અને ઉંચાઈ
C. નજીકના ઝાડથી વ્યાસ અને અંતર
D. વ્યાસ અને ઉંચાઈ

73. છોડની પ્રજાતિ કે જેમાં નર અને માદા ફૂલ જુદા વૃક્ષો પર હોય છે તેને ________ તરીકે ઓળખવામાં આવે છે.

A. મોનોસિઅસ B. ડાયોસિઅસ
C. ટ્રાઇસિઅસ D. બાયોસિઅસ

74. બીટિંગઅપ એ સીલ્વીકલ્ચર ની કઈ પ્રક્રિયા ને કહેવામાં આવે છે?

A. થિનિંગ B. રેસ્ટોકીંગ
C. પ્રુનિંગ D. પોલાર્ડીંગ

75. ભારતમાં કયા વર્ષે પ્લાન્ટેશન ફોરેસ્ટ્રીની શરૂઆત કરવામાં આવી હતી?

A. 1842 B. 1942
C. 1847 D. 1947

69. Girdling is a method used to

A. Kill unwanted trees
B. Harvest mature trees
C. Properly space veneer quality trees
D. Produce knot free Timber

70. A non-grass like herbaceous plant is called________.

A. Forb B. Forage
C. Forest D. Tree

71. The primary purpose of the flower is ____________.

A. Reproduction
B. Respiration
C. To attract beneficial insects
D. Transpiration

72. Which two measurement are needed to determine site index?

A. Age and Diameter
B. Age and Height
C. Diameter and Distance from the closest tree
D. Diameter and Height

73. The species of plant which have a male and female flower on separate trees are termed as ________.

A. Monoecious B. Dioecious
C. Triecious D. Bioecious

74. Beating up refers which silvicultural operation?

A. Thinning B. Restocking
C. Pruning D. Pollarding

75. In which year Plantation Forestry was started in India?

A. 1842 B. 1942
C. 1847 D. 1947

76. પ્રોવેન્સ શબ્દનો અર્થ ________ છે.
A. જમીનની રેસ
B. બીજ સ્ત્રોત
C. ભૌગોલિક સ્ત્રોત
D. ભૌતિક સ્ત્રોત

77. હાઈ ફોરેસ્ટ નો અર્થ શું થાય છે?
A. બીજ દ્વારા વન નવજીવન
B. કોપાઇસ દ્વારા વન નવજીવન
C. પેશી સંવર્ધન દ્વારા વન નવજીવન
D. કાપણી દ્વારા વન નવજીવન

78. શુદ્ધ વનનો અર્થ શું છે?
A. 30% ભાગ એક પ્રજાતિનો હોય
B. 50% પાક એક પ્રજાતિનો હોય
C. 70% પાક એક પ્રજાતિનો હોય
D. 80% પાક એક પ્રજાતિનો હોય

79. પ્રથમ વખત "ફોરેસ્ટ સકસેશન" શબ્દનો ઉપયોગ કોણે કર્યો?
A. ક્લેમેન્ટ B. ડોસન
C. થોરો D. ડાર્વિન

80. એક સજીવ બીજા સજીવ ની અંદર જીવન જીવે તે ________ તરીકે ઓળખાય છે.
A. પરોપજીવી
B. હોસ્ટ
C. એપિફાઇટ્સ
D. મોરીબુંદ

81. પોપ્યુલેશન ઇકોલોજીને ________ તરીકે પણ ઓળખવામાં આવે છે.
A. ઓટેકોલોજી
B. સિનેકોલોજી
C. ઇકોસિસ્ટમ ઇકોલોજી
D. સેન્સસ ઇકોલોજી

76. The term provenance means ________.
A. Land race
B. Seed source
C. Geographical source
D. Physical source

77. What do you mean by High Forests?
A. Forest regenerated through seeds
B. Forest regenerated through coppice
C. Forest regenerated through tissue culture plant material
D. Forest regenerated through cuttings

78. What do you mean by Pure Forest?
A. 30% of the crop consist of a single species
B. 50% of the crop consist of a single species
C. 70% of the crop consist of a single species
D. 80% of the crop consist of a single species

79. by whom "Forest Succession" term is used first time?
A. Clement B. Dawson
C. Thoreau D. Darwin

80. An organism on or within which another organism feeds and develops is known
A. Parasite
B. Host
C. Epiphytes
D. Moribund

81. Population ecology is also known as ________.
A. Autecology
B. Synecology
C. Ecosystem ecology
D. Census ecology

82. છોડનો પોલી ક્લાઇમેક્સ સિદ્ધાંત કોને આપ્યો હતો?
A. ક્લેમેન્ટ B. ટૅન્શલી
C. ડોસન D. થોરો

83. છોડ દ્વારા નિકાલ કરવામાં આવતું મોટાભાગનું પાણી ___________ દ્વારા થાય છે.
A. છિદ્રો B. સ્ટોમાટા
C. ક્યૂટિન D. એક્ઝીટ કોષો

84. ઝાડના પાંદડામાંથી પાણીનું બાષ્પીભવન ___________ તરીકે ઓળખાય છે.
A. ટ્રાન્સ્પિરેશન B. રેસ્પિરેશન
C. પ્રેસિપિટેશન D. ઇવેપોરેશન

85. સિંગલ એપીકલ સ્ટેમવાળા વૃક્ષને ___________ તરીકે ઓળખવામાં આવે છે.
A. મોનોએક્સિયલ
B. ડાયક્સિયલ
C. ટ્રાઇએક્સિયલ
D. ટેટ્રાક્સિયલ

86. ___________ એક એવું ક્ષેત્ર છે જ્યાં છોડને વાવેતર કરી પ્રત્યારોપણ માટે ઉછેરવામાં આવે છે.
A. સીડ ઓર્ચાર્ડ B. ફોરેસ્ટ નર્સરી
C. જિન બેન્ક D. ઓર્ચાર્ડ

87. સ્ટમ્પ વાવેતર ___________ માં સામાન્ય પ્રથા છે.
A. સાગ B. સીસું
C. બંને A અને B D. દેવદાર

88. ___________ એ એક હેક્ટર વિસ્તારના વાવણી માટે બીજની જરૂરિયાત છે.
A. સીડ રેટ B. ફર્ટિલાઇઝર રેટ
C. પ્લાન્ટ રેટ D. સ્ટમ્પ રેટ

89. એક ___________ પ્રજાતિ જે તેની કુદરતી શ્રેણીની મર્યાદાની બહાર ઉગાડવામાં આવે છે.
A. ઇન્ડિજીનીયસ B. એકઝોટિક
C. એન્ડેમિક D. ઇમ્પોર્ટન્ટ

82. Who gave the poly climax theory in plant succession.
A. Clement B. Tansley
C. Dawson D. Thoreau

83. Most of the water vapor given off by a plant escapes through structures called.
A. Pores B. Stomata
C. Cutin D. Exit cells

84. The loss of water vapor from the leaf of a tree is known as ___________.
A. Transpiration B. Respiration
C. Precipitation D. Evaporation

85. Tree with single apical meristem are termed as ___________ trees.
A. Monoaxial
B. Diaxial
C. Triaxial
D. Tetraxial

86. ___________ is an area where plants are raised for eventual planting out has ordinarily both seedlings and transplants.
A. Seed orchard B. Forest Nursery
C. Gene bank D. Orchard

87. Stump planting is common practice in.
A. Teak B. Sissoo
C. Both A and B D. Deodar

88. ___________ is the requirement of seeds for sowing a hectare of area.
A. Seed rate B. Fertilizers rate
C. Plant rate D. Stump rate

89. An ___________ species is one of which is grown outside the limits of its natural range.
A. Indigenous B. Exotic
C. Endemic D. Important

90. ભારતમાં *Acacia auriculiformis* કયા દેશ માંથી પ્રસ્થાપિત કરવામાં આવ્યું ?
A. ક્વીન્સલેન્ડ, ઓસ્ટ્રેલિયા
B. જાપાન
C. ચાઇના
D. યુરોપ

91. વન શબ્દ લેટિન શબ્દ "ફોરિસ" પરથી આવ્યો છે જેનો અર્થ ____________ છે.
A. ગામની સીમાની બહાર
B. ગામની સીમમાં
C. વિસ્તારમાં
D. વિસ્તાર બહાર

92. આરક્ષિત જંગલ એ એક ક્ષેત્ર છે જેમાં સંપૂર્ણ રીતે રક્ષિત છે, જે ભારતીય વન અધિનિયમ - 1927 ના ____________ મુજબ રચાય છે.
A. પ્રકરણ I B. પ્રકરણ II
C. પ્રકરણ III D. પ્રકરણ IV

93. સેન્ટ્રલ બોર્ડ ઓફ ફોરેસ્ટ્રી (CBF) ની રચના ____________ માં કરવામાં આવી હતી.
A. 1948 B. 1950
C. 1952 D. 1954

94. એક પ્રજાતિ જે ____________ તરીકે ઓળખાતી છાયા હેઠળ વિકાસ કરવામાં સક્ષમ છે.
A. શેડ બેરીયર B. શેડ ડિમાન્ડર
C. શેડી D. શેડ લવર

95. નીચેનામાંથી કઈ પ્રજાતિને વિકાસ માટે શેડની જરૂર નથી?
A. અબિસ
B. કેરેયા
C. સિઝિજિયમ
D. ઉપર્યુક્તમાંથી એક પણ નહિ

96. એક દિવસ ____________ મીમી અથવા વધુ વરસાદ પડે છે જેને વરસાદી દિવસ કહેવામાં આવે છે.
A. 2.5 મીમી B. 1.5 મીમી
C. 0.5 મીમી D. 0.2 મીમી

90. *Acacia auriculiformis* is introduced in India from its native place?
A. Queensland, Australia
B. Japan
C. China
D. Europe

91. The word forest is derived from Latin word "foris" meaning ____________.
A. outside the village boundary
B. inside the village boundary
C. Inside land
D. outside land

92. A reserved forest is an area with complete protection, constituted according to ____________ of Indian forest Act-1927.
A. Chapter I B. Chapter II
C. Chapter III D. Chapter IV

93. The Central Board of Forestry (CBF) was constituted in
A. 1948 B. 1950
C. 1952 D. 1954

94. A species which are capable of persisting and developing under shade called
A. Shade bearer B. Shed demander
C. Shady D. Shade lover

95. Which of the following is not a shade demander species?
A. Abis
B. Careya
C. Syzygium
D. None of the above

96. A day in which _____ mm or more rain falls called rainy day.
A. 2.5 mm B. 1.5 mm
C. 0.5 mm D. 0.2 m

97. આબોહવા, વનસ્પતિ અને ઉત્પાદકતા સૂચકાંક CVP index ________ દ્વારા 1956 માં આપવામાં આવી હતી.
A. ટેન્સલી
B. પેટરસન
C. હસ્ટન
D. થોરો

98. ________ પ્રકારના છોડ જે સપોર્ટ માટે અન્ય છોડની ડાળીઓ પર ઉગે છે.
A. એપીફાઇટ્સ B. પેરાસાઇટ્સ
C. પોપ અપ્સ D. ન્યૂમેટોફોર

99. 1 મીટર જેટલી ઉંચાઈ ધરાવતા છોડ ને ____ કહેવામાં આવે છે.
A. રોપા B. છોડ
C. પોલ D. વૃક્ષ

100. સિલિવિકલ્ચર શબ્દ___ભાષા પરથી આવ્યો છે.
A. ગ્રીક B. લેટિન
C. અરબી D. ફ્રેન્ચ

101. __________ રોટેશન જંગલોમાં મુખ્યત્વે સૌંદર્યલક્ષી અને મનોરંજન હેતુ માટે ઉપયોગી છે.
A. ટેક્નિકલ રોટેશન
B. સિલ્વીકલ્ચર રોટેશન
C. મહત્તમ વોલ્યુમ ઉત્પાદનનું રોટેશન
D. ફાઇનાન્સિયલ રોટેશન

102. ઇમારતી લાકડું અને અન્ય વન પેદાશોના ઉત્પાદન માટે અથવા તેના પરોક્ષ લાભો માટે જાળવવામાં આવેલ વિસ્તાર ને શું કહેવામાં આવે છે?
A. વન ફાર્મ
B. વન
C. હોમ ગાર્ડન
D. સામાજિક વનીકરણ

97. CVP (climate, vegetation and productivity index) index were given by ________ in 1956.
A. Tansley
B. Paterson
C. Huston
D. Thoreau

98. ________ are plants which grow on the trunk and branches of other plants for the purpose of supports.
A. Epiphytes B. Parasites
C. Pop ups D. Pneumetoh

99. From recruit stage up to a height of 1 m of the young plant called________.
A. Seedling B. Sapling
C. Pole D. Tree

100. Silviculture came from________ word.
A. Greek B. Latin
C. Arabic D. French

101. ________ rotation is useful in forests managed primarily for aesthetic and recreational purpose.
A. Technical rotation
B. Silviculture rotation
C. Rotation of maximum volume production
D. Financial Rotation

102. What is called an area whcich is set aside for the production of timber and other forest produce or maintained under woody vegetation for certain indirect benefits which it provides?
A. Forest farm
B. Forest
C. Home gardens
D. Social forestry

103. રીજનરેશનના આધારે વનને _______ માં વર્ગીકૃત કરવામાં આવ્યું છે.

A. હાઈ ફોરેસ્ટ
B. કોપાઇસ ફોરેસ્ટ
C. ઉષ્ણકટિબંધીય વન
D. A અને B બંને

104. વન વ્યવસ્થાપનને આધારે વનને _______ માં વર્ગીકૃત કરી શકાય છે.

A. સામાજિક વન
B. ઉત્પાદન વન
C. સંરક્ષણ વન
D. ઉપરોક્ત તમામ

105. રચનાને આધારે વનને _______ માં વર્ગીકૃત કરવામાં આવ્યું છે.

A. શુદ્ધ વન B. અશુદ્ધ વન
C. મિશ્રિત વન D. A અને C બંને

106. ભારતીય વન અધિનિયમ -1927 (પ્રકરણ III) ની જોગવાઈ હેઠળ ગામ સમુદાયને સોંપેલ જંગલ કયા નામે ઓળખાય છે?

A. ગામનું વન
B. અનામત વન
C. રક્ષણ વન
D. અસામાન્ય વન

107. ભારતીય વનનો પ્રથમ ઇન્સ્પેક્ટર જનરલ (IGF) _______ કોણ હતો?

A. ડૉ ડિયેટ્રીક બ્રાન્ડિસ
B. એલ. એસ. ખન્ના
C. ડસ્ટન
D. પ્રકાશ રામ

108. વન મહોત્સવની શરૂઆત કે. એમ. મુનશીએ _______ વર્ષમાં કરી હતી.

A. 1857 B. 1950
C. 1957 D. 1997

103. forest is classified into _______ On the basis of regeneration

A. High forest
B. Coppice forest
C. Tropical forest
D. Both A and B

104. Forest can be classified into _______ on the basis of management forest

A. Social forest
B. Production forest
C. Protection forest
D. All of the above

105. Forest is classified into _______ On the basis of composition.

A. Pure forest B. Impure forest
C. Mixed forest D. Both A and C

106. A forest which is stable forest assigned to a village community under the provisions of Indian forest Act-1927 (Chapter III) is called-

A. Village forest
B. Reserved forest
C. Protection forest
D. Abnormal forest

107. The first Inspector General (IGF) of Indian forest was _______

A. Dr. Dietrich Brandis
B. L. S. Khanna
C. Duston
D. Prakash Ram

108. Van Mahotsav was started by K. M. Munshi in the year of _______

A. 1857 B. 1950
C. 1957 D. 1997

109. CVP સૂચકાંક નું પૂર્ણ સ્વરૂપ ________________ છે.

A. Climate, Vegetation and Precipitation index
B. Climate, Vegetation and Procedure index
C. Climate, Vegetation and Productivity index
D. Climate, Vegetation and Photosynthesis index

110. થેલોફાઇટ્સમાં શું શામેલ છે?

A. ફૂગ
B. બેક્ટેરિયા
C. લાઇકન
D. ઉપર્યુક્ત તમામ

111. ફૂલો, પર્ણસમૂહ અને ફળના વિકાસમાં મોસમી ફેરફારને ________ તરીકે ઓળખવામાં આવે છે.

A. મોર્ફોલોજી B. ફિનોલોજી
C. ગ્રોથ D. ડેવલપમેન્ટ

112. વન નર્સરીમાં સ્ટાન્ડર્ડ બેડનું માપ કેટલું હોય છે?

A. 11.5 m × 1.2 m
B. 12 m × 2 m
C. 12.5 m × 1.2 m
D. 15 m × 1 m

113. નીચેનામાંથી કયું કોપિસર છે ?

A. અલ્બીઝિયા *(Albizia spp.)*
B. દેવદાર *(Cedrus deodora)*
C. હલદૂ *(Adina cordifolia)*
D. ખેર *(Acacia catechu)*

114. ગાદી ક્યારા જમીનથી કેટલી ઉંચાઈ સુધી બનાવામાં આવે છે?

A. 1-5 સે.મી. B. 5 - 10 સે.મી.
C. 10 - 15 સે.મી. D. 15 - 25 સે.મી.

109. full form of CVP index is ________________

A. Climate, Vegetation and Precipitation index
B. Climate, Vegetation and Procedure index
C. Climate, Vegetation and Productivity index
D. Climate, Vegetation and Photosynthesis index

110. Thallophytes includes-

A. Fungi
B. Bacteria
C. Lichens
D. All of the above

111. The seasonal changes in the development of flowering, foliage and fruiting is termed as ________.

A. Morphology B. Phenology
C. Growth D. Development

112. what is the size of Standard bed in a forest nursery?

A. 11.5 m × 1.2 m
B. 12 m × 2 m
C. 12.5 m × 1.2 m
D. 15 m × 1 m

113. Which of the following is a strong coppicer?

A. *Albizia* spp.
B. *Cedrus deodara*
C. *Adina cordifolia*
D. *Acacia catechu*

114. Raised beds are prepared at the height of ________ cm above the ground level-

A. 1 – 5 cm B. 5 – 10 cm
C. 10 – 15 cm D. 15 – 25 cm

115. ચોરસ વાવેતરમાં છોડ ની ગણતરી કરવા માટે જેનો ઉપયોગ થઈ શકે છે તે યોગ્ય સૂત્ર પસંદ કરો.

A. N = (100 × 100) / (રોપણી અંતર (X))
B. N= (100 × 100) / (રોપણી અંતરનો વર્ગ (X))
C. N = (2 × 100 × 100) / (રોપણી અંતરનો વર્ગ (X))
D. N = (2 × 100 × 100) / (રોપણી અંતર (X))

116. ઉષ્ણકટિબંધીય વન મા સરેરાશ વાર્ષિક તાપમાન _________ હોય છે.

A. < 7 °સી B. 7 - 17 °સી
C. 17 - 24 °સી D. > 24 °સી

117. ભીના જંગલમાં સરેરાશ વાર્ષિક વરસાદ કેટલો થાય છે?

A. < 750 મીમી
B. 750 - 1250 મીમી
C. 1250 - 2500 મીમી
D. > 2500 મીમી

118. સિલ્વકલ્ચરલ મેનેજમેન્ટ અનુસાર વનને _________ માં વર્ગીકૃત કરવામાં આવ્યું છે.

A. વર્કિંગ સર્કલ B. ફેલિંગ સિરીઝ
C. કટીંગ સેક્શન D. ઉપર્યુક્ત તમામ

119. પ્યોર ફોરેસ્ટ સામાન્ય રીતે કેટલી પ્રજાતિઓથી બનેલા હોય છે?

A. એક B. બે
C. ત્રણ D. ત્રણ કે તેથી વધુ

120. _________ એ ચોક્કસ પ્રજાતિઓ માટેની સાઇટની સંબંધિત ઉત્પાદક ક્ષમતા તરીકે ઓળખાય છે .

A. સાઇટ ગુણવત્તા
B. રિજનરેશન
C. સાઇટ ઈન્ડેક્સ
D. સાઈટ મેટ્રિક્સ

115. Select appropriate formula which can be used to calculate number of plant (N) in square planting plantation

A. N= (100 × 100)/ (planting distance (X))
B. N = (100 × 100)/ (Square of planting distance (X))
C. N = (2 × 100 × 100)/ (Square of planting distance (X))
D. N = (2 × 100 × 100)/ (planting distance (X))

116. Mean annual temperature for the tropical forest is _________.

A. < 7 °C B. 7 - 17 °C
C. 17 - 24 °C D. > 24 °C

117. In the wet forest how much Mean annual rainfall occurs-

A. < 750 mm
B. 750 - 1250 mm
C. 1250 - 2500 mm
D. > 2500 mm

118. According to Silvicultural Management classification forest are classified into

A. Working Circles B. Felling Series
C. Cutting Sections D. All of the above

119. Pure forests are composed almost entirely of _________ species (not less than 50%).

A. One B. Two
C. Three D. Three or more

120. _____________ is defined as the relative productive capacity of a site for particular species.

A. Site quality
B. Regeneration
C. Site Index
D. Site matrix

121. નીચેની કઇ પ્રજાતિના વિકાસ માટે સૂર્યપ્રકાશ ની જરૂર પડે છે?
A. ડિપ્ટોરોકાર્પસ
B. કેલોફિલમ
C. સાગ
D. ઉપર્યુક્ત તમામ

121. **Which of the following species are Light demander species?**
A. Dipterocarpus
B. Calophylum
C. Teak
D. All of the above

122. નીચેનામાંથી કયા છોડોને ને શેડ બિઅરર પ્રજાતિ તરીકે ગણવામાં આવે છે ?
A. આર્ટોકાર્પસ
B. ટેરોકાર્પસ
C. સીસમ
D. ઉપર્યુક્ત તમામ

122. **Which of the following species is considered as Shade bearer species?**
A. Artocarpus
B. Pterocarpus
C. *Dalbergia latifolia*
D. All of the above

123. નીચેની કઈ પ્રજાતિ શેડ ડીમાન્ડર પ્રજાતિ છે?
A. મેસુઆ
B. શ્કેસેશેરા
C. એબીઝ
D. ઉપર્યુક્ત તમામ

123. **Which of the following species are Shade demander species?**
A. Mesua
B. Schleichera
C. Abies
D. All of the above

124. નીચેનીમાંથી કઈ ડ્રોટ હાર્ડી પ્રજાતિ નથી?
A. *Acacia nilotica,*
B. *Boswellia serrata*
C. *Hardwickia binate*
D. *Madhuca indica*

124. **Which of the following is not a Drought hardy species.**
A. *Acacia nilotica*
B. *Boswellia serrata*
C. *Hardwickia binata*
D. *Madhuca indica*

125 . નીચેનીમાંથી કઇ પ્રજાતિ ડ્રોટ સેન્સિટિવ નથી?
A. *Mangifera indica*
B. *Pterocarpus marsupium*
C. *Hardwikia binate*
D. *Madhuca indica*

125. **Which of the following is not a Drought sensitive species.**
A. *Mangifera indica*
B. *Pterocarpus marsupium*
C. *Hardwickia binata*
D. *Madhuca indica*

126. નીચેનામાંથી મોડરેટલી હાર્ડી પ્રજાતીઓ કઈ છે?
A. ખેર (*Acacia catechu*)
B. હલદ્દ (*Adina cordifolia*)
C. સીસું (*Dalbergia sissoo*)
D. ઉપર્યુક્ત તમામ

126. **Which of the following is a Moderately hardy Species.**
A. *Acacia catechu*
B. *Adina cordifolia*
C. *Dalbergia sissoo*
D. All of the above

127. નીલગિરી માં પિન્ક ડિસીઝ શેના કારણે થાય છે?

A. Corticium salmonicolor

B. *Ganoderma* spp.

C. *Mycoplasma*

D. ઉપર્યુક્તમાંથી એકપણ અહીં

128. ચંદન માં સ્પાઇક ડિસીઝ શેના કારણે થાય છે?

A. Corticium salmonicolor

B. *Ganoderma* spp.

C. *Mycoplasma*

D. ઉપર્યુક્તમાંથી એકપણ અહીં

129. સામાન્ય રીતે ગોળાકાર ક્રાઉન ______ ઝાડમાં જોવા મળે છે.

A. કેરી B. લીમડો

C. આંબલી D. ઉપર્યુક્ત તમામ

130. સામાન્ય રીતે બ્રૂમ આકારનો ક્રાઉન ______ ઝાડમાં જોવા મળે છે.

A. બબુલ B. લીમડો

C. આંબલી D. ઉપરોક્ત તમામ

131. જ્યારે સ્ટેમ અનિયમિત ઈનવોલ્યૂશન થાય તે ______ તરીકે ઓળખાય છે.

A. ફ્લુટિન્ગ B. ફ્રુટીન્ગ

C. ફોર્કિંગ D. ફ્લાવરિંગ

132. રીંગ પોરસ લાકડા કઈ પ્રજાતિમાંથી મળે છે?

A. ટેક્ટોના

B. મોરસ

C. ડાલબર્ગિયા

D. ઉપર્યુક્ત તમામ

133. ટેંડિંગ ઓપરેશનમાં ______ શામેલ છે.

A. વીડિન્ગ

B. કલીનિંગ

C. થિનિંગ

D. ઉપર્યુક્ત તમામ

127. Which of the following is a causal organism for Pink disease in Eucalyptus.

A. *Corticium salmonicolor*

B. *Ganoderma* spp.

C. Mycoplasma

D. None of the above

128. Which of the following is a causal organism for Spike disease in Sandal tree.

A. *Corticium salmonicolor*

B. *Ganoderma* spp.

C. *Mycoplasma*

D. None of the above

129. Generally Spherical Crown shape found in ______ trees.

A. Mango B. Neem

C. Emli D. All of the above

130. Generally Broom Crown shape found in ______ trees.

A. Babul B. Neem

C. Emli D. All of the above

131. When a stem shows irregular involution and swellings termed as ______.

A. Fluting B. Fruiting

C. Forking D. Flowering

132. Which of the followings is Ring porus wood species?

A. Tectona

B. Morus

C. Dalbergia

D. All of the above

133. Tending operations includes ______.

A. Weeding

B. Cleaning

C. Thinning

D. All of the above

134. ઓર્ડિનરી થિનિંગ ને કેટલા ગ્રેડમાં વર્ગીકૃત કરવામાં આવે છે?

A. 4 B. 5
C. 6 D. 7

135. નીચેનામાંથી કયો ઓર્ડિનરી થિનિંગ નો ગ્રેડ નથી.

A. લાઈટ થિનિંગ
B. મધ્યમ થિનિંગ
C. ભારે થિનિંગ
D. ઉપર્યુક્તમાંથી એકપણ અહીં

136. કોપીસર પ્રજાતિઓ કઈ છે?

A. ખેર (*Acacia catechu*)
B. સીસું (*Dalbergia sissoo*)
C. ખાખરો (*Butea monospermum*)
D. ઉપર્યુક્ત તમામ

137. નોન-કોપીસર પ્રજાતિઓ છે કઈ છે?

A. દેવદાર
B. સ્પ્રુસ
C. A અને B બંને
D. ઉપર્યુક્તમાંથી એકપણ અહીં

138. ___________ એક એવું ક્ષેત્ર છે જ્યાં છોડને વાવેતર માટે ઉભા કરવામાં આવે છે જેમાં સામાન્ય રીતે રોપાઓ અને ટ્રાન્સપ્લાન્ટ બંને હોય છે.

A. વન નર્સરી
B. બીજ ઓર્ચાર્ડ
C. SPA
D. ઉપર્યુક્તમાંથી એકપણ અહીં

139. ટ્રાન્સપ્લાન્ટ નર્સરી એક એવી નર્સરી છે જેમાં રોપાઓ રોપવા માટે ફક્ત _____ કયારા છે.

A. સીડલીંગ
B. ટ્રાન્સપપ્લાન્ટીંગ
C. A અને B બંને
D. ઉપર્યુક્તમાંથી એકપણ અહીં

134. Ordinary thinning is classified into ___________ grades.

A. Four B. Five
C. Six D. Seven

135. Which of the followings is not a grade of Ordinary thinning.

A. Light thinning
B. Moderate thinning
C. Heavy thinning
D. None of the above

136. Strong coppicers Species are

A. *Acacia catechu*
B. *Dalbergia sissoo*
C. *Butea monosperma*
D. All of the above

137. Non- coppicers Species are

A. *Cedrus deodara*
B. *Picea smithiana*
C. Both A and B
D. None of the above

138. ___________ is an area where plants are raised for eventual planting out has ordinarily both seedlings and transplants.

A. Forest Nursery
B. Seed orchard
C. SPA
D. None of the above

139. Transplant Nursery is a nursery which has only ________ beds for transplanting seedlings.

A. Seedling
B. Transplanting
C. Both A and B
D. None of the above

140. વધુ સારી રીતે મૂળ અને અંકુરની વૃદ્ધિ માટે રોપણી એક નર્સરીના કયારા માંથી બીજા કયારામાં સ્થાનાંતરિત થઈ રહી છે. આ પ્રક્રિયાને ________ તરીકે પણ ઓળખવામાં આવે છે.

A. પ્રીકીંગ આઉટ B. પ્રીકીંગ ઇન
C. પ્રીકીંગ સાઈડ D. પ્રીકીંગ

141. કઈ પ્રજાતિ ના બીજનો વાયબીલીટી પિરિયડ સામાન્ય તાપમાને 2 અઠવાડિયા સુધીનો છે?

A. *Dipterocarpus* spp. B. Shorea
C. Azadiracta D. ઉપર્યુક્ત તમામ

142. નીચેની કઇ પ્રજાતિ ના સામાન્ય સંગ્રહની સ્થિતિમાં બીજનો એક વર્ષનો વાયબિલીટી પિરિયડ છે?

A. Atrocarpus B. Toona
C. Tectona D. ઉપર્યુક્ત તમામ

143. ટેટ્રાઝોલિયમ પરીક્ષણ ________ ચકાસવા માટે કરવામાં આવે છે.

A. બીજનું વજન)
B. બીજ કદ
C. બીજક્ષમતા
D. બીજ નિષ્ક્રિયતા

144. લાઈન પ્લાન્ટિંગમાં છોડની સંખ્યા (N) ગણવામાટેનું સૂત્ર કયું છે?

A. N = 100 * 100 / X* Y
B. $N = 100 * 100 / (X)^2$
C. $N = 100 * 100 * 1.155 / (X)^2$
D. $N = 2 * 100 * 100 / (X)^2$

145. ટ્રાયેન્ગ્યુલર પ્લાન્ટિંગમાં છોડની સંખ્યા ગણવામાટેનું સૂત્ર કયું છે

A. N = 100 * 100 / X*Y
B. $N = 100 * 100 / (X)^2$
C. $N= 100 * 100 * 1.155 / (X)^2$
D. $N = 2 * 100 * 100 / (X)^2$

140. Transplanting is moving of plant from one nursery bed to another for better root and shoot growth. This process is also called as ________.

A. Pricking out B. Prickin in
C. Pricking side D. Pricking

141. Which of the following species has two-week Viability Period of seed in in normal storage conditions.

A. *Dipterocarpus* spp. B. Shorea
C. Azadirachta D. All of the above

142. Which of the following species has One year Viability Period of seed in in normal storage conditions?

A. Artocarpus B. Toona
C. Tectona D. All of the above

143. Tetrazolium test is performed to test.

A. Seed weight
B. Seed Size
C. Seed viability
D. Seed dormancy

144. What is the formula to calculate number of plants (N) for Line planting pattern.

A. N = 100*100/ X*Y
B. $N = 100*100 / (X)^2$
C. $N = 100*100*1.155/ (X)^2$
D. $N = 2*100*100/ (X)^2$

145. What is the formula to calculate number of plants (N) for Triangular Planting pattern.

A. N = 100*100/ X*Y
B. $N = 100*100 / (X)^2$
C. $N = 100*100*1.155/ (X)^2$
D. $N = 2*100*100/ (X)^2$

146. $N = 2 * 100 * 100 / (X)^2$ **ફોર્મ્યુલા ________ પ્રકારના પ્લાન્ટિંગ ની છે.**

A. લાઇન B. ચોરસ
C. ત્રિકોણાકાર D. ક્વિંકન્ક્સ

147. *Eucalyptus globulus* ________ દેશનું મૂળભૂત વૃક્ષ છે.

A. ઓસ્ટ્રેલિયા B. બર્મા
C. યુરોપ D. ચાઇના

148. ચેમ્પિયન અને શેઠ મુજબ મોન્ટેન ના સમશીતોષ્ણ જંગલોમાં તેના પેટા પ્રકારનાં જંગલો શેમાં શામેલ છે?

A. મોન્ટેન ભીના સમશીતોષ્ણ જંગલો
B. હિમાલયના ભેજવાળા જંગલો
C. હિમાલય શુષ્ક જંગલો
D. ઉપર્યુક્ત તમામ

149. *Azadiracta indica* કઈ લાક્ષણિકતાઓ ધરાવે છે?

A. ડ્રોટ હાર્ડી
B. ફ્રોસ્ટ ટેન્ડર
C. ડ્રોટ ઇન્ટોલરન્ટ
D. A અને B બંને

150. રોગો વાળા વૃક્ષો કયા વર્ગમાં આવે છે?

A. II B. III
C. IV D. V

146. Number of plants (N) = 2*100*100/ $(X)^2$ forula used for ________ Planting pattern.

A. Line B. Square
C. Triangular D. Quincunx

147. *Eucalyptus globulus* is a native of ________ country.

A. Australia B. Burma
C. Europe D. China

148. Montane temperate forests includes its sub types of forest (according to Champion and seth) as

A. Montane wet temperate forests
B. Himalayan moist forests
C. Himalayan dry forests
D. All of the above

149. *Azadirachta indica* is having a characterisics like;

A. Drought hardy
B. Frost tender
C. Drought intolerant
D. Both A and B

150. Diseased trees are classed in

A. II B. III
C. IV D. V

Answers | જવાબો

Que.	Ans.	Que.	Ans.	Que.	Ans.	Que.	Ans.	Que.	Ans.	Que.	Ans.	Que.	Ans.
1.	B	24.	B	47.	B	70.	B	93.	A	116.	D	139.	B
2.	B	25.	D	48.	D	71.	A	94.	A	117.	D	140.	A
3.	D	26.	B	49.	B	72.	B	95.	D	118.	D	141.	D
4.	B	27.	C	50.	A	73.	B	96.	A	119.	A	142.	D
5.	C	28.	B	51.	A	74.	B	97.	B	120.	A	143.	C
6.	D	29.	C	52.	D	75.	A	98.	A	121.	D	144.	A
7.	C	30.	C	53.	A	76.	C	99.	A	122.	D	145.	C
8.	C	31.	A	54.	B	77.	A	100.	B	123.	D	146.	D
9.	C	32.	A	55.	A	78.	D	101.	B	124.	D	147.	A
10.	D	33.	C	56.	B	79.	C	102.	B	125.	C	148.	D
11.	C	34.	D	57.	C	80.	B	103.	D	126.	D	149.	D
12.	C	35.	A	58.	D	81.	B	104.	D	127.	A	150.	D
13.	A	36.	D	59.	D	82.	B	105.	D	128.	C		
14.	B	37.	C	60.	B	83.	B	106.	A	129.	D		
15.	A	38.	B	61.	A	84.	A	107.	A	130.	A		
16.	A	39.	A	62.	C	85.	A	108.	B	131.	A		
17.	A	40.	B	63.	D	86.	B	109.	C	132.	D		
18.	C	41.	A	64.	A	87.	C	110.	D	133.	D		
19.	A	42.	A	65.	D	88.	A	111.	B	134.	B		
20.	A	43.	C	66.	B	89.	B	112.	C	135.	D		
21.	B	44.	C	67.	C	90.	A	113.	D	136.	D		
22.	C	45.	D	68.	D	91.	A	114.	C	137.	C		
23.	A	46.	D	69.	A	92.	B	115.	B	138.	A		

Chapter 11

પ્રકરણ ૧૧

Agroforestry, Social Forestry and Joint Forest Management

કૃષિ વાનિકી, સામાજિક વનીકરણ અને સહભાગી વન વ્યવસ્થાપન

Agroforestry Systems

Agri – silvi – horticulture	forest trees + fruit trees + crops
Agri – horticulture	fruit trees + crops
Agri – silvi – pasture	forest trees + crops + pasture or animals
Agri – silviculture	forest trees + crops
Alley cropping	hedges + crops
Aqua – forestry	forest trees + fishes
Block Plantation	block of trees + block of crops
Boundary Plantation	trees on boundary + crops
Energy Plantation	trees + crops during initial years
Forage forestry	forage trees + pasture
Homestead	combinations of forest trees, fruit trees, vegetables etc.
Horti – apiculture	fruit trees + honeybee
Live fences	shrub and trees on boundary
Shelter belts	trees for wind protection + crops
Silvi – olericulture	forests trees + vegetables
Silvi – pasture	fruit trees + pasture and / or animals
Silvi or horti – sericulture	forest trees or fruit trees + silkworm
Wind breaks	rows of trees + crops

Dhyani 2009

એગ્રોફોરેસ્ટ્રી સિસ્ટમ્સ

એગ્રી - સિલ્વી - હોર્ટિકલ્ચર	વન વૃક્ષો + ફળના ઝાડ + પાક
એગ્રી - હોર્ટિકલ્ચર	ફળના ઝાડ + પાક
એગ્રી - સિલ્વી - પાશ્ચર	વન વૃક્ષો + પાક + ચારો અથવા પ્રાણીઓ
એગ્રી - સિલ્વીકલ્ચર	વન વૃક્ષો + પાક

એલે ક્રોપીંગ	વાડ + પાક
એક્વા - ફોરેસ્ટ્રી	વન વૃક્ષો + માછલીઓ
બ્લોક વાવેતર	વૃક્ષોના બ્લોક + પાકના બ્લોક
બાઉન્ડ્રી પ્લાન્ટેશન	બાઉન્ડ્રી પર વૃક્ષો + પાક
ઊર્જા વાવેતર	પ્રારંભિક વર્ષો દરમિયાન ઝાડ + પાક
ફોરેજ - ફોરેસ્ટ્રી	ઘાસચારાના વૃક્ષો + ગોચર
હોમસ્ટેડ	વન વૃક્ષો, ફળના ઝાડ, શાકભાજી વગેરે નુ સંયોજન
હોર્ટી - એપિકલ્ચર	ફળ ઝાડ + મધમાખી
જીવંત વાડ	સીમા પર નાના છોડ અને વૃક્ષો
શેલ્ટર બેલ્ટ	પવન રક્ષણ માટેના વૃક્ષો + પાક માટે વૃક્ષો
સિલ્વી - ઓલેરીકલ્ચર	વન વૃક્ષો + શાકભાજી
સિલ્વી - પાશ્ચર	ફળના ઝાડ + ગોચર અને / અથવા પ્રાણીઓ
સિલ્વી અથવા હોર્ટી - સેરીકલ્ચર	વન વૃક્ષો અથવા ફળના ઝાડ + રેશમ ના કિડાં
વિન્ડ બ્રેક	ઝાડની હરોળ + પાક

ધ્યાની 2009

MPTs for Indian Agroforestry | ભારતીય એગ્રોફોરેસ્ટ્રી માટે બહુહેતુક વૃક્ષો

Species	Uses
Acacia nilotica	Fuelwood
Acacia tortillis	Fuelwood
Albizia lebbek	Fuelwood, timber
Artocarpus heterophyllus	Fruit, fodder, bark dye, windbreak
Azadirecta india	Timber, oil, insecticidal and medicinal, fuelwood, windbreak
Casuarina equisetifolia	Fuelwood, timber
Cocos nucifera	Nut, nutritional drink, fuel, crafts, housing thatch
Dalbergia sissoo	Timber, fuelwood
Eucalyptus camaldulensis	Fuelwood, timber
Eucalyptus tereticornis	Fuelwood, pole timber, wind break
Gliricidia sepium	Food, fuelwood, poles, fodder, nitrogen fixing, living fence
Gmelina arborea	Timber, fuelwood, furniture
Grevillea robusta	Timber, fuelwood, building material
Leucaena leucocephala	Fuelwood, pole, fodder, windbreak, soil improver
Mangifera indica	Fruit, medicine, leaves as mulch, timber, windbreak
Moringa oleifera	Fruit, medicine, fodder
Prosopis cineraria	Windbreak
Sesbania grandiflora	Fodder, fuelwood, food
Tamarindus indica	Shade, fruit, light timber, medicinal uses
Ziziphus mauritiana	Food, shade

Raj and Lal 2018

Agroforestry Models for Agro-Climatic Zones of India

ભારતના એગ્રો-ક્લાઇમેટ ઝોન માટે એગ્રોફોરેસ્ટ્રી મોડેલ્સ

Sr. No.	Agro-Climatic Zones	Agroforestry System	Tree component	Crop/grass	Description
1	Western Himalayan Region	Silvipasture	*Grewia optiva*	*Setaria* spp.	Ladakh, Kashmir, Punjab, jammu etc.
			Morus alba	*Setaria* spp.	
		Agrihorticulture	*Malus pumila*	Millets, Wheat	Brown soil & silty loam, steep slopes.
		Agrihorticulture	*Prunus persica*	Maize, soyabean	
2	Eastern Himalayan Region	Agrisilviculture	*Anthocephalus cadamba*	Paddy	Arunachal Pradesh, Sikkim and Darjeeling, Manipur etc. High -rainfall and high forest covers heavy soil erosion, Floods.
		Agrihorticulture	*Alnus nepalensis*	Large cardamom/ coffee	
		Silviculture	*Bamboos*		
		Silvipasture	*Alnus nepalensis*	*Setaria* spp.	
			Pinus roxburghii	Napier grass, Native grass	
3	Lower Gangatic Plains Regions	Agrisilviculture (irr)	*Eucalyptus*	Paddy, Wheat	West Bengal, Soils mostly alluvial & are prone to floods.
		Agrihorticulture (irr)	Mango/ Banana, Litchi	Maize	
			Acacia auriculiformis		
		Silvipasture			
4	Middle Gangatic Plains Regions	Agrisilviculture (irr)	*Populus deltoides*	sugarcane-wheat Rice-wheat	Bihar, Uttar Pradesh, High Rainfall, 39% irrigation, cropping intensity 142%
		Agrisilviculture (irr)	*Eucalyptus*	Sesamum	
		Agrisilviculture	*Dalbergia sisoo*	Rice-wheat	
		Agrihorticulture (irr)	Mango/Citrus		
5	Trans Gangatic Plains Regions	Agrihorticulture (irr)	*Emblica officinalis*	Black gram/green gram	Punjab, Haryana, Union territory of Delhi.
		Agrisilviculture	*Azadirachta indica*	Black gram-wheat/mustard	Highest sown area. Irrigated high
6	Upper Gangatic Plains Regions	Agrisilviculture (irr)	*Populus deltoids*	Wheat, Bajra fodder	North region of U.P. (32 Dists) irrigated by canal & tube wells good ground water
		Agrisilviculture (irr)	*Eucalyptus*	Rice-wheat	
		Silvipasture			

Sr. No.	Agro-Climatic Zones	Agroforestry System	Tree component	Crop/grass	Description
7	Eastern Plateaus & Hills Regions	Agrisilviculture	*Gmelina arborea*	Paddy, linseed	Chhota Nagpur, Garhjat hills, M.P.W. bundelkhand plateau. Odisha. Soils shallow to medium sloppy, undulating. Irrigation tank & tube wells.
		Agrisilviculture	*Acacia nilotica*	Paddy	
		Silviculture	*Acacia mangium*		
		Silviculture	*Bamboo*		
8	Central Plateau & Hills Regions	Agrihorticulture (irr)	*Psidium gujava*	Bengal gram	Madhya Pradesh, Rajasthan, UP, climate hot to sub-hunid, rainfall varies from 500mm-1000mm with mixed red, yellow and black soils
		Agrihorticulture (irr)	*Emblica officinalis*	Black gram/green gram	
		Agrisilviculture	*Acacia nilotica/Azadirachta indica*	Soybean, black gram-mustard/wheat	
		Silvipasture (RF-Degraded lands)	*Albizia amara, L. leucocephala, P. cineraria, S. scabra, Jatropha curcas*	*Chrysopogon fulvus, Stylo santhes*	
		TBOs (RF)			
9	Western plateau & Hills Regions	Agrihortisilviculture (irr)	*Tectona grandis, Archus sapota*	Paddy, Maize	Maharashtra, MP, Rainfall 904mm Sown area 65%, forest 11%, irrigation 12.4%
		Agrihoticulture	*Areca catechu*	Black pepper, Cardamom	
		Silviculture	*P.juliflora*		
		Silviculture	*Acacia mangium*		
		Silvipasture	*Zyziphus mauritiana, D.sisoo, T.grandis*		
10	Southern Plateau & Hills Region	Agrisilviculture (rf)	*Eucalyptus*	Cotton	T. Nadu, Andhra Pradesh, Kamataka, Typically semi-arid zone, Dry land Farming 81% Cropping Intensity 11%
		Agrisilviculture (irr)	*Eucalyptus*	Chilli	
		Silviculture (rf)	*Lleucocephala*	—	
		Silviculture (rf)	*Eucalyptus*	—	
		Agrihorticulture (irr)	*Tamarindus indica*	Chilli	
		TBOs	*Pongamia pinnata*	—	
11	East Coast Plains & Hills Region	Agrisilviculture (rf)	*Ailanthus excelsa*	Cowpea	Tamil Nadu, Andhra Pradesh, Odisha. Soils-
		Silviculture	*Casuarina equisetifolia*	—	

Sr. No.	Agro-Climatic Zones	Agroforestry System	Tree component	Crop/grass	Description
		TBOs	*Pongamia pinnata*	—	alluvial, coastal sand, Irigation
		Silvipasture	*A. leucocephala*	*C.ciliaris*	
12	West Coast Plains & Hills Region	Agrisilviculture (rf)	*Acacia auriculiformis* *Artocarpus heterophyllus*	Black pepper	Sourashtra, Maharashtra, Goa, Karnataka, T. Nadu, Variety of cropping Pattern, rainfall & soil types.
		Agrihorticulture (rf)	*Acacia auriculiformis*	Black pepper	
		Agrisilviculture (rf)	*Cocos nucifera / Areca catechu*	Paddy	
		Agrihorticulture	*Casuarina equisetifolia*	Paddy	
		Agrisilviculture	*Casuarina equisetifolia*	Paddy	
		Silvipasture		Hybrid Napier	
13	Gujarat Plains & Hills Region	Agrisilviculture	*Azadirachta indica* *Prosopis juliflora*	Cowpea	Gujarat (19 dists) Low rainfall arid zone. Irrigation 32% well and tube wells.
		Silvipasture	*Ailanthus excelsa*	Green gram	
		Agrisilviculture			
14	Western Dry Region	Agrisilviculture	*Prosopis cineraria, A. indica lama*	Pearl millet	Rajasthan (9 Dists) Hot sandy desert, rainfall erratic, high evaporation. Scanty vegetation, famine, droughts.
		TBOs	*Jatropha curcas* *Zizyphus numularia*		
		Silvipasture	*Capparis deciduas* *Panicum turgidum*		
15	The Island Region	Agrihorticulture	*Cocos nucifera*	Paddy	Eastern Andaman, Nicobar, Westerm Lakshdweep. Typical equatorial, rainfall 3000 mm (9 months) forest zone undulating.
		Silvipasture	*Areca catechu*	Hybrid Napier, Stylo	

Pathak *et al.* 2000

Society Forestry Plantations | સામાજિક વનીકરણમાં વાવેતર

Abandoned Fields	*Eucalyptus hybrid, Bombax ceiba, Dalbergia sissoo, Terminalia tomentosa, Acacia catechu, Prosopis juliflora* (Plains of North India)
Arid Areas	*Prosopis juliflora, P. spicigera, Acacia arabica, A. senegal, Ricinus communis, A. catechu, Zizyphus* spp., *Albizia lebbeck, Zizyphus mumularia, Cassia auriculata, Lasiurus hirsutus, Tamarix articulata, Calligonum polygonoides, Panicu antidotale, Sachharum munja*
Canal Banks	*Acacia arabica, Delonix regia, Ficus bengalensis, Jacaranda ovalifolia, Robinia pseudoacacia, Swietenia macrophylla, Cassia javanica, Eucalyptus* spp., *Populus* spp., *Dalbergia sissoo, Parkinsonia aculeata, Vateria indica*
Coastal Sandy Area	*Casuarina equisetifolia, Acacia auriculiformis, Eucalyptus hybrid, Pongamia pinnata*
Dry Areas with Irrigation	*Tamariz* spp., *Acacia catechu* (Poor kankar soil), *Prosopis juliflora, P. spicigera, Cassia* spp., *Eucalyptus* (soils free of salts), *Acacia catechu, A. arabica, Bombax ceiba, Morus alba, Syzygium cumini, Eucalyptus, Populus ciliata, Dalbergia sissoo* (Mixed plantation)
Dry Areas without Irrigation	*Acacia arabica, A. leucophloe, Dalbergia sissoo, Albizia lebbeck, Zizyphus jujuba, Cassia siamea, Casuarina equisetifolia, Pongamia* spp., *A. catechu, A. senegal, Eucalyptus, Ailanthus excelsa, Holoptelia integrifolia, Cassia fistula, Prosopis juliflora, Anacardium* spp.
Grasslands	*Acacia catehcy, A. arabica, Ailanthus excelsa, Eucalyptus hybrid, Dalbergia sissoo, Bombax ceiba (U.P. terai), Cupressus torulosa, Populus ciliata, Robina psuedoacacia, Grevillia oppositifolia, Pinus roxburghii, P. wallichiana* (Western Himalays), *Morus laevigata, Terminalia myriocarpa* (NE India)
Inland River Sands	*Acacia catechu, Albizia procera, Melia azedarach, Ailanthus excelsa, Ricinus communis, A. arabica, Azadiracta indica, Pongamia pinnata, Prosopis juliflora, Haplophragma* spp.
Lateritic Soils	*Acacia arbica, Alstonia scholaris, Anacardium occidentale, Eucalyptus* spp., *Casuarina equisetifolia, Dendroclamus strictus*
Railway Line Strips	*Acacia catechu, Azadiracta indica, Eucalyptus* sp., *Cassia fistula, Ficus bengalensis, Jacaranda ovalifolia, Madhuca indica, Prosopis juliflora, A. arabica, Albizia procera, Delonix regia, Cassia javanica, Ficus infectoria, Populus* spp., *Mangifera indica, Terminalia arjuna*
Ravines	*Acacia catechu, Melia azedarach, Prosopis juliflora, Dalbergia sissoo, Pongamia pinnata, Albizia* spp. (Sandy soil), *Acacia arabica, A. leucophloea, Terminalia arjuna, Prosopis spicigera* (Clayey soil)
Roadside Strips	*Acacia arabica, Azadiracta indica, Albizia procera, Ficus bengalensis, Mangifera indica, Cassia fistula, Populus* spp., *Bauhinia variegata, Eucalyptus* spp., *Ficus glomerata, Tamarix articulata, Cassia javanica, Delonix regia, Jacaranda ovalifolia*
Saline and Alkaline soil	*Azadiracta indica, Eucalyptus hyb., Terminalia arjuna, Acacia catechu, Albizia procera, Pongamia pinnata, Prosopis juliflora, Acacia arabica*

Severely Eroded Mountain Slopes	*Pinus roxburghii, Acacia catechu* (Moist sub-tropical); *Acacia catechu, A. modesta, Prosopis juliflora, Robinia psedocacia* (Dry sub-tropical), *Pinus wallichiana, Cedrus deodara* (Temperate), *Eucalyptus, Prosopis* spp. (Penninsular India)
Wind Breaks and Shelterbelt	Grasses: *Saccharum spontaneum, Panicum antidotale, S. munja, Cenchrus* spp. Shrubs: *Calotropis procera, Clerodendron phlomoides, Cassia auriculata, Dodonaea viscosa* Trees: *Acacia arabica, Capparis aphylla, A.leucophloea, Dalbergia sissoo, Lannea coromandelica, Tamarix articulata, Eucalyptus* spp., *Parkinsonia aculeata, P. spicigera*

Raj and Lal 2018

Social Forestry Schemes of India | ભારતની સામાજિક વનીકરણ યોજનાઓ

1.	Afforestation of Barren hillocks
2.	Biogas/Solar cooker Distribution
3.	Different Prize Schemes
4.	Environmental Plantation
5.	Improved Crematoria
6.	Kisan Nursery
7.	Planting of Fruit trees on the sides of public roads with participation of Farmers and Land Laborers
8.	Rehabilitation of degraded lands
9.	Special Component Plan
10.	Strip Plantation
11.	Van Mohotsav
12.	Village Woodlot

Multiple Choice Questions | હેતુલક્ષી પ્રશ્નો

1. **ભારતની રાષ્ટ્રીય એગ્રોફોરેસ્ટ્રી નીતિના મુખ્ય ઉદ્દેશ શું છે?**
   A. પાક અને પશુધન સાથે વળતર અને એકીકૃત રીતે ઝાડના વાવેતરને પ્રોત્સાહન અને વિસ્તૃતિકરણ
   B. એગ્રોફોરેસ્ટ્રી ઉત્પાદનોની ઉપલબ્ધતા સુનિશ્ચિત કરવી
   C. હેબીટાટનો વિસ્તાર વધારવો અને પરંપરાગત વન વિસ્તારોની બહાર વન્યપ્રાણીઓનું સંરક્ષણ કરવું
   D. A અને B

2. **સૌથી જૂની જાણીતી એગ્રોફોરેસ્ટ્રી સિસ્ટમ્સ કઇ છે ?**
   A. એલે ક્રોપિંગ
   B. શિફ્ટિંગ કલ્ટિવેશન
   C. ટોન્ગયા સિસ્ટમ
   D. હોમ ગાર્ડન

3. **લીમડાના બીજ સમયગાળો ક્યો છે ?**
   A. જાન્યુઆરીથી માર્ચ
   B. માર્ચ-મે
   C. મે-જૂન
   D. જૂન-જુલાઇ

4. **કેરળના વધુ વરસાદવાળા વિસ્તારોમાં સૌથી પ્રચલિત એગ્રોફોરેસ્ટ્રી સિસ્ટમ્સ કઇ છે ?**
   A. સિલ્વી-પાશ્ચર
   B. એગ્રી - સિલ્વીકલ્ચર
   C. એગ્રી - સિલ્વી - પાશ્ચર
   D. હોમ ગાર્ડન

5. **ભારતીય ગ્રાસલેન્ડ અને વન સંશોધન સંસ્થા ક્યાં આવેલ છે ?**
   A. દહેરાદૂન B. ઝાંસી
   C. જોધપુર D. નાઇજેરિયા

1. **The objective of India's National Agroforestry Policy is**
   A. Encourage and expand tree plantation in compensatory and integrated manner with crops and livestock
   B. Supplement the availability of Agroforestry products
   C. Increase the habitat area and conserve wildlife outside traditional forest areas
   D. A and B

2. **The oldest known agroforestry systems are**
   A. Alley cropping
   B. Shifting cultivation
   C. Taungya systems
   D. Home gardens

3. **The collection period for Neem seed is**
   A. January-March
   B. March-May
   C. May-June
   D. June-July

4. **The most prevalent agroforestry systems in high rainfall areas of Kerala is**
   A. Silvipasture
   B. Agrisilviculture
   C. Agri-silvi-pasure
   D. Home gardens

5. **Indian Grassland and Forest research Institute located at.**
   A. Dehradun B. Jhansi
   C. Jodhpur D. Nigeria

6. ભારતમાં પ્લાન્ટેશન ફોરેસ્ટ્રીની શરૂઆત ક્યા વર્ષ માં થઈ હતી ?

A. 1942 B. 1842
C. 1947 D. 1847

7. ભારતમાં વન મોહોત્સવની શરૂઆત કયા વર્ષ માં કરવામાં આવી હતી ?

A. 1847 B. 1947
C. 1850 D. 1950

8. ઝુમિંગ નું બીજું નામ શું છે ?

A. શિફ્ટિંગ કલ્ટિવેશન
B. ટોન્ગયા
C. A અને B
D. હોમ ગાર્ડન

9. સહભાગી વન વ્યવસ્થાપન રાષ્ટ્રીય વન નીતિ પછી કયા વર્ષમાં વિકસિત થયેલ છે?

A. 1950 B. 1960
C. 1988 D. 1980

10. પર્યાવરણીય વ્યવસ્થાપનમાં બેંચમાર્કિંગ એટલે

A. સંભવિત જોખમ આકારણી.
B. પર્યાવરણીય કામગીરીની જાણ કરવી.
C. કામગીરીમાં સુધારો લાવવા માટે શ્રેષ્ઠ કામગીરી સામે સંસ્થાની વ્યવસાયિક પ્રક્રિયાઓનું મૂલ્યાંકન.
D. પર્યાવરણીય સંચાલકો દ્વારા અનુસરવામાં આવતા પર્યાવરણીય ધોરણોની સ્થાપના

11. છાંયડા નો સામનો કરવા માટે એક વૃક્ષની ક્ષમતા _________ કહેવામાં આવે છે.

A. ઇન્ટોલરન્સ B. ટોલરન્સ
C. હાર્ડીનેશ D. શેડનેશ

12. વુડલોટનો બેઝલ એરિયા શું દર્શાવે છે?

A. ઝાડનો વિસ્તાર ચોરસ ફૂટમાં
B. કાઉનની ડેન્સીટી
C. સ્ટેન્ડ ડેન્સીટી
D. ટ્રંકનો ટેપર

6. **The plantation forestry in India started in**

A. 1942 B. 1842
C. 1947 D. 1847

7. **In India, Van Mohotsav was started in**

A. 1847 B. 1947
C. 1850 D. 1950

8. **Jhuming is the other name for**

A. Shifting cultivation
B. Taungya
C. A and B
D. Home Garden

9. **Joint Forest Management is evolved after National Forest Policy in**

A. 1950 B. 1960
C. 1988 D. 1980

10. **Benchmarking in environmental management refers to**

A. Potential risk assessment.
B. Reporting of environmental performance.
C. Assessment of organization's business processes against the best-in-class operations to improve the performance.
D. Setting of environmental standards to be followed by environmental managers.

11. **The ability of a tree to withstand shade is called as**

A. Intolerance B. Tolerance
C. Hardiness D. Shadiness

12. **Basal area of a woodlot represents**

A. Area occupied by the tree in square feet
B. Crown density
C. Stand density
D. Taper of trunk

13. ઝાડના પાંદડા અને ડાળીઓ ________ તરીકે ઓળખાય છે.

A. ડોમિનન્ટ B. બોલ

C. ક્રાઉન D. ડાયામીટર

14. વધુ સારી ઉપજ લાવવા માટે સમયાંતરે ઝાડ દૂર કરવું તેને શું કહેવામાં આવે છે?

A. ક્લિયર કટીંગ

B. સીડ-ટ્રી કટીંગ

C. સિલેક્ટિવ કટીંગ

D. શેલ્ટર-વૂડ કટીંગ

15. બીઓનોમિલ નોમીનકલચર સિસ્ટમ માટે કોને શ્રેય આપવામાં આવે છે ?

A. કેરોલસ લિનીઅસ

B. ચાર્લ્સ ડાર્વિન

C. ચાર્લ્સ ઇ. બેસી

D. જ્યોર્જ બેન્ટહામ

16. તાજેતરમાં જંગલવાળા વિસ્તારમાં જંગલ અથવા સ્ટેન્ડની સ્થાપના ________ તરીકે ઓળખાય છે.

A. ડેફી ડક

B. સ્લોગેટિંગ

C. ડીફોરેસ્ટેશન

D. વનીકરણ

17. જંગલોના કાપને અટકાવવાનો શ્રેષ્ઠ માર્ગ ક્યો છે ?

A. વનીકરણ B. રિસાયક્લિંગ

C. ફ્લોરિકલચર D. એગ્રોફોરેસ્ટ્રી

18. " સામાજિક વનીકરણ " શબ્દની રજૂઆત કોનાં દ્વારા કરવામાં આવી હતી ?

A. ટેન્સલી B. વેસ્ટોબી

C. બ્રાન્ડિસ D. બ્લેક

19. રસ્તાની બાજુના વાવેતરમાં પ્રથમ હરોળના વાવેતર માટે કઈ પ્રજાતિ યોગ્ય છે?

A. કેરી B. આમલી

C. સાગ D. ઉપરોક્ત તમામ

13. The leaves and branches of a tree are referred to as ________ of the tree.

A. Dominant B. Bole

C. Crown D. Diameter

14. The periodic removal of trees in order to produce better yield is

A. Clear cutting

B. Seed-tree cutting

C. Selective cutting

D. Shelter wood cutting

15. The system of bionomial nomenclature is credited to:

A. Carolus Linnaeus

B. Charles Darwin

C. Charles E. Bessey

D. George Bentham

16. Establishment of forest or stand in an area recently forested is termed as

A. Daffy Duck

B. Sologating

C. Deforestation

D. Afforestation

17. Which is the best way to curb deforestation

A. Afforestation B. Recycling

C. Floriculture D. Agroforestry

18. The term" Social forestry" was introduced by:

A. Tansley B. Westoby

C. Brandis D. Black

19. Species which is suitable for first row planting in roadside planting?

A. Mango B. Tamarind

C. Teak D. All of the above

20. એલેલોકેમિકલ્સ કોના દ્વારા રીલીઝ થાય છે ?
A. લીચિંગ B. એક્સુડેશન
C. વોલેટિલાઇઝેશન D. ઉપરોક્ત તમામ

20. Allelochemicals are released by
A. Leaching B. Exudation
C. Volatilization D. All of the above

21. નાઇટ્રોજન ફિક્સિંગ ફ્રી લિવિંગ બેક્ટેરિયા કયુ છે?
A. એઝોટોબેક્ટર
B. ક્લોસ્ટ્રિડિયમ
C. A અને B
D. ઉપરોક્ત એક પણ નહિ

21. Which is Nitrogen fixing free living bacteria?
A. Azatobacter
B. Clostridium
C. A & B
D. None

22. એગ્રોફોરેસ્ટ્રી સિસ્ટમ જર્નલ ક્યાં થી પ્રકાશિત થાય છે ?
A. કેન્યા B. નેધરલેન્ડ્ઝ
C. ભારત D. ન્યૂયોર્ક

22. Agroforestry system journal is published from
A. Kenya B. Netherlands
C. India D. New York

23. સામાજિક વનીકરણ પ્રવૃત્તિઓને શેમાં વહેંચી શકાય છે?
A. શેલ્ટર બેલ્ટ પ્લાન્ટેશન
B. ફાર્મ પ્લાન્ટેશન
C. બ્લોક પ્લાન્ટેશન
D. ઉપરોક્ત તમામ

23. Social forestry activities can be divided into
A. Shelter belt plantations
B. Farm forestry plantation
C. Block plantations
D. All of the above

24. શિફ્ટિંગ કલ્ટીવેશન સૌથી વધુ ક્યાં પ્રચલિત છે?
A. દક્ષિણ ભારત
B. પશ્ચિમ ભારત
C. ઉત્તર ભારત
D. ઉત્તર પૂર્વ ભારત

24. Shifting cultivation is most prevalent in?
A. South India
B. Western India
C. North India
D. North East India

25. ઉષ્ણકટિબંધીય વિસ્તારીમાં જોવા મળતા હોમ ગાર્ડન્સ _____નું ઉત્તમ ઉદાહરણ છે.
A. એગ્રીસિલ્વિકલ્ચરલ સિસ્ટમ
B. સિલ્વીકલ્ચરલ સિસ્ટમ
C. એગ્રિસિલ્વિપાશ્ચરસ સિસ્ટમ
D. પશુપાલન

25. Home gardens found in tropics are one of the classic examples of
A. Agrisilvicultural system
B. Silvicultural system
C. Agrisilvipastural system
D. Pastoralsilviculture

26. વિશ્વ એગ્રોફોરેસ્ટ્રી સેન્ટર નું મુખ્ય ક્વાર્ટર ક્યાં છે ?
A. નૈરોબી B. ઝાંસી
C. ન્યૂયોર્ક D. હૈદરાબાદ

26. Head Quarter of World Agroforestry center (WAC) is at:
A. Nairobi B. Jhansi
C. New York D. Hyderabad

27. રાજસ્થાનના સુકા ભાગ નું કયું વૃક્ષ કલ્પવૃક્ષ તરીકે ઓળખાય છે.?

A. *Acacia nilotica*
B. *Calligonum polygonoides*
C. *Prosopsis cineraria*
D. *Euphorbia tirucalli*

28. નીલગિરીના પાંદડા માં _______ ની હાજરી એલેલોપેથી માટે જાણીતી છે?

A. યુજેનોલ
B. યુટેક્ટોનોલ
C. મલાનિન
D. ઉપરોક્ત એક પણ નહિ

29. ઉત્તર ભારતીય એગ્રોફોરેસ્ટ્રીમાં નીચેનામાંથી કયુ કોમ્બિનેશન સામાન્ય છે?

A. સાગ સાથે ઘઉં
B. સાગ સાથે ચોખા
C. ચોખા સાથે પોપ્લર
D. ઘઉં સાથે પોપ્લર

30. ચાના એસ્ટેટમાં શેડ માટે સામાન્ય રીતે ઉપયોગમાં લેવામાં આવતી વૃક્ષની પ્રજાતિ કઈ છે ?

A. *Gmelina arborea*
B. *Celtis australis*
C. *Morus alba*
D. *Grevillea robusta*

31. ભારતમાં અગ્રોઇકોનોમિકલ ઝોન કેટલા છે?

A. 14 B. 16
C. 18 D. 20

32. વૃક્ષની પ્રજાતિઓ જે 5Fs આપે છે એટલે કે fuel, fodder, fiber, food and fertilizers તેને શું કહેવામાં આવે છે?

A. ઘાસચારા ના વૃક્ષો
B. ખાતરના વૃક્ષો
C. લેગ્યુમિનસ વૃક્ષો
D. બહુહેતુક વૃક્ષો

27. Kalpavriksha of dry part of Rajasthan?

A. *Acacia nilotica*
B. *Calligonum polygonoides*
C. *Prosopsis cineraria*
D. *Euphorbia tirucalli*

28. The presence of _______ in Eucalyptus leaves is known to cause allelopathy?

A. Eugenol
B. Eutectonol
C. Malanin
D. None

29. In North Indian Agroforestry which of the following combination is common?

A. Wheat with teak
B. Rice with teak
C. Rice with poplar
D. Wheat with poplar

30. Most commonly used tree species for shade in tea estates is

A. *Gmelina arborea*
B. *Celtis australis*
C. *Morus alba*
D. *Grevillea robusta*

31. On the basis of Agroeconomical zone India is divided in to how many zones?

A. 14 B. 16
C. 18 D. 20

32. Tree species which gives 5Fs i.e. fuel, fodder, fiber, food and fertilizers are called:

A. Fodder trees
B. Fertilizer trees
C. Leguminous tress
D. Multipurpose trees

33. કૃષિ પાક સાથે ઝાડ ઉગાડવું એ કઈ રીતે ઓળખાય છે ?
A. એગ્રી-સિલ્વી પાશ્ચર સિસ્ટમ
B. એગ્રી-સિલ્વીકલ્ચર
C. એગ્રી હોર્ટિકલચર
D. સીલ્વીપાશ્ચર સિસ્ટમ

34. એલે ક્રોપીંગમાં, ઝાડની હરોળ ____માં હોવી જોઈએ.
A. N- S દિશા B. N-E દિશા
C. E- W દિશા D. S- W દિશા

35. ઇન્ડિયન જર્નલ ઓફ એગ્રોફોરેસ્ટ્રી ક્યાં થી પ્રકાશિત થાય છે?
A. રાંચી B. ઝાંસી
C. નવી દિલ્હી D. લખનૌ

36. નીચેનામાંથી કયુ એલેલોકેમિકલ્સનું ઉદાહરણ છે?
A. ગેલિક એસિડ
B. એસીટીક એસિડ
C. હાઇડ્રોક્લોરિક એસિડ
D. ઉપરોક્ત એક પણ નહિ

37. રાષ્ટ્રીય એગ્રોફોરેસ્ટ્રી નીતિ____દ્વારા ઘડવામાં આવેલ હતી.
A. કૃષિ અને ખેડૂત કલ્યાણ મંત્રાલય
B. પર્યાવરણ, વન અને જલવાયુ પરિવર્તન મંત્રાલય
C. વિજ્ઞાન અને ટેકનોલોજી વિભાગ
D. આર્થિક બાબતોના વિભાગ

38. રાષ્ટ્રીય એગ્રોફોરેસ્ટ્રી દિવસ ક્યારે ઉજવવામાં આવે છે?
A. 8 મે B. 5 જૂન
C. 21 માર્ચ D. 8 ઓક્ટોબર

39. નીલગિરી + ઘઉં સિસ્ટમ ક્યાં રાજ્યમાં સૌથી પ્રચલિત છે?
A. મધ્ય પ્રદેશ B. રાજસ્થાન
C. પંજાબ D. બિહાર

33. Growing of trees with agricultural crops is known as?
A. Agri-silvi pasture system
B. Agri-silviculture
C. Agri horticulture
D. Silvopasture system

34. In alley cropping, row of trees should be positioned in
A. N-S direction B. N-E direction
C. E-W direction D. S-W direction

35. Indian Journal of agroforestry is published from:
A. Ranchi B. Jhansi
C. New Delhi D. Lucknow

36. Which of the following is an example of allelochemicals?
A. Gallic acid
B. Acetic acid
C. Hydrochloric acid
D. None

37. National Agroforestry Policy formulated by
A. Ministry of Agriculture & Farmers Welfare
B. Ministry of Environment, Forests and Climate Change
C. Department of Science and Technology
D. Department of Economic Affairs

38. National Agroforetry day is celebrated on
A. 8 May B. 5 June
C. 21 March D. 8 Oct

39. Eucalyptus + Wheat system is most prevalent in
A. MP B. Rajasthan
C. Punjab D. Bihar

40. હોમગાર્ડન્સ ક્યાં રાજ્યમાં વધુ પ્રચલિત છે?

A. કર્ણાટક B. તમિલનાડુ
C. કેરળ D. આંધ્ર પ્રદેશ

41. એગ્રોફોરેસ્ટ્રીમાં પ્રકાશની સ્પર્ધા દ્વારા શું ઘટાડી શકાય છે ?

A. કેનોપી મેનેજમેન્ટ
B. રુટ મેનેજમેન્ટ
C. વોટર મેનેજમેન્ટ
D. ન્યુટ્રીઅન્ટ મેનેજમેન્ટ

42. રસાયણોના ઉત્પાદન દ્વારા એક છોડ પર બીજા છોડની સીધી અથવા આડકતરી અસર કઈ રીતે ઓળખાય છે?

A. એલેલોપેથી B. કમ્પિટિશન
C. સીમ્બાયોસીસ D. એલેલોસ્પોલી

43. એલેલોપેથી એક પ્રકારની ક્રિયાપ્રતિક્રિયા છે

A. સકારાત્મક
B. નકારાત્મક
C. બંને A અને B
D. ઉપયુક્ત માંથી એક પણ નહી

44. એલેલોપેથી નું શ્રેષ્ઠ ઉદાહરણ કયું છે ?

A. સીમ્બાયોસીસ B. એસોસિએશન
C. એમેન્સાલિઝમ D. મ્યુચ્યુઅલિઝમ

45. જાણીતું પુસ્તક "ઈન્ટ્રોડકશન ઓફ એગ્રોફેરેસ્ટ્રી" કોના દ્વારા લખાયેલ છે ?

A. પી જી જાર્વિસ B. પી કે આર નાયર
C. એન્ટની યંગ D. એપી દ્વિવેદી

46. અખરોટમાં હાજર એલેલોકેમિકલ કયું છે?

A. મીમોઝિન B. જુગલોન
C. ફિનોલિક્સ D. સિનોલ

47. એલેલોપેથી શબ્દ ક્યાંથી આવ્યો છે?

A. ગ્રીક B. લેટિન
C. ઇંગલિશ D. જર્મન

48. બ્રિટીશ સામ્રાજ્ય દ્વારા શેનો ઉપયોગ કરીને પ્રથમ ટોન્ગયા વાવેતરની સ્થાપના કરવામાં આવી હતી?

A. સાલ B. વાંસ
C. સાગ D. દેવદાર

40. Homegardens are more prevalent in:

A. Karnataka B. Tamilnadu
C. Kerala D. Andhra Pradesh

41. Competition of light in Agroforestry can be reduced through

A. Canopy management
B. Root management
C. Water management
D. Nutrient management

42. The direct or indirect effect of one plant on another by production of chemicals is known as:

A. Allelopathy B. Competition
C. Symbiosis D. Allelospoly

43. Allelopathy is a type of interaction

A. Positive
B. Negative
C. Neutral
D. Both and B

44. Allelopathy is the best example of:

A. Symbiosis B. Association
C. Ammensalism D. Mutualism

45. A renowned book "An Introduction to Agroforestry" is written by:

A. P G Jarvis B. PKR Nair
C. Antony Young D. AP Dwivedi

46. Allelochemical present in Walnut is:

A. Mimosine B. Juglone
C. Phenolics D. Cineole

47. The term allelopathy is derived from:

A. Greek B. Latin
C. English D. German

48. First Taungya plantation was established by British Empire by using?

A. Sal B. Bamboo
C. Teak D. Deodar

49. ઇન્ટરનેશનલ કાઉન્સિલ ફોર રિસર્ચ ઇન એગ્રોફોરેસ્ટરી (ICRAF) ની સ્થાપના ક્યારે કરવામાં આવી હતી?

A. 1975 B. 1977
C. 1978 D. 1980

50. હોમગાર્ડન્સ શેનો ભાગ છે ?

A. એગ્રિસિલવિકલ્ચર
B. સિલ્વીપાશ્ચર
C. એગ્રોસિલોવોપાશ્ચારોલ
D. એક્વાફોરેસ્ટ્રી

51. એગ્રોફોરેસ્ટ્રી ટર્મ __________ વર્ષમાં આવી

A. 1977 B. 1988
C. 1999 D. 2009

52. રાષ્ટ્રીય કૃષિ આયોગે કઈ પંચવર્ષીય યોજનામાં એગ્રોફોરેસ્ટ્રી શિક્ષણ પર ભાર મૂક્યો?

A. સાતમી B. આઠમી
C. નવમી D. દસમી

53. 1985 માં ભારતમાં પ્રથમ એગ્રોફોરેસ્ટ્રી સેમિનારનું આયોજન __________ માં કરવામાં આવ્યું હતું.

A. ઇમ્ફાલ B. ગાંધીનગર
C. બેંગલુરુ D. કોલકાતા

54. હોમ ગાર્ડનને __________ સિસ્ટમ તરીકે પણ ઓળખવામાં આવે છે.

A. સિંગલ લેયર
B. ડબલ લેયર
C. મલ્ટિલિયેર
D. ઉપરોક્ત તમામ

55. પર્યાવરણમાં મુક્ત થતા રાસાયણિક અવરોધકોના ઉત્પાદન દ્વારા એક છોડની સીધી અથવા આડકતરી અસરોને શું કહેવાય છે?

A. એલેલોપેથી B. સિમ્બાયોસિસ
C. બંને A અને B D. ઉપરોક્ત કંઈ નથી

49. International Council for Research in Agroforetry (ICRAF) was established in:

A. 1975 B. 1977
C. 1978 D. 1980

50. Homegardens are part of

A. Agrisilviculture
B. Silvipasture
C. Agrosilvopastoral
D. Aquaforestry

51. Term agroforestry was coined in

A. 1977 B. 1988
C. 1999 D. 2009

52. The National Commission on Agriculture emphasized agroforestry education in the __________ five-year plan period.

A. Seventh B. Eighth
C. Ninth D. Tenth

53. First agroforestry seminar in India was organized at __________ in 1985.

A. Imphal B. Gandhinagar
C. Bengaluru D. Kolkata

54. Home garden also called as __________ system.

A. Single tier
B. Double tier
C. Multitier
D. All of the above

55. __________ is the direct or indirect effects of one plant to another through the production of chemical inhibitors that are released into the environment.

A. Allelopathy B. Symbiosis
C. Both D. None of the above

56. માટીમાંથી છોડ અને જમીનમાં પોષક તત્વોના સાયકલિંગની પ્રક્રિયાને ________ કહેવામાં આવે છે.

A. ન્યુટ્રીઅન્ટ સાયક્લિંગ
B. ન્યુટ્રીઅન્ટ અપટેક
C. ન્યુટ્રીઅન્ટ લિન્ચ
D. ન્યુટ્રીએન્ટ ઇન્ટેક

57. સારી એગ્રોફોરેસ્ટ્રી ડિઝાઇનમાં ________ માપદંડ પૂરા થવો જોઈએ.

A. ઉત્પાદકતા
B. ટકાઉપણું
C. સ્વીકાર્યતા
D. ઉપરોક્ત તમામ

58. વાવેતર કરવાની કઈ રીત કે જેમાં વૃક્ષોની હરોળ પાકની હરોળ સાથે છૂટી પડે છે, માટી અને પાકને પોષક તત્વો સુધરે છે, ખાસ કરીને નાઇટ્રોજન પૂરૂં પાડે છે.

A. એલે ક્રોપીંગ
B. હેજરો અને ઇંટરક્રોપિંગ
C. બંને A અને B
D. ઉપરોક્ત કંઈ નથી

59. ________ એ વૃક્ષોની હરોળ અને / અથવા છોડને ખેતરો, ઘરો, નહેરો અથવા અન્ય વિસ્તારોને ફૂંકાતા પવન અને માટી અથવા રેતીથી બચાવવા માટે વાવેતર છે.

A. એલે ક્રોપીંગ
B. વિન્ડબ્રેક
C. શેલ્ટરબેલ્ટ
D. બેલ્ટસ

60. ________ એ એક એવી ખેતી પદ્ધતિ તરીકે વ્યાખ્યાયિત કરી શકાય છે ઘરની આજુબાજુના જમીનના ક્ષેત્ર પર શારીરિક, સામાજિક અને આર્થિક કાર્યોને જોડે છે.

A. હોમ ગાર્ડન
B. સીડ ઓર્ચાર્ડ
C. એલે ક્રોપીંગ
D. હોર્ટિકુલ્ટર

56. The process of cycling of nutrients from soil to the plants and back to the soil is called as

A. Nutrient cycling
B. Nutrient uptake
C. Nutrient lynch
D. Nutrient intake

57. A good agroforestry design should fulfil the ________ criteria.

A. Productivity
B. Sustainability
C. Adoptability
D. All of the above

58. A method of planting in which rows of trees are interspersed with rows of crops, improving the soil and providing nutrients, particularly nitrogen to the crops is known as ________.

A. Alley cropping
B. Hedgerow Intercropping
C. Both A and B
D. None of the above

59. ________ are the strips of the trees and/or shrub planted to protect fields, homes, canals or other areas from wind and blowing soil or sand.

A. Alley cropping
B. Wind-breaks
C. Shelterbelts
D. Belts

60. A ________ can be defined as a farming system that combines physical, social and economic functions on the area of land around the family home.

A. Home garden
B. Seed orchard
C. Alley cropping
D. Horticulture

61. સોશિયલ ફોરેસ્ટ્રી શબ્દ કોણે આપ્યો?

A. વેસ્ટોબી B. ડોવસન

C. ચૌધરી D. ખન્ના

62. "પરંપરાગત જંગલોની બહારનું વનીકરણ જેનો હેતુ મુખ્યત્વે લોકોના ફાયદા માટે સામાન અને સેવાઓનો સતત પ્રવાહ પૂરો પાડવાનો છે" જે ____________ તરીકે ઓળખાય છે.

A. સામાજિક વનીકરણ

B. ફાર્મ ફોરેસ્ટ્રી

C. એક્સ્ટેંશન ફોરેસ્ટ્રી

D. રૂરલ ફોરેસ્સ્ટ્રી

63. ખેતરમાં અથવા ગામડાની આસપાસની જમીનની અન્ય કામગીરી સાથે સંકળાયેલ વનીકરણ __________ છે .

A. સામાજિક વનીકરણ

B. ફાર્મ ફોરેસ્ટ્રી

C. એક્સ્ટેંશન ફોરેસ્ટ્રી

D. રૂરલ ફોરેસ્સ્ટ્રી

64. શહેરી અને ગ્રામીણ વસ્તી માટે મનોરંજન જંગલો તરીકે સેવા આપવા માટે ફૂલોના છોડ સાથે ઉછેરવાના વનીકરણની પ્રથા શું કહેવામાં આવે છે ?

A. રિએકશન ફોરેસ્ટ્રી

B. ફાર્મ ફોરેસ્ટ્રી

C. એક્સ્ટેંશન ફોરેસ્ટ્રી

D. રૂરલ ફોરેસ્સ્ટ્રી

65. સ્થળાંતર ખેતી ____________ તરીકે પણ ઓળખાય છે.

A. ઝુમ કલ્ટીવેશન

B. સ્લેશ અને બર્ન કલ્ટીવેશન

C. બંને A અને B

D. ઉપરોક્ત કંઈ નથી

61. The word Social forestry was coined by

A. Westoby B. Dowson

C. Chaudhary D. Khanna

62. "Forestry outside the conventional forests which primarily aims at providing continuous flow of goods and services for the benefit of people". What this forestry is known as

A. Social Forestry

B. Farm Forestry

C. Extension Forestry

D. Rural Forestry

63. ____________ is the practice of forestry in all its aspects in and the around the farms or village lands integrated with other farm operations.

A. Social Forestry

B. Farm Forestry

C. Extension Forestry

D. Rural Forestry

64. ____________ is the practice of forestry with the object of raising flowering trees and shrubs mainly to serve as recreation forests for the urban and rural population.

A. Recreation forestry

B. Farm Forestry

C. Extension Forestry

D. Rural Forestry

65. Shifting cultivation is also known as

A. Jhuming cultivation

B. Slash and burn Cultivation

C. Both

D. None of the above

66. ભારતમાં, સંસહભાગી વન આયોજન અને વ્યવસ્થાપન પ્રથમ કયા રાજ્ય માં શરૂ કરવામાં આવ્યું હતું ?
A. ગુજરાત B. કેરળ
C. હિમાચલ પ્રદેશ D. પશ્ચિમ બંગાળ

67. એગ્રોફોરેસ્ટ્રીની શબ્દ કયા વર્ષમાં આવ્યો હતો?
A. 1937 B. 1957
C. 1977 D. 1997

68. ICRAF નું પૂરું નામ શું છે?
A. ઇન્ટરનેશનલ સેન્ટર ફોર રિસેર્ચ ઈન એગ્રોફોરેસ્ટ્રી
B. ઇન્ટરનેશનલ કાઉન્સિલ ફોર રિસેર્ચ ઈન એગ્રીકલ્ચર
C. ઇન્ડિયન કાઉન્સિલ ફોર રિસર્ચ ઇન એગ્રોફોરેસ્ટ્રી
D. ઈન્ડિયન કાઉન્સિલ ફોર રિસર્ચ ઇન એગ્રોનોમી

69. નાયરે એગ્રીકલ્ચર સિસ્ટમ કેટલા વર્ગોમાં ________ વર્ગીકૃત કર્યું છે ?
A. 3 B. 4
C. 5 D. 6

70. વર્ષો સુધી પાક વિનાની જમીનને ________ તરીકે ઓળખવામાં આવે છે.
A. વાણિજ્યિક જમીન
B. પડતર જમીન
C. ગોચર જમીન
D. ઉપરોક્ત તમામ

71. શેલ્ટરબેલ્ટની આદર્શ પહોળાઈ શું હોવી જોઈએ ?
A. 40 મી B. 50 મી
C. 60 મી D. 70 મી

72. શેલ્ટરબેલ્ટ માટે ઉંચાઇ અને પહોળાઈ નો ગુણોત્તર શું હોવું જોઈએ ?
A. 10:1 B. 1:10
C. 1:15 D. 15:1

66. In India, the concept of joint forest planning and management was first started in ________.
A. Gujarat B. Kerala
C. Himachal Pradesh D. West Bengal

67. In which year term agroforestry was coined?
A. 1937 B. 1957
C. 1977 D. 1997

68. Full abbreviation of ICRAF-
A. International Centre for Research in Agroforestry
B. International Council for Research in Agriculture
C. Indian Council for Research in Agroforestry
D. Indian Council for Research in Agronomy

69. Nair has classified agroforestry system into ________ categories?
A. 3 B. 4
C. 5 D. 6

70. The crops-land left without crops for periods from one season to several years is termed as
A. Commercial land
B. Fallow land
C. Pasture land
D. All of the above

71. What should be ideal width of shelterbelts?
A. 40 m B. 50 m
C. 60 m D. 70 m

72. What should be the ratio of height and width for shelterbelts?
A. 10:1 B. 1:10
C. 1:15 D. 15:1

73. એક સામાન્ય શેલ્ટરબેલ્ટનો આકાર __________ હોવો જોઈએ.

A. ચોરસ
B. લંબચોરસ
C. ત્રિકોણાકાર
D. વર્તુળાકાર

74. ગોચર સાથે જોડાયેલા લાકડાવાળા વૃક્ષોના ઉત્પાદન માટે નીચેની કઈ સિસ્ટમનો ઉપયોગ થાય છે ?

A. એગ્રી-સિલ્વીકલ્ચર
B. હોર્ટી-સિલ્વી-પાશ્ચર
C. એગ્રી-હોર્ટિકલચર
D. સીલ્વીપાશ્ચર

75. કેટલી ઊભી કેનોપી સામાન્ય હોમ ગાર્ડન માં સ્થિત હોય છે ?

A. 1-2 B. 2-4
C. 3-4 D. 5-6

76. નીચેનામાંથી __________ ઇમારતી લાકડાની રાણી તરીકે ગણવામાં આવે છે.

A. સાગ B. સીસું
C. વાંસ D. ચંદન

77. નીચેનામાંથી કોને ગરીબ નું ઇમારતી લાકડું ગણવામાં આવે છે ?

A. સાગ B. સીસું
C. વાંસ D. ચંદન

78. ફેબ્રુઆરી, 2014 માં કયા શહેરમાં ત્રીજી વર્લ્ડ એગ્રોફોરેસ્ટ્રી કોંગ્રેસ યોજાઇ હતી?

A. નવી દિલ્હી B. ગાંધીનગર
C. કોલકાતા D. મુંબઇ

79. એગ્રોફોરેસ્ટ્રીમાં કોણે D & D સર્વેની દરખાસ્ત કરી?

A. નાયર
B. ચેમ્પિયન અને શેઠ
C. જેબી રેન્ટ્રી
D. પીટરસન

73. A typical shelter belt usually having a __________ shape.

A. Square shape
B. Rectangular shape
C. Triangular shape
D. Circle shape

74. Which of the following system used for production of woody plants combined with pasture.

A. Agri-silviculture
B. Horti-silvi-pasture
C. Agri-horticulture
D. Silvi-pasture

75. How many vertical canopy strata usually has in Home garden?

A. 1-2 B. 2-4
C. 3-4 D. 5-6

76. Which of the following is considered as Queen of timber?

A. Teak B. Sissoo
C. Bamboo D. Sandal

77. Which of the following is considered as poor's man timber?

A. Teak B. Sissoo
C. Bamboo D. Sandal

78. In which city 3rd World Agroforestry Congress was held in Feb 2014.

A. New Delhi B. Gandhinagar
C. Kolkata D. Mumbai

79. Who proposed D&D survey in agroforestry?

A. Nair
B. Champion & Seth
C. J.B Raintree
D. Paterson

80. ભારતના કયા રાજ્ય એ સામાજીક વનીકરણ પ્રથમ વખત સફળતાપૂર્વક અપનાવ્યું?

A. ગુજરાત B. કેરળ
C. હિમાચલ પ્રદેશ D. પશ્ચિમ બંગાળ

81. વૃક્ષોને કાપવા અને તેને આગ લગાડવાની અને પરિણામી રાખ પર પાક ઉભા કરવાની પ્રથાને ________ શું કહેવામાં આવે છે?

A. ઝૂમ કલ્ટીવેશન
B. શિફટિંગ કલ્ટીવેશન
C. સ્લેશ અને બર્ન કલ્ટીવેશન
D. ઉપરોક્ત તમામ

82. સામાજીક વનીકરણ કેટલા ઘટકો ધરાવે છે ?

A. 2 B. 3
C. 4 D. 0.5

83. નીચેના માંથી શું સામાજિક વનીકરણ નો ભાગ નથી ?

A. ફાર્મ ફોરેસ્ટ્રી
B. વિસ્તરણ વનીકરણ
C. મનોરંજન વનીકરણ
D. સૌંદર્યલક્ષી વનીકરણ

84. ટોન્ગયા માં 'ટોન્ગ' નો અર્થ ________ છે, અને 'યા' એટલે ની ખેતી.

A. હોમ ગાર્ડન B. જંગલ
C. પ્રાણી D. ટેકરી

85. ટોન્ગયા પ્રથમ કોના દ્વારા શરૂ કરવામાં આવી હતી ?

A. એમ.એસ. સ્વામિનાથન
B. કે.એમ. મુન્શી
C. એ.ટી. ચૌધરી
D. બ્રાન્ડિસ

86. ભારતમાં સૌ પ્રથમ ટોન્ગયાની શરૂઆત કયા વર્ષમાં થઈ હતી?

A. 1850 B. 1890
C. 1950 D. 1990

80. The concept of Social forestry was first successfully adopted in ________ State of India.

A. Gujarat B. Kerala
C. Himachal Pradesh D. West Bengal

81. The practice of cutting down trees and setting them or fire and raising crops on the resulting ash is termed as ________.

A. Jhuming cultivation
B. Shifting cultivation
C. Slash and burn cultivation
D. All of the above

82. How many components Social forestry has.

A. 2 B. 3
C. 4 D. 5

83. Which of the following is/are not a component of social forestry.

A. Farm forestry
B. Extension forestry
C. Recreation forestry
D. Aesthetic forestry

84. In taungya meaning of 'taung' is ________, and of 'ya' is Cultivation.

A. Home garden B. Forest
C. Animal D. Hill

85. Taungya was first introduced by

A. M S Swaminathan
B. K. M. Munshi
C. A T Chaudhary
D. Brandis

86. In which Year Taungya was first introduced in India?

A. 1850 B. 1890
C. 1950 D. 1990

87. કયા વર્ષમાં સહભાગી વન વ્યવસ્થાપન શરૂ થયું ?

A. 1890 B. 1990
C. 1999 D. 2000

88. સેન્ટર ફોર સોશ્યલ ફોરેસ્ટ્રી એન્ડ ઇકો-રિહેબિલીટીશન ભારત માં ક્યાં સ્થિત છે ?

A. દહેરદૂન B. ગાંધીનગર
C. નવી દિલ્હી D. અલ્હાબાદ

89. શબ્દ ટોન્ગયા ________ ભાષાનો શબ્દ છે.

A. લેટિન B. અરબી
C. ગ્રીક D. બર્મીઝ

90. શિફ્ટિંગ કલ્ટીવેશનને કેરળમાં ________ તરીકે પણ ઓળખાય છે .

A. હનુમો B. શંબા
C. પૂનમ D. જુમર

91. ભારતમાં કયા વર્ષે ટોન્ગયા ના વાવેતર થયા હતા?

A. 1896 B. 1996
C. 1886 D. 1986

92. નીચેનામાંથી કયું ટોન્ગયાના પ્રકાર તરીકે ગણવામાં આવે છે?

A. ડિપાર્ટમેન્ટલ ટોન્ગયા
B. વિલેજ ટોન્ગયા
C. લીઝ્ડ ટોન્ગયા
D. ઉપરોક્ત તમામ

93. એગ્રી-સિલ્વી-પાશ્યોરસ એટલે ________.

A. લાકડાના વૃક્ષોનું ગોચર સાથે મળીને ઉત્પાદન
B. પાક સાથે જોડાયેલા ફળોના વૃક્ષોનું ઉત્પાદન
C. કૃષિ પાકો સાથે લાકડાવાના વૃક્ષોનું ઉત્પાદન
D. કૃષિ પાકો સાથે લાકડાના વૃક્ષોનું અને ગોચરનું ઉત્પાદન

87. In which Year Joint forest management (JFM) was started was started?

A. 1890 B. 1990
C. 1999 D. 2000

88. Centre for Social Forestry and Eco-Rehabilitation is located in

A. Dehra Dun B. Gandhinagar
C. New Delhi D. Allahabad

89. Term taungya is ________ word.

A. Latin B. Arabic
C. Greek D. Burmese

90. Shifting cultivation is also known as ________ in Kerala.

A. Hanumo B. Shamba
C. Poonam D. Jumar

91. In which of the following year first taungya plantation was raised in India.

A. 1896 B. 1996
C. 1886 D. 1986

92. Which of the following is/are considered as a type of taungya.

A. Departmental taungya
B. Village taungya
C. Leased taungya
D. All of the above

93. Agri-silvi-pastoral means ________.

A. production of woody plants combined with pasture
B. production of fruits plants combined with crops
C. production of woody plants combined with agricultural crops
D. production of woody plants & pasture combined with agricultural crops

94. હોમગાર્ડનમાં ____________ ઉપરના સ્તર માં પ્રભુત્વ ધરાવે છે.

A. ઘાસ
B. શાકભાજી
C. ફળના વૃક્ષો
D. ઇમારતી લાકડા માટેના વૃક્ષો

95. નીચેની ઝાડમાંથી કઈ પ્રજાતિઓનો ઉપયોગ એલે ક્રોપીંગ માટે સૌથી વધુ થાય છે?

A. લેગ્યુમ વૃક્ષોની પ્રજાતિઓ
B. નોન-લેગ્યુંમિનશ વૃક્ષોની પ્રજાતિઓ
C. A અને B બંને
D. ઉપર્યુક્તમાંથી કોઈ નહિ

96. વિન્ડબ્રેક કેટલાં ઔંશના ખૂણે વાવવા જોઈએ જેથી રક્ષણ મળી રહે?

A. 30° B. 45°
C. 60° D. 90°

97. ભારતમાં ક્યાં પ્રદેશોમાં હોમ ગાર્ડન્સનો સૌથી વધુ ઉપયોગ થાય છે?

A. શુષ્ક
B. અર્ધ શુષ્ક
C. સમશીતોષ્ણ
D. ભેજયુક્ત ઉષ્ણકટિબંધીય

98. ભારતમાં નીચેનામાંથી ક્યાં શિફ્ટિંગ કલ્ટીવેશનનો વ્યાપકપણે અભ્યાસ કરવામાં આવે છે?

A. ઉત્તરીય B. ઉત્તર-પૂર્વ
C. દક્ષિણ D. પશ્ચિમી

99. ટોન્ગયા સિસ્ટમ ____________ તરીકે પણ ઓળખાય છે.

A. ઝુમ વાવેતર
B. સ્લેશ અને બર્ન વાવેતર
C. ટેકરી પર ખેતી
D. ઉપરોક્ત તમામ

94. The uppermost layer is dominated by ____________ in home garden.

A. Grass
B. Vegetables
C. Fruit plants
D. Timber trees

95. Which of the following tree species is most commonly used for alley cropping.

A. Legumes tree species
B. Non-leguminous tree species
C. Both A and B
D. None of the above

96. At which angles wind-breaks should be planted to the wind from which protection is needed

A. 30° B. 45°
C. 60° D. 90°

97. In which region home garden is most commonly practiced in India

A. Arid
B. Semi- arid
C. Temperate
D. Humid Tropical

98. Which of the following region shifting cultivation is practiced extensively in India.

A. Northern B. North-eastern
C. Southern D. Western

99. Taungya system also knowno as

A. Jhum cultivation
B. slash and burn cultivation
C. Hill cultivation
D. All of the above

100. વાંસ ________ તરીકે પણ ઓળખાય છે.

A. લીલું સોનું
B. ગરીબોનું ઇમારતી લાકડું
C. A અને B બંને
D. ઉપરોક્ત કંઈ નથી

101. આપણા દેશમાં વૃક્ષોના આવરણને વધારવા માટે જંગલ વિસ્તાર વધારવા સિવાયનો એકમાત્ર કુદરતી અભિગમ ________ છે.

A. ભીના મેદાનોમાં વાવેતર
B. જળચર વિસ્તાઓમાં વાવેતર
C. સામાજિક વનીકરણ વાવેતર
D. ઘાસના મેદાનોમાં વાવેતર

102. ગુજરાતમાં રાજ્ય કક્ષાના વન-મહોસ્તવની ઉજવણી કેવી રીતે કરવામાં આવે છે?

A. વિધાનસભા ખાતે વિધિપૂર્વકનું વાવેતર કરીને
B. દરેક સરકારી કચેરી ખાતે વૃક્ષારોપણ કરીને
C. શૈક્ષણિક સંસ્થાઓમાં વૃક્ષારોપણ કરીને
D. દર વર્ષે અમુક થીમ પર સાંસ્કૃતિક વન વિકસાવીને

103. ગુજરાત ના ખેડા જિલ્લામાં કયું વૃક્ષ જે કૃષિ વનીકરણના ભાગ રૂપે ઉગાડવામાં આવે છે?

A. *Azardirecta indica*
B. *Eucalyptus* spp.
C. *Tectona grandis*
D. *Acacia nilotica*

104. ઉત્તર ગુજરાતના, ખેડૂતો કૃષિ ક્ષેત્ર બંડ પર કયું વૃક્ષ ઉછેર છે?

A. *Azardirecta indica*
B. *Butea monosperma*
C. *Ailanthus excelsa*
D. *Acacia nilotica*

100. Bamboo also known as

A. Green gold
B. Poor man's timber
C. Both A and B
D. None of the above

101. The only natural approach to enhance the tree cover India, other than recognized forests is;

A. Plantations in wetlands
B. Plantations in waterbodies
C. Social forestry plantations
D. Plantations in grasslands

102. In Gujarat, the State level Van-mahostava is being celebrated as

A. Ceremonial plantation at Vidhan sabha
B. Plantations at every Govt office
C. Plantations in Educational institutions
D. Development of Cultural forest (Sanskritic Van) on certain theme every year.

103. In Kheda district of Gujarat, the main tree which is being grown as a part of agro forestry

A. *Azardirecta indica*
B. *Eucalyptus* spp.
C. *Tectona grandis*
D. *Acacia nilotica*

104. In North Gujarat, farmers are raising the fodder tree on the bund of the agriculture field

A. *Azardirecta indica*
B. *Butea monosperma*
C. *Ailanthus excelsa*
D. *Acacia nilotica*

105. ગાંધીનગરમાં હરિયાળી સામાજિક વનીકરણના વાવેતર દ્વારા કરવામાં આવી છે, તેમાં જે વૃક્ષો વાવવામાં આવ્યા છે તે મુખ્ય વૃક્ષ કયું છે?
A. *Azardirecta indica*
B. *Tectona grandis*
C. *Ailanthus excelsa*
D. *Mangifera indica*

106. વૃક્ષની મુખ્ય કઈ પ્રજાતિઓ ગુજરાતના દરિયાકાંઠાના વિસ્તારોમાં વાવેતર માટે વપરાય છે?
A. *Azardirecta indica*
B. *Casuarina equisetifolia*
C. *Ailanthus excelsa*
D. *Acacia nilotica*

107. *Casuarina equisetifolia* દરિયા કિનારે ક્યારે વાવેતર કરવામાં આવે છે?
A. વરસાદ પહેલા
B. વરસાદમાં
C. વરસાદ પછી તરત જ
D. શિયાળા દરમિયાન

108. અમદાવાદ-વડોદરા એક્સપ્રેસ હાઇવે પર ઉનાળાની દરમિયાન રસ્તાના કાંઠે પીળા ફૂલના વૃક્ષો જોવા મળે છે, તે કયા છે?
A. *Butea monosperma*
B. *Delonix regia*
C. *Acacia Tabebuia*
D. *Acacia auriculiformi*

109. અમદાવાદના કાલિદાસ ભાઈ પટેલ શેના માટે જાણીતા હતાં ?
A. દૂધ ઉત્પાદનમાં તેમનું યોગદાન
B. કોટન મિલોમાં તેમનું યોગદાન
C. નીલગિરીના વાવેતર વધારવામાં તેમનું યોગદાન
D. ગાંધીવાદી ફિલસૂફીના જાણીતા ઉપદેશક

105. In Gandhinagar, the greenery has been developed through social forestry plantations, the main tree spp which has been planted is
A. *Azardirecta indica*
B. *Butea monosperma*
C. *Ailanthus excelsa*
D. *Acacia nilotica*

106. On Coastal areas of Gujarat, the main tree species are being used for plantations;
A. *Azardirecta indica*
B. *Casuarina equisetifolia*
C. *Ailanthus excelsa*
D. *Acacia nilotica*

107. The *Casuarina equisetifolia* on coastal is being planted
A. Before rainy season
B. During rainy season
C. Just after rainy season
D. During winter season

108. On Ahmedabad-Vadodara Express Highway, yellow flower trees are seen along roadside during summer season, which are these plants?
A. *Butea monosperma*
B. *Delonix regia*
C. *Acacia Tabebuia*
D. *Acacia auriculiformis*

109. Ahmedabad's Kalidas Bhai Patel was known for?
A. His contribution in Milk production
B. His contribution in Cotton mills
C. His contribution in raising *Eucalyptus* spp plantations
D. Well known preacher on Gandhian philosophy

110. દક્ષિણ ગુજરાતમાં નીલગિરીને એગ્રોફોરેસ્ટ્રી તરીકે ઉગાડવામાં આવે છે, કારણ કે

A. કોઈ અન્ય ઝાડની પ્રજાતિઓ ઉગતી નથી

B. જળ પ્રવેશને કાબૂમાં રાખવા માટે વધુ પાણીનો વપરાશ કરવા

C. CPM માટે સપ્લાય કરવા માટે અને પોલ તરીકે વપરાય છે.

D. સારા ઘાસચારાના વૃક્ષ હોવાને કારણે

111. કઇ ઝાડની પ્રજાતિને તેના પોષક મૂલ્યને કારણે શાળાઓમાં વાવેતર કરવા પ્રોત્સાહન આપવામાં આવે છે?

A. જામફળ

B. દાડમ

C. લીંબુ

D. સરગવો

112. કેન્દ્ર સરકાર દ્વારા ક્યા વર્ષ માં સહભાગી વન વ્યવસ્થાપન માર્ગદર્શિકા જારી કરવામાં આવી છે?

A. 1952 B. 1960

C. 1988 D. 1990

113. સહભાગી વન વ્યવસ્થાપન 1980 ના દાયકામાં ક્યાં રાજ્યમાં ઉદભવ્યું હતું?

A. પશ્ચિમ બંગાળ

B. મધ્યપ્રદેશ

C. ઉત્તર પ્રદેશ

D. ગુજરાત

114. સહભાગી વન વ્યવસ્થાપન ની વિભાવના નો અર્થ શું થાય છે ?

A. સરકાર દ્વારા જંગલોનું વ્યવસ્થાપન અને ઉદ્યોગો.

B. વન અને અન્ય વિભાગો દ્વારા જંગલોનું વ્યવસ્થાપન

C. સ્થાનિક લોકોના સહયોગથી સરકાર દ્વારા જંગલોનું વ્યવસ્થાપન

D. વન અને પંચાયત વિભાગો દ્વારા જંગલોનું વ્યવસ્થાપન

110. In South Gujarat, *Eucalyptus* spp are being grown as agroforestry, because

A. No other tree species can come up

B. To consume more water to control water logging

C. To supply for CPM and used as poles

D. Because of good fodder tree

111. Which of tree species is promoted to be planted in schools because of its nutritional value

A. Guava

B. Anar

C. Lemon

D. Drumstick

112. The Joint Forest Management guidelines have been issued by the Central government in

A. 1952 B. 1960

C. 1988 D. 1990

113. Joint Forest Management was originated in which state during 1980's

A. West Bengal

B. Madhya Pradesh

C. Utter Pradesh

D. Gujarat

114. The concept of Joint Forest management means

A. Management of forests by Govt. and Industries.

B. Management of forests by Forest and other Departments

C. Management of forests by Govt. in collaboration with local peoples

D. Management of forests by Forest and Panchayat Departments

115. તાજેતરમાં ગુજરાત રાજ્ય વન મહોત્સવ કાર્યક્રમ અંતર્ગત, કચ્છ ખાતેના સાંસ્કૃતિક વનનો વિકાસ કરવામાં આવ્યો છે, તેનું શું નામ છે?

A. પુનિત વન B. જૈવવિવિધતા વન
C. રક્ષક વન D. શક્તિ વન

115. Recently under Gujarat State Van mahostava programme, the Cultural forest at Kutch has been developed, known as

A. Punit Van B. Biodiversity Van
C. Rakshak Van D. Shakti van

116. ભારત સરકાર દ્વારા કયા વર્ષમાં રાષ્ટ્રીય એગ્રોફોરેસ્ટ્રી નીતિ શરૂ કરવામાં આવી?

A. 2014 B. 2016
C. 1992 D. 2010

116. National Agroforestry Policy has been launched by Govt. of India in

A. 2014 B. 2016
C. 1992 D. 2010

117. ગુજરાતમાં, ઝાડનું આવરણ લગભગ 11% ની મર્યાદામાં છે, જેમાં તેના ભૌગોલિક ક્ષેત્રના જંગલ ની બહાર કેટલું વૃક્ષ આવરણ છે?

A. 1% B. 3.5%
C. 8% D. 10%

117. In Gujarat, the tree cover is in the extent of around 11%, in which how much tree cover outside forest (TOF) of its geographical area?

A. 1% B. 3.5%
C. 8% D. 10%

118. ગુજરાત વન વિભાગે ક્યાં વર્ષ માં નોન-ફોરેસ્ટ લેન્ડ પર વૃક્ષારોપણ માટે "સામાજિક વનીકરણ કાર્યક્રમ" શરૂ કર્યો હતો

A. 1965-66 B. 1969-70
C. 1975-76 D. 1980-81

118. Gujarat Forest Department launched a "Social Forestry Programme" for planting trees on non-forest lands in the year

A. 1965-66 B. 1969-70
C. 1975-76 D. 1980-81

119. સામાજિક વનીકરણ હેઠળ, શાળા નર્સરીઓ ઉભી કરવામાં આવી રહી છે જેનો અર્થ શું થાય છે ?

A. વન વિભાગ શાળા કેમ્પસમાં નર્સરી ઉભી કરી રહ્યું છે
B. વન વિભાગ શાળાના નામે દરેક ગામમાં નર્સરી ઉભી કરી રહ્યું છે
C. વિદ્યાર્થી પોતે વન અધિકારીઓના માર્ગ-દર્શન હેઠળ નર્સરી ઉભી કરી રહ્યા છે.
D. શિક્ષકો વન વિભાગના સહયોગથી નર્સરી ઉભી કરી રહ્યા છે

119. Under Social forestry, school nurseries are being raised that means

A. Forest department is raising nurseries in school campus
B. Forest department is raising nurseries in every village in the name of school
C. The student themselves are raising nurseries under guidance of forest officers.
D. The teachers are raising nurseries in collaboration with forest department.

120. ગુજરાત માં કોના દ્વારા વિકેન્દ્રિત નર્સરી ઉભી કરવામાં આવી રહી છે?

A. શાળાઓ B. ખેડુતો
C. સ્ત્રીઓ D. ઉપરોક્ત તમામ

120. In Gujarat, the decentralized nurseries are being raised in social forestry by

A. Schools B. Farmers
C. Women D. All of above

121. ગુજરાત વન વિભાગની સામાજિક વનીકરણ યોજના અંતર્ગત પંચાયતની જમીનમાં ગ્રામ વન ઉભા કરવામાં આવી રહ્યા છે, પરિપક્વતા પછી ઝાડની હરાજી કરવામાં આવે છે અને ચોખ્ખી આવકનો કેટલા% નફો વિકાસ કાર્યોમાં ઉપયોગ માટે ગ્રામ પંચાયતને સોંપવામાં આવે છે?

A. 75% B. 50%
C. 25% D. 100%

122. ગુજરાત વન વિભાગ દ્વારા કયા પર્યાવરણીય વાવેતરો કરવામાં આવી રહ્યા છે?

A. સરકારી પરિસર
B. જાહેર મહત્વની સંસ્થાઓ
C. પર્યટક અને ધાર્મિક મહત્વના સ્થાનો
D. ઉપરોક્ત તમામ

123. ગુજરાત વન વિભાગની સામાજિક વનીકરણ યોજના અંતર્ગત રસ્તાની બાજુના વાવેતર ઉભા કરવામાં આવી રહ્યા છે, પરિપક્વતા પછી વૃક્ષોની હરાજી કરવામાં આવે છે અને ચોખ્ખી આવકનો કેટલા% વિકાસના કામોમાં તાલુકા પંચાયતને સોંપવામાં આવે છે ?

A. 75% B. 50%
C. 25% D. 100%

124. ગુજરાત વન વિભાગની સામાજિક વનીકરણ યોજના અંતર્ગત, મોટા પ્રમાણમાં ઉપયોગ માટે સામાન્ય સ્થળે વનકુટીર બનાવવામાં આવે છે, વન વિભાગ દ્વારા આ વનકુટીરના નિર્માણનો મુખ્ય શો ઉદ્દેશ છે?

A. આરામ કરવાની જગ્યા માટે
B. સ્થાનિકો દ્વારા બસ સ્ટેન્ડ તરીકે વાપરવા માટે
C. બાળકોના પિકનિક માટે
D. પર્યાવરણ, વનીકરણ અને વૃક્ષોની ખેતી વિશેની જાગૃતિને પ્રોત્સાહન આપવા

121. Under Social Forestry scheme of Gujarat forest department, the Gram van are being raised on panchayat land, after maturity, the trees are auctioned and how much% of the net income is handed over to Gram Panchayat for use in development works?

A. 75% B. 50%
C. 25% D. 100%

122. Environmental plantations are being raised by Gujarat forest department on

A. Government premises
B. institutions of public importance
C. places of tourist and religious importance
D. all above

123. Under Social Forestry scheme of Gujarat forest department, the road side plantations are being raised, after maturity, the trees are auctioned and how much% of the net income is handed over to Taluka Panchayat for use in development works?

A. 75% B. 50%
C. 25% D. 100%

124. Under Social Forestry scheme of Gujarat forest department, the Vankutir are being constructed on common place for mass utilization, the main objective of construction of these vankutir by forest department is

A. Just use as resting place
B. For use as Bus stand by the locals
C. For picnic of children
D. To promote the awareness about environment, forestry and tree farming

125. ગુજરાત વન વિભાગની સામાજિક વનીકરણ યોજના અંતર્ગત ગ્રામીણ વિસ્તારોમાં સુધારેલ સ્મશાન સ્થાપવામાં આવે છે, વન વિભાગ દ્વારા આ સ્થાપનાનો મુખ્ય શો ઉદ્દેશ છે?

A. આરોગ્ય સ્થિતિ સુધારવા માટે
B. સપાટીની સ્થિતિ સુધારવા માટે
C. બળતણ લાકડાનો વપરાશ ઓછો કરવો
D. પર્યાવરણ અંગે જાગૃતિ લાવવા

126. "વન પંડિત એવોર્ડ" ગુજરાત વન વિભાગ દ્વારા કોને આપવામાં આવે છે?

A. જંગલો વિશે જાણકાર વ્યક્તિને
B. જંગલો વિશે ઉચ્ચ શિક્ષણ પ્રાપ્ત કરેલ વ્યક્તિને
C. ખાનગી અને સામાન્ય જમીન પર મોટા વાવેતરમાં ફાળો આપનાર વ્યક્તિને
D. વન સંરક્ષણ માં ફાળો આપનાર વ્યક્તિને

127. ગુજરાતના વન વિભાગે ક્યા જિલ્લામાં આદિવાસી સમુદાયોનીની ખેતીની જમીન પર સાગના વાવેતર માટે એક વિશેષ યોજના શરૂ કરી છે?

A. છોટા ઉદેપુર
B. વલસાડ
C. ડાંગ
D. પંચમહાલ

128. ગુજરાત વન વિભાગ દ્વારા 2015 માં એગ્રોફોરેસ્ટ્રીને પ્રોત્સાહન આપવા માટે સૌરાષ્ટ્ર ફેલિંગ એક્ટ હેઠળ વૃક્ષની કેટલી જાતોને પરવાનગી માંથી મુક્તિ કરવામાં આવી છે?

A. 67 B. 77
C. 87 D. 97

125. Under Social Forestry scheme of Gujarat forest department, the improved crematoria are being installed in rural areas, the main objective of installation of these by forest department is

A. To improve the hygienic conditions
B. To improve surface conditions
C. To reduce the consumption of fuelwood
D. To promote the awareness about environment.

126. "Van Pandit Award" is being given by the Gujarat forest department to the person, who

A. having deep knowledge about the forests
B. having higher education about the forests
C. contributed in mass plantations on private and common lands
D. contributed in natural forest protection

127. Gujarat forest department has launched a special scheme to raise teak plantations on tribal's farmland in the district

A. Chota udepur
B. Valsad
C. Dangs
D. Panchmahal

128. In 2015, how many tree species have been freed for permission under Saurastra Felling of Tree Act 1951 to promote the Agroforestry by the Gujarat forest department.

A. 67 B. 77
C. 87 D. 97

129. ગુજરાત (જાહેર સેવા માટે નાગરિકોનો અધિકાર) અધિનિયમ, 2013 હેઠળ અનામત વૃક્ષો કાપવાની મંજૂરી કેટલા સમયમાં આપવાની હોય છે?

A. 3 મહિના B. 1 મહિનો
C. 2 મહિના D. 4 મહિના

130. ગુજરાત (જાહેર સેવા માટે નાગરિકોનો અધિકાર) અધિનિયમ, 2013 હેઠળ વન પેદાશોના પરિવહન માટેની પરવાનગી કેટલા સમયમાં આપવાની હોય છે?

A. 3 મહિના B. 1 મહિનો
C. 2 મહિના D. 4 મહિના

131. માઈક્રો પલાનના અમલીકરણ વિસ્તારમાંથી કાપવામાં આવેલા પોલ અને વાંસના કેટલા% ભાગ ગુજરાતમાં સંયુક્ત વન વ્યવસ્થાપન સમિતિ ના સભ્યોને આપવામાં આવે છે?

A. 50% B. 40%
C. 30% D. 20%

132. 1991 ની શરૂઆતમાં, ગુજરાત વન વિભાગ દ્વારા સહભાગી વન વ્યવસ્થાપન હેઠળ જંગલોની ઘનતા _______ થી ઓછી હોય તેવા જંગલોને મંજૂરી આપવામાં આવી છે.

A. 0.4 B. 0.5
C. 0.2 D. 0.3

133. ગુજરાત વન વિભાગમાં કેટલી સહભાગી વન વ્યવસ્થાપન સમિતિઓ કાર્યરત છે?

A. 3414 B. 6414
C. 8414 D. 1414

134. ગુજરાત વન વિભાગની સહભાગી વન વ્યવસ્થાપન કમિટી દ્વારા કેટલો વન વિસ્તાર સંચાલિત કરવામાં આવે છે?

A. 4.5 લાખ હેક્ટર B. 6.5 લાખ હેક્ટર
C. 7.5 લાખ હેક્ટર D. 1.5 લાખ હેક્ટર

129. Under Gujarat (Right to citizens to public service) Act, 2013, the permission for felling of reserved trees has to be given within

A. 3 months B. 1 month
C. 2 months D. 4 months

130. Under Gujarat (Right to citizens to public service) Act, 2013, the permission for transit of forest produce has to be given within

A. 3 months B. 1 month
C. 2 months D. 4 months

131. How much% of poles and bamboos, harvested from the implementation area of micro plan is given to members of Joint forest management committee in Gujarat?

A. 50% B. 40%
C. 30% D. 20%

132. Initially in 1991, the forests which are having the density less then ____has been permitted under Joint Forest management by the Gujarat Forest department

A. 0.4 B. 0.5
C. 0.2 D. 0.3

133. How many Joint Forest Management Committees are working in Gujarat Forest department;

A. 3414 B. 6414
C. 8414 D. 1414

134. How much forest area is being managed by Joint Forest Management Committees in Gujarat Forest department?

A. 4.5 lakh ha B. 6.5 lakhs ha
C. 7.5 lakhs ha D. 1.5 lakhs ha

135. એગ્રોફોરેસ્ટ્સમાં પણ લાઇટ ડિમાન્ડર પ્રજાતિઓનું વાવેતર ________ પછી કરવામાં આવે છે.

A. સિલેક્શન ફેલિન્ગ

B. ક્લિયર ફેલિન્ગ

C. છાંયડાવાળાં ઊભા ઢોળાવ પર

D. ગાઢ જંગલોમાં ખુલ્લી જગ્યામાં

136. વાંસની સફાઇ એટલે

A. પાણી દ્વારા વાંસની સફાઇ

B. સુકા અને તૂટેલા વાંસને ક્લમ્પથી દૂર કરવા

C. વાંસની શાખાઓ દૂર કરવી

D. વાંસના મૂળ દૂર કરવા

137. ISFR 2019 મુજબ, ગુજરાતમાં સૌથી વધુ વૃક્ષ આવરણ ધરાવતો જિલ્લો ________ છે.

A. ડાંગ B. વલસાડ

C. આનંદ D. જુનાગઢ

138. ISFR 2019 મુજબ, ગુજરાતમાં સૌથી ઓછા વૃક્ષ આવરણ ધરાવતો જિલ્લો ________ છે.

A. સુરેન્દ્રનગર

B. કચ્છ

C. બનાસકાંઠા

D. રાજકોટ

139. સાગ વાવેતર કેવી રીતે કરવામાં આવે છે?

A. એક વર્ષ રોપાઓ વાવીને

B. ખાડાઓમાં રોપાઓ વાવીને

C. પ્રથમ વરસાદ પછી સ્ટમ્પ વાવેતર

D. પરિપક્વ વૃક્ષની શાખાઓનું વાવેતર

140. ચેર નું વાવેતર ક્યારે કરવામાં આવે છે ?

A. વરસાદમાં

B. ઉનાળામાં

C. વરસાદ પછી શિયાળામાં

D. સાંજે

135. Even in Agroforests, the plantation of light demander species is being carried out after

A. Selection felling

B. Clear felling

C. On vertical slopes having shade

D. At openings of dense forests

136. Bamboo cleaning means

A. Cleaning of Bamboo by water

B. Removing dried and broken bamboos from clump.

C. Removing branches of bamboo

D. Removing roots of bamboo

137. As per last ISFR 2019, the district having the highest tree cover in Gujarat is

A. Dangs B. Valsad

C. Anand D. Junagadh

138. As per last ISFR 2019, the district having the lowest tree cover in Gujarat is

A. Surendranagar

B. Kutch

C. Banaskatha

D. Rajkot

139. How does the Teak plantation is being carried?

A. Planting of one-year seedlings

B. Planting in bug size of pits

C. Stump planting just after first rain

D. Planting of branches of mature tree

140. Mangroves plantations are being carried out in

A. Rainy season

B. Summer season

C. Winter season after rains

D. Evening

141. કેન્ડિડેટ પ્લસ વૃક્ષો શાને માટે પસંદ કરવામાં આવે છે?

A. સારા લાકડા મેળવવા માટે
B. નર્સરી માટે બીજ એકત્રિત કરવા
C. ગમ એકત્રિત કરવા માટે
D. તે વૃક્ષનું જતન કરવા

142. ISFR 2019 મુજબ, કયા વૃક્ષની પ્રજાતિઓ ગુજરાતમાં જંગલની બહાર સૌથી વધુ સંખ્યા ધરાવે છે?

A. દેશી બાવળ B. આંબો
C. લીમડો D. નીલગીરી

143. ISFR 2019 મુજબ, ગુજરાતમાં લીમડાના વૃક્ષની અંદાજીત ટકાવારી ____ છે.

A. 25.17% B. 20.17%
C. 35.17% D. 40.17%

144. વૃક્ષ સુધારણા કાર્યક્રમમાં શું શામેલ છે?

A. સીડલીંગ સીડ ઓર્ચાર્ડ
B. ક્લોનલ બીજ વાવેતર
C. કેન્ડિડેટ પ્લસ વૃક્ષોની પસંદગી
D. ઉપરોક્ત તમામ

145. ISFR 2019 મુજબ ડાંગ જિલ્લામાં ____% વૃક્ષના આવરણ હેઠળ ભૌગોલિક ક્ષેત્ર ધરાવે છે.

A. 50.16%
B. 55.16%
C. 77.16%
D. 33.16%

146. "તમારા છોડને અવલોકન કરવું તેના વીશેશીબયું ઉગાડવું અને તેનું પાલન કરવું" એ શાનો મુખ્ય ઉદ્દેશ છે?

A. શાળા નર્સરી
B. કિસાન નર્સરી
C. મહિલા નર્સરી
D. વિભાગ નર્સરી

141. Candidate Plus trees are selected for

A. To get good timber
B. To collect seeds for nursery
C. To collect gum
D. To conserve that tree

142. As per last ISFR 2019, the dominant tree species is having the highest number outside the forest in Gujarat is;

A. Desi Baval B. Mango
C. Neem D. Nilgiri

143. As per last ISFR 2019, the Neem tree species is having the highest number outside the forest in Gujarat, the percentage estimated in Neem trees is

A. 25.17% B. 20.17%
C. 35.17% D. 40.17%

144. Tree improvement programme includes

A. Seedling Seed orchard
B. Clonal seed orchard
C. Selection of plus trees
D. All above

145. As per last ISFR 2019, the Dangs district is having ________% of its geographical area under tree cover

A. 50.16%
B. 55.16%
C. 77.16%
D. 33.16%

146. "To observe, learn, grow and nurture your plant" is main objective of

A. School Nursery
B. Kisan Nursery
C. Mahila Nursery
D. Department Nursery

147. સંયુક્ત વન વ્યવસ્થાપન વિસ્તારોમાંથી વૃક્ષો કાપવાથી જે આવક પેદા થાય, તેમાંથી ________% JFMમાં જાય છે અને બાકી ગુજરાત સરકાર પાસે.

A. 20% B. 30%
C. 50% D. 40%

147. In Joint Forest Management areas, income generated from tree harvesting, ____________ goes to the JFMCs and remaining with the Government of Gujarat.

A. 20% B. 30%
C. 50% D. 40%

148. બિન અનામત ઘાસના મેદાનને કઈ સ્થાનિક સંસ્થાઓને ફાળવવામાં આવે છે

A. સ્થાનિક પશુ સંભાળ કેન્દ્રો, પાંજરાપોળ
B. સ્થાનિક માલધારી સહકારી મંડળી
C. સ્થાનિક દૂધ ઉત્પાદક સમાજ/સમિતિ
D. ઉપરોક્ત તમામ

148. The Non-reserved grasslands are allotted to Which organizations or local self-government bodies

A. Local cattle care centers, panjarapols
B. Local Maldhari Sahkari Mandali
C. Local milk producing society/ committee
D. All the above

149. કયું વૃક્ષ સામાન્ય રીતે માનવ વસવાટોની નજીક જોવા મળે છે?

A. પીપળો B. મહુડો
C. આંબો D. લીમડો

149. Which tree is generally found near human habitations?

A. Pipal B. Mahuda
C. Mango D. Neem

150. કયા વૃક્ષને "જંગલોની જ્યોત" તરીકે ઓળખવામાં આવે છે?

A. ખાખરો B. ગુલમોહર
C. શીમળો D. રૂખડો

150. Which tree is known as "the flame of forests"?

A. Palash B. Gulmohar
C. Semal D. Rukhedo

Answers | જવાબો

Que.	Ans.	Que.	Ans.	Que.	Ans.	Que.	Ans.	Que.	Ans.	Que.	Ans.	Que.	Ans.
1.	D	24.	D	47.	A	70.	B	93.	D	116.	A	139.	C
2.	D	25.	C	48.	C	71.	B	94.	D	117.	B	140.	C
3.	D	26.	A	49.	C	72.	B	95.	A	118.	B	141.	B
4.	D	27.	C	50.	C	73.	C	96.	D	119.	C	142.	C
5.	B	28.	A	51.	A	74.	D	97.	D	120.	D	143.	A
6.	B	29.	D	52.	A	75.	C	98.	B	121.	A	144.	D
7.	D	30.	D	53.	A	76.	A	99.	C	122.	D	145.	C
8.	A	31.	D	54.	C	77.	C	100.	C	123.	B	146.	A
9.	C	32.	D	55.	A	78.	A	101.	C	124.	D	147.	A
10.	C	33.	B	56.	A	79.	C	102.	D	125.	C	148.	D
11.	B	34.	A	57.	D	80.	A	103.	D	126.	C	149.	D
12.	C	35.	B	58.	C	81.	C	104.	C	127.	C	150.	A
13.	C	36.	A	59.	B	82.	C	105.	A	128.	C		
14.	C	37.	A	60.	A	83.	D	106.	B	129.	C		
15.	A	38.	A	61.	A	84.	D	107.	C	130.	B		
16.	D	39.	C	62.	A	85.	D	108.	C	131.	D		
17.	A	40.	C	63.	B	86.	B	109.	C	132.	A		
18.	B	41.	A	64.	A	87.	B	110.	C	133.	A		
19.	B	42.	A	65.	C	88.	D	111.	D	134.	A		
20.	D	43.	D	66.	D	89.	D	112.	D	135.	B		
21.	C	44.	C	67.	C	90.	C	113.	A	136.	B		
22.	B	45.	B	68.	A	91.	A	114.	C	137.	A		
23.	D	46.	B	69.	B	92.	D	115.	C	138.	D		

Chapter 12

પ્રકરણ ૧૨

Forest Utilization

વન વપરાશ

Timber classification | ઇમારતી લાકડાનું વર્ગીકરણ

According to refraction	
Highly refractory wood **(Class A)**	*Shorea robusta*
Moderately refractory wood **(Class B)**	*Tectona grandis, Dalbergia sissoo*
Non-refractory wood **(Class C)**	*Mangifera indica*
According to strength	
Exceptionally strong	*Mesua ferrea, Shorea robusta*
Extremely strong	*Acacia nilotica, Grewia tiliaefolia, Terminalia alata*
Very strong	*Tectona grandis, Anogeissus latifolia*
Strong	*Dalbergia sissoo, Dalbergia latifolia, Albizia lebbeck*
Moderately strong	*Adina cordifolia, Cedrus deodara, Chukrasia tabularis*
Somewhat weak	*Abies pindrew, Pinus roxburghii*
Weak	*Picea smithiana, Lannea coromandelicas*
Very weak	*Hymenodicton excelsum, Bombax ceiba, Cryptomeria japonica*
According to hardness	
Extremely hard	*Kusum, Mesua ferrea, Sal, Babul, Rosewood*
Very hard	*Terminalia alata, Dalbergia sissoo, Morus alba*
Hard	*Chukrasia tabularis, Tectona grandis*
Moderately soft	*Mangifera indica*
Soft	*Lannea coromandelica, Deodar, Silver fir*
Very soft	*Boswellia serrata, Chirpine, Spruce*
Extremely soft	*Semal, Cryptomeria japonica*

According to elasticity	
Very elastic	*Grewia tiliaefolia, Anogeissus latifolia, Mesua ferrea*
Elastic	*Chloroxylon swietenia, Sal, Babul, Sissoo, Rosewood*
Moderately elastic	*Mangifera indica, Toona ciliata, Tectona grandis*
According to fissility	
Extremely fissile	*Tectona grandis*
Easily fissile	*Acacia catechu, Adina cordifolia, Toona ciliata*
Moderately fissile	*Bombax ceiba, Anogeissus latifolia*
Difficult to split	*Dalbergia sissoo & Shorea robusta*
Very difficult to split	*Dalbergia latifolia, Ougenia oojeinensis*

Important Wood Uses | લાકડાનાં મહત્વપૂર્ણ ઉપયોગો

Used in	Wood qualities	Suitable species
Agricultural implements	Strongest, toughest and hardest	*Acacia nilotica* (Babul), *Acacia catechu* (Khair), *Anogeissus latifolia* (Axlewood), *Shorea robusta* (Sal), *Ougenia oojeinensis (Sandan), Mesua ferrea* (Mesua), *Schleichera oleosa* (Kusum), *Xylia xylocarpa* (Irul) etc.
Aircrafts timber	Sound. Knot free. Straight grain, light weight. High strength/weight ratio.	*Picea sitchensis* (Sitka spruce) *Abies pindrew* (Silver fir) *Picea smithiana* (Spruce)
Batons	A very heavy and strong wood.	*Olea ferruginea* (Olive) *Dalbergia latifolia* (Rosewood) *Heritiera minor* (Sundri) *Mesua ferrea* (Mesua)
Battery separators	Good matching property, permeable, straight grained, strong, knot free, free from volatile acids, tannins & resinous matters, resistance to the action of electrolyte & mineral acids.	*Mostly conifers.* *Abies pindrew* (Silver fir) *Picea smithiana* (Spruce) *Cedrus deodara* (Deodar) *Cupressus torulosa* (Cypress) *Pinus wallichiana (*Kail) etc.
Bearings	Extreme hardness, very great strength	Lignum vitae obtained from *Guaiacum officinalis & G. sanctum* *Manilkara littoralis* (Andman bullet wood), *Hardwickia binata* (Anjan), *Soymida fabrifuga* (Rohini) etc.
Bent-wood articles (sports goods)	Straight, grained, strong, knot free and well-seasoned	*Morus alba* (Mulberry) *Fraxinus species* *Celtis australis* (Celtis) *Salix alba*

Used in	Wood qualities	Suitable species
Boat & ship	Strong, elastic, more durable, knot free, non-corrosive, low expansion & contraction coefficient and free from tannic acid.	**Ship-** *Tectona grandis* (Teak) *Admirality teak* (Burma teak) **Boat-** *Quercus* spp. (European Oakwood) *Ailanthus excelsa* (Mahogany) *Acer species* (Maple)
Boot-lasts & Shoe-heels	Medium weight, moderately hard but tough, fine textured, smooth finishing etc.	*Fagus species & Acer species* (Maple) – *shoe-lasts* *Dalbergia sissoo* (Shisham) *Adina cordifolia* (Haldu) *Abies pindrew* (Silver fir) *Picea smithiana* (Spruce) *Acer species* (Maple) for Shoe-heels
Brushes & Brooms	Medium hardness, fine to medium fine texture, smooth finishing, easy to season, light colour etc. For ornamental brushes a good figure & colour are also desirables.	**Toilet & hair brushes-** *Chloroxylon swietenia* (Satin wood) *Dalbergia latifolia* (Rosewood) *Dalbergia sisso* (Shisham) *Diospyros ebenum* (Ebony) **Shaving brushes-** *Mitragyna parviflora* (Kaim/Kadamb) **Horse brushes-** *Lannea coromandelica* (Jhingan) **Paint & varnish brushes-** *Acer species* (Maple) *Adina cordifolia* (Haldu) *Gmelina arborea* (Ghamari) *Grewia tiliaefolia* (Dhaman) **Brooms & general utility brushes-** *Cedrus deodara* (Deodar) *Holoptelea integrifolia* (Kanju) *Mitragyna parviflora* (Kaim/Kadamb) *Mangifera indica* (Mango)
Furniture, cabinet-making & Panelling	Good colour, strong, elastic & tough, strain resistance, good grain & figure, non-liability to crack, non-split, warp or move excessively and easy to work & finish.	*Tectona grandis* (Teak) *Dalbergia latifolia* (Rosewood) *Dalbergia sisso* (Shisham) *Pterocarpus marsupium* (Bijasal) *Toona ciliata* (Toon) *Albizia* spp.(Siris) *Betula alnoides* (Birch) *Gmelina arborea* (Ghamari) *Ougenia oojeinensis* (Sandan) *Aesculus indica* (Horse chestnut)

Used in	Wood qualities	Suitable species
Matchwood industry	Straight grain, good fissility, strength & toughness, good white colour, free from knots, easily peelable, light weight, capacity to absorb paraffin and burn smokeless, efficient to burn.	***Populus tremula*** (European aspen) & ***Tilia japonica*** best matchwood. *Sideroxylon longipetiolatum* (Lambapatti), *Bombax ceiba* (Semal), *Sterculia villosa* (Udal), *Trewia nudifera* (Gutel), *Mangifera indica* (Mango), *Alstonia scholaris* (Chatian), *Anthocephalus cadamba* (Kadam), *Endospermum malaccense* (Bakota).
Mathematical instruments	Fine, even textured, non-liable to crack & wrap.	*Pinus wallichiana* (Kail), *Boxus sempervirens* (Boxwood), *Juglans regia* (Walnut), *Tectona grandis* (Teak) *Toona ciliata* (Toon), *Abies pindrew* (Fir) *and Adina cordifolia* (Haldu)
Musical instruments	Fine, even textured, non-liable to crack & wrap, free from knots, having uniform annual rings, regular structure.	Conifer wood- ***Picea smithiana*** **Violins-** Bridge- *Acer species* (Maple) Keys- *Tectona grandis* (Teak) *Diospyros spp.* (Ebony) Bows- *Dalbergia latifolia* (Rosewood) *Heritiera minor* (Sundri) **Sitars-** Body- *Toona ciliata* (Toon) Neck- *Tectona grandis* (Teak) Keys- *Cedrus deodara* (Deodar) *Dalbergia sisso* (Shisham) **Vina & Tambora-** *Artocarpus hetrophyllus* (Kathal) *Pterocarpus marsupium* (Bijasal) *Gmelina arborea* (Ghamari) **Guitar-** *Canrium euphyllum* (Ghite dhup) **Harmonium-** *Tectona grandis* (Teak) **Drums-** *Fraxinus* spp. (Ash) *Dalbergia sisso* (Shisham) *Albizia* spp. (Siris) **Piano-** *Juglans regia* (Walnut) *Swietenia* spp. (Mahogany)
Packing case industry	Light weight, reasonably strong, good nailing properties, preferably	*Picea smithiana* (Spruce). *Pinus wallichiana* (Kail), *Abies pindrew* (Fir),

Used in	Wood qualities	Suitable species
	whitish, easily worked, cheap & plentiful, free from warping & shrinkage free from bed odour.	*Bombax ceiba* (Semal), *Mangifera indica* (Mango), *Trewia nudifera* (Gutel), *Ailanthus excelsa* (Maharukh), *Alstonia scholaris* (Devil tree), *Tetrameles nudifera* (Maina).
Pencil industry	Light weight, pink in colour, have good whittling & matching properties, fine even textured, straight & closed grain, free from knots.	*Juniperus virginiana* (Pencil cedar) *Juniperus procera* (Wood balsa) *Juniperus macropoda* (Juniper) *Alnus nepelensis* (Alder)
Plywood industry	Wood with closed firm grain, round, free from knots & resin, interlocked fibres and free from excessive bulges, taper & buttresses.	*Dipterocarpus macrocarpus* (Hollong) *Terminalia myriocarpa* (Hollock), *Mangifera indica* (Mango), *Ailanthus spp. Toona ciliata* (Toon). *Populus* spp., *Holoptelea integrifolia* (Kanju), *Swietenia spp.* (Mahogany), *Terminalia bellerica* (Bahera)
Railway sleepers	Strength, resist to all pressure, good seasoned, shock resistance, good absorber of coal tar creosote.	*Shorea robusta* (Sal) *Mesua ferrea* (Mesua) *Cedrus deodara* (Deodar) *Pterocarpus marsupium* (Bijasal)
Railway carriages	Sufficient strength, free from seasoning defects, availability & retention of shape.	*Tectona grandis* (Teak) *Shorea robusta* (Sal)
Rifle parts & Gun stocks	Light moderately hard, closed grain, round. Non-split, seasoned wood, capable of standing up to the high-speed machine used in manufacturing process, warp or move in service.	*Juglans regia* (Walnut) *Juglans nigra* (Wild walnut) *Acer pictum* (Maple) *Prunus padus* (Cherry)
Sports good industry	Sound, knot free, straight grain, light weight. High strength/weight ratio, pliable, tough, shock resisting, easy to work, preferably light in colour and taking a good polish capacity.	Europe & America- *Fraxinus* spp. (Ash) In India most commonly used – *Morus alba* (Mulberry)
	Billiard cues-light weight, Knot free. Straight grain, white colour or nearly so.	*Grewia tiliaefolia* (Dhaman), *Diospyros melanoxylon* (Ebony) & *Polyalthia fragrans* (Debdharu).
	Bows & Arrows- Very strong, tough & an elastic wood,	Bow-*Taxus baccata* (Yew), *Grewia tiliaefolia* (Dhaman), *Parrotia jacquemontiana* (Parrotia), *Acacia catechu* (Khair). Arrows- reeds & bamboos.

Used in	Wood qualities	Suitable species
	Cricket bats, stumps & bails- light weight, preferably light in colour, reasonably strong, shock resistance.	*Salix* spp. (Willow) in India. The best cricket bats are made of the English willow (*Salix alba* var. Caerulea). Stumps & bails- Morus alba (Mulberry) & Celtis.
	Fishing rod	*Ocotea rodioei* (Greenheart)
	Golf clubs	*Carya* spp. (South American Hickory)
	Hockey sticks- Light weight, preferably light in colour, reasonably strong, shock resistance amenable to steam bending	*Morus alba* (Mulberry) *Fraxinus spp.* (Ash) *Celtis australis* (Celtis)
	Skis (skiing)- toughness, flexibility, even-textured, straight grain,	*Morus alba* (Mulberry) *Fraxinus* spp. (Ash) *Dalbergia sissoo* (Shisham)
	Tennis, Badminton & squash rackets-straight grain, long fibre wood with good shock resistance.	*Morus alba (Mulberry)* *Fraxinus* spp. (Ash)
Shoulder poles	Strong, tough & an elastic wood free from knots.	*Anogeissus latifolia (Axlewood)* *Anogeissus acuminata* (Yon) These are usually made from bamboo.
Shingles	Durable, free from knots, non-splitting, light in weight, adequate strength, taking paint & preservative capacity.	*Pinus wallichiana* (Kail) *Pinus roxburghii* (Chirpine) *Cedrus deodara* (Deodar) *Abies pindrew* (Fir) *Hopea odorata* (Thingan)
Tents poles & tents pegs	Round or squid wood, strong. Mostly made of bamboos. For tent pegs-hard, durable & tough.	*Grewia tiliaefolia* (Dhaman), *Shorea robusta* (Sal), *Acacia nilotica* (Babul), *Tamarinds indica* (Imli), *Schleichera oleosa* (Kusum), *Ougenia oojeinensis* (Sandan), *Heritiera minor* (Sundri).
Tobacco pipes	High heat withstand capacity, non-split, strong wood.	*Dalbergia sisso* (Shisham) *Dalbergia latifolia* (Rosewood) *Tectona grandis* (Teak) knots, **Country pipes & hookahs-** *Amoora cucullata, Boswellia serrate* *Dalbergia sisso* (Shisham)
Toys	Light weight, fine even textured, having turning & carving qualities.	*Adina cordifolia* (Haldu), *Artocarpus hetrophllus* (Kathal), *Dysoxylum malabaricum* (White Cedar), *Miliusa tomentosa* (Hoom), *Wrightia tinctoria (Dhudhi), Santalum album* (Sandal wood), *Aegle marmelos* (Bel).

Used in	Wood qualities	Suitable species
Umbrella handle	Extensively bamboos.	*Arundinaria* spp. *Dendrocalamus strictus*
Walking sticks	Straightness & strength wood	Bamboo & canes are extensively used. *Celtis australis* (celtis), *Dalbergia latifolia* (rosewood), *Dalbergia sisso* (Shisham) & *Santalum album.*

Insect defects in timber | ઇમારતી લાકડામાં કીટકોની અસરો અને ઉપાયો

Insects	Insect attacks on	Preventive measures
Bark beetles	Bark of felled tree	Debarking & quick seasoning of felled wood and chemical treatment
Carpenter bees	Hole in timber	Spraying wood with chemical
Flatheaded borers	Bark & sapwood	Debarking of felled wood soon
Longhorn borers	Bark, sapwood & heartwood	Spraying wood with chemical
Powder-post borer	Sapwood	Steam sterilization & chemical soaking or spraying of greenwood

Fibres of India | ભારતના ફાઇબર્સ

Species	Fibre color	Use
Fibres obtained from leaves		
Agave sislana (Sisal fibre)	—	Cordage, rope making
Caryota urens (Kittul)	Brown to black	Rope making, fishing nets
Musa textilis	—	Rope making,
Pandanus species	—	Cordage, Rope making, fishing nets
Fibres obtained from stems		
Acacia leucophloea (Hiwar)	—	Fishing nets & cordage
Bauhinia vahlii	—	Tying purpose
Boehmeria nivea ('Rhea' or 'Ramie' fibre)	White	Rope making
Calotropis gigantean	—	Fishing nets & cordage
Cannabis sativa	White	Ropes, cable, twines, mats, canvas,
Ficus species	—	Rope making
Girardinia hetrophylla	—	Rope making, twines & coarse cloths
Grewia tiliaefolia	Yellow	Rope making
Hardwickia binata (Anjan)	Red	Cordage & rope making
Sterculia villosa (Udal)	Whitish-pink	Elephant harness, drags ropes, raft tying
Trema orientalis	Light brown	Twines and coarse cloth & rope making
Flosses yielding plants		
Bombax ceiba (Indian kapok)	—	Cotton industry

Species	Fibre color	Use
Calotropis gigantea	—	Stuffing mattresses, pillows,
Ceiba patendra (kapok)	—	Spinning, weaving or stuffing
Cochlospermum religiosum	Yellow	Stuffing mattresses, pillows, cushions and life saving belts

Dyes of India | ભારતનાં ડાઇઝ

Species name	Known as	Dye colour	Uses
Dyes obtained from woods			
Pterocarpus santalinus	Santalinus dye	Red	Salmon pink colour to cloth
Caesalpinia sappan	Brazilin dye	Red	Dying of cotton, silk, wool
Atrocarpus hetrophyllus	Kathal dye	Yellow	Silk cloth dying
Acacia catechu	Cutch dye	Black	Tanning of lather
Dyes obtained from barks			
Acacla species	Acacla dye	Black	Dying agent
Alnus species	Alnus dye	Red	Cloth dying
Terminalia alata	Alata dye	Black	Dying as black colour
Dyes obtained from flowers & fruits			
Mallotus philippensis	Kamela dye	Orange	Dying silk
Bixa orellana	Annatto dye	Yellow	Dying of silk, wool & calico
Butea monosperma	Dhak dye	Orange	Dying cloth
Nyctanthes arbortristis	Harsringar	Orange	Liquors colouring
Wrightia tinctoria	Dudhi dye	Blue	Colouring agent
Dyes obtained from leaf			
Indigofera species	Indigo dye	Blue	Cloth dying
Lawsonia inermis	Heena dye	Red	Hair, palms & fabric dying

Gum yielding trees of India | ભારતનાં ગુંદર આપનારા વૃક્ષો

Gum	Species	Use
Acacia gum	*Acacia catechu*	Dying and colouring
Dhaura gum	*Anogeissus latifolia*	Paper sizing & calico print.
Gum Arabic (Indian)	*Acacia nilotica*	Calico-printing, sizing material for silk and cotton
Gum Arabic (True)	*Acacia senegal*	Medicinal purpose
Gum kino	*Pterocarpus marsupium*	As astringent in medicine
Jhingan gum	*Lannea coromandelica*	Making ink ,confectionary
Katira gum	*Sterculia urens*	Used in medicine for diarrhoea and dysentery
Moringa gum	*Moringa pterygosperma*	Medicinal use
Salai gum	*Boswellia serrata*	As astringent in medicine
Semla gum	*Bauhinia retusa*	Eating & sizing cloth, paper

Plant based oils of India | ભારતના વનસ્પતિ આધારિત તેલ

Species	Obtained	Colour	Source of	Use
Grass oils				
Palmrosa grass *Cymbopogon martini*	Infloresc-ence	Pale yellow liquid	Geraniol	Cosmetics industry, mosquito repelling ointments
Lemon grass *Cymbopogon Flexosus*	Whole grass	Reddish yellow	Citral (41-85%)	Scents, Cosmetics industry mosquito repelling ointments
Ginger grass *Cymbopogon martini*	—	—		Soap industry
Citronella grass *Cymbopogon nardus*	—	—	—	Low cost perfume
Khus grass *Vetiver zizaniodes*	—	—	—	Perfume, soap and cosmetics
Root oils				
Costus oil *Sassuria lappa*	Roots	Pale yellow	—	Perfume blending
Valerian oil *Valerian wallichii*	Dry Rhizome	—	—	Flavoring and blending
Flower oils				
Keora oil *Pandanus tectorius*	Flower	—	Keora Attar	Hair oils, perfumery and cosmetics
Acacia farnesiana	Flower	—	Cassie	Perfumery
Champ oil *Michelia champa*	Flower	—	Attar	Perfumery
Leaf oils				
Eucalyptus oil *Eucalyptus globulus*	Leaves, branch	Colour Less	—	Toothpaste, perfume, vermin-repellents etc.
Citriodora oil	Leaves		citronellal	Soap & perfumery
Camphor oil *Cinnamomum camphora*	Leaves and wood	—	—	Cold remedies, insecticides and perfumery
Mint oil *Mentha arvensis*	Leaves	—	Menthol	Medicines and confectionary
Wintergreen oil *Gaultheria fragrantissima*	Leaves	—	—	Perfumery and pharmaceutical
Wood oils				
Sandalwood oil *Santalum album*	Heart Wood	—	Santalol	Perfume industry
Agar oil *Aquilaria agallocha*	Heart Wood	—	—	Perfume industry as a perfume retainer

Species	Obtained	Colour	Source of	Use
Deodar wood oil *Cedrus deodara*	Wood	Reddish-brown		Perfume, soap and cosmetics
Pine oil	Heart Wood	—	—	Medicine, veterinary disinfectant sprays, soap

Seed oils of India | ભારતના સીડ ઓઇલ્સ

Species	Seed yield	Use
Aleurites fordii	Tung oil	Paints and varnish
Azadirachta indica	Margo oil	Medicinal and soap making
Madhuca butyracea	Phulwara Butter	Soap, candles and chocolate manufacture
Garcinia indica	Kokam Butter	Edible fat
Hydnocarpus kurzii	Chaulmugra	Leprosy & skin disease
Madhuca indica	Mahua Butter	Cooking purpose, hair oil, soap & medicine
Pongamia pinnata	Kanju oil	Soap and leather tanning
Schleichera oleosa	Kusum oil	Cooking, lightning and soap making
Shorea robusta	Sal Butter	Cooking, lightning and soap making
Vateria indica	Piney tallow	Medicinal and lightning purpose

Resins of India | ભારતના રેઝીન્સ

Resin type	Species
True resins	
Black dammar	*Canarium stricum*
Green dammar	*Shorea umbuggaia*
Rock dammar	*Hopea odorata*
White dammar	*Vateria indica*
Amber	*Pinus succinifera*
Lacquer	*Rhus vernicifera*
Hellackm	*Laccifera lacca* (Insect)
Mastic	*Pistacia leutiscus*
Sand arac	*Callitiris quadrivalvis*
Oleoresin (Resin mixed with essential oils)	
Salai gum	*Boswellia serrata*
Gum resins	
Asafoetid	*Ferula asofoetida*
Gambage	*Garcinia* spp.
Galbanum	*Ferula galbaniflua*
Myrrah	*Commiphera* spp.

Important Medicinal Plants of India | ભારતની મહત્વપૂર્ણ ઔષધીય વનસ્પતિઓ

Medicinal Plant	Use
Abrus precatorius	Seed used for skin diseases
Adhatoda vasica	Dry cough
Aegle marmelos	Ripe fruits use for stomach disorder
Ageratum conyoides	Leaf powder used for headache
Aloe barbadensis	Wet cough, constipation, cut and wounds
Alstonia scholaris	Stem bark used for fever
Annona squomosa	Leaf juice is antiseptic
Argemone mexicana	Latex used in eye disorders
Azadirachta indica	Antiseptic, Skin diseases, Stomach ache
Cassia fistula	Fruit pup for liver disorder
Cassia tora	Skin diseases
Celastrus panicularus	Seed oil used for hair tonic
Centella asiatica	Improvement of memory
Cissampelos pareira	Stomach pain
Crisium arvense	Root paste for diuretic purpose
Cynodon dactylon	Heal wounds
Emblica officinalis	Constipation
Flemingia strobolifera	Leaves extract used as animal tonic
Helicteres isora	Dysentery and stomach pain
Hibiscus rosa-sinensis	Staminal column used for diuretic in kidney trouble
Holarrhenct antidysentrica	Diarrhea
Lerwsonia inermis	Fungal infection
Mallotus philippensis	Powder from fruit used for bone factures
Mimosa pudica	Cut and wounds
Moringa oleifera	Fungal infection
Ocimum sanctum	Decoction of leaves used in fever
Piper longum	Cold and cough
Pterocarp usmarsupium	Chest pain
Punica granatum	Acidity
Rauvolfia serpentina	Roots used for fever
Shorea robusta	Dysentery
Vanda tasellata	Bone facture
Vetiveria zizanoides	Relief pain
Withania somnifera	Chronic disorders

Species for Drugs | વિવિધ પ્રકારની દવાઓ માટેની પ્રજાતિઓ

Root Drugs	*Aconitum species* *Acorous calamus* *Asparagus adscendens* *Berberis aristata* *Dioscorea deltoidea* *Glycyrhiza glabra* *Picrorhiza kurrooa* *Podophyllum hexandrum* *Rauwolfia serpentine* *Rheum emodi* *Saussuria lappa* *Valerian wallichii*
Bark Drugs	*Cinchona hybrida* *Holarrhena antidysentrica* *Soymida febrifuga*
Flower/fruit/seed Drugs	*Aegle marmelos* *Artemisia species* *Caesalpinia erista* *Cassia fistula* *Chenopodium species* *Plantago ovata* *Pyrethrum species* *Ricinus communis* *Strychnos nux-vomica* *Tamarindus indica*
Leaf Drugs	*Atropa acuminata* *Azadirachta indica* *Cannabis sativa* *Datura species* *Ephedra gerardiana* *Gaultheria fragrantissima* *Mentha species* *Ocimum killimandscharicum* *Swertia chirata* *Vitex peduncularis*

Species for Industries | ઉદ્યોગો માટે પ્રજાતિઓ

Pulpwood Species	*Bamboo* *Casuarina* spp. *Eucalyptus* spp. *Leucaena* spp. *Acacia* spp. *Populus* spp.

Matchwood Species	*Ailanthus excelsa* *Albizia falcataria* *Ailanthus triphysa* *Albizia lebbeck* *Anthocephalus cadamba* *Erythrina indica* *Populus* spp.
Timber Species	*Tectona grandis* *Terminalia* spp. *Albizia* spp. *Gmelina* spp. *Azadirachta indica* *Pterocarpus* spp. *Mangifera indica* *Artocarpus* spp. *Dalbergia* spp. *Dipterocarpus* spp. *Eugenia* spp.
Plywood Industries	*Populus* spp. *Melia dubia* *Paulownia* spp. *Eucalyptus* spp. *Ailanthus* spp.
Energy Crop	*Prosopis* spp. *Acacia* spp. *Albizia* spp. *Dalbergia sissoo* *Leucaena leucocephala* *Casuarina equisetifolia* *Eucalyptus* spp. *Gliricidia* spp. *Ceasalpinea* spp. *Chuckrassia* spp. *Other hard woods*

Multiple Choice Questions | હેતુલક્ષી પ્રશ્નો

1. સ્ટર્ડી ટ્રી માં ____________ ખામી "કુદરતી ખામી" છે
   A. ગાંઠો B. બ્રેશનેશ
   C. રિએક્શન વુડ્સ D. ઉપરોક્ત તમામ
2. રિએક્શન અથવા અસામાન્ય વૂડ્સ કયુ છે?
   A. કમ્પ્રેશન લાકડુ
   B. ખામીયુક્ત લાકડું
   C. ટેંશન લાકડું
   D. બંને A અને C
3. ગોન્ડ વન સમુદાય નીચેનામાંથી કયા રાજ્યનો છે?
   A. ઝારખંડ B. જમ્મુ અને કાશ્મિર
   C. ગુજરાત D. છત્તીસગઢ
4. કયા સમુદાયો કુશળ વૃક્ષ કાપનારા છે?
   A. આફ્રિકાના માસાઈ
   B. ઓરિસ્સાના ગોન્ડ
   C. જાવા ના કલન્ગ
   D. છોટા નાગપુર ના મુંડા
5. દક્ષિણ પૂર્વ એશિયામાં સ્થાનાંતરિત ખેતીનું સ્થાનિક નામ __________ છે.
   A. પોડુ B. મિલ્પા
   C. નેરાડ D. લોડાંગ
6. જાવા માં "સ્કોર્ચડ અર્થ" નીતિ શેના પરથી આવી?
   A. જાપાનીઝ B. ડચ
   C. બ્રિટિશ D. ફ્રેન્ચ
7. કેલ (*Pinus wallichiana*) નાં લાકડાનો ઉપયોગ શેમાં કરવામાં આવે છે?
   A. જોડાણના લાકડા
   B. બાંધકામના કામો
   C. હલકા ફર્નિચર
   D. ઉપરોક્ત તમામ

1. **The defects "Natural defect" in sturdy tree are**
   A. Knots B. Brashness
   C. Reaction woods D. All of the above
2. **Reaction or abnormal woods are**
   A. Compression wood
   B. Defective woods
   C. Tension wood
   D. Both A & C
3. **The Gond forest community belongs to which of the following State?**
   A. Jharkhand B. J&K
   C. Gujarat D. Chattisgarh
4. **Which communities are of skilled forest cutters?**
   A. Maasais of Africa
   B. Gonds of Orissa
   C. Kalangs of Java
   D. Mundas of Chotanagpur
5. **The vernacular name of shifting cultivation in SE Asia is**
   A. Podu B. Milpa
   C. Nerad D. Lodang
6. **The "Scorched earth" policy in Java was followed by**
   A. Japanese B. Dutch
   C. British D. French
7. **Kail (*Pinus wallichiana*) utilization of timber**
   A. Joinery wood
   B. Construction works
   C. Light furniture
   D. All of the above

8. અખરોટ (*Juglans regia*) લાકડાનો ઉપયોગ _________ માં થાય છે.
A. કોતરકામ અને રાઇફલ સ્ટોક્સ
B. કેબિનેટ્સ
C. બંને A અને B
D. ઉપર્યુક્તમાંથી એક પણ નહીં

9. અસામાન્ય લાકડા કયા કારણોસર બને છે?
A. ગુરુત્વાકર્ષણ
B. તણાવ અને સંકોચન
C. પ્લાન્ટના હોર્મોન્સમાં ખલેલ
D. ઉપરોક્ત તમામ

10. લાકડામાં મહત્તમ રેઝિનસ પદાર્થ કયા હેતુ માટે સારા છે?
A. ફર્નિચર
B. બળતણ
C. વાસણો
D. રમતગમત

11. ક્યા પ્રિઝર્વેટિવ્સ નોન-ફિક્સ્ડ વોટર-સોલ્યુબલ પ્રકારના લાકડાના પ્રિઝર્વેટિવ્સના ઉદાહરણ છે?
A. ZMA
B. ACA
C. ACC
D. ઉપરોક્ત તમામ

12. લેમિન કોર બોર્ડની પહોળાઈ _________ છે.
A. 2.5 સે.મી. કરતાં ઓછી
B. 3.5 સે.મી. કરતાં ઓછી
C. 4.0 સે.મી. કરતાં વધુ
D. 5.5 સે.મી. કરતાં વધુ

13. _________ લાકડુ "મોડીફાઇડ વૂડ્સ" તરીકે પણ ઓળખાય છે
A. સંયુક્ત લાકડુ
B. લેમિનેટ લાકડુ
C. સુધારેલ લાકડુ
D. પ્લાયવુડ

8. Walnut (*Juglans regia*) wood is utilize as_____.
A. Carving and rifle stocks
B. Cabinets
C. Both A & B
D. None of these

9. What causes of abnormal wood?
A. Gravity
B. Tension and compression
C. Disturbance in plant hormones
D. All of the above

10. Woods have maximum resinous substance are good for which purpose?
A. Furniture
B. Fuel
C. Utensils
D. Sports

11. Which preservatives are the example of non-fixed water-soluble type wood preservatives?
A. ZMA
B. ACA
C. ACC
D. All of the above

12. Lamin core board has width size of....
A. less than 2.5 cm
B. less than 3.5 cm
C. more than 4.0 cm
D. more than 5.5 cm

13. Modified woods are also known as
A. Composite wood
B. Laminate wood
C. Improved wood
D. Plywood

14. હિસ્ટોજન સિદ્ધાંત ________ દ્વારા આપવામાં આવ્યો હતો.

A. ન્યૂમન
B. કાર્લ નાગેલી અને હોફમિસ્ટર
C. શ્મિટ
D. હેન્સ્ટન

15. નીચેનામાંથી કયું વન વ્યવસ્થાપનનો ભાગ નથી?

A. સાઇટ સમાનતા
B. વય-વૃદ્ધિ
C. સામાન્ય વૃદ્ધિ
D. ફેલીંગ

16. ભારતીય લાખ સંશોધન સંસ્થા ક્યાં આવેલ છે?

A. બિહાર B. મહારાષ્ટ્ર
C. ઝાંસી D. રાંચી

17. નીચેનામાંથી કયા આંશિક મૂળના પરોપજીવી છોડ છે?

A. વિસ્કમ B. લોરેન્થસ
C. ચંદન D. બંને A અને C

18. દર વર્ષે 60 સે.મી.થી વધુ વૃદ્ધિ કરનાર છોડની પ્રજાતિ ______ તરીકે ઓળખાય છે.

A. ઝડપી વિકસતી પ્રજાતિઓ
B. મધ્યમ વધતી પ્રજાતિઓ
C. ધીમી ગતિએ પ્રજાતિઓ
D. ખૂબ ઝડપી વિક્સની પ્રજાતિઓ

19. "ઇંધણના લાકડાની ઘનતા" નું પ્રમાણ કઈ પદ્ધતિ દ્વારા નક્કી કરી શકાય છે?

A. સ્પેસિફિક ગ્રેવિટી પદ્ધતિ
B. ઝાયલોમેટ્રિક પદ્ધતિ
C. બંને A અને B
D. ઉપર્યુક્તમાંથી કોઈ નહિ

20. ક્યુબિક કન્ટેન્ટમાં સોલિડ વોલ્યુમ કયા નામે ઓળખાય છે?

A. સ્ટોક વોલ્યુમ B. સ્ટેક્ડ વોલ્યુમ
C. ટોટલ વોલ્યુમ D. એકચ્યુઅલ વોલ્યુમ

14. Histogen theory was proposed by

A. Newman
B. Karl Nageli & Hofmeister
C. Schmidt
D. Hanstein

15. Which from the following is not a part of norms in forest management?

A. Site equality
B. Age-gradation
C. Normal increment
D. Felling

16. Indian Lac Research Institute is situated at

A. Bihar B. Maharashtra
C. Jhansi D. Ranchi

17. Which of the following is a partial root parasite plant?

A. Viscum B. Loranthus
C. Sandal D. Both B & C

18. The height increment of not less than 60 cm per year during growth, such species are called as ______

A. Fast growing Species
B. Moderate growing Species
C. Slow growing Species
D. Very fast growing Species

19. The volume of solid firewood can determine by which method?

A. Specific gravity method
B. Xylometric method
C. Both A & B
D. None of the above

20. Solid volume in cubic content of wood is also called as

A. Stock volume B. Stacked volume
C. Total volume D. Actual volume

21. ઉભા વૃક્ષની ઘનતા નક્કી કરવા માટે પરોક્ષ માપન પદ્ધતિ કઈ છે?
A. સ્પીગેલ રેલાસ્કોપ
B. ટેલી રેલાસ્કોપ
C. વ્હીલર પેન્ટા પ્રિઝમ
D. ઉપરોક્ત તમામ

22. વોલ્યુમ કોષ્ટક કયા ચલ પર આધારિત છે?
A. સ્થાનિક વોલ્યુમ કોષ્ટક
B. સામાન્ય વોલ્યુમ કોષ્ટક
C. પ્રાદેશિક વોલ્યુમ કોષ્ટક
D. પ્રમાણભૂત વોલ્યુમ કોષ્ટક

23. શાહી પ્રવેશ માટે કાગળને વધુ કે ઓછા અભેદ્ય બનાવવા માટે કયા પદાર્થ ઉપયોગ થાય છે?
A. રેઝિન B. મીણ
C. ગુંદર D. રોઝિન

24. કઈ પ્રજાતિનો ઉપયોગ હાથીના સજ્જા બનાવવા માટે થાય છે?
A. *Grewia tiliaefolia*
B. *Hardwickia binata*
C. *Sterculia villosa*
D. *Bauhinia vahlii*

25. કયા પ્રકારનાં લાકડાને રેપીડ સીઝનિંગની પ્રક્રિયા સામે ઓછા રક્ષણની જરૂર હોય છે?
A. પરાવર્ત્ય લાકડું
B. સાધારણ પરવર્ત્ય લાકડું
C. બિન પરાવર્ત્ય લાકડું
D. ઉપર્યુક્તમાંથી કોઈ નહિ

26. સુકા અને લીલા લાકડાનાં સંગ્રહ માટે નીચેનીમાંથી કઈ પદ્ધતિ અપનાવવામાં આવી છે?
A. સેપ ડિસ્પ્લેસમેન્ટ
B. પલાળીને
C. ગરમી અને ઠંડક
D. સૂકવણી

21. Which is the indirect measurement method to determine the volume of standing tree?
A. Spiegel Relaskop
B. Tele Relaskop
C. Wheeler Penta Prism
D. All of these

22. On which one variable the volume table is based on?
A. Local volume table
B. General volume table
C. Regional volume table
D. Standard volume table

23. Which produce is use to make paper more or less impervious to ink penetration in sizing?
A. Resin B. Wax
C. Gum D. Rosin

24. Which species is use to make elephant harness rope?
A. *Grewia tiliaefolia*
B. *Hardwickia binata*
C. *Sterculia villosa*
D. *Bauhinia vahlii*

25. Which type of wood require little protection against rapid seasoning?
A. Refractory wood
B. Moderately refractory wood
C. Non-refractory wood
D. None of the above

26. Which of the following method is adopted for preservation of dry and green woods?
A. Sap displacement
B. Soaking
C. Heating and cooling
D. Drying

27. વોલ્યુમ કોષ્ટક માટે સૌથી મહત્વપૂર્ણ વેરિયેબલ કયું છે?
A. ડીબીએચ B. વિસ્તાર
C. ઊંચાઈ D. ઝાડની ત્રિજ્યા

27. **Which is the most important variable for volume table?**
A. DBH B. Area
C. Height D. Radius of tree

28. ફેલીન્ગ કાપ ________ ની વીરુદ્ધ બનાવવામાં આવે છે
A. 20 થી 25 સે.મી. અન્ડરકટ ઉપર
B. 15 થી 25 સે.મી. અન્ડરકટ નીચે
C. 10 થી 25 સે.મી. અન્ડરકટ ઉપર
D. 5 થી 10 સે.મી. અન્ડરકટ ઉપર

28. **The felling cut is made opposite to the**
A. 20 to 25 cm above the undercut
B. 15 to 25 cm below the undercut
C. 10 to 25cm above the undercut
D. 5 to 10 cm above the undercut

29. લાકડાના પાટિયા ની મહત્તમ જાડાઈ કેટલી હોય છે?
A. 4.5 સે.મી. B. 5.5 સે.મી
C. 6.0 સે.મી. D. 5.0 સે.મી.

29. **Wooden Planks having maximum thickness of?**
A. 4.5 cm B. 5.5 cm
C. 6.0 cm D. 5.0 cm

30. લાકડાના વજન દ્વારા સુધારેલી પદ્ધતિથી અનવેસ કાથાની ઉપજનું પ્રમાણ આશરે કેટલું હોય છે?
A. 3-4% B. 0.5-1.0%
C. 2.5-3.5% D. 5.0-6.0%

30. **What is the yield of katha from modified method by weight of wood?**
A. 3-4% B. 0.5-1.0%
C. 2.5-3.5% D. 5.0-6.0%

31. બબૂલના લાકડા ની કેલરીફિક મૂલ્ય કેટલું છે?
A. 5094 કેલરી B. 4870 કેલરી
C. 4948 કેલરી D. 4939 કેલરી

31. **What is the calorific value of babul wood?**
A. 5094 cal B. 4870 cal
C. 4948 cal D. 4939 cal

32. "ડ્યુરામેન" ________ તરીકે ઓળખાય છે.
A. સેપ વુડ B. હાર્ડવુડ
C. હાર્ટવુડ D. સોફ્ટવુડ

32. **What is the Duramen known as?**
A. Sapwood B. Hardwood
C. Heartwood D. Softwood

33. ઝાડની ડાળીના વ્યાસને સચોટ રીતે માપવા માટે કયા સાધનનો ઉપયોગ કરવામાં આવે છે?
A. વ્હીલર પેન્ટા પ્રિઝમ
B. સ્પીગેલ રેલાસ્કોપ
C. ડિજિટલ કેલિપર
D. ક્લીનોમીટર

33. **Which instrument is used to measure stem diameter accurately?**
A. Wheeler Penta Prism
B. Spiegel Relaskop
C. Digital calliper
D. Clinometer

34. વૃક્ષ કાપવા માટેની આદર્શ સિઝન કઈ છે?
A. વસંત B. ઉનાળો
C. શિયાળો D. ચોમાસુ

34. **The ideal season for tree felling is**
A. Spring B. Summer
C. Winter D. Monsoon

35. નીચેનામાંથી કયું ઝાડ મૂળ થી જ કાપવામાં આવે છે?
A. ખેર B. અખરોટ
C. ચંદન D. ઉપરોક્ત તમામ

36. ભારતમાં લોગિંગ કાર્ય કોણે શરૂ કર્યું છે?
A. સર ડાયટ્રીક બ્રાન્ડિસ
B. ડો. એ હ્યુબર અને એ. કોરોલેફ
C. હાંવલી
D. હર્ષબર્ગર

37. રેઝિન ટેપિંગ માં કઈ ટેપીંગ કરવામાં આવતી નથી?
A. ભારે ટેપીંગ
B. મધ્યમ ટેપીંગ
C. ઓછી સતત ટેપીંગ
D. ઉપર્યુક્તમાંથી કોઈ નહિ

38. ભારે ટેપીંગમાં ઝાડ નો ઘેરાવો ________ ઉપર હોવો જોઈએ.
A. 60 સે.મી. B. 50 સે.મી.
C. 20 સે.મી. D. 100 સે.મી.

39. લાખના કીટકોની કઈ પ્રજાતિ મુખ્યત્વે ભારતમાં જોવા મળે છે?
A. મૌસમી
B. રંગીની
C. બંને A અને B
D. ઉપર્યુક્તમાંથી કોઈ નહિ

40. 15 સે.મી.ની જાડાઈ ધરાવતું ચોરસ અથવા લંબચોરસ આકાર નું લાકડું શું કહેવાય છે?
A. સ્કેટલિંગ્સ
B. હકીઝ
C. પોલ
D. પ્લેન્ક

41. ટેનિન અને કેટેચિન બાવળના લાકડાના કયા ભાગમાંથી કાઢવામાં આવે છે?
A. નવી ડાળીઓ B. છાલ
C. સેપવૂડ D. હાર્ટવુડ

35. Which of the following tress is felled by roots?
A. Khair B. Walnut
C. Sandal D. All of the above

36. Who have started logging work in India?
A. Sir Dietrich Brandis
B. Dr. A. Huber & A. Koroleff
C. Hawley
D. Hershberger

37. Which tapping is not done in resin taping?
A. Heavy tapping
B. Moderate tapping
C. Light continuous tapping
D. None of the above

38. In heavy tapping tree girth should be above
A. 60 cm B. 50 cm
C. 20 cm D. 100 cm

39. Which of the following strains of Lac insects are mainly found in India?
A. Mausami
B. Rangeeni
C. Both A & B
D. None of the above

40. What is the square or rectangular shape having thickness of 15 cm under section is called?
A. Scatlings
B. Hakries
C. Poles
D. Pank

41. Tannis and catechin are abstracted from which wood part of Acacia?
A. Early wood B. Bark
C. Sapwood D. Heartwood

42. ભારતમાં કયા વર્ષે કાગળ બનાવવાની શરૂઆત કરવામાં આવી?
A. 1830 B. 1825
C. 1930 D. 1815

43. કાગળની ભૌતિક ગુણધર્મો કઈ પ્રક્રિયા પર આધારીત છે?
A. સાઈઝીંગ B. બીટિંગ
C. કલરીંગ D. લોડિંગ

44. ઝાડના રેસીનમાંથી ટર્પેન્ટાઇન તેલ અને રોસિન મેળવવા માટે કઈ પ્રક્રિયાનો ઉપયોગ કરવામાં આવે છે?
A. ફ્રેર્મેન્ટેશન B. મેસરેશન
C. નિસ્યંદન D. ઉપરોક્ત તમામ

45. ભારતમાં લાખના ઉછેર માટેના મુખ્ય યજમાન છોડ કયા છે ?
A. ઝિઝિફસ *મૌરીસીઆના* (બોર)
B. *સ્ક્લિચેરા ઓલિઓસા* (કુસુમ)
C. *બુટિયા મોનોસ્પેર્મા* (પલાસ)
D. ઉપરોક્ત તમામ

46. કાગળ બનાવતી વખતે નીચેનીમાંથી કઈ પ્રક્રિયા કરવામાં આવતી નથી?
A. વેક્સિંગ
B. બ્લીચિંગ
C. પલ્પિંગ
D. સ્ટોક તૈયારી

47. લીફ ફાઇબર કઈ છોડ પ્રજાતિમાંથી મેળવવામાં આવે છે?
A. *પેંડેનસ પ્રજાતિઓ*
B. *કેરીઓટા યુરેન્સ*
C. એગેવ *પ્રજાતિઓ*
D. ઉપરોક્ત તમામ

48. કયા છોડના રેસાને રેયા અથવા રેમિ ફાઇબર તરીકે ઓળખાય છે?
A. *બોહેમેરિયા નિવિયા*
B. *સ્ટર્ક્યુલિયા વિલોસા*
C. *બૌહિનીયા વહાલી*
D. *બોમ્બેક્સ સીબા*

42. In which year the paper making is started in India?
A. 1830 B. 1825
C. 1930 D. 1815

43. Physical properties of paper depends on which crucial process?
A. Sizing B. Beating
C. Colouring D. Loading

44. Which process is use to obtain turpentine oil and rosin from tree resins?
A. Fermentation B. Maceration
C. Distillation D. All of the above

45. Which is the major commercial host plants for lac insect in India?
A. *Zizyhus mauritiana* (ber)
B. *Schleichera oleosa* (Kusum)
C. *Butea monosperma* (palas)
D. All of the above

46. Which of the following process is not performed during paper making?
A. Waxing
B. Bleaching
C. Pulping
D. Stock preparation

47. Leaf fibre is obtained from which species?
A. *Pandanus* spp.
B. *Caryota urens*
C. *Agave spp.*
D. All of the above

48. Which of the plant fibre is known as Rhea or Ramie fibre?
A. *Boehmeria nivea*
B. *Sterculia villosa*
C. *Bauhinia vahlii*
D. *Bombax ceiba*

49. કોઇર પિથ મુખ્યત્વે કઈ વનસ્પતિમાંથી મેળવવામાં આવે છે?
A. બોમ્બેક્સ સીબા
B. કોકોસ ન્યુસિફેરા
C. સીઇબા પતેન્દ્ર
D. કોકલોસ્પાર્મમ રિલિજીઓસમ

50. *Cymbopogon flexosus* નું તેલ ________ તરીકે ઓળખાય આવે છે.
A. લેમન તેલ B. કસ્તુરી તેલ
C. જીરાનીઓલ D. ઉપરોક્ત તમામ

51. નીચેનામાંથી કઈ જાતનો ઉપયોગ મોટાભાગે પાંદડાના તેલ માટે થાય છે?
A. સાન્તાલમ આલ્બમ
B. સેડ્રસ દિયોદરા
C. નીલગિરી
D. ઉપર્યુક્તમાંથી કોઈ નહિ

52. "ફૂલવારા બટર" મેળવવા માટે નીચેની કઈ પ્રજાતિનો ઉપયોગ કરવામાં આવે છે ?
A. મધુકા બ્યુટરેસી
B. ગાર્સિનિયા ઇન્ડીકા
C. અલેઉરાઇટ્સ ફોર્ડીઆઈ
D. પોંગામિયા પિનાટા

53. નીચેનામાંથી પ્રત્યવર્તી લાકડાની પ્રજાતિઓ કઈ છે?
A. ટર્મિનલિયા અલાટા
B. શોરિયા રોબસ્ટા
C. બંને A અને B
D. ઉપર્યુક્તમાંથી કોઈ નહિ

54. 5 સે.મી.ની જાડાઈ ધરાવતા ઉચ્ચ પ્રત્યાવર્તી લાકડાને સૂકવવા માટે કેટલો સમય લાગે છે?
A. 3 મહિના B. 6 મહિના
C. 8 મહિના D. 12 મહિના

55. લાકડાના પરિવહનની કઈ પદ્ધતિને "સ્વીપિંગ મેથડ" પણ કહેવામાં આવે છે?
A. રિવર ફ્લોટિંગ B. ફ્રી ફ્લોટિંગ
C. વેટ સ્લાઇડિંગ D. ઉપરોક્ત તમામ

49. **Coir pith are mainly obtained from which plant species?**
A. *Bombax ceiba*
B. *Cocos nucifera*
C. *Ceiba patendra*
D. *Cochlospermum religiosum*

50. **Oil abstracted from *Cymbopogon flexosus* is termed as**
A. Lemon oil B. Musk oil
C. Geraniol D. All of the above

51. **Which of the following species is mostly use for leaf oil?**
A. *Santalum album*
B. *Cedrus deodara*
C. *Eucalyptus* spp.
D. None of the above

52. **Which of the following species is use to obtain Phulwara Butter?**
A. *Madhuca butyraceae*
B. *Garcinia indica*
C. *Aleurites fordii*
D. *Pongamia pinnata*

53. **Which of the following is/are refractory wood species?**
A. *Terminalia alata*
B. *Shorea robusta*
C. Both A & B
D. None of the above

54. **How much time does 5 cm thickness highly refractory wood takes for drying?**
A. 3 months B. 6 months
C. 8 months D. 12 months

55. **Sweeping method of wood transportation is also called as?**
A. River floating B. Free floating
C. Wet sliding D. All of the above

56. ***Gaultheria fragrantissima*** **નું તેલ ________ તરીકે ઓળખવામાં આવે છે.**

A. વેલેરીયન તેલ
B. વિન્ટરગ્રીન તેલ
C. ફિક્સ તેલ
D. જીંજર તેલ

57. ***Anogeissus latifolia*** **ના નવા પર્ણ માં ટેનિન નું પ્રમાણ કેટલું હોય છે?**

A. 35% B. 45%
C. 50% D. 55%

58. નીચેનામાંથી કયું લાકડું ખુબ ધૂમાળો કરે છે તેથી બળતણ અને અગ્નિ માટે વપરાતું નથી?

A. સાગનું લાકડું B. વાંસનું લાકડું
C. સીસુ લાકડું D. આંબાનું લાકડું

59. લાકડાના કોરપિથથી મધ્યસુધી વિસ્તરેલી તિરાડો ________ તરીકે ઓળખાય છે?

A. રિંગ શેક્સ B. રેડિયલ શેક્સ
C. હાર્ટ શેક્સ D. ઉપરોક્ત તમામ

60. ઝાડના થડમાં વેવીનેસ અને સડો થવાના મુખ્ય કારણો કયા છે?

A. ફૂગ
B. બેક્ટેરિયા
C. ફીઝીયોલોજિકલ ફેરફારો
D. ઉપરોક્ત તમામ

61. નીચેનામાંથી ઝાડની કઈ પ્રજાતિનું લાકડું સૌથી મજબૂત તરીકે ગણવામાં આવે છે?

A. *બોમ્બેક્સ સીઇબા*
B. *ટેક્ટોના ગ્રાન્ડિઝ*
C. *મેસુઆ ફેરીઆ*
D. ઉપરોક્ત તમામ

62. કાગળ બનાવવા માટે કાચા માલ તરીકે નીચેની કઈ પ્રજાતિનો ઉપયોગ થાય છે?

A. લીમડો B. વાંસ
C. પોપ્યુલસ D. મહુડો

56. Oil obtained from *Gaultheria fragrantissima* is known as

A. Valerian oil
B. Wintergreen oil
C. Fix oil
D. Ginger oil

57. What is the tannin content of young leaves of *Anogeissus latifolia*?

A. 35% B. 45%
C. 50% D. 55%

58. Which of the following wood gives heavy smokes an is not use for fuel and fire?

A. Teak wood B. Bamboo wood
C. Sissoo wood D. Mango wood

59. Cracks extended from pith to center of the wood is called as?

A. Ring shakes B. Radial Shakes
C. Heart shakes D. All of the above

60. Waviness and decay are main defects in woods which is due to?

A. Fungal attack
B. Bacterial attack
C. Physiological changes
D. All of the above

61. Which of the following is considered as a the most hardwood species?

A. *Bombax ceiba*
B. *Tectona grandis*
C. *Mesua ferrea*
D. All of the above

62. Which of the following tree species is use as a raw material for paper making?

A. Neem B. Bamboo
C. *Populus* spp. D. Mahua

63. કયા ઝાડનાં લાકડામાંથી બટનો, ચેસ અને કોતરણી ની બનાવટો બનાવવામાં આવે છે?
A. સોપારી B. નાળિયેર
C. ટાગુઆનટ D. ઉપરોક્ત તમામ

64. નીચેનામાંથી કઇ જાતિના ફળનો ઉપયોગ ફ્લોસિસ અને રેશમી કોટન માટે વપરાય છે?
A. *બોમ્બેક્સ સીઇબા*
B. *એરિઓડેંડ્રોન અનફેક્ટ્યુઓઝમ*
C. *કોકલોસ્પર્મમ ગોસિપિયમ*
D. ઉપરોક્ત તમામ

65. ઝાડની કઈ પ્રજાતિમાંથી "ટ્રુ ડામર" પ્રાપ્ત થાય છે?
A. *અગાથીસ લોરેન્થિફોલિઆ*
B. *હોપી ઓડોરેટા*
C. *શોરિયા રોબસ્ટા*
D. *કેનેરિયમ સ્ટ્રિક્ટમ*

66. સબમરીન કેબલ્સના ઉત્પાદનમાં નીચેનામાંથી કયા ઉત્પાદનનો ઉપયોગ કરવામાં આવે છે?
A. ગુટ્ટાપર્ચા
B. રેઝિન
C. ડામર
D. ઉપરોક્ત તમામ

67. ડુંગરાળ પ્રદેશોમાં વૂડ્સના પરિવહન માટે કઈ પરિવહન પદ્ધતિનો ઉપયોગ થાય છે?
A. વેટસ્લાઇડ્સ
B. ટેલિસ્કોપિક ફ્લોટિંગ
C. રાફટીંગ
D. ઉપરોક્ત તમામ

68. કોમ્પ્રિગ્નેટેડ વૂડ્સનો મુખ્ય ઉપયોગ શું છે ?
A. એરપ્લેન પ્રોપેલર બ્લેડ
B. કોર બોર્ડ
C. પાર્ટીકલ બોર્ડ
D. રેલ્વે સ્લીપર્સ

63. From which nut tree buttons, chess and carvings are made?
A. Arecanut B. Coconut
C. Taguanut D. All of the above

64. Which of the of the following species fruits is used for flosses and silk cottons?
A. *Bombax ceiba*
B. *Eriodendron anfractuosum*
C. *Cochlospermum Gossypium*
D. All of the above

65. From which of the fooling species a true dammar is obtained?
A. *Agathis loranthifolia*
B. *Hopea odorata*
C. *Shorea robusta*
D. *Canarium strictum*

66. Which of the following produce is used to employ in the manufacture of submarine cables?
A. Guttapercha
B. Resin
C. Dammar
D. All of the above

67. Which transport method is used to transport woods in hilly regions?
A. Wetslides
B. Telescopic floating
C. Rafting
D. All of the above

68. What is the main use of Compregnated woods?
A. Aeroplanes propeller blades
B. Core boards
C. Particle boards
D. Railway sleepers

69. સુતરાઉ રેસાની મુખ્ય રચના શું છે?
A. લિગ્નીન B. લિપિડ
C. સેલ્યુલોઝ D. ઉપરોક્ત તમામ

70. કયા પ્રકારનાં લાકડાની વાહિનીઓ અસમ પ્રમાણ રીતે ગોઠવાય છે?
A. છિદ્રાળુ રિંગ લાકડું
B. વિખરાયેલ છિદ્રાળુ લાકડું
C. વિખરાયેલ બિન-છિદ્રાળુ લાકડું
D. રિંગ બિન-છિદ્રાળુ લાકડું

71. ઝાડ ને લાકડાના લોગમાં કાપવાને _______ કહેવાય છે?
A. બકિંગ B. સોવિંગ
C. ફેલિન્ગ D. ઉપરોક્ત તમામ

72. નીચે આપેલામાંથી કયામાં ટૅડિંગ ઓપરેશન શામેલ નથી?
A. કાપણી
B. નવીનીકરણ છટણી
C. સુધારણા કાપણી
D. વેલાની કાપણી

73. નીચેનામાંથી ક્યા વેલા વાળા ઝાડ નો ઉપયોગ કાથો કાઢવા થાય છે?
A. *પેનિસ્ટમ પર્પ્યુરિયમ*
B. *એકેસિયા કેટેચુ*
C. *અનકેરિયા ગેમ્બિયર*
D. ઉપર્યુક્તમાંથી કોઈ નહિ

74. નીચેનામાંથી કયા લાકડામાં ડીફ્યુસનની ટ્રીટમેન્ટની આપવામાં આવે છે?
A. પાણીમાં અદ્રાવ્ય લાકડા
B. લીલા દ્રાવ્ય લાકડા
C. લીલા વાંસ
D. ઉપર્યુક્તમાંથી કોઈ નહિ

75. કત્ય કયા પદાર્થમાં દ્રાવ્ય છે?
A. ઠંડુ પાણી B. તેજાબ
C. ગરમ પાણી D. બંને A અને C

69. What is the main composition of cotton fibre?
A. Lignin B. Lipid
C. Cellulose D. All of the above

70. In which type of woods vessels are arranged asymmetrically?
A. Ring porous woods
B. Diffused porous woods
C. Diffused non-porous woods
D. Ring non-porous woods

71. The cutting of trees into wooden logs is called as
A. Bucking B. Sawing
C. Felling D. All of the above

72. Which of the following does not include in Tending operation?
A. Pruning
B. Regeneration felling
C. Improvement felling
D. Climber cutting

73. Which woody climber is also use to abstract Katha?
A. *Pennistum purpureum*
B. *Acacia catechu*
C. *Uncaria gambier*
D. None of the above

74. In Which of the following timber the diffusion treatment is given?
A. Green timber without water soluble wood
B. Green timber with water soluble wood
C. Green bamboo only
D. None of the above

75. Cutch is soluble in which solution?
A. Cold water B. Acid
C. Hot water D. Both A & C

76. નીચેની કઇ જાતિ વાદળી ગમ વૃક્ષ તરીકે ઓળખાય છે?

A. *યુકેલિપ્ટ્સ ગ્લોબ્યુલસ*
B. *યુકેલિપ્ટ્સ ટ્રાઈકારપા*
C. *યુકેલિપ્ટ્સ સીટ્રિઓડોરા*
D. ઉપરોક્ત તમામ

77. ક્રિકેટ બેટ બનાવવા માટે ક્યા ઝાડનું લાકડુ શ્રેષ્ઠ છે?

A. *સેલ્ટિસ ઓસ્ટ્રેલિસ*
B. *ફ્રેક્સીનસ* પ્રજાતિ
C. *એકેસિયા કેટેચુ*
D. ઉપરોક્ત તમામ

78. ચામડામાં સ્થાનાંતરિત કરવા માટે પ્રાણીની ચામડીની સારવાર અને ટેનીન પ્રક્રિયા શું કહેવામાં છે?

A. નિસ્યંદન
B. રેટિંગ
C. ટેનિંગ
D. લેધરિંગ

79. નીચેનામાંથી કઈ પ્રજાતિના મૂળમાંથી મૂલ્યવાન લાલ રંગ મેળવવામાં આવે છે?

A. *રૂબિયા કોર્ડીફોલીઆ*
B. *લાવસોનીયા ઇનરમીસ*
C. *મેલોટ્સ ફિલિપેન્સિસ*
D. *નિક્ટેન્થેસ આર્બોટ્રિસ્ટિસ*

80. નીચેની કઈ પ્રજાતિમાંથી ટન્ગ તેલ મળે છે?

A. *ગાર્સિનિયા ઇન્ડીકા*
B. *મધુકા બુટીરશિ*
C. *એઝાડિરેક્ટા ઇન્ડિકા*
D. *અલેઉરાઇટ્સ ફોર્ડીઆઈ*

81. મલ્ટી સ્ટેજ બ્લીચિંગની પ્રક્રિયા દરમિયાન કયા રસાયણોનો ઉપયોગ થાય છે?

A. Cl_2 અને ClO_2 B. SO_2 અને SO_3
C. $CaOCl_2$ D. $CaCl_2$ અને CaO

76. **Which of the following species is/are known as a blue gum tree?**

A. *Eucalyptus globulus*
B. *Eucalyptus tricarpa*
C. *Eucalyptus citriodora*
D. All of the above

77. **Which of the following species is best for making cricket bat?**

A. *Celtis australis*
B. *Fraxinus* spp.
C. *Acacia catechu*
D. All of the above

78. **What is the process of treating animal skins and hides with tannins to transfer it to leather?**

A. Distillation
B. Retting
C. Tanning
D. Leathering

79. **From which of the following species roots, the valuable red dye is obtained?**

A. *Rubia cordifolia*
B. *Lawsonia inermis*
C. *Mallotus philippensis*
D. *Nyctanthes arbotristis*

80. **From which of the following species the Tung oil is obtained?**

A. *Garcinia indica*
B. *Madhuca butyracea*
C. *Azadiracta indica*
D. *Aleurites fordii*

81. **Which chemicals are use during the process of multi stage bleaching?**

A. Cl_2 & ClO_2 B. SO_2 & SO_3
C. $CaOCl_2$ D. $CaCl_2$ & CaO

82. સિટ્રોનેલા તેલ મેળવવા માટે કઈ પ્રજાતિનો ઉપયોગ થાય છે?
A. *સિમ્બોપોગન નાર્ડસ*
B. *સિમ્બોપોગન ફ્લેક્સોસસ*
C. *સિમ્બોપોગન માર્ટીની*
D. *સિમ્બોપોગન જાવા*

83. લાકડા સિવાયના વન ઉત્પાદનો ______ તરીકે ઓળખાય છે?
A. ગૌણ વન ઉત્પાદન
B. મુખ્ય વન ઉત્પાદન
C. એનટીએફપી
D. ઉપરોક્ત તમામ

84. લેમન ગ્રાસના તેલમાં સિટ્રલ કેટલું છે?
A. 35 - 50% B. 40 - 85%
C. 35 - 45% D. 20 - 35%

85. કયા પ્લાન્ટનો ઉપયોગ પિની ટેલો તેલ મેળવવા માટે થાય છે?
A. પાઈન B. લીમડો
C. વેલાપીન D. કુસુમ

86. બેટ્સ ગુઆનો એ શેનો એક સમૃદ્ધ સ્રોત છે?
A. નાઇટ્રેટ
B. ફોસ્ફેટ
C. ફેરસ
D. બંને A અને B

87. છોડની કઈ પ્રજાતિનો ઉપયોગ ચૌલમુગરા મેળવવા માટે થાય છે?
A. *ગાર્સિનિયા ઇન્ડીકા*
B. *અલેઉરાઇટ્સ ફોર્ડીઆઈ*
C. *મધુકા ઈન્ડિકા*
D. *હાઇડનોકાર્પસ કુર્ઝી*

88. કયા ઝાડની છાલ માછલીના ઝેર તરીકે વપરાય છે?
A. બાવળ B. સુબાવળ
C. સીરસ D. ઉપરોક્ત તમામ

82. Which species is use to obtained Citronella oil?
A. *Cymbopogon nardus*
B. *Cymbopogon flexosus*
C. *Cymbopogon martini*
D. *Cymbopogon java*

83. Forest products other than timber is known as
A. Minor Forest Produce
B. Major Forest Produce
C. NTFP
D. All of the above

84. What is the citral content of lemon grass oil?
A. 35 - 50% B. 40 - 85%
C. 35 - 45 % D. 20 - 35 %

85. Which plant is use to obtain Piney tallow solid oil?
A. Pine B. Neem
C. Vellapine D. Kusum

86. Bats guano is a rich source of
A. Nitrate
B. Phosphate
C. Ferrous
D. Both A & B

87. Which species use to obtained Chaulmugra?
A. *Garcinia indica*
B. *Aleurites fordii*
C. *Madhuca indica*
D. *Hydnocarpus kurzii*

88. Which tree's bark is used as a fish poison?
A. Babul B. Subabul
C. Siris D. All of the above

89. "ટ્રેસીડ્સ" કયા છોડના ભાગની મિનિટ સ્ટ્રક્ચર નું ઉદાહરણ છે?
A. મૂળ B. લાકડું
C. પાંદડા D. ફળ

90. ઉચ્ચ ગુણવત્તાના બીડી પાંદડા મેળવવા માટે કઈ જાતિનો ઉપયોગ થાય છે?
A. *બૌહીનીયા વહાલી*
B. *ડિઓસપયરોસ મેલાનક્સયલોન*
C. *બોરેસસ ફ્રેબિલિફેર*
D. ઉપરોક્ત તમામ

91. રેઝિનની લાક્ષણિકતા શું છે?
A. બરડ B. આકારહીન
C. ઓછા પારદર્શક D. ઉપરોક્ત તમામ

92. કઈ પ્રજાતિનો ઉપયોગ "ડીવી-ડીવી" ટેન મેળવવા માટે થાય છે?
A. *કેસિઆ ઓરિક્યુલાટા*
B. *સીઝાલ્પિનિયા કોરીરિયા*
C. *સેરીઓપ્સ રોક્સબર્ગિગિઆના*
D. *અલેઉરાઇટ્સ ફ્રોડીઆઈ*

93. ખૂબ જ હલકા લાકડાની ઘનતા કેટલી હોય છે?
A. 200 કિગ્રા / મીટર^3
B. 400 કિગ્રા / મીટર^3
C. 300 કિગ્રા / મીટર^3
D. 100 કિગ્રા / મીટર^3

94. સર્પાકાર બંદૂક ________ તરીકે પણ ઓળખાય છે?
A. સ્પ્લિટ B. ક્રોસ ગ્રેઈન
C. ટવીસ્ટ ઓફ ધ ટ્રી D. ટેપર

95. "ધરુઆ ગુંદર" માટે કઈ પ્રજાતિનો ઉપયોગ થાય છે ?
A. *અલ્બીઝિયા પ્રોસેરા*
B. *બોમ્બેક્સ સીબા*
C. *બોસ્વેલિયા સેરાટા*
D. એનોજેસસ *લાટીફોલીઆ*

89. "Tracheids" is the example of minute structure of which plant part?
A. Root B. Wood
C. Leaves D. Fruit

90. Which of the following species is use to obtain premium bidi leaves?
A. *Bauhini vahlii*
B. *Diospyros melanxylon*
C. *Borasus flabellifer*
D. All of the above

91. What is the characteristic of Resin?
A. Brittle B. Amorphous
C. Less transparent D. All of the above

92. Which of the following species is used to obtain "Divi-Divi" tans?
A. *Cassia auriculata*
B. *Caesalpinia coriaria*
C. *Ceriops roxburghiana*
D. *Aleurites fordii*

93. What is the density of very light timber?
A. 200 kg/m^3
B. 400 kg/m^3
C. 300 kg/m^3
D. 100 kg/m^3

94. Spiral gun is also called as?
A. Split B. Cross grain
C. Twist of the tree D. Taper

95. Which of the following species is used to abstract dharua gum?
A. *Albizia procera*
B. *Bombax ceiba*
C. *Boswellia serata*
D. *Anogeissus latifolia*

96. કિતુલ ફાઇબર કઈ પ્રજાતિમાંથી મેળવવામાં આવે છે?
A. *એગાવ પ્રજાતિ*
B. *કેરીઓટા યુરેનસ*
C. *કોર્ચરસ કેપ્સ્યુલરિસ*
D. ઉપર્યુક્તમાંથી કોઈ નહિ

97. કયા પ્રકારનાં ડેપો અસ્થાયી છે?
A. વન ડેપો
B. વેચાણ ડેપો
C. પરિવહન ડેપો
D. ઉપરોક્ત તમામ

98. નીચેનામાંથી પ્રત્યવર્તી લાકડાની પ્રજાતિઓ કઈ છે?
A. *હોપ ઓડોરેટા*
B. *શોરિયા રોબસ્ટા*
C. *ટર્મિનાલિયા અલાતા*
D. ઉપરોક્ત તમામ

99. નીચેનામાંથી કયા પ્રકારનું તેલ લાકડાંના પ્રિઝર્વેટિવનું ઉદાહરણ છે?
A. $ZnSO_4$
B. Lindane
C. Creosote
D. PCP

100. ફૂગના હુમલામાં સેપવુડની સંવેદનશીલતાનું કારણ શું છે?
A. સ્ટાર્ચ
B. કાર્બોહાઇડ્રેટ
C. પ્રોટીન
D. બંને B અને C

101. સપાટીમાંથી પ્રકાશને પ્રતિબિંબિત કરવાની કોષ દિવાલની ક્ષમતા________તરીકે ઓળખાય છે.
A. ચમક
B. તેજ
C. બંને A અને B
D. ઉપર્યુક્તમાંથી કોઈ નહિ

96. Kittul fibre is obtained from which species?
A. *Agave* spp.
B. *Caryota urenus*
C. *Corchorus capsularis*
D. None of the above

97. Which types of depots are temporary in nature?
A. Forest depots
B. Sale depots
C. Transit depots
D. All of the above

98. Which of the following is/are refractory wood species?
A. *Hopea odorata*
B. *Shorea robusta*
C. *Terminalia alata*
D. All of the above

99. Which of the following is an example of oil type wood preservative?
A. $ZnSO_4$
B. Lindane
C. Creosote
D. PCP

100. What is the reason for susceptibility of sapwood to fungal attacks?
A. Starch
B. Carbohydrate
C. Protein
D. Both B & C

101. An ability of cell wall to reflect the light from the surface is called as
A. Sheen
B. Luster
C. Both A & B
D. None of the above

102. "સેલક્યુર" એ લાકડાનું કયા પ્રકારનું પ્રિઝર્વેટિવ છે?

A. તેલ
B. પાણીમાં દ્રાવ્ય
C. અકાર્બનિક દ્રાવક
D. કાર્બનિક દ્રાવક

103. લાકડાના વરાળની પ્રક્રિયા એને વધુ શેના તરફ દોરી જાય છે?

A. સ્થિતિસ્થાપકતા
B. વ્યવહારક્ષમતા
C. ચળકાટ
D. સુગમતા

104. દેવદારના લાકડાનું કેલરીફિક મૂલ્ય કેટલું છે?

A. 4500 B. 5190
C. 5294 D. 5390

105. લાકડાની જાળવણીની કઈ પદ્ધતિમાં પ્રિઝર્વેટિવને પ્રારંભિક વેક્યૂમ કરવાની જરૂર નથી?

A. ફૂલ સેલ પ્રક્રિયા
B. લોવરી પ્રક્રિયા
C. રુપીંગ પ્રક્રિયા
D. ઉપર્યુક્તમાંથી કોઈ નહિ

106. કયા છોડ નોંધપાત્ર ગંધ ધરાવે છે?

A. કપૂર B. દેવદાર
C. સાગ D. ઉપરોક્ત તમામ

107. કોને વિશ્વના શ્રેષ્ઠ લાકડા માનવામાં આવે છે?

A. બાલસા B. સાગ
C. સાલ D. દેવદાર

108. કઈ પ્રજાતિનાં લાકડાનો ઉપયોગ કૃષિ સાધનો માટે કરવામાં આવે છે?

A. સાગ B. ખેર
C. બાવળ D. બંને B અને C

102. Celcure is a which type of wood preservative?

A. Oil type
B. Water soluble type
C. Inorganic solvent type
D. Organic solvent type

103. Steaming process of woods leads to more?

A. Elasticity
B. Feasibility
C. Shinning
D. Flexibility

104. What is the calorific value of deodar wood?

A. 4500 B. 5190
C. 5294 D. 5390

105. Which method of wood preservation initial vacuum for applying preservative is not needed?

A. Full cell process
B. Lowry process
C. Rueping process
D. None of the above

106. Which of the following have unmistakable remarkable odour?

A. Camphor B. Deodar
C. Teak D. All of the above

107. Which of the following is considered as the world best timber?

A. Balsa B. Teak
C. Sal D. Deodar

108. Which of the following species woods are utilized for agricultural implements?

A. Teak B. Khair
C. Babul D. Both B & C

109. વીણા અને તંબુરા બનાવવા માટે કયા ઝાડના લાકડાનો ઉપયોગ થાય છે?
A. *મેલીના આર્બોરીયા*
B. *આર્ટોકાર્પસ હેટેરોફિલસ*
C. બંને A અને B
D. *કેનેરીયમ યુફાઈલમ*

110. ડચ શબ્દ "બૂમ" નો અર્થ ________ થાય છે.
A. ફ્રી ફ્લોટિંગ
B. ફ્લોટિંગ ચેનલ
C. ફ્લોટિંગ ઓબસ્ટ્રક્શન
D. ઉપરોક્ત તમામ

111. લાકડાના પોલાણમાં રહેલા પાણીને ________ તરીકે ઓળખવામાં આવે છે.
A. હાઇગ્રોસ્કોપિક પાણી
B. ફ્રી પાણી
C. શોષિત પાણી
D. બાઉન્ડ પાણી

112. ઇમારતી લાકડાનું ગ્રેડિંગ કયા સૂત્ર પર આધારિત છે?
A. ક્વાર્ટર ગર્થ સૂત્ર
B. પ્રિઝમોઇડલ સૂત્ર
C. ન્યુટનનું સૂત્ર
D. ઉપર્યુક્તમાંથી કોઈ નહિ

113. કયા ઝાડની પ્રજાતિઓ રેલ્વે સ્લીપર્સ માટે સૌથી વધુ યોગ્ય છે?
A. દેવદાર B. ઇરુલ
C. સાલ D. ઉપરોક્ત તમામ

114. *બ્રુસોનેટીયા પેપિરિફેરાનું* સામાન્ય નામ શું છે ?
A. કાગળ શેતૂર
B. રેશમ શેતૂર
C. સામાન્ય શેતૂર
D. ઉપર્યુક્તમાંથી કોઈ નહિ

115. ભારતમાં પ્રથમ પેપર મિલ ક્યાં સ્થપાઇ હતી ?
A. આંધ્રપ્રદેશ B. પશ્ચિમ બંગાળ
C. ગુજરાત D. પંજાબ

109. Which of the tree woods are used to make Vina and Tambura?
A. *Gmelina arboria*
B. *Atrocarpus heterophyllus*
C. Both A & B
D. *Canarium euphyllum*

110. The Dutch word "Boom" means
A. Free floating
B. Floating channel
C. Floating obstruction
D. All of the above

111. The water present in the cavity of wood is termed as
A. Hygroscopic water
B. Free water
C. Absorbed water
D. Bound water

112. Grading of the timber is based on which formula?
A. Quarter girth formula
B. Prismoidal formula
C. Newton's formula
D. None of the above

113. Which tree species are most suitable for railway sleepers?
A. Deodar B. Irul
C. Sal D. All of the above

114. What is the common name of *Broussonetia papyrifera*?
A. Paper Mulberry
B. Silk mulberry
C. Common Mulberry
D. None

115. Where was the first paper mill in India established?
A. Andhra Pradesh B. West Bengal
C. Gujarat D. Punjab

116. વૂડ્સને સૂકવવા માટે શું મહત્વપૂર્ણ છે?

A. FSP (Fibre Saturation Point)
B. AWP (Absorbed water point)
C. FWP (Free water point)
D. EMP (Equilibrium moisture point)

117. રાસાયણિક રીતે સુધારેલા લાકડા માટે કયા રાસાયણો નો ઉપયોગ થાય છે?

A. યુરિયા હાઇડ્રોક્સિલિન
B. સલ્ફૂરીક એસિડ
C. એસિટિક એનિહાઇડ્રાઇડ
D. ઉપરોક્ત તમામ

118. પ્લાયવુડ વિનીર ની સંખ્યા હંમેશા ______ રહે છે

A. બેકી
B. એકી
C. બંને A અને B
D. ઉપર્યુક્તમાંથી કોઈ નહિ

119. નિષ્ક્રિય કળીઓ ના વિકાસને લીધે કઈ જટિલ રચના રચાય છે?

A. કપિંગ B. ગાંઠ
C. બર D. ફાઈબર

120. નીચેની કયી પ્રજાતિમાંથી રેયોન પ્રાપ્ત થાય છે?

A. શેતૂર B. બાલસમ
C. વાંસ D. ઉપરોક્ત તમામ

121. નીચેનામાંથી કઈ પ્રજાતિના બીજમાં ઝેરી "અબ્રિન" છે?

A. *ઓસીમમ કિલિમંડ્સચરિકમ*
B. *એબ્રસ પ્રિક્ટોરિયસ*
C. *માયરીકા એસક્યુલેન્ટા*
D. *ઓજેઈના ઓજેનેન્સીસ*

122. નીચેનામાંથી કઈ પ્રજાતિમાંથી રેડ ડામર પ્રાપ્ત થાય છે?

A. સાલ B. ગુગળ
C. ચંદન D. ગુર્જન

116. Which is critical for drying of woods?

A. FSP (Fibre Saturation Point)
B. AWP (Absorbed water point)
C. FWP (Free water point)
D. EMP (Equilibrium moisture point)

117. Which chemical is used for chemically modified woods?

A. Urea hydroxylin
B. Sulphuric acid
C. Acetic anhydride
D. All of the above

118. The number of veneers in plywoods is always

A. Even
B. Odd
C. Both A & B
D. None of the above

119. Which complex structure is formed due to the vigorous growth of dormant buds?

A. Cupping B. Knot
C. Burr D. Fibre

120. From which of the following species Rayon is obtained?

A. Mulberry B. Balsam
C. Bamboo D. All of the above

121. Which of the following species contains poisonous "Abrin" in their seeds?

A. *Ocimum killimandscharicum*
B. *Abrus precatorius*
C. *Myrica esculanta*
D. *Ougeina oojeinensis*

122. From which of the following species "Ral Dammar" is obtained?

A. Sal B. Guggal
C. Sandal D. Gurjan

123. *કોસ્ટસ ફોર્સકોહલી*નો કયો ભાગ તેલ કાઢવા માટે ઉપયોગમાં લેવાય છે?
A. પાંદડા B. મૂળ
C. ફૂલો D. ડાળી

124. વૂડ્સ કંડીશનીંગ કરાવાથી કેટલા પ્રમાણ સુધી ભેજ ઘટે છે?
A. 5% B. 10%
C. 12% D. 15%

125. લાકડાની જાળવણી કઈ પદ્ધતિમાં પ્રિઝર્વેટિવ્સ દૂરવર્તી છેડે ચોક્કસ દબાણ હેઠળ લગાવવામાં આવે છે?
A. સ્ટેપિંગ પ્રક્રિયા
B. બૂચેરી અને મોડીફાઇડ બૂચેરી પ્રક્રિયા
C. સેપ ડિસ્પ્લેસમેન્ટ પ્રક્રિયા
D. સોકિંગ પ્રક્રિયા

126. સૂકવણીનો દર કયા પરિબળથી વિપરિત પ્રમાણસર છે?
A. સંબંધિત ભેજનું પ્રમાણ%
B. તાપમાન
C. હવાનું પરિભ્રમણ
D. ઉપરોક્ત તમામ

127. પેન્સિલ બનાવવા માટે કઇ પ્રજાતિનો વ્યાપકપણે ઉપયોગ થાય છે?
A. *જ્યુનિપરસ પ્રોસેરા*
B. *એલ્સ્ટોનીયા સ્કોલારીસ*
C. *કપ્રેસસ ટ્રોલોસા*
D. ઉપરોક્ત તમામ

128. થર્મોસેટિંગ ફાઇબર પેનિટ્રેટીંગ રેઝિન સાથે ઉપચાર કરાયેલા વિનીર ________ તરીકે ઓળખાય છે.
A. કોમ્પ્રેગ B. લેમિનેટેડ
C. ઇમપ્રેગ D. સ્થિર

129. સોફ્ટવુડ્સ સામાન્ય રીતે________માં મળી આવે છે.
A. જિમ્નોસ્પર્મ્સ B. એન્જીયોસ્પર્મ્સ
C. બ્રાયોફાઇટ્સ D. ટેરીડોફાયઇટ્સ

123. Which part of the *Costus forskohli* is utilized for oil extraction?
A. Leaves B. Roots
C. Flowers D. Stem

124. Conditioning of woods can decrease the moisture content upto
A. 5% B. 10%
C. 12% D. 15%

125. In which wood preservation method, preservatives are applied at the distal end at certain pressure?
A. Stepping Process
B. Boucherie & modified Boucherie process
C. Sap displacement process
D. Socking process

126. Rate of the drying is inversely proportional to which factor?
A. RH %
B. Temperature
C. Air circulation
D. All of the above

127. Which of the following species is widely use to make pencils?
A. *Juniperous procera*
B. *Alstonia scholaris*
C. *Cupresus tourolosa*
D. All of the above

128. Veneers treated with thermosetting fibre penetrating resins are called as
A. Compreg B. Laminated
C. Impreg D. Stabilized

129. Softwoods are generally found in which of the following?
A. Gymnosperms B. Angiosperms
C. Bryophytes D. Pteridophytes

130 . કયું લાકડું હળવા રંગનું છે અને ઓછી ગીચતા દર્શાવે છે?

A. ઓટમ લાકડું B. સ્પ્રિંગ લાકડું
C. બંને A અને B D. લેટ લાકડું

131. નીચેનામાંથી કયા ઝાડના લાકડાનો ઉપયોગ ટેનિસ રેકેટ બનાવવા માટે થાય છે?

A. *સેલ્ટિસ ઓસ્ટ્રેલીસ*
B. *મોરસ આલ્બા*
C. *લેજેસ્ટ્રોમિયા લેન્સોલેટા*
D. *સેલિક્સ આલ્બા*

132. પોરડ વૂડ્સ ને __________પણ કહેવામાં આવે છે.

A. હેટેરોક્ષાયલસ લાકડું
B. પેરાટ્રેસીલ લાકડું
C. સિન્ટ્રેસીલ લાકડું
D. હોમોક્ષાયલસ લાકડું

133. કાગળ બનાવવાની પ્રક્રિયામાં સાઈઝીંગ શેના માટે કરવામાં આવે છે?

A. પારદર્શકતા B. ચળકાટ
C. શાહીથી અભેદ્ય D. મજબૂતાઈ

134. લાકડાના વૈજ્ઞાનિક અભ્યાસને શું કહેવામાં આવે છે?

A. વુડ એનોટોમી
B. ક્ષાયલોટોમી
C. ડેંડ્રોલોજી
D. હિસ્ટોલોજી

135. લોગના વોલ્યુમની ગણતરી કરવા માટે હોપ્પસ નિયમ નું સૂત્ર શું છે?

A. $V = (g/4)^3 X l$ B. $V = (g/4) X l$
C. $V = (g/4)^2 X l$ D. $V = (g/2)^2 X l$

136. "ફ્રેન્ચ થિનિંગ" એ __________તરીકે પણ ઓળખાય છે.

A. ક્રાઉન થિનિંગ
B. ઓર્ડિનરી થિનિંગ
C. સાઈડ થિનિંગ
D. ટોપ થિનિંગ

130. Which wood is lighter in colour and exhibits lower density?

A. Autum wood B. Spring wood
C. Both A & B D. Late wood

131. Which of the following tree woods are used to make Tennis rackets?

A. *Celtis australis*
B. *Morus alba*
C. *Lagestroemia lanceolata*
D. *Salix alba*

132. Poroud woods are also called as

A. Heteroxylous wood
B. Paratracheal wood
C. Syntracheal wood
D. Homoxylous wood

133. In the process of paper making, sizing is done to render

A. Transparent B. Brightness
C. Impervious to ink D. Strengthening

134. Scientific study of wood is called as

A. Wood anatomy
B. Xylotomy
C. Dendrology
D. Histology

135. What is the formula of Hoppus rule to calculate volume of logs

A. $V = (g/4)^3 X l$ B. $V = (g/4) X l$
C. $V = (g/4)^2 X l$ D. $V = (g/2)^2 X l$

136. "French thinning" is also known as a

A. Crown thinning
B. Ordinary thinning
C. Side thinning
D. Top thinning

137. ઇન્ડિયન ઇન્સ્ટિટ્યૂટ ઓફ પ્લાન્ટેશન મેનેજમેન્ટ ક્યાં સ્થિત છે?
A. નવી દિલ્હી B. બેંગ્લોર
C. હૈદરાબાદ D. કાલિકટ

138. બર્માનો ચોરસ સાગ ________ તરીકે પણ ઓળખાય છે.
A. અફઘાન સાગ
B. દક્ષિણ પૂર્વ એશિયા સાગ
C. ચાઇનીઝ સાગ
D. એડમિરલિટી સાગ

139. કયા પ્રકારનું લાકડું વાર્ષિક રિંગ્સના મુખ્ય ભાગની રચના કરવામાં મદદ કરે છે?
A. વસંત લાકડું B. પાનખર લાકડું
C. A અને B બંને D. કંઈ નહીં

140. પાનખર લાકડું ક્યારે જોવા મળે છે?
A. વાર્ષિક રિંગ્સના અંતમાં
B. વાર્ષિક રિંગ્સની શરૂઆતમાં
C. વાર્ષિક રિંગ્સની મધ્યમાં
D. બાહ્ય સ્તરમાં

141. હાર્ટવુડ ને ________ તરીકે પણ ઓળખવામાં આવે છે.
A. સેપ લાકડું B. ડ્યુરામેન
C. છિદ્રાળુ લાકડું D. બિન છિદ્રાળુ લાકડું

142. હાર્ટવુડમાં કયો ભાગ આછા રંગનો છે?
A. આંતરિક ભાગ
B. પેરિફેરલ ભાગ
C. મધ્ય ક્ષેત્ર ભાગ
D. બાહ્ય ભાગ

143. કયા પ્રકારના લાકડામાં વેસેલ નો સંપૂર્ણ અભાવ હોય છે?
A. નરમ લાકડું B. સખત લાકડું
C. છિદ્રાળુ લાકડું D. બિન છિદ્રાળુ લાકડું

144. કયો મોનોકોટ અસંગત ગૌણ વૃદ્ધિ દર્શાવે છે?
A. ઘઉં B. ડ્રેકૈના
C. મકાઈ D. ચોખા

137. Indian Institute of Plantation Management is situated at
A. New Delhi B. Bangalore
C. Hyderabad D. Calicutt

138. The Teak square of Burma is known as
A. Afghan Teak
B. SE Asia Teak
C. Chinese Teak
D. Admirality Teak

139. Which type of wood helps in formation of Major part of annual rings?
A. Spring wood B. Autumn wood
C. Both A & B D. None

140. Autumn wood is present at
A. End of annual rings
B. Beginning of annual rings
C. Middle of annual rings
D. Outer layer

141. Heartwood is also called as
A. Sap wood B. Duramen
C. Porous wood D. Non-porous wood

142. Which region in heartwood is light in colour?
A. Inner region
B. Peripheral region
C. Middle region
D. Outer region

143. The wood that entirely lacks vessel
A. Soft wood B. Hard wood
C. Porous wood D. Non porous wood

144. Which Monocot show anomalous secondary growth?
A. Wheat B. Dracaena
C. Maize D. Rice

145. જૂના થડમાં, લાકડાનો કયો મોટો ભાગ ટેનીન, રેઝિન અને ગુંદરથી ભરેલો છે?

A. સેપ વૂડ B. સખત લાકડું

C. નરમ લાકડું D. હાર્ટ વૂડ

146. 90 – 95% ટ્રેકીડ પ્રાકૃતિક રીતે ________ છે.

A. સેપ વૂડ B. સખત લાકડું

C. નરમ લાકડું D. હાર્ટ વૂડ

147. કયું લાકડું પ્રમાણમાં વધુ ટકાઉ હોય છે?

A. સેપ વૂડ B. સખત લાકડું

C. નરમ લાકડું D. હાર્ટવૂડ

148. વાર્ષિક રિંગ્સની ગણતરી કરીને ઝાડની ઉંમર જાણવાની પ્રક્રિયા ને શું કહેવામાં આવે છે?

A. કાઉન્ટોલોજી B. ડેંડ્રોલોજી

C. ડેંડ્રોકાઉન્ટોલોજી D. ડેંડ્રોક્રોનોજી

149. કયા છોડ પાસે વેસેલ્સનો અભાવ હોય છે?

A. કેરી B. ઘઉં

C. પાઇનશ D. જાસ્મિન

150. ઝાયલેમ માં શેનો સમાવેશ થાય છે?

A. tracheids, fibres and vessels

B. tracheids, fibres and parenchyma

C. tracheids, vessels and companion cells

D. tracheids, vessels, fibres and parenchym

151. ફ્લોઇમ માં શેનો સમાવેશ થાય છે?

A. vessels, tracheids, sieve tube cells, companion cells

B. vessels, tracheids, phloem parenchyma, phloem fibres

C. vessels, sieve tube cells, phloem parenchyma, phloem fibres

D. sieve tube cells, companion cells, phloem parenchyma, phloem fibres

145. In the old stem, Which major part of wood is filled with tannins, resins and gums

A. Sap wood B. Hard wood

C. Soft wood D. Heart wood

146. The percentage tracheid is 90-95% in

A. Sap wood B. Hard wood

C. Soft wood D. Heart wood

147. The wood which is comparatively more durable

A. Sap wood B. Hard wood

C. Soft wood D. Heart wood

148. The age of tree by counting the annual rings is called

A. Countology B. Dendrology

C. Dendrocountology D. Dendrochronogy

149. Which plant doesn't have vessels?

A. Mango B. Wheat

C. Pinus D. Jasmine

150. Xylem contains

A. tracheids, fibres and vessels

B. tracheids, fibres and parenchyma

C. tracheids, vessels and companion cells

D. tracheids, vessels, fibres and parenchyma

151. Phloem contains

A. vessels, tracheids, sieve tube cells, companion cells

B. vessels, tracheids, phloem parenchyma, phloem fibres

C. vessels, sieve tube cells, phloem parenchyma, phloem fibres

D. sieve tube cells, companion cells, phloem parenchyma, phloem fibres

152. કયું ફ્લોઇમ રેસાઓ નું સારું ઉદાહરણ છે?
A. હાર્પ B. જૂટ
C. ફિઆક્ષ D. ઉપરના બધા

153. પાણીના વહનમાં કઈ પેશીઓ સમાવેશ થાય છે?
A. ઝાયલેમ અને ફ્લોઇમ
B. ટ્રેચેઇડ્સ અને ટ્રેકેઈ
C. ઝાયલેમ રેસા અને બાસ્ટ ફાઇબર
D. પેરેન્કાયમા અને સ્ક્લેરંચાઇમા

154. નીચેના છોડના કયા જૂથોમાં ફ્લોઇમ પેરેન્કાયમા નો અભાવ હોય છે?
A. જિમ્નોસ્પર્મ્સ B. મોનોકોટ્સ
C. ડાયકોટ્સ D. ઉપરોક્ત તમામ

155. કોઈ વિશિષ્ટ સંસ્કૃતિ અને પ્રદેશના લોકો દેશી છોડનો ઉપયોગ કેવી રીતે કરે છે તેનો અભ્યાસ ____________ ઓળખાય છે.
A. વર્ગીકરણ B. પેથોલોજી
C. પ્લાન્ટ ઇકોલોજી D. ઈથનોબોટની

156. *એકેશિયા જેકમોનટાઇ* ના પાંદડા કયા પ્રકાર પશુ ઇલાજ માટે આપવામાં આવે છે?
A. દૂધ તાવ
B. પેશાબનો અવરોધ
C. એન્થ્રેક્સ
D. ટિક ફિવર

157. *મોરિંગા રિવા* શેના માટે વપરાય છે?
A. સ્નાયુઓમાં નબળાઇ
B. આથ્રાઇટીસ
C. તાવ
D. A અને B બંને

158. *રોઝા ઈન્ડીકા* શેના ઈલાજ માટે વાપરી શકાય છે?
A. અસ્થમા
B. કિડનીના પથ્થરને દૂર કરવા
C. પિત્તાશય પથ્થર માટે
D. બધા

152. Which is a good example of phloem fibres
A. Hemp B. Jute
C. Fiax D. All

153. The tissues involved in water conduction are
A. Xylem and Phloem
B. Tracheids and tracheae
C. Xylem fibres and bast fibre
D. Parenchyma and sclerenchyma

154. Which of the following plant groups lack in phloem parenchyma?
A. Gymnosperms B. Monocots
C. Dicots D. All of above

155. The study of how people of a particular culture and region make of use of indigenous plants is known as.
A. Taxonomy B. Pathology
C. Plant Ecology D. Ethnobotany

156. Leaves of *Acacia jacquemontii* are given to livestock for the cure of
A. Milk Fever
B. Block of Urine
C. Anthrax
D. Tick Fever

157. *Moringa rivae* is used for
A. Weakness in muscles
B. Arthritis
C. Fever
D. A & B

158. *Rosa indica* can be used for
A. Asthma
B. Removal of kidney stone
C. Gall bladder stone
D. All

159. *ડેટુરા ફાસ્ટુઓસા* **શેના ઈલાજ માટે વપરાય છે?**

A. જૂ વિરોધી B. અસ્થમા વિરોધી

C. કંઈ નહીં D. A અને B બંને

160. થડ નું વિશ્લેષણ શું નક્કી કરવા માટે કરવામાં આવે છે?

A. ઝાડની ઉંમર B. ઝાડની ઉંચાઈ

C. ઝાડનું વજન D. ઝાડની ઘનતા

161. બિલ્ટમોર સ્ટીક શું માપવા માટે વપરાય છે?

A. વૃક્ષનું વજન

B. વૃક્ષનો વ્યાસ

C. વૃક્ષની ઉંમર

D. વૃક્ષનો બેસલ વિસ્તાર

162. લોગના અંતનું ક્ષેત્ર ________ ને વિસ્તાર કહેવામાં આવે છે

A. બેસલ B. અંત

C. સ્ટમ્પ D. બટ

163. પર્ણ પતન કરતા પહેલા પાંદડાના રંગમાં પરિવર્તન આવે છે તેને શું કહેવાય છે?

A. પાનખર ટિન્ટ

B. પાંદડાનું રંગબદલ

C. પાંદડાની દૂષિતતા

D. પાંદડામાં ઉણપ

164. ભારતમાં ફોરેસ્ટ પેથોલોજીના પિતા કોણ છે?

A. ડો .કે. ડી. બગચી

B. સર એડવિન જ્હોન બટલર

C. પેટ્રિક રસેલ

D. સર ડાયેટ્રિક બ્રાન્ડિસ

165. કેન અને વાંસ ટેકનોલોજી સેન્ટર ક્યાં આવેલું છે?

A. ગુવાહાટી, આસામ

B. રુદ્રપુર, ઉત્તરાખંડ

C. ઝાંસી, ઉત્તર પ્રદેશ

D. જબલપુર, મધ્યપ્રદેશ

159. ***Datura fastuosa* is used for**

A. Anti-lice B. Anti-asthma

C. None D. A & B

160. Stem analysis is done to determine

A. Age of a tree B. Height of a tree

C. Weight of a tree D. Density of a tree

161. The Biltmore stick is used to measure

A. Tree Weight

B. Tree Diameter

C. Tree Age

D. Tree basal area

162. The area of the end of a log is called the ________ area

A. Basal B. End

C. Stump D. Butt

163. The leaf color change before the leaf fall is called

A. Autumn tint

B. Discoloration of leaf

C. Malformation of leaf

D. Deficiency in leaf

164.. Who is the father of Forest Pathology in India?

A. Dr. K.D. Bagchee

B. Sir Edwin John Butler

C. Patrick Russell

D. Sir Dietrich Brandis

165. Where is the Cane and Bamboo Technology Centre (CBTC)?

A. Guwahati, Assam

B. Rudrapur, Uttarakhand

C. Jhansi, Uttar Pradesh

D. Jabalpur, Madhya Pradesh

166. આમાંથી કયો છોડ બાયોડિઝલ પ્લાન્ટ છે?
A. જાટ્રોપા B. મહુડો
C. લીમડો D. કુંવારપાઠું

167. *અકેશિયા સેનેગલ* ગુણવત્તામાં શ્રેષ્ઠ માનવામાં આવે છે કેમકે તેમાં ટેનીનની માત્રા ________ હોય છે.
A. ઉચ્ચ B. ઓછી
C. સમાન D. સંતુલિત

168. કયો દેશ *અકેશિયા સેનેગલ* ના ગુંદરમાં વિશ્વભરમાં અગ્રણી ઉત્પાદક છે?
A. જાપાન B. ભારત
C. સુદાન D. ઓસ્ટ્રેલિયા

169. અરબી ગુંદર પાણીમાં ________ દ્રાવ્યતા છે અને અન્ય ગુંદરોની તુલનામાં પ્રમાણમાં ઓછી સ્નિગ્ધતા છે.
A. ઉચ્ચ
B. નીચી
C. સમાન
D. ઉપરોક્તમાંથી એક પણ નહિ

170. *અકેશિયા સેનેગલ* નું અરબી ગમ સ્થાનિક રીતે તરીકે ક્યા નામે ઓળખાય છે?
A. હશાબ B. તલ્હ
C. સલાઇ D. ગુગલ

171. સ્ટેર્ક્યુલિયા પ્રજાતિના ઝાડમાંથી મેળવેલ સૂકા ગુંદર એક્ઝુડેટ ________ તરીકે ઓળખાય છે.
A. અરબી ગમ B. કડાયા ગમ
C. ટ્રગાકાંથ ગમ D. ગુગળ

172. ________ ગુંદર કેટલીકવાર ટ્રગાકાંથના હલકી ગુણવત્તાવાળા વિકલ્પ તરીકે ઉપયોગમાં લેવામાં આવતો હતો, અને તેને ભારતીય પણ કહેવામાં આવે છે.
A. અરબી ગમ
B. કારાયા ગમ
C. ટ્રગાકાંથ ગમ
D. ગુગળ

166. Biodiesel plant is
A. Jatropa B. Madhuca
C. Azadirachta D. Aloevera

167. *Acacia senegal* is considered the best in quality due to a ______ quantity of tannins as
A. Higher B. Lower
C. Equal D. Balanced

168. Which Country is the leading producer of Acacia Gums worldwide?
A. Japan B. India
C. Sudan D. Australia

169. Gum Arabic has _____ water solubility and a relatively low viscosity compared with other Gums.
A. High
B. Low
C. Same
D. None of the above

170. Gum Arabic from *Acacia senegal* is locally known as?
A. Hashab B. Talha
C. Salai D. Gugal

171. The gum dried exudate obtained from trees of *Sterculia species* is known as.
A. Arabic Gum B. Karaya Gum
C. Tragacanth gum D. Gugal

172. ____________ Gum was sometimes used as an inferior substitute for tragacanth, and also called Indian Tragacanth.
A. Arabic Gum
B. Karaya Gum
C. Tragacanth gum
D. Guggul

173. કયો દેશ વિશ્વભરમાં કડાયા ગુંદરનું અગ્રણી ઉત્પાદક છે?
A. જાપાન B. ભારત
C. સુદાન D. ઓસ્ટ્રેલિયા

174. ટ્રગકાંથ ગુંદર એ ___________ની કેટલીક ઝાડવાળા જાતિના મૂળ અને શાખાઓને ટેપ કરીને ઉત્પન્ન થયેલ સૂકા એક્ઝુડેટ છે.
A. *એસ્ટ્રાગાલસ ગમ્મિફર*
B. *કેલોટ્રોપિસ પ્રોસેરા*
C. *બોમ્બેક્સ સીબા*
D. ઉપરોક્ત તમામ

175. પ્રખ્યાત કુદરતી વાદળી રંગ (ઈન્ડિગો) કયા છોડના પાંદડામાંથી મેળવવામાં આવે છે?
A. *રુબિયા* પ્રજાતિઓ
B. *ઈન્ડિગોફેરા ટિંક્ટોરીયલ*
C. *મોરિંડા સાઇટ્રિફોલીઆ*
D. *કાર્થેમસ ટિંક્ટોરિયસ*

176. "પ્રાકૃતિક રંગોનો રાજા" ક્યાંથી પ્રાપ્ત થાય છે?
A. રુબિયા પ્રજાતિઓ
B. *ઈન્ડિગોફેરા ટિંક્ટોરીયલ*
C. *મોરિંડા સાઇટ્રિફોલીઆ*
D. *કાર્થેમસ ટિંક્ટોરિયસ*

177. "કુદરતી રંગોની રાણી" ક્યાંથી મળે છે?
A. *રુબિયા ટીંક્ટોરમ*
B. *ઈન્ડિગોફેરા ટિંક્ટોરીયલ*
C. *મોરિંડા સાઇટ્રિફોલીઆ*
D. *કાર્થેમસ ટિંક્ટોરિયસ*

178. *સીસલપિના સપ્પન*ના લાકડામાંથી લાલ રંગ મેળવવામાં આવે છે. જેને ___________ તરીકે ઓળખાય છે.
A. ઈન્ડિગો B. પટાંગ
C. એન્નાટ્ટો D. કચ

173. Which Country is the leading producer of Karaya Gums worldwide?
A. Japan B. India
C. Sudan D. Australia

174. Tragacanth gum is the dried exudate produced by tapping the tap root and branches of certain shrubby species of ___________
A. *Astragalus gummifer*
B. *Calotropis procera*
C. *Bombax ceiba*
D. All of the above

175. The famous natural blue dye, indigo is obtained from the leaves of the which plant?
A. *Rubia* species
B. *Indigofera tinctorial*
C. *Morinda citrifolia*
D. *Carthamus tinctorius*

176. "King of natural dyes" is obtained from?
A. *Rubia species*
B. *Indigofera tinctorial*
C. *Morinda citrifolia*
D. *Carthamus tinctorius*

177. "Queen of natural dyes" is obtained from?
A. *Rubia tinctorum*
B. *Indigofera tinctorial*
C. *Morinda citrifolia*
D. *Carthamus tinctorius*

178. A red dye is obtained from the wood of *Caesalpina sappan*, known as
A. Indigo B. Patang
C. Annatto D. Cutch

179. નીચેનામાંથી કયા છોડ લાલ રંગ નથી આપતા?
A. *રૂબિયા પ્રજાતિઓ*
B. *ઈન્ડિગોફેરા ટિંકટોરીયલ*
C. *મોરિંડા સાઇટ્રિફોલીઆ*
D. ઉપરોક્ત તમામ

180. નીચેનામાંથી કયુ પીળી ડાઈ નથી?
A. હળદર ડાઈ
B. કેસર ડાઈ
C. એન્નાટ્ટો
D. ઉપરોક્ત કંઈ નથી

181. એન્નાટ્ટો (પીળો રંગ) ________ ના બીજમાંથી મેળવવામાં આવે છે?
A. *ટર્મિનલિયા ચેબ્યુલા*
B. *બિકસા ઓરેલાના*
C. *ટર્મિનલિયા અર્જુના*
D. *બર્બેરિસ એરિસ્ટેટ*

182. માયરોબોલા (પીળો રંગ) ________ ના ફળમાંથી મેળવવામાં આવે છે?
A. *ટર્મિનલિયા ચેબ્યુલા*
B. *બિકસા ઓરેલાના*
C. *ટર્મિનલિયા ચેબ્યુલા*
D. *બર્બેરિસ એરિસ્ટેટ*

183. *હેમેટોક્સીલોન કેમ્પેચિયનમ* ________ રંગ આપે છે.
A. પીળો રંગ B. લાલ રંગ
C. કાળા રંગ D. વાદળી રંગ

184. નીચેનામાંથી કયા ફૂલ છોડમાંથી તેલ કાઢી શકાય છે?
A. *જાસ્મિનમ ઓફિસિનાલ*
B. *બવાયોલા ઓડોરેટા*
C. *મીમોસોપ્સ એલેંગી*
D. ઉપરોક્ત તમામ

179. Which of the following plant does not yields a Red Dye?
A. *Rubia* species
B. *Indigofera tinctorial*
C. *Morinda citrifolia*
D. All of the above

180. Which of the followings is not a Yellow dye?
A. Turmeric dye
B. Saffron dye
C. Annatto
D. None of the above

181. Annatto (Yellow Dye) is obtained from the seed of ________?
A. *Terminalia chebula*
B. *Bixa orellana*
C. *Terminalia chebula*
D. *Berberis aristate*

182. Myrobolan (Yellow Dye) is obtained from the fruits of ________?
A. *Terminalia chebula*
B. *Bixa orellana*
C. *Terminalia chebula*
D. *Berberis aristate*

183. *Haematoxylon campechianum* yields ________ dye.
A. Yellow Dye B. Red Dye
C. Black Dyes D. Blue Dye

184. Which of the followings plants yields Flower oil.
A. *Jasminum officinale*
B. *Viola odorata*
C. *Mimosops elengi*
D. All of the above

185. તેલ જે હવાના સંપર્કમાં લાંબી અવધિ માટે પ્રવાહી સ્થિતિમાં રહે છે તે ________તેલ તરીકે ઓળખાય છે.
A. સૂકવેલ તેલ
B. સૂકવણી વિનાનાં તેલ
C. અર્ધ સૂકવણી તેલ
D. ઉપરોક્ત કંઈ નથી

185. **The oils which remains in liquid state for prolonged period on exposure to air are known as ________oils.**
A. Dried Oils
B. Non- Drying Oils
C. Semi- Drying Oils
D. None of the above

186. નીચેનામાંથી કયાને ડ્રાયિંગ ઓઇલ માનવામાં આવતું નથી?
A. ટંગ તેલ B. અળસીનું તેલ
C. વોલનટ તેલ D. નાળિયેર તેલ

186. **Which of the followings is not considered as Drying Oils?**
A. Tung Oil B. Linseed Oil
C. Walnut Oil D. Coconut Oil

187. રેઝિનને કેટલા ભાગમાં વર્ગીકૃત કરી શકાય છે?
A. સાચા રેઝિન B. ગમ રેઝિન
C. ઓલિઓરેઝિન D. ઉપરોક્ત તમામ

187. **Resin can be classified into**
A. True resins B. Gum resins
C. Oleoresins D. All of the above

188. રેઝિન ટેપિંગ માટે નીચેનામાંથી કઈ પદ્ધતિ ઉપયોગી છે?
A. બોક્ષ પદ્ધતિ
B. કપ અને લિપની પદ્ધતિ
C. રીલ પદ્ધતિ
D. ઉપરોક્ત તમામ

188. **Which of the followings are useful for Resin tapping?**
A. Box method
B. Cup and lip method
C. Rill method
D. All of the above

189. ઇન્ડિયન ઈન્સ્ટિટ્યૂટ ઓફ નેચરલ રેઝિન એન્ડ ગમ ક્યાં આવેલું છે?
A. રાંચી B. ભોપાલ
C. જયપુર D. દિલ્હી

189. **Indian Institute of Natural Resins and Gums located at**
A. Ranchi B. Bhopal
C. Jaipur D. Delhi

190. કેન્સની ભારતમાં કઇ પ્રજાતિઓ છે?
A. કાલામસ ઇરેક્ટસ
B. કાલામુસ ગુરુબા
C. *કાલામસ લાફિટોલીઅસ*
D. ઉપરોક્ત તમામ

190. **Which of the following species of canes founds in India?**
A. Calamus erectus
B. Calamus guruba
C. *Calamus latifolius*
D. All of the above

191. આમાંથી કયા વૃક્ષો/છોડ રેશમી ફ્લોસ ઉત્પન્ન કરતા નથી?
A. *સીઇબા પેન્ટાન્ડ્રા*
B. *બોમ્બક્સ સીબા*
C. *કોકલીસ્પર્મમ રીલીજીયોસમ*
D. ઉપરોક્ત કંઈ નથી

191. **Which of these trees/ plants not produce silky flosses?**
A. *Ceiba pentandra*
B. *Bombax ceiba*
C. *Cochlospermum religiosum*
D. None of the above

192. એક લોકપ્રિય ફૂલ ડાઈ જે *બુટિયા મોનોસ્પરમા* માંથી મેળવવામાં આવે છે તે __________ તરીકે ઓળખાય છે.
A. ધક, પલાશ
B. અન્નાટ્ટો
C. કામેલા
D. કામેલી

193. નીચેનામાંથી કયા વન વૃક્ષો વિવિધ વ્યાવસાયિક મહત્વના બીજ તેલ ઉત્પન્ન છે?
A. એઝાડીરેક્ટા ઈન્ડીકા
B. મધુકા ઈન્ડીકા
C. *પોંગામિયા પિનાટા*
D. ઉપરોક્ત તમામ

194. નીચેનામાંથી કયા પ્રાણી આધારિત કુદરતી રેસા નથી?
A. અંગોરા B. પશ્મિના
C. અલ્પાકા D. હેમ્પ

195. નીચેનામાંથી કયું ફળ ટેન આપનાર છોડ માનવામાં આવે છે?
A. ટર્મિનલિયા ચેબુલા
B. ટર્મિનલિયા બેલિરિકા
C. *એમ્બ્લિકા ઓફિસિનાલિસ*
D. ઉપરોક્ત તમામ

196. બીડી પત્તા __________ ઝાડમાંથી આવે છે.
A. ટર્મિનલિયા ચેબુલા
B. ડાયસ્પોરોસ મેલાનોક્સિલોન
C. *કોર્ડિયા ડિકોટોમા*
D. ઉપરોક્ત તમામ

197. જંગલી મધમાખીની બે પ્રકારની સૌથી સામાન્ય જાતો ________ અને ________ છે.
A. એપિસ ડોર્સાટા, એપિસ ઇન્ડિકા
B. એપિસ ઇન્ડેકા, એપિસ મેલીફેરા
C. *એપીસ મેલીફેરા, એપિસ ડોર્સાટા*
D. ઉપરોક્ત કંઈ નથી

192. A popular flower dye which is obtained from *Butea monosperma* is known as __________.
A. Dhak, Palash
B. Arnotto
C. Kamela
D. Kameli

193. Which of the followings forest trees bear seeds oils of varying commercial importance.
A. *Azadirachta indica*
B. *Madhuca indica*
C. *Pongamia pinnata*
D. All of the above

194. Which of the following is not a Animal based natural fibers.
A. Angora B. Pashmina
C. Alpaca D. Hamp

195. Which of the followings is considered to be a Fruit Tans yielding plants?
A. *Terminalia chebula*
B. *Terminalia bellirica*
C. *Emblica officinalis*
D. All of the above

196. Tendu leaf or bidi patta comes from the tree __________.
A. *Terminalia chebula*
B. *Diospyros melanoxylon*
C. *Cordia dichotoma*
D. All of the above

197. The two commonest of wild honey bees are __________ and __________.
A. *Apis dorsata, Apis indica*
B. *Apis indica, Apis mellifera*
C. *Apis mellifera, Apis dorsata*
D. None of the above

198. ત્રિફળા ચુર્ણ નીચેના કયા છોડમાંથી બને છે?

A. *ટર્મિનલિયા ચેબ્યુલા*

B. *ટર્મિનલિયા બેલિરિકા*

C. એમ્બ્લિકા ઓફિસિનાલિસ

D. ઉપરોક્ત તમામ

199. ઇન્ટરનેશનલ ટ્રોપિકલ ટીમ્બર ઓર્ગેનાઇઝેશન ક્યાં સ્થિત છે?

A. કાનાગાવા, જાપાન

B. જિનેવા, સ્વિટ્ઝલેન્ડ

C. ન્યૂ યોર્ક, યુનાઇટેડ સ્ટેટ્સ

D. ગ્લેન્ડ, સ્વિટ્ઝલેન્ડ

200. ભારતીય વૃક્ષ (લાકડું) એનાટોમીના પિતા તરીકે કોણ જાણીતું છે?

A. એચ.વાય. મોહન

B. ડાયેટ્રિક બ્રાન્ડિસ

C. સર એડવિન જ્હોન બટલર

D. કે. એ. ચૌધરી

201. ભારતીય પ્લાયવુડ ઉદ્યોગ સંશોધન અને તાલીમ સંસ્થા __________ માં સ્થિત છે.

A. બેંગ્લોર B. પુણે

C. નવી દિલ્હી D. કોલકાતા

202. ચંદનનું લાકડું ______________ માંથી મેળવવામાં આવે છે.

A. સાન્તાલમ આલ્બા

B. ટેક્ટોના ગ્રાન્ડિઝ

C. શોરિયા રોબસ્ટા

D. *રાઈટિયા ટિંક્ટોરીયલ*

203. ઝાડની મૂળ અને છાલ *મોરિંડા સિટ્રિફોલિયા* ________ રંગ આપે છે.

A. વાદળી B. પીળો

C. લાલ D. કાળો

204. મેંગોસ્ટીન તેલ ________________ માંથી મેળવી શકાય છે.

A. *ગાર્સિનિયા ઈન્ડીકા*

B. *એઝાડીરેક્ટા ઇન્ડિકા*

C. *મીમોસોપ્સ એલેંગી*

D. *એમ્બ્લિકા ઓફિસિનાલિસ*

198. Triphala Churna made up from which of the followings plants?

A. *Terminalia chebula*

B. *Terminalia bellirica*

C. *Emblica officinalis*

D. All of the above

199. International Tropical Timber Organization (ITTO) is located at _________.

A. Kanagawa, Japan

B. Geneva, Switzerland

C. New York, United States

D. Gland, Switzerland

200. Who is known as the Father of Indian Plant (wood) anatomy?

A. H. Y. Mohan

B. Dietrich brandis

C. Sir Edwin John Butler

D. K. A. Chaudhary

201. Indian Plywood Industries Research and Training Institute is located at ______.

A. Bangalore B. Pune

C. New Delhi D. Kolkata

202. Sandal wood is obtained from ______________.

A. *Santalum alba*

B. *Tectona grandis*

C. *Shorea robusta*

D. *Wrightia tinctorial*

203. The root and bark of the tree *Morinda citrifolia* yield _________ (colour) dye.

A. Blue B. Yellow

C. Red D. Black

204. Mangosteen Oil can be obtained from

A. *Garcinia indica*

B. *Azadirachta Indica*

C. *Mimosops elengii*

D. *Emblica officinalis*

205. શંકુદ્રુપ વૃક્ષના લાકડા સામાન્ય રીતે ________ તરીકે ઓળખાય છે.
A. સોફ્ટવુડ B. હાર્ડવુડ
C. બંને D. ઉપરોક્ત કંઈ નથી

206. હાર્ડવુડ્સના ઝાડમાં ફાઇબર સોફ્ટ વુડઝાડની તુલનામાં ________ હોય છે.
A. ટૂંકા B. લાંબા
C. સમાન D. ખુબ લામ્બા

207. *Dendrocalamous strictus* ના ફાઇબરની લંબાઈ આશરે ________ થી ________ મીમી હોય છે.
A. 1.34 થી 1.43 B. 3.34 થી 3.43
C. 5.34 થી 5.43 D. ઉપરોક્ત કંઈ નથી

208. લોગને રોલ કરવા, બંધ કરવા અને ચાલુ કરવા માટે લિવર તરીકે કામ કરવા માટેના હેતુ માટે ________ ઉપયોગી છે.
A. છાલ કાઢવાની કુહાડી B. કુહાડી
C. કેન્ટ-હૂક D. કરવત

209. લાકડું જેની લંબાઈ 4 મીટર કરતા વધુ છે અને લંબચોરસ અથવા ચોરસ ક્રોસ સેકશનની કોઈપણ બાજુ જે 15 સે.મી. થી ઓછી નથી તેને ________ તરીકે ઓળખવામાં આવે છે.
A. બીમ B. સ્કેન્ટલિંગ
C. બ્લોક D. પ્લેન્ક

210. એક એવી જગ્યા જ્યાં લાકડા અથવા અન્ય વનપેદાશો નિકાલ માટે ભેગી કરવામાં આવે છે?
A. ડેપો B. સ્ટેશન
C. જંકશન D. સ્ટેન્ડ

211. ________ નો અર્થ એ છે કે તેની લાકડાની મૂળ સ્થિતિમાં હાજર ભેજને દૂર કરવું?
A. રેટિન્ગ
B. રોપિંગ
C. લાકડાનું બાઈન્ડીંગ
D. લાકડાનું સીઝનિંગ પ્રક્રિયા

205. **Wood of coniferous trees is generally termed as ________.**
A. Softwood B. Hardwood
C. Both D. None of the above

206. **Hardwoods trees have ________ fiber as compared to Softwood trees.**
A. Short B. Long
C. Equal D. Very long

207. **Fiber length of *Dendrocalamous strictus* varies rom ________ to ________ mm.**
A. 1.34 to 1.43 B. 3.34 to 3.43
C. 5.34 to 5.43 D. None of the above

208. **________ is useful for the Purpose to act as lever to roll, stop and turn logs**
A. Debarking Spade B. Axe
C. Cant-hook D. Saw

209. **A Sawn Timber like Length more than 4 metre and any side of rectangular or square cross section not less than 15 cm is known as ________.**
A. Beam B. Scantling
C. Block D. Plank

210. **A ________ is a place where wood or other forest produce is stored pending its disposal.**
A. Depot B. Station
C. Junction D. Stand

211. **________ means removal of moisture present in the timber in its green state.**
A. Retting
B. Roping
C. Wood binding
D. Wood seasoning

212. નીચેના માંથી કઈ પ્રક્રિયામાં લાકડું ભેજ ગુમાવે છે?

A. શ્રીંક B. સ્વેલિંગ
C. બંને A અને B D. સીઝનીંગ

213. નીચેના માંથી કઈ પ્રક્રિયામાં લાકડું ભેજને શોષી લે છે?

A. શ્રીંક B. સ્વેલિંગ
C. બંને A અને B D. સીઝનીંગ

214. સૂકવણીનો દર __________ દ્વારા નિયંત્રિત થાય છે?

A. ગરમી B. હવામાં ભેજ
C. હવા પરિભ્રમણ D. ઉપરોક્ત તમામ

215. નીચેનામાંથી કયુ બાર્ક ટેન્સ નું ઉદાહરણ નથી?

A. *એકેસીયા નિલોટિકા*
B. *કેસિયા ફિસ્ટુલા*
C. *ટર્મિનાલિયા અર્જુન*
D. ઉપરોક્ત તમામ

216. સાન્તાલીન ડાઇ લાકડાની રંગોથી જાણીતો છે. __________ ના લાકડાથી તેજસ્વી લાલ રંગ સાંતાલિન રંગ મળે છે.

A. *ટેરોકાર્પસ સાન્ટાલિનસ*
B. *સાન્તાલમ આલ્બમ*
C. *આર્ટોકાર્પસ ઈન્ટેગ્રિફોલીયા*
D. *એકેસીયા કેટેચુ*

217. ખેર હાર્ટવુડના મુખ્ય ઘટકો ______ (કાથા) છે.

A. કેટેચિન
B. એઝાડીરેક્ટીન
C. ક્વેર્સિટિન
D. ઉપરોક્ત તમામ

218. કયા ફળને ચીનના ચમત્કારિક ફળ તરીકે ઓળખવામાં આવે છે?

A. કિવિ B. લીચી
C. લોકટ D. અનાનસ

212. Wood __________ as it loses moisture.

A. Shrinks B. Swells
C. Both A and B D. Seasoning

213. Wood __________ as it absorbs moisture.

A. Shrinks B. Swells
C. Both A & B D. Seasoning

214. The rate of drying is controlled by __________?

A. Heat B. Humidity in the air
C. Air circulation D. All of the above

215. Which of the following is not an example of Bark Tans.

A. *Acacia nilotica*
B. *Cassia fistula*
C. *Terminalia arjuna*
D. All of the above

216. Santaline dye is probably the best known of wood dyes. The wood of __________ yields a bright red dye Santaline dye.

A. *Pterocarpus santalinus*
B. *Santalum album*
C. *Artocarpus integrifolia*
D. *Acacia catechu*

217. The principal constituents of Khair heartwood are __________ (Katha).

A. Catechin
B. Azadirachtin
C. Quercetin
D. All of the above

218. Which fruit is known as miracle fruit of China?

A. Kiwi B. Litchi
C. Loquat D. Pineapple

219. કયા ઔષધીય ઉપયોગ માટે *કોમીફેરા વાઈટી*નું ગુંદર જાણીતું છે?

A. મુખ્યત્વે એલિવેટેડ રક્ત કોલેસ્ટરોલનું સ્તર ઘટાડવા માટે વપરાય છે
B. ઘાને મટાડવા માટે
C. તાવને કાબૂમાં રાખવા
D. તીવ્ર ઠંડીમાં વપરાય છે

220 ગુગળ ગુજરાતમાં મુખ્યત્વે કયા જિલ્લામાં જોવા મળે છે?

A. ડાંગ B. વલસાડ
C. ગોધરા D. કચ્છ

221. મહુડાના ફળનો ઉપયોગ ________ માટે થાય છે.

A. આદિવાસીઓના રસોઈ તેલની તૈયારી
B. પશુઓનો ખોરાક
C. ઉનાળામાં રસ કાઢવા
D. ખોરાકમાં

222. બીલી ના ફળનો ઉપયોગ ________ માટે થાય છે.

A. આદિવાસીઓના રસોઈ તેલની તૈયારી
B. પશુઓના ખોરાક
C. ઉનાળામાં રસ કાઢવા
D. ખોરાકમાં

223. પલાસ ના મૂળિયા ________ માટે વપરાય છે.

A. રોપ્સની તૈયારી
B. પશુઓના ખોરાક
C. કોલસો તૈયાર કરવા માટે
D. આભૂષણ

224. કયા લાકડાનો કોલસો વધુ કેલરીફિક મૂલ્ય ધરાવે છે?

A. *Prosopis juliflora*
B. *Bombex ceiba*
C. *Ailenthus excela*
D. *Butea monosperma*

219. For which medicinal value the gum of *Commiphora wightii* is known?

A. used primarily to reduce elevated blood cholesterol levels.
B. To heal up wounds
C. To control fever
D. Used in severe cold

220. In which district of Gujarat *Commiphora wightii* is found naturally?

A. Dangs B. Valsad
C. Godhra D. Kutch

221. The fruits of *Madhuca indica* are used for

A. Preparation of cooking oil of tribals.
B. Used as Cattle feed
C. Used to extract juice in summer
D. Consumption

222. Fruits of *Aegle Marmelos* (Billi) are used for

A. Preparation of cooking oil of tribals.
B. Used as Cattle feed
C. Used to extract juice in summer
D. Consumption

223. Roots of *Butea Monosperma* (Palas) are used for

A. Preparation of Ropes
B. Used as Cattle feed
C. Used to prepare coal
D. Ornament

224. Which of the wooden coal is having higher calorific value?

A. *Prosopis juliflora*
B. *Bombex ceiba*
C. *Ailenthus excela*
D. *Butea monosperma*

225. ગુજરાતમાં *ગાંડો બાવળ* મુખ્યત્વે ________ માટે વપરાય છે.

A. કોલસા માટે B. બળતણના લાકડા
C. ઇમારતી લાકડા D. A અને B બંને

225. In Gujarat, *Prosopis juliflora* is mainly used for

A. Preparation of coal B. Firewood
C. Timber D. Both A & B

Answers | જવાબો

Que.	Ans.	Que.	Ans.	Que.	Ans.	Que.	Ans.	Que.	Ans.	Que.	Ans.	Que.	Ans.
1.	D	**34.**	C	**67.**	B	**100.**	D	**133.**	C	**166.**	A	**199.**	A
2.	D	**35.**	D	**68.**	A	**101.**	C	**134.**	B	**167.**	B	**200.**	D
3.	D	**36.**	B	**69.**	C	**102.**	B	**135.**	C	**168.**	C	**201.**	A
4.	C	**37.**	B	**70.**	B	**103.**	D	**136.**	A	**169.**	A	**202.**	A
5.	D	**38.**	A	**71.**	A	**104.**	C	**137.**	B	**170.**	A	**203.**	C
6.	B	**39.**	C	**72.**	B	**105.**	B	**138.**	D	**171.**	B	**204.**	A
7.	D	**40.**	A	**73.**	C	**106.**	D	**139.**	B	**172.**	B	**205.**	A
8.	C	**41.**	D	**74.**	B	**107.**	A	**140.**	A	**173.**	B	**206.**	A
9.	D	**42.**	A	**75.**	D	**108.**	D	**141.**	B	**174.**	A	**207.**	B
10.	B	**43.**	B	**76.**	A	**109.**	C	**142.**	B	**175.**	B	**208.**	C
11.	D	**44.**	C	**77.**	A	**110.**	C	**143.**	D	**176.**	B	**209.**	A
12.	A	**45.**	D	**78.**	C	**111.**	B	**144.**	B	**177.**	A	**210.**	A
13.	C	**46.**	A	**79.**	A	**112.**	A	**145.**	D	**178.**	B	**211.**	D
14.	B	**47.**	D	**80.**	D	**113.**	D	**146.**	C	**179.**	B	**212.**	A
15.	A	**48.**	A	**81.**	A	**114.**	A	**147.**	D	**180.**	D	**213.**	B
16.	D	**49.**	B	**82.**	A	**115.**	B	**148.**	D	**181.**	B	**214.**	D
17.	D	**50.**	A	**83.**	C	**116.**	A	**149.**	C	**182.**	A	**215.**	D
18.	A	**51.**	C	**84.**	B	**117.**	D	**150.**	D	**183.**	C	**216.**	A
19.	C	**52.**	A	**85.**	A	**118.**	A	**151.**	D	**184.**	D	**217.**	A
20.	D	**53.**	C	**86.**	D	**119.**	C	**152.**	D	**185.**	B	**218.**	A
21.	D	**54.**	D	**87.**	D	**120.**	D	**153.**	B	**186.**	D	**219.**	A
22.	A	**55.**	A	**88.**	C	**121.**	B	**154.**	B	**187.**	D	**220.**	D
23.	D	**56.**	B	**89.**	B	**122.**	A	**155.**	D	**188.**	D	**221.**	A
24.	C	**57.**	D	**90.**	B	**123.**	B	**156.**	B	**189.**	A	**222.**	C
25.	B	**58.**	A	**91.**	D	**124.**	C	**157.**	D	**190.**	D	**223.**	A
26.	B	**59.**	C	**92.**	B	**125.**	B	**158.**	D	**191.**	D	**224.**	A
27.	A	**60.**	A	**93.**	C	**126.**	A	**159.**	D	**192.**	A	**225.**	D
28.	C	**61.**	C	**94.**	C	**127.**	D	**160.**	A	**193.**	D		
29.	D	**62.**	B	**95.**	D	**128.**	C	**161.**	B	**194.**	D		
30.	A	**63.**	C	**96.**	B	**129.**	A	**162.**	A	**195.**	D		
31.	B	**64.**	D	**97.**	A	**130.**	B	**163.**	A	**196.**	B		
32.	C	**65.**	A	**98.**	D	**131.**	C	**164.**	A	**197.**	A		
33.	A	**66.**	A	**99.**	C	**132.**	A	**165.**	A	**198.**	D		

Chapter 13
પ્રકરણ ૧૩

Wildlife Management
વન્યજીવ વ્યવસ્થાપન

National Board for Wildlife (NBWL)

Earlier name	Indian Board for Wildlife (IBWL)
Established in	1952
Reconstituted in	2003
Rename	Under Section 5 of the Wildlife (Protection) Act of 1972, IBWL was replaced by a statutory board renamed as National Board for Wildlife (NBWL).
Governed by	MoEFCC, Govt of India

નેશનલ બોર્ડ ફોર વાઇલ્ડલાઇફ (NBWL)

જૂનું નામ	ઇન્ડિયન બોર્ડ ફોર વાઇલ્ડલાઇફ (IBWL)
સ્થાપના	1952
પુનઃસ્થાપના	2003
નવું નામ	1972 ના વન્યજીવ (સંરક્ષણ) અધિનિયમની કલમ 5 હેઠળ, IBWLને બદલીને નેશનલ બોર્ડ ફોર વાઇલ્ડલાઇફ (NBWL) નામ આપવામાં આવ્યું.
દ્વારા સંચાલિત	MoEFCC, ભારત સરકાર

A.	Prime Minister	Chairman
B.	Minister in-charge of Forests and Wildlife	Vice Chairperson
C.	Three Ministers of Parliament of whom two shall be from the House of the People and one from the Council of States	Member
D.	Planning Commission in-charge of Forests and Wildlife	Member
E.	Five persons to represent non-governmental organizations to be nominated by the Central Government	Member
F.	Ten persons to be nominated by the Central Government from amongst eminent conservationists, ecologists and environmentalists	Member
G.	Secretary to the Government of India in-charge of the Ministry or Department of the Central Government dealing with Forests and Wildlife	Member
H.	Chief of the Army Staff	Member
I.	Secretary to the Government of India in-charge of the Ministry of Defence	Member

J.	Secretary to the Government of India in-charge of the Ministry of the Information and Broadcasting	Member
K.	Secretary to the Government of India in-charge of the Department of Expenditure, Ministry of Finance	Member
L.	Secretary to the Government of India, Ministry of Tribal welfare	Member
M.	Director General of Forests in the Ministry or Department of the Central Government dealing with Forests and Wildlife	Member
N.	Director General of Tourism, Government of India	Member
O.	Director General, Indian Council for Forestry Research and Education, Dehradun	Member
P.	Director, Wildlife Institute of India, Dehradun	Member
Q.	Director, Zoological Survey of India	Member
R.	Director, Botanical Survey of India	Member
S.	Director, Indian Veterinary Research Institute	Member
T.	Member-Secretary, Central Zoo Authority	Member
U.	Director, National Institute of Oceanography	Member
V.	One representative each from ten States and Union territories by rotation, to be nominated by the Central Government	Member
W.	Director of Wildlife Preservation	Member Secretary

A.	પ્રધાન મંત્રી	અધ્યક્ષ
B.	વન અને વન્યપ્રાણી પ્રભારી મંત્રી	ઉપાધ્યક્ષ
C.	સંસદના ત્રણ મંત્રીઓ, જેમનામાંથી બે ગૃહમાંથી અને એક રાજ્યની પરિષદમાંથી	સભ્ય
D.	વન અને વન્ય જીવનના પ્રભારી આયોજન પંચ	સભ્ય
E.	કેન્દ્ર સરકાર દ્વારા નિયુક્ત કરવા માટે બિન-સરકારી સંસ્થાઓ (NGOs) નું પ્રતિનિધિત્વ કરવા માટે પાંચ વ્યક્તિઓ	સભ્ય
F.	કેન્દ્ર સરકાર દ્વારા નામાંકિત થનાર પ્રખ્યાત સંરક્ષણવાદીઓ, અને પર્યાવરણવિદો	સભ્ય
G.	વન અને વન્યજીવન સાથે કામ કરતા કેન્દ્ર સરકારના મંત્રાલય અથવા ભારત સરકારના જે તે વિભાગના પ્રભારી સચિવ	સભ્ય
H.	ચીફ ઓફ આર્મી સ્ટાફ	સભ્ય
I.	ભારત સરકારના સંરક્ષણ મંત્રાલયના પ્રભારી સચિવ	સભ્ય
J.	ભારત સરકારના માહિતી અને પ્રસારણ મંત્રાલયના પ્રભારી સચિવ	સભ્ય
K.	ભારત સરકારના નાણાં મંત્રાલયના ખર્ચ વિભાગના પ્રભારી સચિવ	સભ્ય
L.	ભારત સરકારના આદિજાતિ કલ્યાણ મંત્રાલયના સચિવ	સભ્ય
M.	કેન્દ્ર સરકારના મંત્રાલય અથવા વન અને વન્યજીવ વિભાગના વન મહાનિયામક	સભ્ય
N.	ભારત સરકારના પર્યટન નિયામક	સભ્ય
O.	મહાનિયામક, ભારતીય વનીકરણ સંશોધન અને શિક્ષણ પરિષદ (ICFRE), દહેરાદૂન	સભ્ય

P.	નિયામક, વાઇલ્ડલાઇફ ઇન્સ્ટિટ્યૂટ ઓફ ઈન્ડિયા (WII), દહેરાદૂન	સભ્ય
Q.	નિયામક, ઝૂઓલોજિકલ સર્વે ઓફ ઈન્ડિયા (ZSI)	સભ્ય
R.	નિયામક, બોટનિકલ સર્વે ઓફ ઇન્ડિયા (BSI)	સભ્ય
S.	નિયામક, ભારતીય વેટરનરી રિસર્ચ ઇન્સ્ટિટ્યૂટ (IVRI)	સભ્ય
T.	સભ્ય-સચિવ, સેન્ટ્રલ ઝૂ ઓથોરિટી (CZA)	સભ્ય
U.	નિયામક, નેશનલ ઇન્સ્ટિટ્યૂટ ઓફ ઓશેનોગ્રાફી	સભ્ય
V.	કેન્દ્ર સરકાર દ્વારા નિયુક્ત દસ રાજ્યો અને કેન્દ્રશાસિત પ્રદેશોમાંથી પ્રતિનિધિ રોટેશન મુજબ	સભ્ય
W.	વન્યપ્રાણી સંરક્ષણ નિયામક	સભ્ય સચિવ

IUCN Red List Categories and Criteria | આઇયુસીએન રેડ લીસ્ટ શ્રેણીઓ અને માપદંડ

- IUCN established in 1964.
- World's most comprehensive and trusted source of information on animal and plant species.
- Divided world's species in 9 categories.

Category	Symbol	Criteria	Example
Extinct	EX	No individual is left on earth, all have died.	Dodo, Black mamo, Passenger pigeon, Golden toad
Extinct wild	EW	Taxa is extinct at wild but surviving in captivity	Guam kingfisher, Franklin tree, She cabbage tree, Kalimantan mango,
Critically Endangered	CR	Taxa facing extremely high risk of extinction • Population size reduced >90% in last 10 years or 3 generations whichever is longer • Extent of occurrence is <500 km^2 • Area of occupancy is <10 km^2 • Population size <250 matured individuals	Hangul, Slender billed Vulture, Great Indian Bustard, Siberian Crane, Gharial
Endangered	EN	Taxa facing very high risk of extinction • Population size reduction of >70% in last 10 years or 3 generations whichever is longer • Extent of occurrence is 5000 km^2 • Area of occupancy is <300 km^2 • Population size <2500 matured individuals	Tiger, Asian elephant, Dhole, Blue whale, Green peafowl, Lesser florican, Greater adjutant, Ganges river dolphin
Vulnerable	VU	Taxa facing high risk of extinction • Population size reduction of >50% in last 10 years or 3 generations whichever is longer	Lion, Leopard, Snow leopard, Clouded leopard, Cheetah, Sloth bear, Sarus crane, Sambar, Lesser

Category	Symbol	Criteria	Example
Vulnerable	VU	• Extent of occurrence is 20,000 km^2 • Area of occupancy is 2,000 km^2 • Population size <1,000 matured individuals	Adjutant, Coffin tree, King Cobra
Near Threatened	NT	A taxon not qualify for CR, EN or VU now, but is close to qualifying for a threatened category in the near future.	Asiatic wild ass, Stripped hyaena, Jaguar, Lesser Flamingo, Red kite, Painted Stork, Bull shark, Tiger shark
Least Concern	LC	Widespread and abundant taxa	Grey wolf, Chital, Blackbuck, Spotted hyaena, Black kite, Yellow sea snake
Data deficient	DD	Insufficient information on taxa	Indian mackerel, Somali pigeon, Cherry plum
Not Evaluated	NE	Taxon not evaluated.	Reticulated python

IUCN 2020

Conservation Projects | સંરક્ષણ પરિયોજનાઓ

Project name	Year	Remarks
Project Tiger	1973	Core-buffer strategy. Core areas were freed from human activities and buffer areas were imposed conservational land use.
Lion Conservation Project	1972	Conservation of lions in their habitat and social upliftment of maldhari tribes.
Crocodile Conservation Project	1975	Intensive captive breeding and rearing programme. For 3 crocodilian species found in India.
Project Elephant	1992	To reduce the threats of loss of habitat, migration path and poaching for tusks. Part of project MIKE of CITES.
Project Snow Leopard	2009	In Jammu and Kashmir, Himachal Pradesh Uttarakhand. Arunachal Pradesh and Sikkim.
Operation Rhino	2005	Phase I- 2005 to 2008, Phase II- 2008-2012. Presently working on IRV-2020 (India Rhino Vision).
Project Hangul	1970	Jammu Kashmir Government with IUCN and WWF implemented the project in Dachigam National Park.
Project Great Indian Bustard	2013	To conserve the remaining population of critically endangered GIB.
Species Recovery Plan of Bengal Florican	1995	To ensure to manage Bengal florican scientifically in their natural habitats.
Project Snow Leopard	2009	To safeguard and conserve India's Snow leopards.
Sea turtle Project	1999	MoEF started the project with UNDP and WII for Olive Ridley Turtles.

MoEFCC 2020

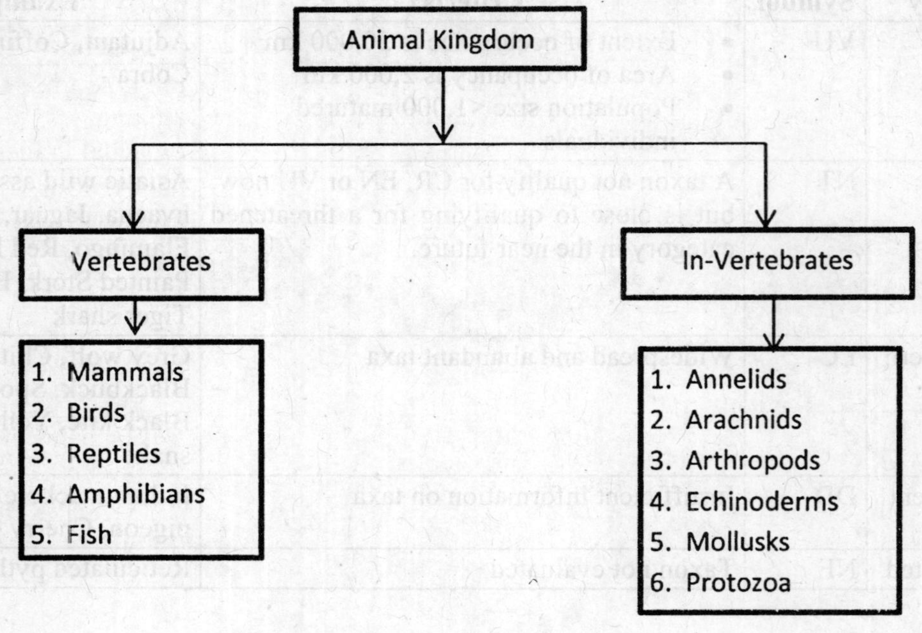
Animal Kingdom
Vertebrates
In-Vertebrates
1. Mammals
2. Birds
3. Reptiles
4. Amphibians
5. Fish
1. Annelids
2. Arachnids
3. Arthropods
4. Echinoderms
5. Mollusks
6. Protozoa

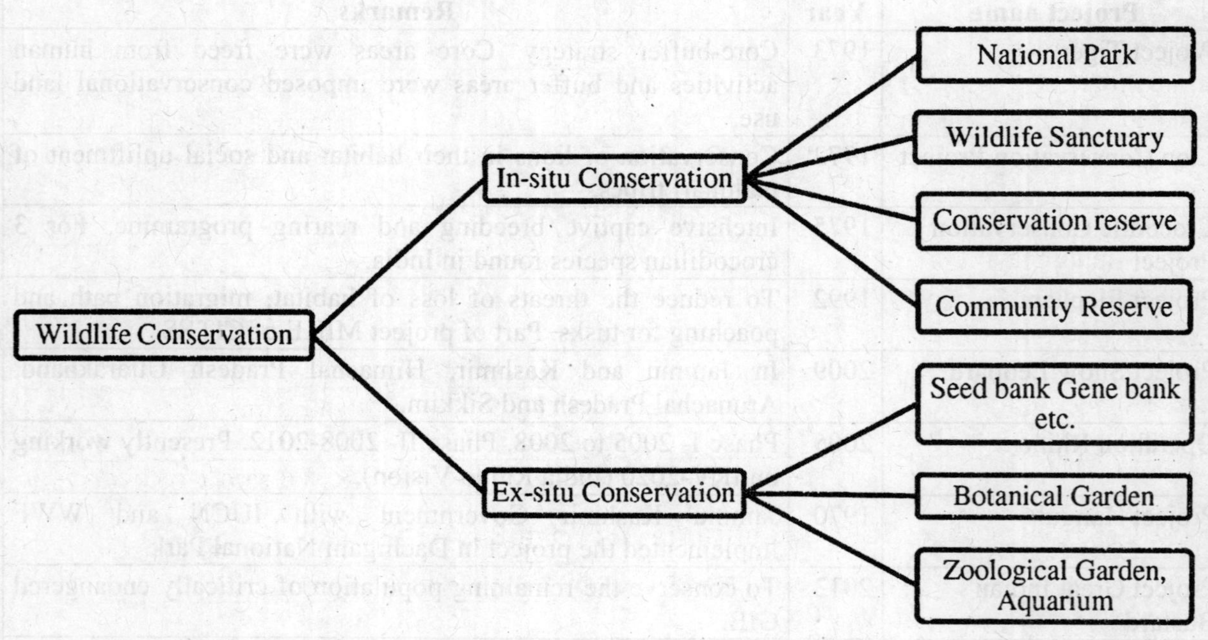
Wildlife Conservation
In-situ Conservation
Ex-situ Conservation
National Park
Wildlife Sanctuary
Conservation reserve
Community Reserve
Seed bank Gene bank etc.
Botanical Garden
Zoological Garden, Aquarium

Important Conservation Areas of India | ભારતના મહત્વપૂર્ણ સંરક્ષણ ક્ષેત્રો

Conservation area	Nos.	First	Year	Latest	Year
Tiger Reserves	50	Jim Corbett, Uttarakhand	1973	Orang, Asam Kamlang, Arunachal Pradesh	2016
Elephant Reserves	32	Singhbhum, Jharkhand	2001	Uttar Pradesh Elephant Reserve	2009
Biosphere Reserves	18	Nilgiri	1986	Panna	2011
RAMSAR Wetland Sites	41	Chilka Lake, Odisha Keoladeo NP, Rajastahn	1981	Sur Sarovar, UP Lonar lake, Maharashtra	2020
Natural World Heritage Sites	7	Kaziranga, Keoladeo and Manas NP	1985	Great Himalayan National Park Conservation Area	2014
Cultural World Heritage Sites	30	Taj Mahal	1983	Jaipur City, Rajasthan	2019
Mixed World Heritage Sites	1	Khangchendzonga National Park	2016	—	—
Important Coastal and Marine Biodiversity Areas	107	—	—	—	—
Marine Protected Areas	131 (25 in penincular India and 106 in Islands)	Point Calimere, TN	1967	Thane Creek Flamingo, Maharashtra	2015
Important Bird Areas	467	—	—	—	—
Potential Important Bird Areas	96	—	—	—	—
Key Biodiversity Areas	531	—	—	—	—
Biodiversity Heritage Sites	18	Nallur Tamarind Grove, Karnataka	2007	Mandasaru, Odisha	Dec.-2019

ENVIS, WII 2020 & UNESCO 2020

Tiger Reserves of India | ભારતના ટાઇગર રિઝર્વ્ઝ

Started in	1973 as a Project Tiger by Govt of India
Tiger reserves when started	9 (1973)
Tiger reserves	50 (2020)
Biggest	Nagarjunsagar Srisailam Tiger Reserve
Smallest	Bor Tiger Reserve
First	Jim Corbett in 1973
Latest	Orang, Asam and Kamlang, Arunachal Pradesh in 2016

ENVIS, WII 2020

1	Bandipur	Karnataka
2	Corbett Amangarh	Uttarakhand Uttar Pradesh
3	Kanha	Madhya Pradesh
4	Manas	Assam
5	Melghat	Maharashtra
6	Palamau	Jharkhand
7	Ranthambore	Rajasthan
8	Similipal	Odisha
9	Sunderbans	West Bengal
10	Periyar	Kerala
11	Sariska	Rajasthan
12	Buxa	West Bengal
13	Indravati	Chhattisgarh
14	Namdapha	Arunachal Pradesh
15	Dudhwa	Uttar Pradesh
16	Kalakad-Mundanthurai	Tamil Nadu
17	Valmiki	Bihar
18	Pench	Madhya Pradesh
19	Tadoba-Andhari	Maharashtra
20	Bandhavgarh	Madhya Pradesh
21	Panna	Madhya Pradesh
22	Dampa	Mizoram
23	Bhadra	Karnataka
24	Pench	Maharashtra
25	Pakke	Arunachal Pradesh
26	Nameri	Assam
27	Satpura	Madhya Pradesh
28	Anamalai	Tamil Nadu
29	Udanti-Sitanadi	Chattisgarh
30	Satkosia	Odisha
31	Kaziranga	Assam
32	Achanakmar	Chattisgarh
33	Dandeli-Anshi	Karnataka
34	Sanjay-Dubri	Madhya Pradesh
35	Mudumalai	Tamil Nadu
36	Nagarahole	Karnataka
37	Parambikulam	Kerala
38	Sahyadri	Maharashtra

39	Biligiri Ranganatha Temple	Karnataka
40	Kawal	Telangana
41	Sathyamangalam	Tamil Nadu
42	Mukandra Hills	Rajasthan
43	Nawegaon-Nagzira	Maharashtra
44	Nagarjunsagar Srisailam	Andhra Pradesh
45	Amrabad	Telangana
46	Pilibhit	Uttar Pradesh
47	Bor	Maharashtra
48	Rajaji Tige Reserve	Uttarakhand
49	Orang Tiger Reserve	Assam
50	Kamlang Tiger Reserve	Arunachal Pradesh

WII 2020

Type of Wetlands | વેટલેન્ડ્સના પ્રકારો

Type	Remarks
Estuaries	Deep water wetlands receiving regular tidal flush.
Floodplains	Wetlands present adjacent to river in riverine ecosystem. Floodplains are inundated regularly or seasonally and are rich in alluvial deposits.
Lakes	Permanent water bodies having steady water.
Man-made Wetlands	Excavated or transformed for human need.
Marshes	Breeding and wintering grounds for ducks, geese, herons, shorebirds and stopover points.
Peatlands	Waterlogged areas saturated by groundwater or rainwater.
Swamps	Wetlands with trees and shrubs.

Ramsar Wetland Sites of India | ભારતની રામસર વેટલેન્ડ સાઇટ્સ

First Ramsar site	Chilka Lake, Odisha and Keoladeo National Park, Rajasthan
Year	1981
Total Ramsar sites	41
Governed by	Ramsar Convension
40th and 41st sites (added in Nov 2020)	Sur Sarovar (commonly known as Keetham Lake) in Uttar Pradesh and Lonar Lake in Maharashtra

Declaration of wetland for Waterbirds | વોટરબર્ડ્સ માટે વેટલેન્ડની ઘોષણા

Important Criteria	Remarks
1	supports vulnerable, endangered, or critically endangered species
2	regularly supports 20,000 or more waterbirds.
3	regularly supports 1% of the individuals in a population of one species or subspecies of waterbird

Ramsar sites of India | ભારતના રામસર સ્થળોની સૂચિ

No.	Name of Site	State	Year of Declaration
1	Asan Conservation Reserve	Uttarakhand	2020
2	Asthamudi Wetland	Kerala	2002
3	Beas Conservation Reserve	Punjab	2019
4	Bhitarkanika Mangroves	Orissa	2002
5	Bhoj Wetlands	Madhya Pradesh	2002
6	Chandertal Wetland	Himachal Pradesh	2005
7	Chilka Lake	Orissa	1981
8	Deepor Beel	Assam	2002
9	East Kolkata Wetlands	West Bengal	2002
10	Harike Lake	Punjab	1990
11	Hokera Wetland	Jammu and Kashmir	2005
12	Kabartal Wetland	Bihar	2020
13	Kanjli Lake	Punjab	2002
14	Keoladeo Ghana NP	Rajasthan	1981
15	Keshopur-Miani Community Reserve	Punjab	2019
16	Kolleru Lake	Andhra Pradesh	2002
17	Loktak Lake	Manipur	1990
18	Lonar Lake	Maharashtra	2020
19	Nalsarovar Bird Sanctuary	Gujarat	2012
20	Nandur Madhameshwar	Maharashtra	2019
21	Nangal Wildlife Sanctuary	Punjab	2019
22	Nawabganj Bird Sanctuary	Uttar Pradesh	2019
23	Parvati Agra Bird Sanctuary	Uttar Pradesh	2019
24	Point Calimere Wildlife and Bird Sanctuary	Tamil Nadu	2002
25	Pong Dam Lake	Himachal Pradesh	2002
26	Renuka Wetland	Himachal Pradesh	2005
27	Ropar Lake	Punjab	2002
28	Rudrasagar Lake	Tripura	2005
29	Saman Bird Sanctuary	Uttar Pradesh	2019

No.	Name of Site	State	Year of Declaration
30	Samaspur Bird Sanctuary	Uttar Pradesh	2019
31	Sambhar Lake	Rajasthan	1990
32	Sandi Bird Sanctuary	Uttar Pradesh	2019
33	Sarsai Nawar Jheel	Uttar Pradesh	2019
34	Sasthamkotta Lake	Kerala	2002
35	Sunderbans Wetland	West Bengal	2019
36	Surinsar-Mansar Lakes	Jammu and Kashmir	2005
37	Sur Sarovar	Uttar Pradesh	2020
38	Tsomoriri Lake	Jammu and Kashmir	2002
39	Upper Ganga River (Brijghat to Narora Stretch)	Uttar Pradesh	2005
40	Vembanad Kol Wetland	Kerala	2002
41	Wular Lake	Jammu & Kashmir	1990

Ramsar Convention, As on November 2020

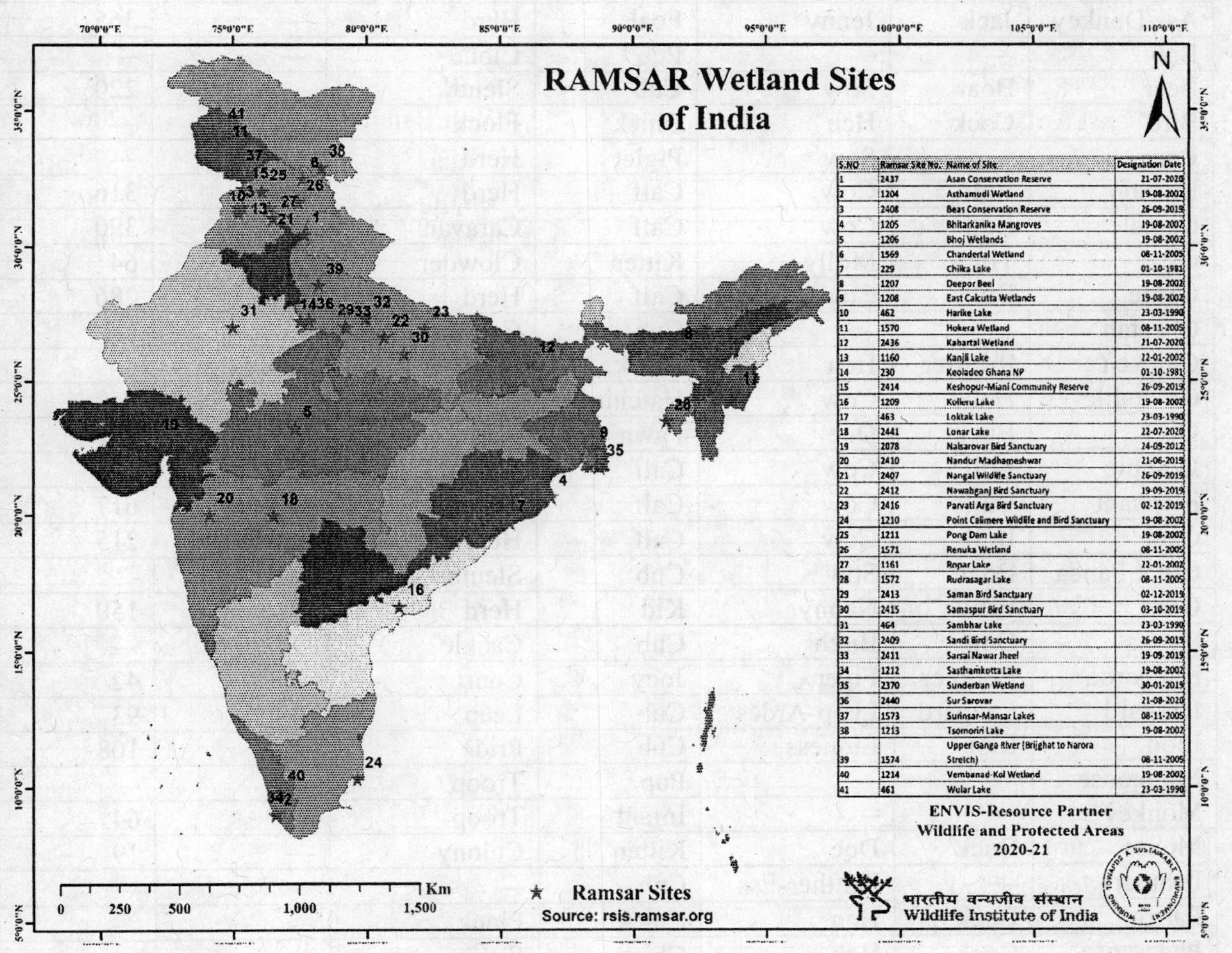

S.NO	Ramsar Site No.	Name of Site	Designation Date
1	2437	Asan Conservation Reserve	21-07-2020
2	1204	Asthamudi Wetland	19-08-2002
3	2408	Beas Conservation Reserve	26-09-2019
4	1205	Bhitarkanika Mangroves	19-08-2002
5	1206	Bhoj Wetlands	19-08-2002
6	1569	Chandertal Wetland	08-11-2005
7	229	Chilka Lake	01-10-1981
8	1207	Deepor Beel	19-08-2002
9	1208	East Calcutta Wetlands	19-08-2002
10	462	Harike Lake	23-03-1990
11	1570	Hokera Wetland	08-11-2005
12	2436	Kabartal Wetland	21-07-2020
13	1160	Kanjli Lake	22-01-2002
14	230	Keoladeo Ghana NP	01-10-1981
15	2414	Keshopur-Miani Community Reserve	26-09-2019
16	1209	Kolleru Lake	19-08-2002
17	463	Loktak Lake	23-03-1990
18	2441	Lonar Lake	22-07-2020
19	2078	Nalsarovar Bird Sanctuary	24-09-2012
20	2410	Nandur Madhameshwar	21-06-2019
21	2407	Nangal Wildlife Sanctuary	26-09-2019
22	2412	Nawabganj Bird Sanctuary	19-09-2019
23	2416	Parvati Arga Bird Sanctuary	02-12-2019
24	1210	Point Calimere Wildlife and Bird Sanctuary	19-08-2002
25	1211	Pong Dam Lake	19-08-2002
26	1571	Renuka Wetland	08-11-2005
27	1161	Ropar Lake	22-01-2002
28	1572	Rudrasagar Lake	08-11-2005
29	2413	Saman Bird Sanctuary	02-12-2019
30	2415	Samaspur Bird Sanctuary	03-10-2019
31	464	Sambhar Lake	23-03-1990
32	2409	Sandi Bird Sanctuary	26-09-2019
33	2411	Sarsai Nawar Jheel	19-09-2019
34	1212	Sasthamkotta Lake	19-08-2002
35	2370	Sunderban Wetland	30-01-2019
36	2440	Sur Sarovar	21-08-2020
37	1573	Surinsar-Mansar Lakes	08-11-2005
38	1213	Tsomoriri Lake	19-08-2002
39	1574	Upper Ganga River (Brijghat to Narora Stretch)	08-11-2005
40	1214	Vembanad-Kol Wetland	19-08-2002
41	461	Wular Lake	23-03-1990

WII 2020

Important Endangered Wildlife of India | ભારતના મહત્વના ભયગ્રસ્ત વન્યજીવો

Taxa	India	Gujarat
Mammals	Lion-tailed macaque, Nilgiri thar, Indian giant squirrel	Indian Wild Ass, Indian Gazelle, Black buck, Asiatic Lion, Dugong, Indian Leopard
Birds	Narcondam hornbill, Grey junglefowl, Great Indian bustard, Forest owlet, Jerdon's courser	Great Indian bustard, Forest owlet, Red-headed Vulture, Adjutant-Stork, Great Indian Bustard, Lesser Florican
Reptile	Assam keelback, Darjeeling kukri snake, Andaman cobra, Nilgiri forest lizard	Leatherback Sea Turtle, Green Sea Turtle
Amphibia	Malabar toad, Nagaland toad	Ramanella frog, Ichthyophis
Butterflies	Nilgiri tiger, Andaman Mormon	

Wildlife of India | ભારતના વન્યજીવો

Wildlife Species	Male	Female	Offspring	Group	Meat	Gestation Period
Antelope	Buck	Doe	Calf	Herd	—	—
Ass/Donkey	Jack	Jenny	Foal	Herd	—	365
Bat	—	—	Pup	Cloud	—	—
Bear	Boar	Sow	Cub	Sleuth	—	220
Bird	Cock	Hen	Chick	Flock	—	—
Boar		Sow	Piglet	Herd	—	—
Buffalo	Bull	Cow	Calf	Herd	Buffin	316
Camel	Bull	Cow	Calf	Caravan	—	390
Cat	Tom	Molly	Kitten	Clowder	—	64
Cattle	Bull	Cow	Calf	Herd	Beef	286
Cheetah	—	—	Cub	Coalition	—	—
Chicken	Rooster	Hen	Chick	Flock	Poultry	—
Crocodile	Bull	Cow	Hatching	Bask	—	—
Deer	Buck	Doe	Fawn	Herd	Venison	201
Dolphin	Bull	Cow	Calf	School	—	—
Elephant	Bull	Cow	Calf	Herd	—	617
Gaur	Bull	Cow	Calf	Herd	Gara-Beef	215
Giant Panda	Boar	Sow	Cub	Sleuth	—	—
Goat	Billy	Nanny	Kid	Herd	Chevon	150
Hyena	Dog	Bitch	Cub	Cackle	—	—
Kangaroo	Boomer	Flyer	Joey	Court	—	42
Leopard	Leopard	Leop-Ardes	Cub	Leap	—	93
Lion	Lion	Lioness	Cub	Pride	—	108
Mongoose	—	—	Pup	Troop	—	—
Monkey	—	—	Infant	Troop	—	64
Mouse	Buck	Doe	Kitten	Colony	—	19
Panther	—	Panther-Ess	Cub	—	—	—
Parrot	Cock	Hen	Chick	Flock	—	—
Pheasant	Cock	Hen	Chick	Flock	—	—

Wildlife Species	Male	Female	Offspring	Group	Meat	Gestation Period
Rabbit	Buck	Doe	Bunny	Nest	—	31
Sheep	Ram	Ewe	Lamb	Flock	Mutton	147
Rat	Buck	Doe	Kitten	Colony		22
Rhinoceros	Bull	Cow	Calf	Herd		450
Sheep	Ram	Ewe	Lamb	Flock	Lamb	
Snake	—	—	Snakelet	Nest	—	
Squirrel	Buck	Doe	Pup	Scurry	—	35
Tiger	Tiger	Tigress	Cub	Ambush	—	109
Walrus	Bull	Cow	Calf	Herd	—	—
Whale	Bull	Cow	Calf	Pod	Blubber	535
Wolf	Dog	Bitch	Pup	Pack	—	64
Yak	Bull	Cow	Calf	Herd		
Zebra	Stallion	Mare	Colt	Herd	—	375

Legal protection to Important Wildlife of India

ભારતના મહત્વપૂર્ણ વન્યપ્રાણીઓને કાનૂની સુરક્ષા

Wildlife Species	WLPA Schedule	IUCN Category	CITES Annexure
Asian Elephant	I	EN	
Asiatic Lion	I	EN	I
Asiatic Wild Ass	I	NT	I
Asiatic Wild Dog/ Dhole	I	EN	II
Bengal Florican	I	CR	I
Blackbuck	I	NT	III
Brow-antlered Deer	I	EN	I
Caracal	I	LC	I
Cheetah	I	VU	I
Clouded Leopard	I	VU	I
Crocodiles	I	VU	I
Dolphins	I	LC	I
Dugong	I	VU	I
Fishing Cat	I	EN	II
Four-Horned Antelope	I	VU	III
Ganges River Dolphin	I	EN	I
Gaur	I	VU	I
Gharial	I	CR	I
Gray Slender Loris	I	LC	II
Great Indian bustard	I	CR	I
Great Indian hornbill	I	VU	I
Greater One-horned Rhinoceros	I	VU	I

Wildlife Species	WLPA Schedule	IUCN Category	CITES Annexure
Himalayan Black Bear	I	VU	I
Himalayan Musk Deer	I	EN	I
Hoolock Gibbon	I	EN	I
Indian Gazelle	I	LC	I
Jerdon's Courser	I	CR	-
Lion-Tailed Macaque	I	EN	I
Malabar Civet	I	CR	III
Olive Ridley Turtle	I	VU	I
Otters	I	NT	I
Peafowl	I	LC	III
Pythons	I	NE	II
Rusty Spotted Cat	I	NT	I/II
Sloth Bear	I	VU	I
Swamp Deer	I	VU	I
Tiger	I	CR	I
Whale Shark	I	EN	II
White-Rumped Vulture	I	CR	II

ENVIS WII 2020; IUCN 2020; CITES 2020

Types of wildlife Corridors | વાઇલ્ડલાઇફ કોરિડોરના પ્રકાર

Type	Example
Regional	Migratory pathways
Sub-regional	Ridgelines
Local	Strips of vegetation connecting remnant patches
Linear Corridors	Hedges, strips of forest and the vegetation growing on banks of rivers and streams.
Stepping Stone Corridors	—
Landscape Corridors	—

Important Deers of India | ભારતના મહત્વના હરણ

Deer Species	Distribution	Best seen at National Park/ Sanctuary
Blue bull or Nilgai	Throughout India	Sariska, Gir
Brow-antlered Deer or Eld's Deer or Sangai	Manipur	Keibul Lamjao
Chital or Spotted Deer	Through Peninsular India	Corbett, Bandipur, Kanha, Gir
Hangul or Kashmir Red deer	Jammu & Kashmir and Himachal Pradesh	Dachigam
Himalayan Musk Deer	Kashmir, Himachal Pradesh, Sikkim	Kedarnath

Deer Species	Distribution	Best seen at National Park/ Sanctuary
Hog Deer	Himachal Pradesh, North East India, Punjab, Uttar Pradesh and West Bengal	Kaziranga, Dudhwa
Indian Muntjac	Almost throughout the country except Jammu & Kashmir and desert region	Bandhavgarh National Park, Bandipur and Kaziranga
Leaf Deer	Arunachal Pradesh	Namdapha Tiger Reserve
Mouse Deer or Indian Chevrotain	Peninsular India, all through South India	Bandipur
Sambar	Throughout India except Himalayas	Sariska, Kanha, Corbett, Ranthambore, Bandhavgarh, Gir, Dudhwa, Kaziranga, Periyar
Swamp Deer	Assam, Madhya Pradesh and Uttar Pradesh	Kanha, Kaziranga,

Menon 2014

Important Antelopes of India | ભારતના મહત્વપૂર્ણ મૃગો

Antelope Species	Distribution	Best seen at National Park/ Sanctuary
Blackbuck	Rajasthan, Gujarat, TN	Velavadar, Point Calimere
Four-horned Antelope or Chousingha	Throughout India	Gir, Mudumalai
Himalayan Tahr	Jammu Kashmir, Sikkim	Kedarnath
Indian gazzle or Chinkara	Arid and Semi-arid parts	All over Rajasthan and Gujarat
Nilgiri Tahr	Western Ghats	Eravikulam

Menon 2014

Deer	Antelope
• Belongs to the family Cervidae.	• Belongs to the famil Bovidae.
• Found in the forests of Asia, Europe and North America.	• Native to Asia and Africa.
• Only male deers have antlers.	• Both male and female antelope have horns.
• Antlers are not permanent and shed annually.	• Horns are permanent and do not shed.
• Antlers are branched.	• Horns are twisted, ridged and curved.
• Life span - 10 to 20 years.	• Life span 10 to 25 years.
• Mule Deer, Elk, Moose, Reindeer, Fallow Deer are deers.	• Gazelle, Oryx, Waterbuck, Impala are antelopes.
• Indian deers are Blue bull, Spotted Deer, Sambar, etc.	• Indian antelopes are Blackbuck, Chousingha, Chinkara, etc.

Important Scientific Rules in Wildlife Science

વન્યજીવ વિજ્ઞાન ના મહત્વપૂર્ણ વૈજ્ઞાનિક નિયમો

Rule	Example	Explantation
Allen's Rule	Ears of Arctic fox are smaller and Desert fox are bigger	The tail, neck and other external parts of the animals living in colder parts are compact as well as shorter in comparison with the animals living in warmer parts
Bergmann's Rule	Polar bear is bigger than sloth bear	Animals living in colder region are much larger than the warmer region.
Gloger's Rule	Polar bears are lighter than other bears	Animals in warm and humid climate are darker than the animals living in cold or dry climate
Jorden's Rule	Species of fishes	Species with low temperature have more vertebrate than those living in warm water.

Territory and Home Range of wild animals | વન્ય પ્રાણીઓની ટેરીટરી અને હોમ રેન્જ

• Territory is an area defended by the species from the individuals of the same species.	• Home range is an area where animal roams around for day to day activity.
• No overlapping in territories.	• Home ranges can overlap.

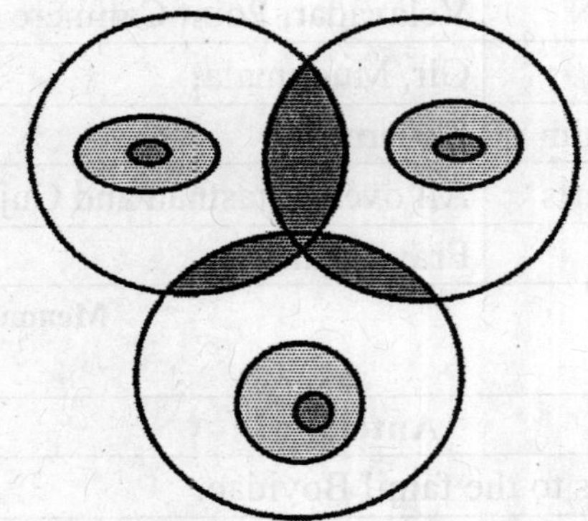

Scent glands in wild animals | વન્ય પ્રાણીઓમાં સુગંધિત ગ્રંથીઓ

Gland	Use	Location
Forehead or sudoriferous	Used to apply scent on branches to express domination and territory.	On the forehead between the antlers and eyes
Interdigital	Emit a yellow waxy chemical which has offensive odour for defense.	Between the hooves of each foot
Metatarsal	Used for alarm.	On the outside of each hind leg between ankle and hooves
Pre-orbital	Rubbed on tree branches for domination and territory marking.	In front of the eye
Tarsal	Used for alarm.	On the upper inside on each hind leg

Migration types of Wild animals | વન્ય પ્રાણીઓના સ્થળાંતરના પ્રકારો

Altitudinal migration	between different terrestrial elevations
Annual migration	Synchronized with the annual cycle.
Complete migration	Migration of a all the individuals of species.
Diadromous migration	Animals move between freshwater and saltwater ecosystems.
Differential migration	Different age or sex of a species migrate at different distance or times.
Drift migration	Individuals are blown off by the winds or water currents
Facultative migration	Migration is a matter of choice for the individuals of species.
Horizontal migrations	Individuals travel a few hundred meters across a space.
Nomadism	Migration does not follow any regular pattern or route
Obligate migration	Migration is a compulsory for all the individuals of species.
One-way migration	Move from a location where they breed.
Partial migration	Migration of a few individuals of species.
Round-trip migration	Animals return to the general breeding area from which they originated.
Seasonal migration	Synchronized with the season of the year.
To-and-fro migration	Animals migrate between two different habitats.
Vertical migrations	Between different water depths.

How Wild animals navigate during migration
સ્થળાંતર દરમિયાન વન્ય પ્રાણીઓ કેવી રીતે શોધખોળ કરે છે

Electric fields	Electric fish
Magnetic field of earth	Sea turtles, Birds
Smell	Salmon, Moths
Sound	Bats, Elephants
Temperature	Snakes
Vibrations	Spiders, Rats, Mole
Visual understanding	Ants, Bees, Birds

Common diseases of Wildlife | વન્યપ્રાણીસૃષ્ટિના સામાન્ય રોગો

Type	Diseases	Causes organism
Viral Diseases	Foot and Mouth Disease	FMD Virus
	Rinderpest	Microbilli virus
	Canine Distemper	—
	Rabies	Rhabdo virus
	Contagious Ecthyma	Causative virus
	Canine Parvovirus	—
	Avian Pox	Avipox viruse
	Trypanosomia	Trypanosomia virus

Type	Diseases	Causes organism
Viral Diseases	Eastern Equine Encephalitis	RNA virus
	Avian Influenza	—
	Duck Plague	herpes virus
	Bluetongue Disease	Midges
Bacterial Diseases	Anthrax	*Bacillus anthraxis*
	Bumble Foot	—
	Chlamydiosis	*Chlamydia psittaci*
	Leptospirosis	*Leptospira spp.*
	Brucellosis	*Brucella spp.*
	Bubonic Plague	*Yersinia pestis*
	Salmonellosis	*Salmonella spp.*
	Tularemia	*Francisella tularensis.*
Protozoan Diseases	Giardiasis	*Giardia intestinalis*
	Toxoplasmosis	*Toxoplasma gondii*
Mycotic Diseases	Aspergillosis	*Aspergillus fumigatus.*
	Histoplasmosis	*Histoplasma capsulatum*
	White-nose Syndrome in Bats (WNS)	*Geomyces destructans*
Tick-borne Diseases	Lyme Disease	*B. burgdorferi, B. afzelii,* and *B. garinii*
	Rocky Mountain Spotted Fever	*Rickettsia rickettsia*
Mite-borne Diseases	Mite-borne Mange	*Sarcoptes scabei*

Wildlife Body Parts in Illegal Trade | ગેરકાયદેસર વેપારમાં વન્યજીવોના બોડી પાર્ટ્સ

Item/ Body Part	Animal	Illegal use
Antlers and Horns	Chital, Sambar, Swamp deer, Barking deer, Blackbuck, Chinkara, Gaur, Markhor	Ornamental and decorative purpose
Bone and Teeth	Tiger, Lion, Leopard	Decorative items and fashion accessories.
Canines and Claws	Tiger, Leopard, Bears	Ornamental
Corals	Stony corals	Ornamental
Frog legs	Ranidae frogs	Food
Gall bladder	Bears	Bear bile is used in oriental medicines in the far East, especially in China, Japan, Korea and Taiwan.
Ivory	Asiatic elephant	To build ornaments and artifacts
Lepidopterans	Butterflies of the swallow-tail family	Ornamental
Live Animals	Snakes, Lizards, Tortoises, Birds, Small mammals, Marine wildlife	Pet, Food and other illegal uses.

Item/ Body Part	Animal	Illegal use
Musk or Kasturi	Males of Himalayan musk deer.	In the perfume industries to produce exquisite fragrance. Also used in Chinese traditional medicines.
Ornamental Fishes	Guppies (*Poecilidae*), Mollies (*Cyprinodontidae*)	Food
Pelt (Skin) and Fur	Tiger, Lion, Leopard, Snow Leopard, Bears, Red panda, Jackals, Foxes, Otters, Deers, Civets, Snakes and Crocodiles	Ornamental and decorative use.
Rhino Horn	Greater one-horned rhino	To treat fever, high blood pressure, paralysis and considered as aphrodisiac in Oriental and Tibetan medicine.
Scale and Spines	Pangolin, Indian Porcupine	Traditional medicine, meat and food, Leather industry
Shells	Turban shell	Ornamental
Spiders	Tarantulas	Brightly colored spiders in decorations.
Swiftlet nests	Swifts	Food

Wildlife (Protection) Act, 1972 | વન્યપ્રાણી (સંરક્ષણ) અધિનિયમ, 1972

Chapters	Remarks
I	Preliminary include short title, extent, commencement and definitions
II	Authorities to be appointed or constituted under the Act
III	Hunting of wild animals
IIIA	Protection of specified plants
IV	Protected areas including sanctuaries, national park, and closed areas
IVA	Central Zoo Authority and recognition zoos
IVB	National Tiger Conservation Authority. New chapter added in Wildlife (Protection) Amendment Act, 2006.
IVC	Tiger and Other Endangered Species Crime Control Bureau. New chapter added in Wildlife (Protection) Amendment Act, 2006.
V	Trade or commerce in wild animals, animal articles and trophies
VA	Prohibition of trade or commerce in trophies, animal articles, etc. derived from certain animals
VI	Prevention and detection of offences
VIA	Forfeiture of property derived from illegal hunting and trade. New chapter added in Wildlife (Protection) Amendment Act, 2002.
VII	Miscellaneous

Schedules of Wildlife (Protection) Act, 1972
વન્યપ્રાણી (સંરક્ષણ) અધિનિયમ, 1972 ની સૂચિઓ

Schedule	Remarks
I.	Rare and threatened animals
II.	Animals that enjoy special game status
III.	Big games wild animals
IV.	Small game wild animals
V.	Vermin animals
VI.	Cultivation of plants

Species in Wildlife Management | વન્યજીવન વ્યવસ્થાપનમાં વિવિધ પ્રજાતિઓ

Type	Example	Remarks
Flagship species	Indian Tiger	Species chosen to represent an environmental cause
Keystone species	Tiger, Lion, Leopard	Special attention from the conservation point of view
Indicator species	Lichens	Species indicates particular environmental conditions
Foundation species	Corals and kelp	Dominant primary producer in an ecosystem
Umbrella species	Spotted owl	Species whose requirements include those of many other species.

Cats *Vs* Dogs | બિલાડી કુળ *Vs.* કુતરા કુળના પ્રાણીઓ

Cats (Felids)	Dogs (Canids)
Solitary (Exception Lion)	Group living, strong family linking
Retractile claws (Exception Cheetah)	Non-retractile claws
Plantigrade movement	Digitigrade movement
Hunt by stealth (Exception Cheetah)	Hunt by run
Tiger, Leopard, Lion, Caracal, etc.	Wild Dog, Wolves, Foxes, Jackals, etc.

The Mammals | સસ્તન પ્રાણીઓ

Fastest	Cheetah
Slowest	Sloth
Tallest	Girrafe
Smallest	Bumblebee Bat (2 g) Pygmy Shrew (1.2 to 2.7 g)
Biggest	Blue Whale (On earth) African Elephant (On land)
Loudest	Blue Whale
Smelliest	Stripped Skunk

Fattest	Ringed Seal Pup (50 % fat)
Largest bovid	Gaur/ Wild Buffalo
Largest deer of India	Sambar
Smallest deer of India	Mouse deer

The Birds | પક્ષીઓ

Fastest	Peregrine Falcon
Tallest	Ostrich
Tallest flying bird	Sarus Crane
Largest	Ostrich
Longest migratory bird	Arctic Tern

The Reptiles | સરીસૃપો

Largest reptile	Estuarine Crocodile
Longest reptile	Reticulated Python
Largest sea turtle	Leatherback Turtle
Largest lizard	Komado Dragon
Nest dwelling Snake	King Cobra

The Amphibians | ઉભયજીવીઓ

Largest amphibian	Chinese Giant Salamander
Largest frog	Goliath Frog

Multiple Choice Questions | હેતુલક્ષી પ્રશ્નો

1. **વન્યજીવોના શિકાર પર સખત પ્રતિબંધ છે , વન્યપ્રાણી (સંરક્ષણ) અધિનિયમ, 1972 મુજબ ગુનેગારને કેટલા વર્ષ કેદની સજા ફટકારી શકાય ?**
   A. 2 વર્ષ B. 7 વર્ષ
   C. 5 વર્ષ D. 9 વર્ષ

1. **Hunting is strictly prohibited for wild animals, how much imprisonment can be imposed to the culprit as per Wildlife (Protection) Act, 1972?**
   A. 2 years B. 7 years
   C. 5 years D. 9 years

2. **નીચેનામાંથી કઈ સંગઠન MoEFCCની સ્વાયત સંસ્થા છે ?**
   A. WWF B. BNHS
   C. WII D. BCSG

2. **Which of the following organizations is an autonomous body of Ministry of Environment Forests and Climate Change, Government of India?**
   A. WWF B. BNHS
   C. WII D. BCSG

3. **કાપણી પછીના પ્રારંભિક તબક્કા દરમિયાન, ઘાસચારાની ઉપલબ્ધતાને લીધે શાકાહારી પ્રાણીઓની વસ્તીમાં તીવ્ર ________ થાય છે.**
   A. ઘટાડો B. વર્ચસ્વ
   C. વધારો D. શરણાગતિ

3. **During the initial stages after felling, there is a sharp ________in the herbivore population owing to the availability of forage and browse.**
   A. Decrease B. Dominance
   C. Increase D. Submission

4. **BNHSનું મુખ્ય મથક કયા નામ તરીકે ઓળખાય છે?**
   A. સ્પેરો હાઉસ B. હોર્નબિલ હાઉસ
   C. સરસ હાઉસ D. પિફોલ હાઉસ

4. **The Headquarter of BNHS is known as**
   A. Sparrow House B. Hornbill House
   C. Sarus House D. Peafowl House

5. **ભારતના રાષ્ટ્રીય વન્યપ્રાણી બોર્ડ (NBWL)ના વડા કોણ છે ?**
   A. કેન્દ્રીય વન મંત્રી
   B. ઉપ પ્રમુખ
   C. વન મહાનિર્દેશક
   D. પ્રધાન મંત્રી

5. **Who is the head of the National Board for Wildlife of India?**
   A. Union Forest Minister
   B. Vice President
   C. Inspector General of Forests
   D. Prime Minister

6. **પ્રજાતિઓની યાદીઓ IUCN ના પુસ્તકમાં શામેલ છે. આ પુસ્તક કયા નામ તરીકે ઓળખાય છે**
   A. પિન્ક ડેટા બુક
   B. ગ્રીન ડેટા બુક
   C. રેડ ડેટા બુક
   D. બ્લેક ડેટા બુક

6. **The lists of species are included in the book of IUCN. This book called**
   A. Pink Data Book
   B. Green Data Book
   C. Red Data Book
   D. Black Data Book

7. WWFનું આંતરરાષ્ટ્રીય સચિવાલય ક્યાં આવેલું છે?

A. ઓસ્ટ્રેલિયા
B. યુનાઇટેડ કિંગડમ
C. યુએસએ
D. સ્વિટ્ઝર્લૅન્ડ

8. ભારતીય સંજોગો માટે વન્યપ્રાણી નુકશાન નિયંત્રણ પદ્ધતિમાંની સૌથી યોગ્ય પદ્ધતિ છે?

A. શૂટિંગ B. પોઇઝનિંગ
C. ટ્રાન્સલોકેશન D. ડ્રાઇવિંગ

9. કઈ સ્થિતિમાં જૈવિક નિયંત્રણ અસરકારક ન હોઈ શકે?

A. નાના વિસ્તારોમાં
B. મોટા વિસ્તારોમાં
C. સ્ક્રબી વિસ્તારોમાં
D. ગાઢ જંગલોમાં

10. નીચેનામાંથી કયા જંગલી પ્રાણીને ઉપદ્રવી જીવાત માનવામાં આવતું નથી?

A. વાગોળ
B. શાહુડી
C. ઝરખ
D. શિયાળ

11. એક ગામની નજીક અથવા શેરડીના ક્ષેત્રમાં માનવી નો શિકાર કરતો એક વાઘ, બધી સંભાવનાઓમાં એક ____________હશે

A. માનવ-ખૂની વાઘ
B. આકસ્મિક હત્યા
C. માનવભક્ષી વાઘ
D. ભૂલથી હુમલો

12. કોર ઝોનમાં, તમામ વનીકરણ અને લણણીની કામગીરી પર પ્રતિબંધ છે પરંતુ કયા અર્થે પ્રવેશ કરી શકાય ?

A. વિદેશી વીવીઆઈપીની ખાસ મુલાકાત
B. વસ્તી ગણતરી માટે
C. સંરક્ષણ સંબંધિત મૂવી શૂટિંગ
D. શાળાઓ દ્વારા વૈજ્ઞાનિક પ્રોજેક્ટ

7. The international secretariat of WWF is located in

A. Australia
B. United Kingdom
C. USA
D. Switzerland

8. Which of the wildlife damage control methods is the most suitable method for Indian conditions?

A. Shooting B. Poisoning
C. Translocation D. Driving

9. In which condition biological control through predators may not be effective?

A. In small areas
B. In large areas
C. In scrubby areas
D. In dense forests

10. Which of the following wild animal is not considered as pest?

A. Flying Fox (Pteroptus giganteus)
B. Porcupine (Hystrix indica)
C. Hyena (Hyaena hyaena)
D. Jackal (canis aureus)

11. A male tiger killing a human being near a village or in sugarcane field will in all probability be a

A. Man-killer tiger
B. Accidental killing
C. Man-eater tiger
D. Mistake attack

12. In Core Zone, all forestry and harvesting operations are prohibited and even entry is restricted except for

A. Special visit of foreign VVIPs
B. Census for population studies
C. Conservation related movie shooting
D. Scientific project by schools

13. ભૂતકાળમાં, ભારતમાં વાઘ ની વસ્તીમાં ઘટાડો થવાનું કારણ શું હતું?
A. વેપાર માટે શિકાર
B. રમત શિકાર
C. ઔષધીય હેતુ માટે શિકાર
D. ઉપરના બધા

14. નીચેનામાંથી કયું પરિબળ પ્રવાસનને પરિસરીય પ્રવાસન તરીકે સિદ્ધ કરે છે?
A. જ્યારે પર્યટકની મુલાકાતીની અસરની તકેદારી રાખવામાં આવે છે
B. જ્યારે સ્થાનિક સમુદાયોને આજીવિકાની તકો પૂરી પાડવામાં આવે છે
C. જ્યારે ઉદ્યોગ-સાહસોની સધ્ધરતા અને કામગીરીની સુવિધા આપવામાં આવે છે
D. જ્યારે પ્રાચીન વન વિસ્તારોની સંભાવનાઓનો ઉપયોગ કરવામાં આવે છે

15. વૃક્ષો, લેન્ડસ્કેપ્સ અને ગીચ ઝાડ જેવા હિમાયતી પરિબળોને લગતું કયું નિવાસસ્થાન છે જે વન્યપ્રાણી વસ્તીને અસર કરે છે?
A. એસ્ટુટ્યુરિન હેબીટાટ
B. માઇક્રોહેબીટાટ
C. ટેરેસ્ટ્રિયલ હેબીટાટ
D. મેક્રોહેબીટાટ

16. લિમ્નોલોજી શું છે?
A. એસ્ચ્યુરીન હેબીટાટનો અભ્યાસ
B. કીચડવાળા પાણીના હેબીટાટનો અભ્યાસ
C. મરીન હેબીટાટનો અભ્યાસ
D. તાજા પાણીના હેબીટાટનો અભ્યાસ

17. બાયોસ્ફિયરમાં સૌથી મોટો હેબીટાટ કયાં છે?
A. એસ્ચ્યુરીન હેબીટાટ
B. ફ્રેશ વોટર હેબીટાટ
C. મરીન હેબીટાટ
D. ટેરેસ્ટ્રિયલ હેબીટાટ

13. In the past, what was the reason for dramatic decline of tiger population in India?
A. Poaching for trade
B. Game hunting
C. Hunting for medicinal purpose
D. All of above

14. Inclusion of which factor advocates a tourism turn out to be an ecotourism
A. When vigilance of visitor impact of tourist is carried out
B. When livelihood opportunities to local communities are provided
C. When the viability and performance of enterprises is facilitated
D. When the potential of pristine forest areas are exploited

15. Which habitat pertains to onsite factors like trees, landscapes and thickets which affect wildlife populations?
A. Estuarine habitat
B. Microhabitat
C. Terrestrial habitat
D. Macrohabitat

16. Limnology is
A. Study of estuarine water habitats
B. Study of muddy water habitats
C. Study of sea water habitats
D. Study of fresh water habitats

17. Which is the largest habitat in biosphere?
A. Estuarine habitat
B. Freshwater habitat
C. Marine habitat
D. Terrestrial habitat

8. ઉષ્ણકટિબંધીય જંગલોમાં વિશ્વની અડધાથી વધુ જાતિઓ પૃથ્વીના કુલ ભૂમિ વિસ્તારના ____% વિસ્તાર ધરાવે છે.

A. 6 B. 7
C. 5 D. 8

9. નીચેના માંથી કયું પ્રાણી પૃથ્વી પરથી લુપ્ત થઇ ગયેલ છે?

A. સ્લેન્ડર બિલ ગીધ
B. ઓલિવ બબૂન
C. ડોડો
D. ઉપરથી કંઈ નહીં

0. નીચેનામાંથી કયું પ્રાણી વન્યપ્રાણી (સંરક્ષણ) અધિનિયમનું શેડ્યૂલ-1 માં નથી?

A. સ્વેમ્પ ડિયર
B. સાંબર
C. મસ્ક ડિયર
D. માર્ખોર

1. વિશ્વમાં રીંછની આઠ પ્રજાતિઓમાંથી કઈ પ્રજાતિ ભારતમાં નથી?

A. બ્લેક રીંછ
B. સ્લોથ રીંછ
C. સ્પેડટેકલ્ડ રીંછ
D. સન રીંછ

2. પાકને નુકસાન એ કયા જંગલી પ્રાણીનો મોટો સંઘર્ષ નથી?

A. હાથી (*Elaphas maximus*)
B. ગોર (*Bos gaurus*)
C. નીલગાય (*Boselaphus tragocamelus*)
D. દીપડો (*Panthera pardus*)

:3. 21 માર્ચ કયા દિવસ તરીકે ઉજવવામાં આવે છે?

A. વિશ્વ પૃથ્વી દિવસ
B. વિશ્વ જૈવવિવિધતા દિવસ
C. વિશ્વ સાપ દિવસ
D. વિશ્વ વન દિવસ

18. The tropical forests have more than half of the world's species and occupying ____ % of the total land area of the earth.

A. 6 B. 7
C. 5 D. 8

19. Which of the following creature is reasonably undoubted that the last individual has died?

A. Slender billed vulture (*Gyps tenuirostris*)
B. Olive Baboon (*Papio anumbis*)
C. Dodo (*Raphus cucullatus*)
D. None of Above

20. Which of the following creatures is not in schedule –I of Wildlife (Protection) Act?

A. Swamp Deer (*Rucervus duvaucelii*)
B. Sambar (*Rusa unicolor*)
C. Musk Deer (*Moschus leucogast*)
D. Markhor (*Capra falconi*)

21. Out of eight bear species in the world, which of the followings does not found in India?

A. Black Bear (*Ursus thibetanus*)
B. Sloth Bear (*Melursus ursinus*)
C. Spectacled Bear (*Tremarctos ornatus*)
D. Sun Bear (*Helarctos malayanus*)

22. Crop damage is not major conflict of this wild animal in India.

A. Elephant (*Elaphas maximus*)
B. Gaur (*Bos gaurus*)
C. Blue bull (*Boselaphus tragocamelus*)
D. Panther (*Panthera pardus*)

23. 21 March is celebrated as?

A. World Earth Day
B. World Biodiversity Day
C. World Snake Day
D. World Forestry Day

24. ભારતના પ્રખ્યાત હર્પેટોલોજિસ્ટ કોણ છે?
A. સાલીમ અલી B. વિવેક મેનન
C. રિષદ નૌરોજી D. જે. સી. ડેનિયલ

25. ________________ એ એક વિસ્તાર છે જે તેની દૈનિક પ્રવૃત્તિમાં જંગલી પ્રાણી દ્વારા આવરી લેવામાં આવે છે
A. ટેરીટરી
B. હોમ રેન્જ
C. કોર એરિયા
D. ઉપર્યુક્તમાંથી કોઈ નહિ

26. નીચેનામાંથી કયા જંગલી પ્રાણીને વાઘ શિકાર માટે સૌથી વધુ પ્રાધાન્ય આપશે?
A. ગૌર (*Bos gaurus*)
B. ભારતીય વાઇલ્ડ પિગ (*Sus scrofa*)
C. રીસસ મકાક (*Macaca mulatta*)
D. સ્વેમ્પ ડિયર (*Cervus duvaucelii*)

27. ગુજરાતના વન માં કયો સાપ મળતો નથી?
A. સિંધ ક્રેટ (*Bungarus sindanus*)
B. સ્ટ્રિપ્ડ કીલબેક (*Amphiesma stolat*)
C. કિંગ કોબ્રા (*Ophiophagus hannah*)
D. મોનોસિલેટ કોબ્રા (*Naja kaouthia*)

28. એક ફિમેલનો બે અથવા વધુ મેલ સાથે વિશિષ્ટ સંબંધ હોય તો તે કયા નામે ઓળખવામાં આવે છે?
A. પોલીજેન્ડ્રી
B. પોલિજયની
C. પોલિએન્ડ્રી
D. મોનોગેમી

29. ભારતનો સૌથી વ્યાપક રીતે ફેલાયેલો રીંછ ક્યો છે?
A. એશિયાટિક બ્લેક રીંછ (*Ursus thibetanus*)
B. સન રીંછ (*Helarctos malayanus*)
C. હિમાલય બ્રાઉન રીંછ (*Ursus arctos*)
D. સ્લોથ રીંછ (*Melursus ursinus*)

24. Who is famous herpetologist of India?
A. Salim Ali B. Vivek Menon
C. Rishad Naoroji D. J.C. Denial

25. ________________ is an area covered b a wild animal in its day to day activity
A. Territory
B. Home range
C. Core area
D. None of above

26. Which of the following wild animals wi tiger prefer the most to stalk?
A. Gaur (*Bos gaurus*)
B. Indian Wild Pig (*Sus scrofa*)
C. Rhesus Macaque (*Macaca mulatta*)
D. Swamp deer (*Cervus duvaucelii*)

27. The snake that doesn't found i wilderness of Gujarat?
A. Sindh Krait (*Bungarus sindanus*)
B. Stripped Keelback (*Amphiesma stolate*)
C. King Cobra (*Ophiophagus hannah*)
D. Monocellate Cobra (*Naja kaouthia*)

28. One female has an exclusive relationshi with two or more males is termed as
A. Polygyndry
B. Polygyny
C. Polyandry
D. Monogymy

29. The most widely spread bear of India is
A. Asiatic Black Bear (*Ursus thibetanus*)
B. Sun Bear (*Helarctos malayanus*)
C. Himalayan Brown Bear (*Ursus arctos*)
D. Sloth Bear (*Melursus ursinus*)

30. નીચેનામાંથી કઈ પદ્ધતિ વન્ય પ્રાણીઓની ગણતરીની સીધી પદ્ધતિ નથી ?

A. એરિયલ પદ્ધતિ
B. ટ્રાંસેક્ટ ગણતરી પદ્ધતિ
C. કિલ એવિડન્સ પદ્ધતિ
D. ટેમ્પોરલ ગણતરી

31. વાઇલ્ડલાઇફ મેનેજમેન્ટના પિતા કોણ છે?

A. કાર્લ લિઓપોલ્ડ
B. જ્યોર્જ શેલર
C. અલ્ડો લિઓપોલ્ડ
D. ડેવિડ એટનબર્ગ

32. ભારત સરકારની એક્સપર્ટ કમિટી મુજબ, 'A', 'B' અને 'C' પ્રાણી સંગ્રહાલય શું સૂચવે છે?

A. 70, 50 અને 50 હેક્ટર થી ઓછો વિસ્તાર
B. 50, 25 અને 25 હેક્ટર થી ઓછો વિસ્તાર
C. 35, 20 અને 20 હેક્ટર થી ઓછો વિસ્તાર
D. 25, 10 અને 10 હેક્ટર થી ઓછો વિસ્તાર

33. ભારતમાં વાઘની કેટલી જાતો જોવા મળે છે?

A. 4 B. 3
C. 2 D. 1

34. નિશ એટલે

A. સજીવની વહન ક્ષમતા
B. સજીવની વસ્તી
C. સજીવની સ્થિતિ અથવા સ્થાન
D. સજીવની વૃદ્ધિ

35. સામાન્ય રીતે ટેરિટોરીમાં સમાન પ્રજાતિના અન્ય પ્રાણીઓ ________ હોય છે.

A. સમાવિષ્ટ B. બાકાત
C. કોઈ સ્પર્ધામાં D. દુશ્મનો

36. કયાં સમયે ટેરીટરી વધુ પ્રબળ હોય છે ?

A. ઉનાળો B. શિયાળો
C. ચોમાસુ D. બ્રીડીંગ

30. Which of the following is not a direct method of wild animal counting?

A. Arial method
B. Transect count method
C. Kill evidence method
D. Temporal count

31. Who is the father of Wildlife Management?

A. Carl Leopold
B. George Schaller
C. Aldo Leopold
D. David Attenborough

32. As per the Expert Committee of Govt of India, What signifies category 'A', 'B' and 'C' zoos?

A. 70, 50 & below 50 ha area
B. 50, 25 & below 25 ha area
C. 35, 20 & below 20 ha area
D. 25, 10 & below 10 ha area

33. How many species of Tigers are found in India?

A. 4 B. 3
C. 2 D. 1

34. Niche is the

A. Carrying capacity of an organism
B. Population of an organism
C. Status or place of an organism
D. Growth of an organism

35. In territories, it is generally observed that other individuals of same species are

A. Included B. excluded
C. in no competition D. enemies

36. Territory becomes more effective at the time of

A. Summer B. Winter
C. Monsoon D. Breeding

37. કયાં વર્ષ માં સેન્ટ્રલ ઝૂ ઓથોરિટી બનાવવામાં આવી હતી?

A. 1972 B. 1968
C. 1992 D. 1998

38. વન્યજીવોના ટેરીટરી વર્તણૂક અંગે નીચેનામાંથી કયું વિધાન સાચું છે ?

A. બધા પ્રાણીઓ ટેરીટરી દર્શાવે છે
B. બધા પ્રાણીઓ ટેરીટરી બતાવતા નથી
C. બધા પ્રાણીઓ તેમના ટેરીટરી નો બચાવ કરે છે
D. બધા પ્રાણીઓ ટેરીટરી માર્કિંગ કરે છે

39. હોમ રેન્જનો કયો વિસ્તાર વન્યજીવો દ્વારા અન્ય વિસ્તારો કરતા વધુ ઉપયોગ કરવામાં આવે છે ?

A. ક્રિટિકલ વિસ્તાર
B. ટેરિટોરીયલ વિસ્તાર
C. સ્પેશિયલ વિસ્તાર
D. કોર વિસ્તાર

40. સોલ્ટ લીક બનાવવા માટે ________ માટી સાથે ભેળવવામાં આવે છે.

A. HCL B. H_2SO_4
C. NaCl D. NH_4

41. પ્રાણી સંગ્રહાલય એ એક જટિલ સંસ્થા છે જેણે ઘણી જરૂરીયાતો પૂરી કરવી પડે છે, નીચેનામાંથી કયું પ્રાણી સંગ્રહાલયનું મુખ્ય કાર્ય નથી?

A. સંશોધન B. મનોરંજન
C. સંરક્ષણ D. શિક્ષણ

42. નીચેનામાંથી કયુ રાજ્ય એશિયાઇ હાથીની જંગલી વસ્તી ધરાવતું નથી?

A. અરુણાચલ પ્રદેશ
B. મિઝોરમ
C. પંજાબ
D. ઝારખંડ

37. The Central Zoo Authority was created in.

A. 1972 B. 1968
C. 1992 D. 1998

38. Which of the following statement is true regarding territorial behavior of wild animals?

A. All animals show territoriality
B. All animals don't show territoriality
C. All animals defend their territoriality
D. All animals mark their territory

39. Which area is an area more frequently used by wild animal with greater regularity than other parts within a home range?

A. Critical area
B. Territorial area
C. Special area
D. Core area

40. For making artificial salt licks, ____________ is mixed with soil.

A. HCL B. H_2SO_4
C. NaCl D. NH_4

41. The zoo is complex organization that has to fulfill many requirements, which of the following is not the prime function of modern zoo?

A. Research B. Recreation
C. Conservation D. Education

42. Which of the following states doesn't harbor wild populations of Asiatic Elephant?

A. Arunachal Pradesh
B. Mizoram
C. Punjab
D. Jharkhand

43. કાળિયાર સામાન્ય રીતે કેવા પ્રકાર ના જંગલો પસંદ કરે છે?
A. સુકા ગાઢ જંગલો
B. ભેજ વાળા ગાઢ જંગલો
C. ઘાસના ખુલ્લા મેદાનો
D. ટુન્ડ્રા

44. ફાયર લાઇન ક્યારે તૈયાર કરવી જોઈએ ?
A. આગ લાગ્યા પછી
B. આગ લાગે તે પહેલાં
C. ઉનાળા પહેલા
D. ચોમાસા પહેલા

45. જંગલના નીચેનામાંથી કયા સંસાધનો કુદરતી ફાયર લાઈનની જેમ કાર્ય કરી શકે છે?
A. સ્ટ્રીમ બેડ
B. ફોરેસ્ટ રોડ
C. રીવેર બેડ
D. બધા A, B અને C

46. અગ્નિ સંબંધિત કમ્યુનિકેશનમાં, વાદળી એટલે
A. આગ ચાલુ હતી પણ હવે ઠીક છે
B. આગ નથી, બધું ઠીક છે
C. નજીકના ક્ષેત્રમાં આગ
D. પાણીથી આગ બુઝાઈ ગઈ છે

47. જમ્મુ અને કાશ્મીરના ડાચીગામ નેશનલ પાર્કનું મુખ્ય પ્રાણી ક્યું છે
A. કસ્તુરી હરણ
B. હંગુલ
C. માર્ખોર
D. કાળું રીંછ

48. વાઘની નીચેની કઈ જાતિ વિશ્વમાંથી લુપ્ત થઈ છે?
A. બાલી ટાઇગર
B. મલયાન ટાઇગર
C. અમુર ટાઇગર
D. દક્ષિણ ચાઇના ટાઇગર

43. Blackbuck (*Antilope cervicapra*) generally prefers _____ type of forests.
A. Dry Dense
B. Moist Dense
C. Open meadow
D. Tundra

44. When should fire lines be prepared?
A. After fire is set
B. Just before fire
C. Before summer
D. Before monsoon

45. Which of the following resources of the jungle can act as natural fire line?
A. Stream bed
B. Forest road
C. River bed
D. All A, B & C

46. In Fire related communication, Blue means
A. Fire was on but now ok
B. No fire, all is clear
C. Fire in adjoining jurisdiction
D. Fire cleared with water

47. The main animal of Dachigam National Park, Jammu and Kashmir is
A. Musk deer (*Moschus chrysogaster*)
B. Hangul (*Cervus canadensis*)
C. Markhor (*Capra falconeri*)
D. Black Bear (*Ursus thibetanus*)

48. Which of the following species of Tiger is extinct from the world?
A. Bali Tiger
B. Malayan Tiger
C. Amur Tiger
D. South china Tiger

49. ભારતના હાથીઓમાં વસ્તી ગણતરી અને આનુવંશિક પરિણામોના આધારે હાથીઓની વસ્તીમાં ફિમેલ નુ પ્રમાણ મેલ હાથીઓ કરતા વધારે છે, કારણ શું છે?

A. માંસ માટે ફિમેલ હાથીઓના શિકાર
B. હાથીદાંત માટે મેલ હાથીઓના શિકાર
C. મેલ હાથીઓની વસવાટની ખોટ
D. ફિમેલ હાથીઓની સંવર્ધન નિષ્ફળતા

50. ગુજરાત રાજ્યના પ્રોટેક્ટેડ એરિયા નેટવર્ક _______ સંરક્ષિત વિસ્તારોનો સમાવેશ કરે છે.

A. 28 B. 23
C. 30 D. 24

51. વાઘને ભારતના રાષ્ટ્રીય પ્રાણીનો દરજ્જો ક્યારે આપવામાં આવ્યો હતો

A. 1947 B. 1965
C. 1950 D. 1973

52. કયું વન્ય પ્રાણી સૌથી પ્રાચીન જાણીતું ભારતીય સસ્તન પ્રાણી છે જેનો લડાઇ દરમિયાન પ્લેટફોર્મ તરીકે ઉપયોગ કરવામાં આવતું, જે શિકારમાં પણ ઉપયોગમાં લેવામાં આવતું અને હડપ્પન સંસ્કૃતિમાં પ્રતિષ્ઠાનું પ્રતીક તરીકે પણ મનાતું હતું ?

A. ઘોડો B. હાથી
C. ઊંટ D. નીલ ગાય

53. 1969 માં દિલ્હીમાં IUCNના તકનીકી સત્રમાં એશિયાટિક સિંહોના નિવાસસ્થાન સાથે તેના સંરક્ષણની ચર્ચા થઈ હતી અને ત્યારબાદ સમસ્યાઓ હલ કરવા માટે વિશ્વભરના સંખ્યાબંધ વાઇલ્ડલાઇફર્સ અને ઇકોલોજીસ્ટ્સની સલાહ લેવામાં આવી હતી. એક પ્રોજેક્ટને તરત જ 'પ્રોજેક્ટ લાયન' નામથી ઘડવામાં આવ્યો. આ પ્રોજેક્ટ ક્યારે શરૂ કરવાંમાં આવેલ હતો?

A. 1970 B. 1971
C. 1972 D. 1973

49. What is the reason for highly skewed male: female ratio of elephant populations which has resulted serious demographic and genetic consequences in elephants of India?

A. Poaching of female elephants for flesh
B. Poaching of male elephants for ivory
C. Habitat loss of male elephants
D. Breeding failure of female elephants

50. The Protected Area (PA) network of the State of Gujarat comprises of _____ Protected Areas.

A. 28 B. 23
C. 30 D. 24

51. The Tiger was accorded the status of National Animal of India in

A. 1947 B. 1965
C. 1950 D. 1973

52. Which wild animal was the earliest known Indian mammal that was employed to be mounted upon during battles, a raised platform used in hunting and also as a status symbol during the age of Harappan Civilization?

A. Horse B. Elephant
C. Camel D. Blue bull

53. Conservation of Asiatic Lion with its habitat was discussed at the technical session of IUCN in Delhi in 1969 and a number of wildlifers and ecologists all over the world were then consulted to solve the problems. A project was immediately formulated named as 'Project Lion'. This project was initiated in

A. 1970 B. 1971
C. 1972 D. 1973

54. MoEFCC એ વન્યપ્રાણી સંરક્ષણ માટે ત્રીજી રાષ્ટ્રીય વન્યપ્રાણી એક્શન પ્લાનનું અનાવરણ કર્યું છે. આ યોજના કેટલા વર્ષો ની છે?

A. 2017-2031 B. 2017-2022
C. 2017-2027 D. 2017-2037

55. WWF ના અહેવાલ મુજબ પાછલા સો વર્ષોમાં વિશ્વ માં વાઘ ની સંખ્યામાં ____ ટકાનો ઘટાડો થયો છે ?

A. 90% B. 95%
C. 80% D. 85%

56. એશિયાઈ સિંહ નું ટ્રાન્સલોકેશન સાસણ ગીર થી ____ મધ્ય પ્રદેશ માટે સૂચન કરવામાં આવે છે.

A. કાન્હા B. ધોની
C. કુનો D. કારેરા

57. IUCN ના મુજબ ____ થ્રેટન્ડ પ્રજાતિઓમાં શામેલ છે

A. EX, CR અને EN
B. CR, EN અને VU
C. EN, VU અને NT
D. VU, NT અને LC

58. ભારતમાં પ્રાણી સંગ્રહાલયો ને કઈ સંસ્થા દ્વારા માન્યતા આપવામાં આવે છે ?

A. FSI B. BSI
C. ASI D. CZA

59. CITESનું આંતરરાષ્ટ્રીય સચિવાલય ક્યાં આવેલું છે

A. ઓસ્ટ્રેલિયા B. યુનાઇટેડ કિંગડમ
C. યૂએસએ D. સ્વિટ્ઝર્લેન્ડ

60. વાઘ ના અંદાજ 2018 મુજબ, ભારતમાં વાઘ ની વસ્તી કેટલી છે ?

A. 2967 B. 1411
C. 1706 D. 2226

54. The MoEFCC has unveiled third National Wildlife Action Plan to chalk out future road map for wildlife conservation. This plan is for how many years?

A. 2017-2031 B. 2017-2022
C. 2017-2027 D. 2017-2037

55. According to WWF report, over the past hundred years the world's tiger population has decreased by ___ per cent.

A. 90% B. 95%
C. 80% D. 85%

56. The translocation of Asiatic lion from Sasan Gir to ____ in Madhya Predesh is suggested.

A. Kanha B. Kheoni
C. Kuno D. Karera

57. According to IUCN, threatened species include

A. EX, CR and EN
B. CR, EN and VU
C. EN, VU and NT
D. VU, NT and LC

58. The Zoos in India are recognized by

A. FSI B. BSI
C. ASI D. CZA

59. The international secretariat of CITES is located in

A. Australia B. United Kingdom
C. USA D. Switzerland

60. In 2018 tiger estimation, the population of Tiger (*Panthera tigris*) in India is

A. 2967 B. 1411
C. 1706 D. 2226

61. બર્ડમેન ઓફ ઇન્ડિયા સાલીમ અલી દ્વારા અપાયેલી ગુજરાતની એક વિશ્વ પ્રસિદ્ધ ફ્લેમિંગો બ્રીડીંગ સાઇટ "ફ્લેમિંગો સિટી" ક્યાં આવેલ છે?

A. કચ્છ રણ અભયારણ્ય
B. થોલ વન્યપ્રાણી અભયારણ્ય
C. ખિજડિયા પક્ષી અભયારણ્ય
D. નલ સરોવર પક્ષી અભયારણ્ય

62. સસ્તન લેટિન શબ્દ " મમ્મા " પરથી ઉતરી આવ્યો છે , મમ્મા એટલે

A. ગરમ લોહી
B. વાળ
C. સ્તન
D. સીધો જન્મ આપવાની ક્ષમતા

63. 1950 માં ______________ ને રાષ્ટ્રીય પ્રતીક તરીકે અપનાવવામાં આવ્યું.

A. સિંહ B. વાઘ
C. હાથી D. ડોલ્ફિન

64. વિશ્વના 65 ટકા જંગલી વાઘ અને એશિયન હાથી ___________ માં જોવા મળે છે.

A. ચીન B. મલેશિયા
C. ભૂટાન D. ભારત

65. NBWLના વાઇસ ચેરમેન કોણ હોય છે?

A. DGF
B. સચિવ, MoEFCC
C. મંત્રી, MoEFCC
D. રાજ્ય કક્ષાના મંત્રી, MoEFCC

66. ગુજરાતના વન્યપ્રાણી સલાહકાર મંડળના વડા કોણ છે?

A. વન મંત્રી B. વન સચિવ
C. મુખ્યમંત્રી D. PCCF & HoFF

67. ટેક્સીડર્મી એટલે શું?

A. પ્રાણીઓ સ્ટફિંગ કરવાની કળા
B. પરિવહનના પ્રાણીઓ
C. પ્રાણીઓની સ્થળાંતર પેટર્ન
D. પ્રાણીઓ ની કલિંગની કળા

61. The bird-man of India Salim Ali gave a world-famous flamingo breeding site of Gujarat the name "Flamingo City" Where is flamingo city located

A. Kutch Desert Sanctuary
B. Thol Wildlife Sanctuary
C. Khijadiya Bird Sanctuary
D. Nal Sarovar Bird Sanctuary

62. The word mammal has derived from latin word "mamma", mamma means

A. Warm blood
B. Hair
C. Breast
D. Ability to give direct birth

63. ______________ was adopted as national emblem in 1950.

A. Lion Capital B. Tiger capital
C. Elephant Capital D. Dolphin

64. per cent of world's wild tigers andAsian Elephants are found in

A. China B. Malaysia
C. Bhutan D. India

65. Who is the Vice Chairman of NBWL?

A. DGF
B. Secretory, MoEFCC
C. Minister, MoEFCC
D. State Minister, MoEFCC

66. Who, by virtue of post, is the head of WildlifeAdvisory Board of Gujarat?

A. Forest Minister B. Forest Secretary
C. Chief Minister D. PCCF & HoFF

67. What is taxidermy?

A. The art of stuffing animals
B. Transportation animals
C. Migratory pattern of animals
D. Culling the animals

68. રીંછ માં પ્રભાવી ઇન્દ્રિય કઈ હોય છે?
A. દૃષ્ટિ B. ગંધ
C. સ્પર્શ D. સ્વાદ

69. પક્ષીઓની નબળી ઇન્દ્રિય કઈ હોય છે?
A. દૃષ્ટિ B. સ્મેલ
C. સ્પર્શ D. સ્વાદ

70. દોડવીર તરીકે કયા પ્રાણીમાં વધુ શક્તિ હોય છે?
A. ચિત્તા B. વાઘ
C. ઘોડો D. દીપડો

71. કાન્હા રાષ્ટ્રીય ઉદ્યાન ક્યાં આવેલું છે?
A. કેરળ
B. મધ્યપ્રદેશ
C. કર્ણાટક
D. મહારાષ્ટ્ર

72. COP માં CITES રિઝોલ્યુશન થયું હતું કે વર્ષ 2003 માં દક્ષિણ એશિયામાં MIKE કાર્યક્રમ શરૂ કરવો. MIKE નો અર્થ
A. Monitoring the Illegal Killing of Elephants
B. Migratory Killing Enforcement
C. Maintenance of the Illicit Killing of Elephants
D. Management of Illegal Killing of Endangered Species

73. પ્રોજેક્ટ ટાઇગર શરૂઆતમાં ________ ટાઇગર રિઝર્વ માં શરૂ કરાયો હતો. હાલમાં તે 18 રાજ્યોને આવરી લેતા 50 ટાઇગર રિઝર્વ સુધી પહોંચી ગયો છે.
A. 9 B. 10
C. 19 D. 15

74. પ્રોજેક્ટ ટાઇગર ____ માં શરૂ કરાયો હતો.
A. 1972 B. 1992
C. 1973 D. 1993

68. Which is dominant sense in bears?
A. Sight B. Smell
C. Touch D. Taste

69. Which is the weakest sense of birds?
A. Sight B. Smell
C. Touch D. Taste

70. Which animal has greater endurance as a runner?
A. Cheetah B. Tiger
C. Horse D. Leopard

71. Kanha National Park is located in
A. Kerala
B. Madhya Pradesh
C. Karnataka
D. Maharashtra

72. Mandated by COP resolution of CITES, MIKE program started in SouthAsia in the year 2003. MIKE means
A. Monitoring the Illegal Killing of Elephants
B. Migratory Killing Enforcement
C. Maintenance of the Illicit Killing of Elephants
D. Management of Illegal Killing of Endangered Species

73. Project Tiger was started in _____tiger reserves initially. Presently it has reached to 50 tiger reserves, covering 18 states.
A. 9 B. 10
C. 19 D. 15

74. Project Tiger was started in
A. 1972 B. 1992
C. 1973 D. 1993

75. વન્યપ્રાણી (સંરક્ષણ) અધિનિયમના કઈ સેક્શન હેઠળ નેશનલ ટાઇગર કન્સર્વેઝન ઓથોરિટી (NTCA) ની સ્થાપના કરવામાં આવી હતી?

A. સેક્શન 38 એલ (1)
B. સેક્શન 380 (1)
C. સેક્શન 38 ઓ (1)
D. સેક્શન 35 (1)

76. ભારતનું સૌથી મોટુ હરણ કયુ છે?

A. સાંબર B. નીલગાય
C. સ્વેમ્પ ડિયર D. સ્પોટેડ ડિયર

77. એલ્ડ્સ ડિયર ક્યાં જોવા મળે છે ?

A. અરુણાચલ પ્રદેશ B. મણીપુર
C. પશ્ચિમ બંગાળ D. આસામ

78. નીચેના માંથી ક્યું દરિયાઇ સસ્તન પ્રાણી નથી?

A. બ્લુ વ્હેલ B. ડોલ્ફિન
C. ડુગોંગ D. વ્હેલ શાર્ક

79. ક્યાર થી વૈશ્વિક સ્તરે વ્હેલના શિકાર પર પ્રતિબંધ મૂકાયો છે ?

A. 1992 B. 1986
C. 1990 D. 1981

80. નીચેનીમાંથી કઈ પ્રજાતિ ગુજરાતમાંથી લુપ્ત હોવાનું માનવામાં આવતું હતું પરંતુ તાજેતરમાં વાંસદા નેશનલ પાર્કમાં ફરીથી શોધ થઈ ?

A. વાઘ B. ગૌર
C. મોટી ખિસકોલી D. ધોલ

81. ગુજરાતનો કયો વિસ્તાર રાજ્યના 70% મેંગ્રોવ્સને આશ્રય આપે છે ?

A. કચ્છ B. સૌરાષ્ટ્ર
C. દક્ષિણ ગુજરાત D. કચ્છનો અખાત

82. ગુજરાતના કયા જિલ્લામાં સૌથી વધુ વન આવરણ છે ?

A. જુનાગઢ B. જામનગર
C. ડાંગ D. કચ્છ

75. **Under which section of Wildlife (Protection) Act, National Tiger ConservationAuthority was established**

A. Section 38 L (1)
B. Section 380 (1)
C. Section 38 O (1)
D. Section 35 (1)

76. **Which is the largest deer of India?**

A. Sambar B. Blue bull
C. Swamp Deer D. Spotted Deer

77. **The endemic Eld's Deer found only in**

A. Arunachal Pradesh B. Manipur
C. Mizoram D. Assam

78. **Which of the following is not a marine mammal**

A. Blue whale B. Dolphin
C. Dugong D. Whale shark

79. **Since when the commercial whaling has been banned globally**

A. 1992 B. 1986
C. 1990 D. 1981

80. **Which of the following species was considered to be extinct from Gujarat but recently rediscovered in Vansda National Park**

A. Tiger B. Gaur
C. Giant squirrel D. Dhole

81. **Which area of Gujarat harbors 70 % of mangroves of the state**

A. Kachchh B. Saurashtra
C. South Gujarat D. Gulf of Kachchh

82. **Which district of Gujarat has the highest forest?**

A. Junagadh B. Jamnagar
C. Dangs D. Kachchh

83. **અગાઉ વ્હેલ શાર્ક આડેધડ કતલ કરવામાં આવી હતી, ક્યાં વર્ષમાં ભારત સરકાર દ્વારા વ્હેલ શાર્ક ને વન્યપ્રાણી (સંરક્ષણ) અનુસૂચિ-I માં સમાવેશ કરાયો?**

A. 2015 B. 2001
C. 2010 D. 2005

84. **ગુજરાતના દરિયાકાંઠે વિશ્વની સૌથી મોટી માછલી કઈ છે ?**

A. ગ્રેટ વાઇટ શાર્ક B. બાસ્કિંગ શાર્ક
C. ટાઇગર શાર્ક D. વ્હેલ શાર્ક

85. **કચ્છ બસ્ટર્ડ અભયારણ્યની મુખ્ય પ્રજાતિ કઈ છે?**

A. ગ્રેટ ઇન્ડિયન બસ્ટાર્ડ
B. લેસર ફ્લોરિકન
C. હૌબરા બસ્ટાર્ડ
D. ચિંકારા

86. **લુપ્ત થવાના જોખમનો સામનો કરી રહેલા એક લુપ્ત થનાર પક્ષી, જેને વન્યપ્રાણી (સંરક્ષણ) અધિનિયમની અનુસૂચિ-I હેઠળ મુકવામાં આવ્યું છે .**

A. મેક કવીન બસ્ટર્ડ
B. લેસર ફ્લોરિકન
C. હૌબરા બસ્ટાર્ડ
D. ગ્રેટ ઇન્ડિયન બસ્ટાર્ડ

87. **કચ્છના નાના રણની પ્રમુખ પ્રજાતિ કઈ છે ?**

A. ઝરખ B. ભારતીય ઘુડખર
C. ચિંકારા D. વરુ

88. **માર્ચ 2020 માં ગુજરાત સરકાર દ્વારા સાપ સંશોધન સંસ્થા ની સ્થાપના ક્યાં કરવામાં આવી?**

A. ગાંધીનગર B. રાજપીપળા
C. આહવા D. ધરમપુર

89. **નલ સરોવરને રામસાર સાઇટ તરીકે ક્યારે જાહેર કરાઈ હતી ?**

A. 2012 B. 2015
C. 2018 D. 2020

83. **Earlier the Whale Shark was slaughtered indiscriminately. In which year Govt of India added Whale Shark in Schedule-I of the Wildlife (Protection) Act?**

A. 2015 B. 2001
C. 2010 D. 2005

84. **Which is the world's largest fish and harbors in Gujarat coasts**

A. Great white shark B. Basking shark
C. Tiger shark D. Whale shark

85. **The main species of Kutch Bustard Sanctuary**

Great Indian Bustard
Lesser Florican
Houbara Bustard
Chinkara

86. **A Critically Endangered bird facing the risk of extinction and put under Schedule-1 of Wildlife (Protection),Act.**

A. MacQueen's Bustard
B. Lesser Florican
C. Houbara bustard
D. Great Indian Bustard

87. **Which is the flagship species of Little Rann of Kachchh**

A. Hyena B. Indian WildAss
C. Chinkara D. Wolf

88. **Govt of Gujarat established Snake Research Institute in March 2020 at**

A. Gandhinagar B. Rajpipla
C. Ahwa D. Dharampur

89. **Nal Sarovar was declared as Ramsar site in**

A. 2012 B. 2015
C. 2018 D. 2020

90. કયો દેશ વિશ્વ માં સૌથી વધુ હાથીઓની વસ્તી ધરાવે છે?

A. ભારત B. થાઇલેન્ડ
C. બોત્સ્વાના D. શ્રીલંકા

91. પ્રજાતિઓ કે જે ભૂતકાળમાં ઉપસ્થિત હતી પરંતુ હવે એક પણ પ્રાણી સંભવિત વિસ્તારોની શોધ કર્યા પછી પણ મળ્યા નથી તો તેવી પ્રજાતિને શું કહી શકાય ?

A. સ્થાનિક પ્રજાતિઓ
B. દુર્લભ પ્રજાતિઓ
C. ભયગ્રસ્ત પ્રજાતિઓ
D. લુપ્ત પ્રજાતિઓ

92. બકસા ટાઇગર રિઝર્વમાં શેનુ ખાણકામ ગંભીર રીતે નુકસાન પોંહચાડે છે ?

A. કોલસો B. કોપર
C. ડોલોમાઇટ D. આયર્ન

93. નીચેના માંથી કયા રાજ્યમાં પેરિયાર ટાઇગર રિઝર્વ સ્થિત છે ?

A. રાજસ્થાન B. આસામ
C. ઉત્તરાંચલ D. કેરળ

94. કઈ શાકાહારી ગરોળી રાજસ્થાન થાર રણમાં જોવા મળે છે ?

A. લિઝાર્ડ
B. મંગૂઝ
C. મોનિટર લિઝાર્ડ
D. સ્પાઇની ટેઈલ લિઝાર્ડ

95. વન વિભાગ દ્વારા કરવામાં આવેલી વસ્તી ગણતરી મુજબ, 2005 થી 2018 ની વચ્ચે ગુજરાતે ____% ગીધ ગુમાવ્યાં છે.

A. 60% B. 70%
C. 80% D. 90%

96. ગુજરાતમાં ગીધની વસ્તી કેટલી છે ?

A. 820 B. 1300
C. 1800 D. 2300

90. Which county possess world's highest elephant population?

A. India B. Thailand
C. Botswana D. Srilanka

91. The species whose last individuals are even not found after searches of known or likely areas where they were present in past

A. Endemic species
B. Rare species
C. Endangered species
D. Extinct species

92. Buxa Tiger Reserve is seriously threatened by the mining of

A. Coal B. Copper
C. Dolomite D. Iron

93. In which one of the following states is Periyar Tiger Reserves Located?

A. Rajasthan B. Assam
C. Uttaranchal D. Kerala

94. Which is the vegetarian Lizard found in Thar Desert of Rajasthan?

A. Lizard
B. Mongoose
C. Monitor Lizard
D. Spiny tail Lizard

95. As per the census conducted by Forest Department, Gujarat has lost ____ % of vultures between 2005 and 2018.

A. 60 % B. 70 %
C. 80 % D. 90 %

96. The population of vultures in Gujarat

A. 820 B. 1300
C. 1800 D. 2300

97. તાજેતરમાં ડિસેમ્બર 2020 માં, ________ ગીધ પાવી જેતપુર, વડોદરા નજીક છ વર્ષ પછી જોવા મળ્યું હતું.
A. રેડ હેડેડ ગીધ
B. વ્હાઇટ રમ્પ્ડ ગીધ
C. લોન્ગ બિલ્ડ ગીધ
D. ઇજિપ્શિયન ગીધ

98. રાષ્ટ્રીય પોર્ટલ 'સુરક્ષા' ક્યાં પ્રાણીના સંરક્ષણ માટે છે?
A. વાઘ
B. ગ્રેટ ઇન્ડિયન બસ્ટાર્ડ
C. એશિયન હાથી
D. ગંગા નદી ડોલ્ફિન

99. વિશ્વ સંરક્ષણ સંઘ IUCN ની સ્થાપના ક્યા વર્ષમાં થઈ?
A. 1948 B. 1980
C. 1992 D. 2002

100. પૃથ્વી પર 50,000 થી 60,000 એશિયન હાથીઓ છે અને તેમાંથી ______ % ભારતમાં છે.
A. 60% B. 70%
C. 80% D. 90%

101. ક્રેનની કઈ પ્રજાતિ ગુજરાતમાં નથી?
A. ડેમોઇઝેલ ક્રેન B. સરસ ક્રેન
C. સાઇબેરીયન ક્રેન D. કોમન ક્રેન

102. ક્રેનની કઈ પ્રજાતિ ગુજરાતમાં સ્થાયી રીતે રહે છે?
A. ડેમોઇઝેલ ક્રેન B. સારસ ક્રેન
C. સામાન્ય ક્રેન D. સાઇબેરીયન ક્રેન

103. IUCN સ્ટેટસ મુજબ ક્રેનની એકમાત્ર પ્રજાતિ જોખમમાં છે?
A. સારસ ક્રેન
B. હૂડેડ ક્રેન
C. સાઇબેરીયન ક્રેન
D. વાદળી ક્રેન

97. Recently in December 2020, ________ vulture was spotted after six years near Pavi Jetpur, Vadodara.
A. Red headed vulture
B. White rumped vulture
C. Long billed vulture
D. Egyptian vulture

98. National Portal 'Suraksha' is for the conservation of
A. Tiger
B. Great Indian Bustard
C. Asian elephant
D. Ganges river dolphin

99. The world conservation union IUCN was found in the year
A. 1948 B. 1980
C. 1992 D. 2002

100. There are 50,000 to 60,000 Asian elephants on earth and out of them ___ % are in India.
A. 60% B. 70%
C. 80 % D. 90 %

101. Which species of Crane is not harbored in Gujarat?
A. Demoiselle crane B. Sarus crane
C. Siberian crane D. Common crane

102. Which species of Crane is resident in Gujarat?
A. Demoiselle crane B. Sarus crane
C. Common crane D. Siberian crane

103. Which is the only species of crane Critically Endangered as per IUCN status?
A. Sarus crane
B. Hooded crane
C. Siberian crane
D. Blue crane

104 . ગુજરાતમાં સારસ ક્રેન ની આશરે 80% વસ્તી ક્યાં જિલ્લાઓમાં છે?: (i) આણંદ, (ii) અમદાવાદ, (iii) ખેડા , (iv) વડોદરા.
A. (i) અને (iv)
B. (iii) અને (iv)
C. (i), (ii) અને (iii)
D. (i), (ii), (iii) અને (iv)

105 . ગુજરાત વન વિભાગ દ્વારા વ્હેલ શાર્ક ના સંરક્ષણ માટે પ્રયાસો ક્યારે શરૂ કરાયા હતા ?
A. 1990 B. 2003
C. 1013 D. 2019

106. ભારતમાં ગીધની કેટલી જાતો જોવા મળે છે?
A. 15 B. 25
C. 5 D. 9

107. બાયોસ્ફિયર રિઝર્વના કયા ક્ષેત્રમાં મર્યાદિત માનવ પ્રવૃત્તિઓને મંજૂરી છે?
A. ટ્રાંઝીશન ઝોન B. બફર ઝોન
C. કોર ઝોન D. A અને B બંને

108. ભારતની સૌથી જૂની પર્યાવરણ સંસ્થા કઈ છે?
A. IIFM B. BNHS
C. WWF D. FRI

109. વન્યજીવનની વસ્તીમાં પ્રાણીઓની સંખ્યામાં થતા ફેરફારના વિશ્લેષણ શુ કહેવાય છે ?
A. વસ્તી ગતિશીલતા
B. વન્યપ્રાણી ગણતરી
C. વન્યજીવ વિજ્ઞાન
D. વસ્તી વ્યવસ્થાપન

110. ક્યાં વૈજ્ઞાનિક વિશ્વમાં હોટ સ્પોટની કલ્પના આપી હતી?
A. જિમ કોર્બેટ B. નોર્મન મેયર
C. આલ્ફ્રેડ મેયર D. મિશેલ ગિલ્સ

104. Around 80 % population of Sarus crane is supported by the districts of Gujarat: (i)Anand, (ii)Ahmedabad, (iii) Kheda, (iv) Vadodara.
A. (i) and (iv)
B. (iii) and (iv)
C. (i), (ii) and (iii)
D. (i), (ii), (iii) and (iv)

105. Gujarat Forest Department initiated efforts for Whale Shark Consevaton in
A. 1990 B. 2003
C. 1013 D. 2019

106. How many species of vulture are found in India?
A. 15 B. 25
C. 5 D. 9

107. In Which region of Biosphere Reserve limited human activities are allowed?
A. Transition zone B. Buffer zone
C. Core zone D. BothA and B

108. Which is the oldest environment organization of India?
A. IIFM B. BNHS
C. WWF D. FRI

109. Which is the study called that deals with the analysis of the changes in the numbers of animals in wildlife population at a particular time?
A. Population dynamics
B. Wildlife census
C. Wildlife science
D. Population Management

110. The Scientist who gave the concept of hot spot in the world?
A. Jim Corbett B. Norman Mayer
C. Alfred Mayer D. Michel Gills

111. ખાદ્ય, પાણી અને આશ્રયના સંપૂર્ણ સુધારાઓ __________ તરીકે ગણવામાં આવે છે
A. શેલ્ટર મેનિપ્યુલેશન્સ
B. હેબીટાટ મેનિપ્યુલેશન્સ
C. સંરક્ષણ
D. રક્ષણ પગલાં

112. વન્યજીવ વ્યવસ્થાપનમાં કૃત્રિમ રીતે ખોરાક કયા સમયે જ પૂરો પાડવો જોઈએ?
A. અનુકૂળ સમયગાળો
B. ઉનાળાની ઋતુ
C. પિન્ચ પેરીએડ
D. શિયાળાની ઋતુ

113. 1936 માં સ્થાપિત ભારતનો સૌથી જૂનો રાષ્ટ્રીય ઉદ્યાન કયો છે?
A. કાન્હા રાષ્ટ્રીય ઉદ્યાન
B. સુંદરવન રાષ્ટ્રીય ઉદ્યાન
C. કોર્બેટ રાષ્ટ્રીય ઉદ્યાન
D. કાઝીરંગા રાષ્ટ્રીય ઉદ્યાન

114. વન્યપ્રાણી (સંરક્ષણ) અધિનિયમની કઈ કલમ હેઠળ રાષ્ટ્રીય ઉદ્યાનની ઘોષણા કરવામાં આવી છે?
A. કલમ 35 B. કલમ 26 એ
C. કલમ 34 D. કલમ 38

115. વન્યપ્રાણી (સંરક્ષણ) અધિનિયમની કઈ કલમ હેઠળ વન્યપ્રાણી અભયારણ્ય જાહેર કરાયા છે ?
A. કલમ 35 B. કલમ 26 એ
C. કલમ 34 D. કલમ 38

116 . ગુજરાતનું સૌથી નાનું વન્યપ્રાણી અભયારણ્ય ક્યુ છે ?
A. પોરબંદર પક્ષી અભયારણ્ય
B. થોલ પક્ષી અભયારણ્ય
C. હિંગોલગઢ પ્રકૃતિ શિક્ષણ અભયારણ્ય
D. ગાગા વન્યપ્રાણી અભયારણ્ય

111. Food, water and shelter improvements altogether are considered as
A. Shelter manipulation
B. Habitat-manipulation
C. Conservation
D. Protection measures

112. In Wildlife Management, the artificial food is supplied in __________ period.
A. Favorable period
B. Summer season
C. Pinch period
D. Winter season

113. Which is the oldest National Park of India established in 1936?
A. Kanha National Park
B. Sundarbans National Park
C. Corbett National Park
D. Kaziranga National Park

114. Under which section of Wildlife (Protection)Act, a national park is declared?
A. Section 35 B. Section 26A
C. Section 34 D. Section 38

115. Under which section of Wildlife (Protection)Act, a Wildlife Sanctuary is declared?
A. Section 35 B. Section 26A
C. Section 34 D. Section 38

116. The smallest Wildlife Sanctuary of Gujarat is
A. Porbandar Bird Sanctuary
B. Thol Bird Sanctuary
C. Hingolgadh Nature Education Sanctuary
D. Gaga Wildlife Sanctuary

117. સ્લોથ રીંછ કયા અભ્યારણ માં જોવા મળે (i) ગીર રાષ્ટ્રીય ઉદ્યાન, (ii) બાલારામ અંબાજી વન્યપ્રાણી અભયારણ્ય, (iii) જાંબુઘોડા વન્યપ્રાણી અભયારણ્ય, (iv) વાંસદા રાષ્ટ્રીય ઉદ્યાન

A. (i) અને (ii)
B. (ii) અને (iii)
C. (i), (ii) અને (iii)
D. (i), (ii), (iii) અને (iv)

118. જેસોર વન્યપ્રાણી અભયારણ્યની મુખ્ય પ્રજાતિ કઈ છે ?

A. સ્લોથ રીંછ B. દીપડો
C. પટ્ટાવાળી ઝરખ D. ચિંકારા

119. ગુજરાતમાં સૌથી નાનો રાષ્ટ્રીય ઉદ્યાન કયો છે?

A. દરિયાઈ રાષ્ટ્રીય ઉદ્યાન
B. ગીર રાષ્ટ્રીય ઉદ્યાન
C. બ્લેકબક રાષ્ટ્રીય ઉદ્યાન
D. વાંસદા રાષ્ટ્રીય ઉદ્યાન

120. ગુજરાતનો સૌથી મોટો રાષ્ટ્રીય ઉદ્યાન કયો છે ?

A. વાંસદા રાષ્ટ્રીય ઉદ્યાન
B. મરીન રાષ્ટ્રીય ઉદ્યાન
C. ગીર રાષ્ટ્રીય ઉદ્યાન
D. બ્લેકબક રાષ્ટ્રીય ઉદ્યાન

121. નીચેનામાંથી કઈ પ્રજાતિઓ મરીન નેશનલ પાર્ક માં મળી આવે છે?

A. સ્પોન્જીસ
B. કોરલ્સ
C. સીહોર્સ
D. ઉપરોક્ત તમામ

122. ગુજરાતનું એકમાત્ર રામસાર સ્થળ કયું છે ?

A. પોરબંદર અભયારણ્ય
B. પાનીયા અભયારણ્ય
C. નલસરોવર વન્યપ્રાણી અભયારણ્ય
D. લાલા પરજન અભયારણ્ય

117. The ProtectedAreas of Gujarat harboring the population of Sloth Bear (i) Gir National Park, (ii) Balaram Ambaji Wildlife Sanctuary, (iii) Jambughoda Wildlife Sanctuary, (iv) Vansda National Park

A. (i) and (ii)
B. (ii) and (iii)
C. (i), (ii) and (iii)
D. (i), (ii), (iii) and (iv)

118. What is the major species of Jessore wildlife sanctuary

A. Sloth bear B. leopard
C. Striped Hyena D. Chinkara

119. Which is the smallest national park in Gujarat?

A. Marine national park
B. Gir national park
C. Black buck national park
D. Vansda national park

120. Which is the biggest national park of Gujarat?

A. Vansda National Park
B. Marine National Park
C. Gir National Park
D. Blackbuck National Park

121. Which of the following species are found in Marine National Park?

A. Sponges
B. Corals
C. Sea hourse
D. All of above

122. The only Ramsar Site of Gujarat is

A. Porbandar Bird Sanctuary
B. Paniya Wildlife sanctuary
C. Nal Sarovar Wildlife Sanctuary
D. Lala parjan Sanctuary

123. નીચેનામાંથી સૌથી મોટું ઉડતું પક્ષી કયું છે?

A. શાહમૃગ
B. પેલિકન્સ
C. ગ્રેટ ઇન્ડિયન બસ્ટર્ડ્સ
D. ફ્લેમિંગો

124. આ નિવેદન કોના દ્વારા આપવામાં આવ્યું - " એશિયાટિક સિંહ ત્યાં ન હોત તો, ગીર રાષ્ટ્રીય ઉદ્યાન દેશ નુ સૌથી વધુ રસપ્રદ પક્ષી અભયારણ્યમાંનુ એક હોત"

A. જિમ કોર્બેટ
B. સાલીમ અલી
C. રિચાર્ડ ડોકિન
D. લવકુમાર ખાચર

125. તાજેતરમાં 2020 માં, નીચેનામાંથી કયા સુરક્ષિત ક્ષેત્રમાં લેસર ફ્લોરિકન ને સેટેલાઇટટ ટેગ કરાયા હતા ?

A. ગીર રાષ્ટ્રીય ઉદ્યાન
B. જામનગર રાષ્ટ્રીય ઉદ્યાન
C. વાંસદા રાષ્ટ્રીય ઉદ્યાન
D. બ્લેકબક રાષ્ટ્રીય ઉદ્યાન

126. ક્યાં રાષ્ટ્રીય ઉદ્યાનમાં નીલગાય જોઇ શકાતી નથી ?

A. વાંસદા રાષ્ટ્રીય ઉદ્યાન
B. વેલાવદર રાષ્ટ્રીય ઉદ્યાન
C. ગીર રાષ્ટ્રીય ઉદ્યાન
D. ગીર રાષ્ટ્રીય ઉદ્યાન

127. રીંછ ની કઈ પ્રજાતિ ગુજરાતમાં જોઇ શકાય છે?

A. ધ્રુવીય રીંછ B. બ્રાઉન રીંછ
C. સન રીંછ D. સ્લોથ રીંછ

128. તાજેતરમાં કયા રક્ષિત ક્ષેત્રમાં, ગ્રેટ ઇન્ડિયન બસ્ટાર્ડ્સ કેપટિવ બ્રીડીંગ કરવામાં આવ્યું હતું?

A. GIB અભયારણ્ય, કચ્છ
B. ગાગા અભયારણ્ય
C. નલ સરોવર પક્ષી અભયારણ્ય
D. ડેઝર્ટ રાષ્ટ્રીય ઉદ્યાન

123. Which is the heaviest flying bird among them?

A. Ostrich
B. Pelicans
C. Great Indian bustard
D. Flamingo

124. Who dictated the statement – "had theAsiatic lion not been there, the area of Gir national park would have been one of the most fascinating bird sanctuaries of the country"

A. Jim Corbett
B. SalimAli
C. Richard Dawkin
D. Lavkumar Khachar

125. Recently in 2020, in which of the following protected area Lesser floricans were Satellite tagged?

A. Geer Nationa Park
B. Jamnagar Nationa Park
C. Vansda Nationa Park
D. Blackbuck Nationa Park

126. Blue Bull can't be seen in this National Park

A. Vansda Nationa Park
B. Velavadar Nationa Park
C. Geer Nationa Park
D. Gir Nationa Park

127. Which of the Beer species can be seen in Gujarat?

A. Polar bear B. Brown bear
C. Sun bear D. Sloth bear

128. Recently in which ProtectedArea, Great Indian Bustards were raised in captive breeding?

A. GIB sanctuary, Kutch
B. Gaga Bird sanctuary
C. Nal Sarovar Bird sanctuary
D. Desert National Park

129. ગાગા પક્ષી અભયારણ્ય કયા જિલ્લામાં આવેલું છે?
A. ભાવનગર
B. દેવ ભૂમિ, દ્વારકા
C. સુરત
D. નવસારી

130. ભારતનો પ્રથમ મરીન નેશનલ પાર્ક કયો છે?
A. મરીન રાષ્ટ્રીય ઉદ્યાન, જામનગર
B. ભીતરકાનિકા રાષ્ટ્રીય ઉદ્યાન
C. ગલ્ફ ઓફ મન્નાર રાષ્ટ્રીય ઉદ્યાન
D. સુંદરવન રાષ્ટ્રીય ઉદ્યાન

131. જંગલ ઓવલેટ ક્યાં રાષ્ટ્રીય ઉદ્યાન માં જોવા મળે છે ?
A. સૂરપાણેશ્વર અભયારણ્ય
B. ગીર રાષ્ટ્રીય ઉદ્યાન
C. બ્લેકબક રાષ્ટ્રીય ઉદ્યાન
D. વાંસદા રાષ્ટ્રીય ઉદ્યાન

132. કયા ક્ષેત્રમાં મોટી ઉડતી ખિસકોલી જોઇ શકાય છે?
A. બાલારામ વન્યપ્રાણી અભયારણ્ય
B. ગીર વન્યપ્રાણી અભયારણ્ય
C. જાંબુઘોડાવન્યપ્રાણી અભયારણ્ય
D. રામપુરા વન્યપ્રાણી અભયારણ્ય

133. નીચેનામાંથી કયું વિધાન યોગ્ય છે?
(i) ગુજરાતમાં 23 રાષ્ટ્રીય ઉદ્યાન અને 4 વન્યપ્રાણી અભયારણ્ય છે.
(ii) ઘુડખર માત્ર ગુજરાતમાં જોવા મળે છે.
(iii) એશિયાઇ સિંહો ફક્ત ગીર રાષ્ટ્રીય ઉદ્યાનમાં જ જોવા મળે છે
A. (i) અને (ii)
B. (i) અને (iii)
C. (ii) અને (iii)
D. (i), (ii) અને (iii)

129. Gaga Bird sanctuary is located in which district?
A. Bhavnagar
B. Dev Bhumi, Dwarka
C. Surat
D. Navsari

130. Which is the first Marine National Park of India?
A. Marine NP, Jamnagar
B. Bhitarkanika NP
C. Gulf of Mannar NP
D. Sundarbans NP

131. Jungle owlet, a threatened species is seen in which of the following parks?
A. Shoolparneshwar Wildfie Sanctuary
B. Gir National Park
C. Black buck National Park
D. Vansda National Park

132. Giant flying squirrel can be observed in which protected area?
A. Balaram Wildlife Sanctuary
B. Gir Wildlife Sanctuary
C. Jambughoda Wildlife Sanctuary
D. Rampura Wildlife Sanctuary

133. Which of the following statement are correct?
(i) Gujarat has 23 national park and 4 wildlife centuries
(ii) Gujarat is the only state to harborAsiatic wildAss
(iii) Asiatic lions are only found in Gir national park
A. (i) & (ii)
B. (i) & (iii)
C. (ii) & (iii)
D. (i), (ii) & (iii)

134. દીપડાની 2018ની ગણતરી મુજબ, ભારતમાં કેટલા દીપડાઓ છે?

A. 10852 B. 12852

C. 20852 D. 25852

135. ભારતમાં દીપડામાં ની ગણતરી, 2018 કોનાં દ્વારા પ્રકાશિત કરવામાં આવી છે?

A. નેશનલ લેપર્ડ ઓથોરિટી

B. નેશનલ ટાઇગર કન્ઝર્વેશન ઓથોરિટી

C. નેશનલ બાયોડાયવર્સીટી ઓથોરિટી

D. નેશનલ ઝૂ ઓથોરિટી

136. કારાકલ ગુજરાતના ક્યાં વન્યપ્રાણી અભયારણ્યમાં જોઇ શકાય છે?

A. પૂર્ણા વન્યજીવન અભયારણ્ય

B. પાનીયા વન્યપ્રાણી અભયારણ્ય

C. નારાયણ સરોવર વન્યપ્રાણી અભયારણ્ય

D. મીતીયાળા વન્યપ્રાણી અભયારણ્ય

137. "ફ્લેમિંગો સિટી" સામૂહિક સંવર્ધન નું સ્થળ ક્યું છે, તે ક્યાં સ્થિત છે?

A. કચ્છનું મોટું રણ

B. કચ્છનું નાનું રણ

C. જામનગર દરિયાઇ અભ્યારણ્ય

D. પોરબંદર પક્ષી અભયારણ્ય

138. ઇકો સેન્સેટિવ ઝોન માટે ઝોનલ માસ્ટર પ્લાનમાં ______ સામેલ હોવું જોઈએ

A. તમામ સંબંધિત રાજ્ય વિભાગોની ભાગીદારી

B. ડીન્યુડેડ વિસ્તારોની ફેરસુધારણા

C. જમીનના ઉપયોગમાં કોઈ ફેરફાર નહીં

D. ઉપરના તમામ

139. ઓર્ગેનિક ફાર્મિંગ સામાન્ય રીતે રક્ષિત વિસ્તારોના ઇકો સેન્સેટિવ ઝોનમાં ______ છે.

A. પ્રતિબંધિત

B. નિયમિત

C. મંજૂર અથવા પ્રોત્સાહિત

D. સંબંધિત નથી

134. As per the Status of Leopard in India, 2018, how many leopards India harbors?

A. 10852 B. 12852

C. 20852 D. 25852

135. Status of Leopard in India, 2018 is published by

A. National LeopardAuthority

B. National Tiger ConservationAuthority

C. National BiodiversityAuthority

D. Central ZooAuthority

136. The Caracal can be seen in the wildlife sanctuary of Gujarat

A. Purna wildlife sanctuary

B. Pania wildlife sanctuary

C. Narayan Sarovar wildlife sanctuary

D. Mitiyala wildlife Sanctuary

137. "Flamingo City" is the site of mass breeding, is located in

A. Great Rann of Kutch

B. Little Rann of Kutch

C. Jamnagar Marine Sanctuary

D. Porbander Bird Sanctuary

138. Zonal master Plan for Eco sensitive zone should include the

A. Involvement of all concerned State Departments

B. Restoration of denuded areas,

C. No change of land use from green uses

D. All above

139. Organic farming is generally ______ in Eco sensitive zone of PAs.

A. Restricted

B. Regulated

C. Permitted or promoted

D. Not relevant

140. સિંહોનું ઝૂંડ (pride) _____________ દ્વારા નિયંત્રિત થાય છે.

A. નર સિંહો
B. માદા સિંહો
C. નિયંત્રણ નથી
D. બચ્ચાં

141. ગુજરાતમાં જંગલીમાં વાઘની વસ્તી કેટલી છે ?

A. 15 B. 10
C. 0 D. 21

142. ખાણ કામ પ્રવૃત્તિ સામાન્ય રીતે રક્ષિત વિસ્તારોના ઇકો સેન્સેટિવ ઝોનમાં _____ છે.

A. પ્રતિબંધિત
B. નિયમિત
C. મંજૂરી અથવા બંધિત
D. સંબંધિત નથી

143. મોટા ભાગના અભયારણ્યો અને રાષ્ટ્રીય ઉદ્યાનોમાં, પર્યટન પ્રવૃત્તિઓ 15 જૂન થી 15 ઓક્ટોબર બંધ હોઈ છે કારણ કે;

A. વન્યપ્રાણી (સંરક્ષણ) અધિનિયમ 1972
B. વરસાદમાં ગતિશીલતા પ્રતિબંધિત છે
C. માંસાહારી પ્રાણીઓનું સંવર્ધન
D. પર્યાવરણ (સંરક્ષણ) અધિનિયમ, 1986

144. રક્ષિત ક્ષેત્રોને કોના દ્વારા સંચાલિત થાય છે ?

A. કાર્યકારી યોજના
B. વ્યવસ્થાપન યોજના
C. સારવાર યોજના
D. રાજ્ય સરકારની માર્ગદર્શિકા

145. વન્યપ્રાણી (સંરક્ષણ) અધિનિયમ, 1972 મુજબ _____ રક્ષિત વિસ્તાર નથી.

A. અભયારણ્ય
B. રાષ્ટ્રીય ઉદ્યાન
C. કનઝર્વેશન રિઝર્વ
D. બાયોસ્ફિયર રિઝર્વ

140. Pride of lions is controlled by

A. Male lions
B. Lionesses
C. No control
D. Cubs

141. The Tiger population in wild in Gujarat is

A. 15 B. 10
C. 0 D. 21

142. Mining activity is generally _____ in Eco sensitive zone of PAs.

A. Restricted
B. Regulated
C. Permitted or promoted
D. Not relevant

143. In most of the Sanctuaries and National Parks, the tourism activities are closed from 15th June to 15th October, because of

A. Wildlife (Protection)Act 1972
B. In rainy season, mobility is restricted.
C. Breeding season of carnivore animals
D. Environment (Protection)Act, 1986

144. The protected areas are to be managed as per the approved

A. Working plan
B. Management plan
C. Treatment plan
D. Guidelines of the State Government.

145. Which is not the Protected area as per Wildlife (Protection)Act 1972;

A. Sanctuary
B. National Park
C. Conservation reserve
D. Biosphere Reserve

146. રક્ષિત વિસ્તારમાં વન્યજીવનનું શ્રેષ્ઠ સંચાલન ________ છે.

A. વન્યપ્રાણીના જીવનમાં ન્યૂનતમ વિક્ષેપ

B. વન્યપ્રાણીના જીવન પર સતત નજર

C. વન્યપ્રાણીઓની નિયમિત સ્વાસ્થ્ય તપાસ

D. વન્યપ્રાણીઓને નિયમિત રસી

147. જંગલ પ્રાણીઓના "બચાવ અને પુનર્વસન" ની કામગીરી વન અધિકારીઓ દ્વારા કેમ કરવામાં આવી રહી છે

A. માણસ-પ્રાણીના તકરાર ને ઓછા કરવા માટે

B. હંમેશા જંગલી પ્રાણીઓની સંભાળ રાખવા

C. જંગલી પ્રાણીઓના આરોગ્યની તપાસ કરવા

D. જંગલી પ્રાણીઓને જંગલોમાં મુક્ત કરવા

148. એશિયાઈ સિંહ એ એક સામાજિક પ્રાણી છે, આનો મતલબ શું થાય?

A. તે માનવ વસવાટ સાથે જીવે છે

B. તે સમૂહ માં શિકાર કરી ખાય છે

C. સિંહની પૂજા કરવામાં આવી રહી છે

D. માનવ સમાજ સિંહોને પસંદ કરે છે

149. ગુજરાતમાં ધુડખરની છેલ્લો વસ્તી અંદાજ ક્યારે આવ્યો?

A. 2020 B. 2019

C. 2018 D. 2015

150. ગુજરાતમાં કેટલા રામસીર વિસ્તારો છે?

A. 1 B. 2

C. 3 D. 4

146. Best management of wildlife in protected area is;

A. Minimum disturbance to wildlife

B. Keep continuous watch on the wildlife

C. Keep regular health checkup of the wildlife

D. Keep regular vaccination to wildlife

147. Why "Rescue and Rehabilitation" operations of wild animals are being done by the forest officials?

A. To avoid man-animal conflicts.

B. To always take care of wild animals

C. To check up the health of Wild animals.

D. To release wild animals in deep forests.

148. Asiatic lion is a social animal; it means

A. Live with human habitations

B. Hunting and eating in groups

C. Lion is being worshipped

D. Human society likes the lions

149. When last estimation of Wild Ass was conducted in Gujarat

A. 2020 B. 2019

C. 2018 D. 2015

150. How many Ramseer Sites in Gujarat?

A. 1 B. 2

C. 3 D. 4

Answers | જવાબો

Que.	Ans.	Que.	Ans.	Que.	Ans.	Que.	Ans.	Que.	Ans.	Que.	Ans.	Que.	Ans.
1.	B	24.	D	47.	B	70.	C	93.	D	116.	A	139.	C
2.	C	25.	B	48.	A	71.	B	94.	D	117.	B	140.	A
3.	C	26.	A	49.	B	72.	A	95.	B	118.	A	141.	C
4.	B	27.	C	50.	A	73.	A	96.	A	119.	D	142.	A
5.	D	28.	C	51.	D	74.	C	97.	C	120.	C	143.	B
6.	C	29.	D	52.	B	75.	A	98.	C	121.	D	144.	B
7.	D	30.	C	53.	D	76.	A	99.	A	122.	C	145.	D
8.	D	31.	C	54.	A	77.	B	100.	A	123.	C	146.	A
9.	B	32.	B	55.	B	78.	D	101.	C	124.	B	147.	A
10.	C	33.	D	56.	C	79.	B	102.	B	125.	D	148.	B
11.	C	34.	C	57.	B	80.	D	103.	C	126.	A	149.	A
12.	B	35.	B	58.	D	81.	A	104.	C	127.	D	150.	A
13.	D	36.	D	59.	D	82.	C	105.	B	128.	D		
14.	B	37.	C	60.	A	83.	B	106.	D	129.	B		
15.	B	38.	B	61.	A	84.	D	107.	D	130.	A		
16.	D	39.	D	62.	C	85.	A	108.	B	131.	D		
17.	C	40.	C	63.	A	86.	D	109.	A	132.	C		
18.	B	41.	B	64.	D	87.	B	110.	B	133.	C		
19.	C	42.	C	65.	C	88.	D	111.	B	134.	B		
20.	B	43.	C	66.	C	89.	A	112.	C	135.	B		
21.	C	44.	C	67.	A	90.	C	113.	C	136.	C		
22.	D	45.	D	68.	B	91.	D	114.	A	137.	A		
23.	D	46.	C	69.	B	92.	C	115.	B	138.	D		

Chapter 14

પ્રકરણ ૧૪

Forest and Environment Acts, Policies and Laws

વન અને પર્યાવરણને લગતા ધારાઓ, નીતિઓ અને કાયદાઓ

Indian Constitution and Environment

Article	Provision	Remarks
48A	Protection and improvement of environment and safeguarding of forests and wild life	Govt of India enacted 42nd Amendment added Article 48-A to the Directive Principles of State Policy in 1976.
51-A (g)	It shall be the duty of every citizen of India to protect and improve the natural environment including forests, lakes, rivers and wildlife and to have compassion for living creatures.	To protect environment is fundamental duty of every citizen of India
253	Legislation for giving effect to international agreements.	Parliament has power to make law for the whole or any part of the country for implementing any treaty, agreement or convention with any other country.
List III 17 A	Forests added to List III-concurrent list	It empowers central Government to make laws on forests and wildlife
List III 17 B	Protection of Wild animals and birds added to List III-concurrent list	

ભારતીય બંધારણ અને પર્યાવરણ

અનુચ્છેદ	જોગવાઈ	વિશેષ નોંધ
48 એ	પર્યાવરણનું રક્ષણ અને સુધારણા અને જંગલો અને વન્યજીવોની સુરક્ષા	ભારત સરકારે 42મો સુધારો કરીને 1976 માં રાજ્ય નીતિના નિર્દેશક સિધ્ધાંતોમાં કલમ 48-એ ઉમેરી .
51-એ (g)	વન, તળાવો, નદીઓ અને વન્યપ્રાણી સૃષ્ટિ સહિતના કુદરતી વાતાવરણની સુરક્ષા અને સુધારણા કરવી અને વન્યજીવો પ્રત્યે કરુણા રાખવી એ ભારતના દરેક નાગરિકની ફરજ રહેશે .	પર્યાવરણની સુરક્ષા કરવી એ ભારતના દરેક નાગરિકની મૂળભૂત ફરજ છે

અનુચ્છેદ	જોગવાઈ	વિશેષ નોંધ
253	આંતરરાષ્ટ્રીય કરારોના અમલ માટે કાયદો.	સંસદ પાસે કોઈ અન્ય દેશ સાથેની કોઈ સંધિ, કરાર અથવા સંમેલનને લાગુ કરવા માટે સમગ્ર અથવા દેશના કોઈપણ ભાગ માટે કાયદો બનાવવાની સત્તા છે.
સૂચિ III 17 A	જંગલોને સૂચિ III-સમવર્તી સૂચિમાં ઉમેર્યું	તે કેન્દ્ર સરકારને જંગલો અને વન્યપ્રાણીસૃષ્ટિ પર કાયદા બનાવવા માટે સત્તા આપે છે
સૂચિ III 17 B	જંગલી પ્રાણીઓ અને પક્ષીઓનું રક્ષણ સૂચિ III-સમવર્તી સૂચિમાં ઉમેર્યું	

Forest policies of India

Detail	1894	1952	1988
Classification of Forests	i) Forest for timber production, ii) Minor forests, iii) Preservation forest, iv) Pasture lands	i) Protection forests ii) National forests iii) Village forests iv) Tree lands	No classification
Forest administration	No guidelines	Trained foresters are recommended	Personal management for professional foresters
Forest legislation	No guidelines	Enactment of forest acts	Appropriate legislation
Forests and tribal	No guidelines	No guidelines	Forest department and tribal's symbiotic relation
Grazing	No guidelines	Grazing is allowed with nominal fee	Grazing according to carrying capacity of the area
Increasing tree cover	No guidelines	Increase green cover through planting on railways, PWD lands.	Afforestation programme emphasized on large scale
Local communities' goodwill	No guidelines	Need for peoples cooperation	Community's participation in forest management
Objective	Focus on agriculture and public interest and rights	Balanced and complementary utilization of forest produce	Focus on conservation of natural heritage, biodiversity and genetic resources
Proportion of forest are	No guidelines	1/3rd of the area, 60% in hills and 20% in plains	33% under forests cover, 66% tree cover on hills
Shifting cultivation	No guidelines	No coercive measures to stop shifting cultivation	Discourage shifting cultivation
Sustainable yield	No guidelines	Scope is limited	Detailed measures are recommended

Detail	1894	1952	1988
Training, education and research in forestry	No guidelines	Through rangers college and universities	Training and scientific forestry research for increasing productivity, afforestation, wildlife and wood substitution
Wildlife protection	No guidelines	Protection of rare and endangered species	Strengthen protection measures network

ભારતની વન નીતિઓ

વિગત	1894	1952	1988
જંગલોનું વર્ગીકરણ	i) લાકડાના ઉત્પાદન માટે વન, ii) નાના વનો, iii) બચાવ વન, iv) ગોચર જમીન	i) સંરક્ષણ જંગલો, ii) રાષ્ટ્રીય વનો, iii) ગામડાનાં જંગલો, iv) વૃક્ષની જમીન	કોઈ વર્ગીકરણ નથી
વન વહીવટ	કોઈ માર્ગદર્શિકા નથી	પ્રશિક્ષિત ફોરેસ્ટર્સની ભલામણ કરવામાં આવે છે	વ્યાવસાયિક ફોરેસ્ટર્સ માટેનું વ્યક્તિગત વ્યવસ્થાપન
વન કાયદાઓ	કોઈ માર્ગદર્શિકા નથી	વન અધિનિયમોનો અમલ	યોગ્ય કાયદો
જંગલો અને આદિજાતિ સમુદાયો	કોઈ માર્ગદર્શિકા નથી	કોઈ માર્ગદર્શિકા નથી	વન વિભાગ અને આદિજાતિ સમુદાયો વચ્ચે સહજીવન
ચરાઈ	કોઈ માર્ગદર્શિકા નથી	નજીવી ફી સાથે ચરાઈને મંજૂરી છે	વિસ્તારની વહન ક્ષમતા અનુસાર ચરાઈ
ઝાડનું આવરણ વધારવું	કોઈ માર્ગદર્શિકા નથી	રેલ્વે, પીડબ્લ્યુડી જમીનો પર વાવેતરો દ્વારા લિલોતરીમાં વધારો	મોટા પાયે વનીકરણ કાર્યક્રમ પર ભાર મૂક્યો
સ્થાનિક સમુદાયોની ઇચ્છા	કોઈ માર્ગદર્શિકા નથી	લોકોના સહયોગની જરૂર છે	વન વ્યવસ્થાપનમાં સમુદાયો ભાગીદારી
ઉદ્દેશ્ય	કૃષિ અને જનઅધિકારો પર ધ્યાન કેન્દ્રિત	વન પેદાશોના સંતુલિત અને પૂરક ઉપયોગ	કુદરતી વારસો, જૈવવિવિધતા અને આનુવંશિક સંસાધનોના સંરક્ષણ પર ધ્યાન કેન્દ્રિત કર્યું
વનનો હિસ્સો	કોઈ માર્ગદર્શિકા નથી	1/3 વિસ્તાર, 60%ટેકરીઓ અને 20%મેદાનોની	33% વિસ્તાર જંગલો, પર્વતોમાં 66% ઝાડ

વિગત	1894	1952	1988
ખેતી સ્થળાંતર	કોઈ માર્ગદર્શિકા નથી	શિફ્ટિંગ કલ્ટીવેશન અટકાવવા માટે કોઈ પગલાં નહીં	શિફ્ટિંગ કલ્ટીવેશન અટકાવવું
ટકાઉ ઉપજ	કોઈ માર્ગદર્શિકા નથી	કાર્યક્ષેત્ર મર્યાદિત છે	વિગતવાર પગલાઓની ભલામણ કરવામાં આવે છે
વનીકરણમાં તાલીમ, શિક્ષણ અને સંશોધન	કોઈ માર્ગદર્શિકા નથી	રેન્જર્સ કોલેજ અને યુનિવર્સિટીઓ દ્વારા	ઉત્પાદકતા, વનીકરણ, વન્યપ્રાણી અને લાકડામાં વધારો કરવા માટે તાલીમ અને વૈજ્ઞાનિક વન સંશોધન
વન્યપ્રાણી સંરક્ષણ	કોઈ માર્ગદર્શિકા નથી	દુર્લભ અને જોખમી પ્રજાતિઓને રક્ષણ	સુરક્ષા નેટવર્કને મજબૂત બનાવવું

Forest and Environment Acts | વન અને પર્યાવરણ અંગેના અધિનિયમો

Law	Year	Chapters	Schedule	Sections	Remarks
Indian Forest Act	1927	13	—	86	Established in 1865 and enacted in 1927
The Saurastra felling of trees (Infliction of punishment) Act, 1951	1951	—	—	11	—
Prevention of Cruelty to Animals Act	1960	6	—	41	Enforced in 1961
Wild Life (Protection) Act	1972	7	6	66	—
Water (Prevention & control of pollution) Act	1974	8	—	64	—
Forest (Conservation) Act	1980	12	—	5	—
Air (Prevention & control of pollution) Act	1981	7	—	54	—
Environment (Protection) Act	1986	4	—	26	—
National Forest Policy	1988	—	—	—	—
The Protection of Plant Varieties and Farmers Rights Act	2001	11	—	97	Enforced in 2005
Biological Diversity Act	2002	12	—	65	—
The scheduled tribes and other traditional forest dwellers (recognition of forest rights) Act	2006	6	—	14	Enforced in 2007

Law	Year	Chapters	Schedule	Sections	Remarks
National Environment Policy	2006	—	—	—	—
National Green Tribunal Act	2010	5	3	38	—
Coastal Regulation Zone	2011	—	—	—	—
National Green Tribunal Act	2010	5	—	38	—
Green highways (Plantation, Transplantation, beautification & maintenance) Policy	2015	—	—	—	—
The Compensatory Afforestation Fund Act	2016	6	—	33	Enforced in 2018

Multiple Choice Questions | હેતુલક્ષી પ્રશ્નો

1. ભારતમાં હાલ કઈ વન નીતિ લાગુ છે?
   A. 1988
   B. 1954
   C. 1894
   D. 2002

1. **Which Forest Policy of India is applicable now?**
   A. 1988
   B. 1954
   C. 1894
   D. 2002

2. જંગલી પ્રાણીઓ રાખવા એ કયા કાયદા હેઠળ ગુનો છે?
   A. જૈવવિવિધતા અધિનિયમ
   B. પ્રાણીઓ પરની ક્રૂરતા નિવારણ અધિનિયમ
   C. વન્યપ્રાણી (સંરક્ષણ) અધિનિયમ
   D. વન (સંરક્ષણ) અધિનિયમ

2. **Under which act keeping wild animal is a crime?**
   A. Biodiversity Act
   B. Prevention of Cruelty to Animals Act
   C. Wild Life (Protection) Act
   D. Forest (Conservation) Act

3. જંગલની જમીનની કેટેગરી કયા એક્ટમાં વર્ણવવામાં આવી છે?
   A. ફોરેસ્ટ (કંઝર્વેશન) એક્ટ 1980
   B. ઇન્ડિયન ફોરેસ્ટ એક્ટ 1927
   C. ફોરેસ્ટ રાઇટ એક્ટ 2006
   D. સૌરાષ્ટ્ર ફેલિંગ ઓફ ટ્રી એક્ટ 1951

3. **The category of forest lands are describing in the act**
   A. Forest (Conservation) Act 1980
   B. Indian Forest Act 1927
   C. Forest Right Act 2006
   D. Saurashtra felling of tree Act 1951

4. વિકાસલક્ષી કામો માટે જંગલની જમીનના પરિવર્તન હેઠળ કયો એક્ટ લાગુ પડે છે?
   A. ફોરેસ્ટ (કંઝર્વેશન) એક્ટ 1980
   B. ઇન્ડિયન ફોરેસ્ટ એક્ટ 1927
   C. ફોરેસ્ટ રાઇટ એક્ટ 2006
   D. સૌરાષ્ટ્ર ફેલિંગ ઓફ ટ્રી એક્ટ 1951

4. **The diversion of forest land for developmental works has to processed under**
   A. Forest (Conservation) Act 1980
   B. Indian Forest Act 1927
   C. Forest Right Act 2006
   D. Saurastra felling of tree Act 1951

5. ફોરેસ્ટ (કંઝર્વેશન) એક્ટ 1980 મુજબ વન જમીનને ફેરબદલ માટે કોની પૂર્વ પરવાનગી મેળવવી જરૂરી છે?
   A. સુપ્રીમ કોર્ટ
   B. હાઇકોર્ટ
   C. રાજ્ય સરકાર
   D. કેન્દ્ર સરકાર

5. **As per Forest (Conservation) Act 1980, the prior permission of the ________ is must for diversion of forest land?**
   A. Supreme Court
   B. High Court
   C. State Government
   D. Central Government

6. કયા ફોરેસ્ટ એક્ટ હેઠળ વન વિસ્તારમાંથી વૃક્ષો કાપવાની મનાઈ છે?
   A. ફોરેસ્ટ (કંઝર્વેશન) એક્ટ 1980
   B. ઇન્ડિયન ફોરેસ્ટ એક્ટ 1927
   C. ફોરેસ્ટ રાઇટ એક્ટ 2006
   D. સૌરાષ્ટ્ર ફેલિંગ ઓફ ટ્રી એક્ટ 1951

7. ઇન્ડિયન ફોરેસ્ટ એક્ટ 1927 મુજબ વન વિસ્તારોનું વર્ગીકરણના કેટલા પ્રકારમાં થાય છે?
   A. રિઝર્વડ ફોરેસ્ટ
   B. પ્રોટેકટેડ ફોરેસ્ટ
   C. વિલેજ ફોરેસ્ટ
   D. ઉપર્યુક્ત તમામ

8. ગુજરાત રાજ્યમાં, વન વિસ્તારના વૃક્ષોના ગેરકાયદેસર કટીંગમાં સામેલ વાહન, વન અધિકારીઓ કયા અધિનિય હેઠળ અર્ધ-ન્યાયિક સુનાવણી સાથે વાહનને જપ્ત કરી શકે છે.
   A. ફોરેસ્ટ (કંઝર્વેશન) એક્ટ 1980
   B. ઇન્ડિયન ફોરેસ્ટ એક્ટ 1927
   C. ફોરેસ્ટ રાઇટ એક્ટ 2006
   D. સૌરાષ્ટ્ર ફેલિંગ ઓફ ટ્રી એક્ટ 1951

9. "સરકારે મંજુર કરેલ મેનેજમેન્ટ પ્લાન સિવાય કોઈપણ જંગલ માં કામ કરવાની છૂટ હોવી જોઈએ નહીં" આ બાબત કયા કાયદા અંતર્ગત છે ?
   A. ફોરેસ્ટ (કંઝર્વેશન) એક્ટ 1980
   B. ઇન્ડિયન ફોરેસ્ટ એક્ટ 1927
   C. ફોરેસ્ટ રાઇટ એક્ટ 2006
   D. નેશનલ ફોરેસ્ટ પોલિસી 1988

10. રાષ્ટ્રીય વન નીતિ 1988 મુજબ, કયા કારણો માટે કુદરતી જંગલોનું વ્યવસ્થાપન કરવું પડશે?
   A. દેશનો આર્થિક લાભ
   B. દેશની ઇમારતી લાકડાની માંગને ધ્યાનમાં રાખીને
   C. પર્યાવરણીય સ્થિરતા અને ઇકોલોજીકલ સંતુલનની જાળવણીની ખાતરી માટે
   D. જંગલોમાંથી લાકડાની નિકાસ માટે

6. The Cutting of trees from the forest area is punishable under the forest act?
   A. Forest (Conservation) Act 1980
   B. Indian Forest Act 1927
   C. Forest Right Act 2006
   D. Saurastra felling of tree Act 1951

7. The classification of forest areas as per Indian Forest Act 1927 are
   A. Reserved Forests
   B. Protected Forests
   C. Village forests
   D. Above all

8. In Gujarat State, the vehicle involved in illicit cutting of trees from forest area, the Forest officer with his quasi-judicial trial may confiscate the vehicle under;
   A. Forest (Conservation) Act 1980
   B. Indian Forest Act 1927
   C. Forest Right Act 2006
   D. Saurastra felling of tree Act 1951.

9. "No forest should be permitted to be worked without - the Government having approved the management plan", this is as per
   A. Forest (Conservation) Act 1980
   B. Indian Forest Act 1927
   C. Forest Right Act 2006
   D. National Forest Policy 1988

10. As per National Forest Policy 1988, the natural forests have to be managed for
   A. Economic benefit of the country
   B. Meet out the timber demand of country
   C. Ensure environmental stability and maintenance of ecological balance.
   D. Export the timber from the forests

11. ઇન્ડિયન ફોરેસ્ટ એક્ટ 1927 મુજબ જંગલોના ક્ષેત્રના કોઈપણ સીમાચિહ્નનો નાશ અથવા વિક્ષેપ કરનાર વ્યક્તિને કેટલા વર્ષની કેદ થઈ શકે છે?

A. એક વર્ષ
B. બે વર્ષ
C. ત્રણ વર્ષ
D. ચાર વર્ષ

12. વન્યપ્રાણી (સંરક્ષણ) અધિનિયમ 1972 મુજબ, "ઓનરરી વાઇલ્ડલાઇફ વોર્ડન્સ" કયા છે?

A. દરેક વન અધિકારી
B. વન્યપ્રાણી વિસ્તારોના વન અધિકારીઓ
C. મહેસૂલ અધિકારી
D. સરકાર દ્વારા નિયુક્ત સ્થાનિક લોકો

13. સલમાનખાન રાજસ્થાનમાં __________ ના શિકારમાં સામેલ હતો.

A. ચિંકારા
B. કાળીયાર
C. ચિતલ
D. ચૌસિંગા

14. સ્ટેટ બોર્ડ ઓફ વાઇલ્ડલાઇફના અધ્યક્ષ કોણ હોય છે?

A. મુખ્યમંત્રી
B. વન મંત્રી
C. મુખ્ય સચિવ
D. અધિક મુખ્ય સચિવ, વન.

15. કોમ્યુનિટી રિઝર્વ અને કંઝર્વેશન રિઝર્વ વચ્ચેનો મુખ્ય તફાવત શું છે?

A. સમુદાય અને વન વિભાગ દ્વારા સંચાલન
B. ખાનગી અથવા સમુદાયની જમીન અને સરકારી પડતર જમીન પર
C. સમુદાયની જમીન અને જાહેર જંગલની જમીન પર
D. JMF વિસ્તાર અને Non-JMF વિસ્તાર પર

11. Destroy or defaces any boundary mark of forests area, as per IFA, 1927, the person can be imprisonment up to

A. one year
B. Two years
C. Three years
D. Four years

12. As per Wildlife (Protection) Act 1972, the 'Honorary Wildlife Wardens" are

A. Every Forest Officer
B. Forest officers posted in Wildlife areas
C. Revenue Officer
D. Any Local people, appointed by Govt.

13. Salman Khan was involved in hunting of __________ in Rajasthan.

A. Chinkara
B. Blackbuck
C. Spotted dear
D. Chausingha

14. Who is the Chairperson of the State Board of Wildlife

A. Chief Minister
B. Forest Minister
C. Chief Secretary
D. Addl. Chief secretary, Forests

15. The main difference between Community Reserve and Conservation Reserve is;

A. Management by Community and by Forest department
B. Private or Community land and Government wasteland
C. Community land and declared forest land
D. JFM area and Non JFM area

16. પ્રાણીસંગ્રહાલય કઈ માર્ગદર્શિકા અનુસાર માન્યતા અને સંચાલન કરવામાં આવી રહ્યું છે?

A. રાજ્ય વન વિભાગ
B. પશુપાલન વિભાગ
C. સેન્ટ્રલ ઝૂ ઓથોરિટી
D. મ્યુનિસિપલ વિભાગ

17. પંચાયતોની જોગવાઈઓ (અનુસૂચિત ક્ષેત્રોમાં વિસ્તરણ) અધિનિયમ, 1996 ની જોગવાઈઓ વન વિસ્તારમાં કયો અધિકાર પૂરો પાડે છે?

A. નાના વન ઉત્પાદકની માલિકી
B. જંગલોમાં લાકડાની માલિકી
C. વન વિસ્તારમાં ખેતીની માલિકી
D. અભયારણ્ય માં માછીમારીનો અધિકાર

18. સૌરાષ્ટ્ર ફેલિંગ એક્ટ 1951 મુજબ ગુજરાતમાં કયું વૃક્ષ રિઝર્વડ કેટેગરીમાં સામેલ નથી?

A. સાગ
B. સીસમ
C. પીપળો
D. ખેર

19. વિકાસના કામો માટે જંગલની જમીનના પરિવર્તન માટે "નેટ પ્રેઝન્ટ વેલ્યુ" કેવી રીતે નક્કી કરવામાં આવે છે?

A. કેન્દ્ર સરકાર દ્વારા નિયત કરેલ જંગલોના પ્રકારનાં આધારે
B. જમીનના વ્યાપારી દર મુજબ
C. કેસના આધારે મહેસૂલ અધિકાર દ્વારા નક્કી કરવું પડે છે
D. વન અધિકારીએ તે વિસ્તારના પ્રવર્તમાન દર મુજબ નક્કી કરેલ મૂલ્ય

20. વન અધિકાર કાયદો, 2006 ક્યાં લાગુ થશે નહી?

A. રાષ્ટ્રીય ઉદ્યાન
B. વન્યપ્રાણી અભયારણ્ય
C. ક્રિટિકલ એરિયા
D. અનામત વનો

16. Zoological park or Zoo are being recognized and managed as per guidelines of;

A. State Forest Department
B. Animal husbandry Department
C. Central Zoo Authority
D. Municipal Department

17. The provisions of the Panchayats (extension to the scheduled areas) Act, 1996 provides rights in the forest area of;

A. Ownership of Minor Forest Produce
B. Ownership of Timber in the Forests
C. Ownership of Cultivation in forest area
D. Right to fishing in Sanctuary area

18. As per Saurastra Felling of Trees Act 1951, which is not the Reserved Tree in Gujarat

A. Teak
B. Shisham
C. Pipal
D. Khair

19. "Net Present Value" in diversion of forest land for development works has to be paid, the "Net Present Value" will be

A. The value as fixed by the Central Govt. on the basis of type of forests,
B. The value as per commercial rate of the land
C. The value that has to be fixed by revenue authority on case to case basis
D. The Forest officer has to decide the value as per prevailing rate of that area

20. Forest right Act, 2006 will not be applicable in

A. National Park
B. Wildlife Sanctuary
C. Critical Areas
D. Reserved Forests

21. વન (સંરક્ષણ) અધિનિયમ 1980 હેઠળ વળતર વનીકરણનો અર્થ શું થાય છે?
A. ડીગ્રેડ થયેલા જંગલોનું નવીનીકરણ
B. ગેરકાયદેસર કટીંગ દ્વારા નાશ પામેલા જંગલોને વનીકરણ
C. બિન-વન પ્રવૃત્તિઓ માટે વન જમીનના ફેરબદલનું નવીનીકરણ
D. વન વિભાગના માળખાકીય વિકાસને વળતર આપવા વનીકરણ

22. દેશના જંગલોના વિસ્તારો પર વન (સંરક્ષણ) કાયદો 1980 ની શું અસરો જોવા મળે છે?
A. સંરક્ષિત વનસ્પતિ અને પ્રાણીસૃષ્ટિ
B. સુરક્ષિત વોટરશેડ વિસ્તારો
C. જંગલની જમીનને બિન-વનીકરણ પ્રવૃત્તિઓમાં મોટા પ્રમાણમાં ફેરવવાની ગતિને ઘણી હદ સુધી ઘટાડી છે.
D. વન વિસ્તારોમાં વધારો કરવામાં આવ્યો છે

23. કયા વર્ષમાં વન (સંરક્ષણ) અધિનિયમ લાગુ કરવામાં આવ્યો હતો?
A. 1988 B. 1972
C. 1980 D. 1927

24. સુપ્રીમ કોર્ટના કયા આદેશથી દેશના વન વિસ્તારોના સંરક્ષણ ઉપર મોટો પ્રભાવ પડ્યો?
A. 202/1995 B. 208/1997
C. 206/1996 D. 85/19805

25. નેશનલ બાયોડાઇવરસીટી ઓથોરિટી નું મુખ્ય મથક ક્યાં આવેલું છે?
A. નવી દિલ્હી B. કોલકાતા
C. ચેન્નાઈ D. મુંબઈ

26. નેશનલ બાયોડાઇવરસીટી ઓથોરિટી ના વર્તમાન અધ્યક્ષ કોણ છે?
A. ડો વી બી માથુર
B. સંજયકુમાર
C. ડો અસદ આર રહેમાની
D. એ એસ રાવત

21. Compensatory afforestation under Forest (Conservation) Act 1980 means;
A. Afforestation to compensate the degraded forests
B. Afforestation to compensate the forests destroyed by illicit cutting
C. Afforestation to compensate the diversion of forest land for non-forest activities
D. Afforestation to compensate the infrastructural development of forest department

22. Impact of Forest (Conservation) Act 1980 on forests areas of Country
A. Conserved the flora & fauna
B. Conserved the watershed areas
C. Reduced the pace of diversion of forest land to non-forestry activities to a great extent.
D. Forest areas have been increased

23. In which year Forest (Conservation) Act was enacted
A. 1988 B. 1972
C. 1980 D. 1927

24. Which order of the Supreme Court had the great impact over conservation of Forest areas in Country
A. 202/1995 B. 208/1997
C. 206/1996 D. 85/1980

25. The Headquarter of National Biodiversity Authority is situated
A. New Delhi B. Kolkata
C. Chennai D. Mumbai

26. The present Chairman of National Biodiversity authority is
A. Dr. V. B. Mathur
B. Sanjay Kumar
C. Dr. Asad R. Rahmani
D. A.S. Rawat

27. જૈવ વિવિધતા અધિનિયમ, 2002 નો મુખ્ય હેતુ શું છે?

A. જૈવિક વિવિધતા જાળવણી
B. તેના ઘટકોનો ટકાઉ ઉપયોગ અને વાજબી અને ન્યાયી વહેંચણી
C. જૈવિક સંસાધનનું પરિવહન અટકાવવું
D. જૈવિક સંસાધનોના ઉપયોગ પર પ્રોત્સાહન

28. કયા કાયદા હેઠળ, સુરક્ષિત વિસ્તારોની આસપાસના ઇકો સેન્સિટિવ ઝોનને સૂચિત કરવામાં આવી રહ્યું છે?

A. વન (સંરક્ષણ) અધિનિયમ 1980
B. ભારતીય વન અધિનિયમ 1927
C. વન અધિકાર અધિનિયમ 2006
D. પર્યાવરણ (સંરક્ષણ) અધિનિયમ 1986

29. ગુજરાતના વન વિભાગને વ્હેલ શાર્ક સંરક્ષણ માટે ઈન્ડિયા બાયોડાયવર્સિટી એવોર્ડ્સ 2014 નો રનર અપ એવોર્ડ આપવામાં આવ્યો તે વિભાગનું નામ શું છે?

A. ગીર (પશ્ચિમ) વિભાગ, જૂનાગઢ
B. ગીર (પૂર્વ) વિભાગ, ધારી
C. પ્રાદેશિક વન વિભાગ, જૂનાગઢ
D. સામાજિક વનીકરણ વિભાગ, ગીર-સોમનાથ

30. વ્હેલ શાર્ક કન્ઝર્વેશનમાં, જેમણે સ્થાનિક માછીમારોને પ્રોત્સાહિત કરવા મહત્વપૂર્ણ ભૂમિકા ભજવી છે.

A. શ્રી મોરારી બાપુ B. શ્રી રામદેવજી
C. સોમનાથ ટ્રસ્ટ D. ગિરનાર ટ્રસ્ટ

31. સેન્ચ્યુરી એશિયા એવોર્ડ દર વર્ષે શ્રેષ્ઠ સંચાલિત અભયારણ્ય અને રાષ્ટ્રીય ઉદ્યાનને આપવામાં આવે છે ગીર રાષ્ટ્રીય અભયારણ્યને કયા વર્ષમાં તેના શ્રેષ્ઠ સંચાલન માટેનો એવોર્ડ મળ્યો છે?

A. 2013 B. 2014
C. 2015 D. 2019

27. National Biological diversity Act, 2002 aims to

A. Preservation of biological diversity
B. sustainable use of its components and fair and equitable sharing
C. No transit of Biological resource
D. Promotion on uses of biological resources

28. Under which Act, the Eco sensitive zone around protected areas are being notified?

A. Forest (Conservation) Act 1980
B. Indian Forest Act 1927
C. Forest Right Act 2006
D. Environment (Protection) Act, 1986

29. The runner up award of India Biodiversity Awards 2014, has been given for the Whale shark conservation to forest division of Gujarat. Name of the forest division?

A. Gir (West) Division, Junagadh
B. Gir (East) Division, Dhari
C. Territorial Forest Division, Junagadh
D. Social Forestry Division, Gir- Somnath

30. In Whale Shark Conservation, who has played an important role to motivate the local fishermen?

A. Shri Morari Bapu B. Shri Ramdev
C. Somnath Trust D. Girnar Trust

31. Sanctuary Asia award is being given every year to the best managed Sanctuary and National Park in the Country, The Gir National & Sanctuary has got the award for its best management in the year?

A. 2013 B. 2014
C. 2015 D. 2019

32. ગુજરાતમાં રાજ્યમાર્ગોના વિસ્તાર કરવા માટે વન (સંરક્ષણ) અધિનિયમ 1980 હેઠળ પરવાનગી લેવાની રહે છે, કારણ કે
A. આ વિસ્તારોમાં વન વિભાગના વાવેતર છે
B. આ વિસ્તારોને સુરક્ષિત વનો તરીકે જાહેર કરવામાં આવ્યા છે
C. આ વિસ્તારોને ગામલોકોના જંગલો તરીકે જાહેર કરવામાં આવ્યા છે
D. આ ક્ષેત્રો ઇકોલોજીકલ સંવેદનશીલ છે

33. નીલગાય ખેડૂતના પાકને મોટે ભાગે નુકસાન પહોંચાડે છે, તેથી તેની વસ્તીને નિયંત્રિત કરવા માટે કયા અધિનિયમ હેઠળ વર્મિન જાહેર કરવામાં આવ્યું છે?
A. વન્યપ્રાણી (સંરક્ષણ) અધિનિયમ 1972
B. ભારતીય વન અધિનિયમ 1927
C. વન અધિકાર અધિનિયમ 2006
D. પર્યાવરણ (સંરક્ષણ) અધિનિયમ 1986

34. જ્યારે કોઈ જંગલી પ્રાણી માનવ જીવન માટે જોખમી બને છે, ત્યારે તે પ્રાણીની હત્યા કરવા માટે કોની પરવાનગી લેવી જરૂરી હોય છે?
A. વિભાગીય વન અધિકારી
B. મુખ્ય વન સંરક્ષક
C. ચીફ વાઇલ્ડલાઇફ વૉર્ડન
D. વન સચિવ

35. જો વન્યપ્રાણી (સંરક્ષણ) અધિનિયમ 1972 અંતર્ગત શેડ્યૂલ-I કેટેગરીના જંગલી પ્રાણીને અન્ય વિસ્તારોમાં સ્થાનાંતરિત કરવા માટે કઈ માર્ગદર્શિકાનું પાલન કરવામાં આવે છે?
A. આઇયુસીએન માર્ગદર્શિકા
B. વન્યપ્રાણી માર્ગદર્શિકા
C. સીઝેડએ માર્ગદર્શિકા
D. રાજ્ય માર્ગદર્શિકા

32. For widening of State highways in Gujarat, the permission has to be taken under Forest (Conservation) Act 1980 because
A. These areas are having the plantations of forest department
B. These areas have been declared as protected forests
C. These areas have been declared as village forests
D. These areas are ecological sensitive

33. Blue bulls are mostly damaging the farmer's crops, so to control its population under which act these animals can be declared as vermin?
A. Wildlife (Protection) Act 1972
B. Indian Forest Act 1927
C. Forest Right Act 2006
D. Environment (Protection) Act 1986

34. When any wild animal becomes dangerous to human life, that individual wild animal can be killed with the permission of?
A. Divisional Forest Officer
B. Chief Conservator of Forests
C. Chief Wildlife Warden
D. Secretary, Forests

35. If any wild animal of schedule I category under Wildlife (Protection) Act 1972 is to be translocated to other areas, which guidelines are to be followed
A. IUCN guidelines
B. Wildlife guidelines
C. CZA guidelines
D. State guidelines

36. 2008 માં કચ્છ બાયોસ્ફિયર રિઝર્વને કયા નિયમ હેઠળ જાહેર કરવામાં આવ્યું?

A. વન્યપ્રાણી (સંરક્ષણ) અધિનિયમ 1972
B. ભારતીય વન અધિનિયમ 1927
C. યુનેસ્કો, એમએબી (MAB) પ્રોગ્રામ
D. પર્યાવરણ (સંરક્ષણ) અધિનિયમ 1986

37. કચ્છ બાયોસ્ફિયર રિઝર્વની જાહેરાત કયા વર્ષમાં કરવામાં આવી હતી?

A. 2009 B. 2007
C. 2008 D. 2010

38. વન (સંરક્ષણ) અધિનિયમ 1980 ના ઉલ્લંઘન હેઠળ, જવાબદાર પ્રભારી વન સંબંધિત અધિકારીને કેટલા સમય માટે સજા થઈ શકે છે?

A. 6 મહિના B. 3 મહિના
C. 15 દિવસ D. એક મહિનો

39. દેશમાં એક પ્રાણી સંગ્રહાલયમાંથી બીજા પ્રાણી સંગ્રહાલયમાં જંગલી પ્રાણીઓના સ્થળાંતર કરવા માટે કોની પરવાનગી લેવી ફરજિયાત છે?

A. પર્યાવરણ અને વન અને જલવાયુ પરિવર્તન મંત્રાલય
B. બંને રાજ્યના મુખ્ય મંત્રી
C. સેન્ટ્રલ ઝૂ ઓથોરિટી
D. વાઇલ્ડલાઇફ ઇન્સ્ટિટ્યૂટ ઓફ ઇન્ડિયા

40. વન્યપ્રાણી (સંરક્ષણ) અધિનિયમ 1972 હેઠળ કઈ વસ્તુઓનું વ્યાપારિક વેચાણ નિષેધ છે?

A. મોરના પીંછા
B. હાથીદાંતના આભૂષણ
C. સાબરના શીંગ
D. ઉપર્યુક્ત તમામ

41. ભારતના બંધારણની 42માં સુધારા મુજબ, ક્યા વર્ષમાં વનો સહવર્તી સૂચિ (concurrent list) માં સમાવેશ કરવામાં આવ્યો?

A. 1976 B. 1978
C. 1980 D. 1972

36. Kutch Biosphere Reserve has been declared in 2008, under

A. Wildlife (Protection) Act 1972
B. Indian Forest Act 1927
C. UNESCO, MAB Programme
D. Environment (Protection) Act 1986

37. Kutch Biosphere Reserve was declared in the Year

A. 2009 B. 2007
C. 2008 D. 2010

38. Under violation of Forest (Conserva-tion) Act 1980, the officer concerned in charge of the forest land, is liable to punish with simple imprisonment, may extends to

A. 6 months B. 3 months
C. 15 days D. One month

39. Transferring of wild animals from one Zoo to another Zoo with in Country, needs the permission of

A. Ministry of Environment & Forests and Climate Change
B. Chief Minsters of both the State
C. Central Zoo Authority
D. Wildlife Institute of India

40. Commercial trading of the following is punishable under the Wildlife (Protec-tion) Act 1972 of

A. Feathers of the Peacock
B. Ivory ornaments
C. Naturally dropped antlers of deer
D. All of above

41. As per 42nd amendment to the constitu-tion of India, the forest has been included in the concurrent list in the year

A. 1976 B. 1978
C. 1980 D. 1972

42. વન્યપ્રાણી (સંરક્ષણ) અધિનિયમ હેઠળ કોને ચીફ વાઇલ્ડલાઇફ વોર્ડન (CWLW) તરીકે નિમણુંક આપવામાં આવે છે ?

A. રાજ્યના મુખ્યમંત્રી
B. વન મંત્રી
C. વરિષ્ઠ વન અધિકારી
D. સચિવ, વન

43. ભારતના બંધારણમાં 51 Aએ સુધારા મુજબ, ભારતના દરેક નાગરિકની મૂળભૂત ફરજ શું છે?

A. કુદરતી પર્યાવરણની સુરક્ષા અને સુધારણા
B. જાહેર સંપત્તિની રક્ષા કરવી અને હિંસાને રોકવી
C. દેશની સમૃદ્ધ વારસાનું મૂલ્ય અને જાળવણી
D. ઉપર્યુક્ત તમામ

44. ગુજરાત બાયોડાયવર્સીટી બોર્ડ ની સ્થાપના કયા વર્ષમાં કરવામાં આવી હતી?

A. 2006 B. 2005
C. 2002 D. 2010

45. ગુજરાત ઇકોલોજીકલ એજ્યુકેશન એન્ડ રિસર્ચ ફાઉન્ડેશન ની સ્થાપના ક્યારે કરવામાં આવી હતી?

A. 1982 B. 1980
C. 1985 D. 1990

46. ગુજરાત પર્યાવરણ વ્યવસ્થાપન સંસ્થા ની સ્થાપના ક્યારે કરવામાં આવી હતી?

A. 1982 B. 1980
C. 1985 D. 1999

47. ગુજરાત ઇકોલોજી કમિશન ની સ્થાપના ક્યારે કરવામાં આવી હતી?

A. 1992 B. 1980
C. 1985 D. 1990

48. લાકડા આધારિત ઉદ્યોગો (સ્થાપના અને નિયમન) વિષેની માર્ગદર્શિકા MoEFCC દ્વારા કયા વર્ષમાં જારી કરવામાં આવી?

A. 2016 B. 2015
C. 2017 D. 2018

42. The Chief Wildlife Warden (CWLW) is the statutory authority, under the Wildlife Protection Act of any State is

A. Chief Minister of the State
B. Forest Minister
C. Senior Rank Forest Officer
D. Secretary, Forests

43. As per 51A amendment to the constitution of India, it is fundamental duty of every citizen of India

A. to protect and improve the natural environment
B. to safeguard public property and to abjure violence
C. to value and preserve the rich heritage of Country
D. All above

44. Gujarat Biodiversity Board (GBB) was established in the year

A. 2006 B. 2005
C. 2002 D. 2010

45. Gujarat Ecological Education and Research Foundation (GEER) was established in;

A. 1982 B. 1980
C. 1985 D. 1990

46. Gujarat Environment Management Institute (GEMI) was established in

A. 1982 B. 1980
C. 1985 D. 1999

47. Gujarat Ecology Commission (GEC) was established in;

A. 1992 B. 1980
C. 1985 D. 1990

48. The Wood based Industries (Establishment and Regulation) Guidelines has been issued by the MoEFCC

A. 2016 B. 2015
C. 2017 D. 2018

49. ગુજરાત સરકારના વ્યવસાયના નિયમો 1990 અનુસાર વન અને પર્યાવરણ વિભાગને કયા વિષયો ફાળવવામાં આવેલા છે?

A. જંગલો અને વન્યજીવનનું વ્યવસ્થાપન
B. જંગલોની બહાર ઝાડનું આવરણ વધારવું
C. જંગલોના મહત્વ વિશે જાગૃતિ
D. ઉપર્યુક્ત તમામ

50. વન પેદાશોના પરિવહનના નિયમનની કાર્યવાહી કયા નિયમ હેઠળ કરવામાં આવે છે?

A. વન્યપ્રાણી (સંરક્ષણ) અધિનિયમ 1972
B. ભારતીય વન અધિનિયમ 1927
C. બોમ્બે ફોરેસ્ટ રૂલ્સ 1942
D. પર્યાવરણ (સંરક્ષણ) અધિનિયમ, 1986

51. રાષ્ટ્રીય વન્યપ્રાણી બોર્ડ ના અધ્યક્ષ કોણ છે?

A. પ્રધાન મંત્રી
B. રાષ્ટ્રપતિ
C. કેન્દ્રીય વન મંત્રી
D. વન મહાનિર્દેશક

52. સ્ટેટ બોર્ડ ફોર વાઇલ્ડલાઇફ ના સભ્ય સચિવ કોણ છે?

A. મુખ્યમંત્રી
B. વન મંત્રી
C. PCCF અને HoFF
D. CWLW

53. વાઇલ્ડ લાઇફ માટે રાજ્ય બોર્ડની કઈ ફરજ છે?

A. સંરક્ષિત વિસ્તારો તરીકે જાહેર કરાયેલ વિસ્તારોની પસંદગી અને સંચાલન
B. જંગલી જીવન અને નિર્દિષ્ટ છોડના સંરક્ષણ માટેની નીતિનું નિર્માણ
C. આદિજાતિ અને અન્ય લોકોની જરૂરિયાતોને ધ્યાનમાં રાખીને જંગલી જીવનની સુરક્ષાના પગલાં
D. ઉપર્યુક્ત તમામ

49. The Government of Gujarat Rules of Business, 1990 subjects allotted to the Forest and Environment Department are

A. management of forests and wildlife,
B. enhance tree cover outside forests
C. awareness about importance of forests
D. All above

50. Regulation of transit of forest produce is being dealt under the

A. Wildlife (Protection) Act 1972
B. Indian Forest Act 1927
C. Bombay Forest Rules 1942
D. Environment (Protection) Act 1986

51. The Chairperson of the National Board for Wildlife is

A. Prime Minister
B. President
C. Union Forest Minister
D. Director General, Forests

52. Who is the member secretary of State Board for Wildlife?

A. Chief Minister
B. Forest Minister
C. PCCF & HoFF, Forests
D. CWLW

53. The Duty of State Board for Wild Life is

A. in the selection and management of areas to be declared as protected areas
B. in formulation of the policy for conservation of the wild life and specified plants
C. measures to be taken for harmonizing the needs of the tribal and other dwellers of the forest with the protec-tion and conservation of wild life
D. All above

54. "કમ્પેનસેન્ટરી એફોરેસ્ટેશન ફંડ એક્ટ, 2016" ક્યા મંત્રાલયે લાગુ કર્યો છે?
A. પર્યાવરણ, વન અને જલવાયુ પરિવર્તન મંત્રાલય
B. કાયદો અને ન્યાય મંત્રાલય
C. આદિજાતિ બાબતોનું મંત્રાલય
D. ગ્રામીણ વિકાસ મંત્રાલય

55. ગ્રીન હાઇવે (વાવેતર, ટ્રાન્સપ્લાન્ટેશન, બ્યુટીફિકેશન એન્ડ મેઇન્ટેનન્સ) ની નીતિનો કાયદો કયા વર્ષમાં કરવામાં આવ્યો?
A. 2014 B. 2015
C. 2016 D. 2017

56. રાષ્ટ્રીય ગ્રીન ટ્રિબ્યુનલ એક્ટ કયા વર્ષમાં લાગુ કરવામાં આવ્યો?
A. 2014 B. 2015
C. 2010 D. 2017

57. વન સંસાધનોના સતત સંચાલન, સંરક્ષણ અને ઉપયોગ માટે અને તેમાં એકરૂપતા લાવવા દેશભરમાં વન વ્યવસ્થાપન આયોજન, MoEFCC એ એક સમાન રાષ્ટ્રીય કાર્યકારી યોજના કોડ માં સુધારો ક્યા વર્ષમાં કર્યો?
A. 2014 B. 2015
C. 2016 D. 2013

58. દેશના જંગલોનું વૈજ્ઞાનિક વ્યવસ્થાપન કરવા માટે દરેક વન વિસ્તારમાં શેની મંજૂરી હોવી જોઈએ?
A. મેનેજમેન્ટ પ્લાન
B. વર્કિંગ પ્લાન
C. ટ્રીટમેન્ટ પ્લાન
D. જૈવવિવિધતા નોંધણી અને નકશો

59. વન વિસ્તારોના વર્કિંગ પ્લાનમાં મેનેજમેન્ટ મેપનો સ્કેલ કેટલો રાખવામાં આવે છે?
A. 1: 12,500 B. 1: 1,25,000
C. 1: 50,000 D. 1: 25,000

54. Which Ministry has enacted the "Compensatory Afforestation Fund Act, 2016"
A. Ministry of Environment, Forests and Climate Change
B. Ministry of Law and Justice
C. Ministry of Tribal affairs
D. Ministry of Rural Development

55. Green Highways (Plantation, Transplantation, Beautification & Maintenance) Policy, is enacted in the year
A. 2014 B. 2015
C. 2016 D. 2017

56. National Green Tribunal Act is enacted in the year
A. 2014 B. 2015
C. 2010 D. 2017

57. To sustainably manage, conserve and utilize the forest resources and to bring uniformity in forest management planning across the country, MoEFCC adopted a uniform National Working Plan Code, revised in the year,
A. 2014 B. 2015
C. 2016 D. 2013

58. For scientific management of forests in country, every forest area should have the approved
A. Management Plan
B. Working Plan
C. Treatment Plan
D. Biodiversity Register and Map

59. In working plan of forest areas, the management maps are of the scale;
A. 1:12,500 B. 1:1,25,000
C. 1: 50,000 D. 1:25,000

60. કયા અધિનિયમ હેઠળ કોઈ પણ જંગલને માન્યતા પ્રાપ્ત યોજના વિના કામ કરવાની મંજૂરી ન હોવી જોઈએ?

A. રાષ્ટ્રીય વન નીતિ 1988
B. ભારતીય વન અધિનિયમ 1927
C. બોમ્બે ફોરેસ્ટ રૂલ્સ, 1942
D. પર્યાવરણ (સંરક્ષણ) અધિનિયમ, 1986

61. લાકડા આધારિત ઔદ્યોગિક એકમનું નવું લાઇસન્સ કોની પૂર્વ મંજૂરી વિના મંજૂર કરવામાં આવતું નથી?

A. વન દળના વડા
B. વન સચિવ
C. વનમંત્રી
D. રાજ્ય કક્ષાની સમિતિ

62. ચીફ વાઇલ્ડલાઇફ વોર્ડન કયા કાર્યો માટે અભયારણ્યમાં પ્રવેશવાની મંજૂરી આપી શકે છે.

A. ફોટોગ્રાફી
B. વૈજ્ઞાનિક સંશોધન
C. પ્રવાસન
D. ઉપર્યુક્ત તમામ

63. રિયો-ડી-જનેરો માં કન્વેનશન ઓન બાયોલોજીકલ ડાઇવર્સિટી અર્થ સમિટ કયા વર્ષમાં યોજાઈ હતી ?

A. 1992 B. 1995
C. 1991 D. 1993

64. જૈવિક સંસાધનો અને વિદેશી નાગરિકો, કંપનીઓ અને બિન-નિવાસી ભારતીયો (NRIs) ને બાયોલોજીકલ રિસોર્સસ અને પરંપરાગત જ્ઞાન મેળવવામાટે કોની પૂર્વ મંજૂરીની જરૂર છે?

A. સ્ટેટ બાયોડાઇવર્સિટી બોર્ડ
B. નેશનલ બાયોડાઇવર્સિટી ઓથોરિટી
C. નેશનલ બાયોડાઇવર્સિટી બોર્ડ
D. વન મહાનિદેશક

60. No forest should be permitted to be worked without an approved working plan by the competent authority, as per

A. National Forest Policy 1988
B. Indian Forest Act 1927
C. Bombay Forest Rules, 1942
D. Environment (Protection) Act, 1986

61. No new license to a wood based industrial unit shall be granted without prior approval of

A. Head of Forest Force
B. Secretary to Forests,
C. Minister, Forests
D. State Level Committee

62. The Chief Wildlife Warden may permit to enter in a sanctuary for

A. Photography
B. Scientific Research
C. Tourism
D. All above

63. The Convention on Biological Diversity (CBD) at Earth Summit in Rio de Janeiro was in the year?

A. 1992 B. 1995
C. 1991 D. 1993

64. Access to biological resources and traditional knowledge to foreign citizens, companies and Non-Resident Indians (NRIs) needs 'prior approval of;

A. State Biodiversity Board
B. National Biodiversity Authority
C. National Biodiversity Board
D. Director General, Forests

65. જૈવિક વિવિધતા અધિનિયમ 2002 મુજબ જૈવિક વિવિધતા નિયમો, પર્યાવરણ અને વન મંત્રાલયે કયા વર્ષ માં ઘડ્યા અને સૂચિત કર્યા?

A. 2004 B. 2003
C. 2002 D. 2005

66. જૈવિક સંસાધનો અને સંકળાયેલ પરંપરાગત જ્ઞાન ના લાભો ની વેહેંચણીનો કાયદો કયા અધિનિયમ હેઠળ ઘડવામાં આવ્યો?

A. વન્યપ્રાણી (સંરક્ષણ) અધિનિયમ 1972
B. ભારતીય વન અધિનિયમ 1927
C. જૈવવિવિધતા અધિનિયમ 2002
D. પર્યાવરણ (સંરક્ષણ) અધિનિયમ, 1986

67. પ્રિવેંશન ઓફ ક્રુએલટી ટુ એનિમલ એક્ટ કયા વર્ષમાં આવ્યો?

A. 1960 B. 1970
C. 1980 D. 1990

68. રાષ્ટ્રીય ગ્રીન ટ્રિબ્યુનલ ક્યાં આવેલ છે?

A. નવી દિલ્હી B. મુંબઈ
C. ચેન્નઈ D. બેંગલોર

69. વાયુ (પ્રદૂષણ નિવારણ અને નિયંત્રણ) અધિનિયમ ક્યારે અમલમાં આવ્યો?

A. 1981 B. 1991
C. 1971 D. 2001

70. પ્રોટેકશન ઓફ પ્લાન્ટ વેરાઈટીસ એન્ડ ફાર્મર્સ રાઈટ એક્ટ 2001 નો મુખ્ય હેતુ શું છે

A. છોડની જાતોના રક્ષણ માટે અસરકારક પ્રણાલી સ્થાપિત કરવી
B. દેશમાં કૃષિ વિકાસને વેગ આપવા
C. દેશમાં બીજ ઉદ્યોગના વિકાસની સુવિધા આપે છે
D. ઉપર્યુક્ત તમામ

71. રાષ્ટ્રીય પર્યાવરણ નીતિ કયારે ઘડવામાં આવી હતી?

A. 2006 B. 2007
C. 2008 D. 2009

65. The Biological Diversity Rules, in accordance with the Biological Diversity Act, 2002 were formulated and notified by the Ministry of Environment & Forests in the Year

A. 2004 B. 2003
C. 2002 D. 2005

66. To access Biological resources and associated traditional knowledge, the benefit sharing concept is being enforced under the Act

A. Wildlife (Protection) Act 1972
B. Indian Forest Act 1927
C. Biological Diversity Act 2002
D. Environment (Protection) Act, 1986

67. Prevention of cruelty to Animals act enacted in the year

A. 1960 B. 1970
C. 1980 D. 1990

68. National Green Tribunal is located in

A. New Delhi B. Mumbai
C. Chennai D. Bangalore

69. The Air (Prevention and Control of Pollution) Act, enacted in

A. 1981 B. 1991
C. 1971 D. 2001

70. Protection of Plant Varieties and Farmers' Rights Act, 2001, the objective of

A. To establish an effective system for the protection of plant varieties
B. To accelerate agricultural development in the country
C. Facilitate the growth of seed industry in the country
D. All above

71. National Environment Policy formulated in

A. 2006 B. 2007
C. 2008 D. 2009

72. આર્થિક અને સામાજિક વિકાસમાં પર્યાવરણીય ચિંતાઓનું એકીકરણ એ શેનો એક મુખ્ય હેતુ છે?

A. રાષ્ટ્રીય પર્યાવરણ નીતિ 2006
B. રાષ્ટ્રીય વન નીતિ 1988
C. પર્યાવરણ (સંરક્ષણ) અધિનિયમ 1986
D. રાષ્ટ્રીય ગ્રીન ટ્રિબ્યુનલ એક્ટ 2010

73. પર્યાવરણ પર યોજનાના પોઝિટિવ અને નેગેટિવ અસરોનું મૂલ્યાંકન એન્વિરોન્મેન્ટલ ઈમ્પેક્ટ એસેસમેન્ટ (EIA) ક્યારે કરવામાં આવે છે?

A. પ્રોજેક્ટ શરૂ થતાં પહેલાં
B. પ્રોજેક્ટની મધ્યમાં
C. પ્રોજેક્ટ પૂર્ણ થયા બાદ
D. ચાલુ પ્રોજેક્ટ પર કોઈપણ સમયે

74. વન અધિકાર અધિનિયમ 2006 કયા કેન્દ્રીય મંત્રાયલ દ્વારા પ્રસિદ્ધ કરવામાં આવેલ હતો?

A. પર્યાવરણ, વન અને જલવાયુ પરિવર્તન મંત્રાલય
B. આદિજાતિ બાબતોનું મંત્રાલય
C. ગ્રામીણ વિકાસ મંત્રાલય
D. કાયદો અને ન્યાય મંત્રાલય

75. વન અધિકાર અધિનિયમ અનુસૂચિત જનજાતિઓ અને પરંપરાગત વન રહેવાસીઓને કયા વન અધિકારોને માન્યતા આપે છે અને હસ્તગત કરે છે?

A. વન આશ્રિત સમુદાયોની આજીવિકા અને ખાદ્ય સુરક્ષાની ખાતરી
B. વન ગામો માટે મૂળભૂત વિકાસ સુવિધાઓ પૂરી પાડવી
C. સમુદાયોને વન રક્ષણ, સંરક્ષણ અને વ્યવસ્થાપન માટે સશક્ત બનાવો
C. ઉપર્યુક્ત તમામ

76. અનુસૂચિ-I ના પ્રાણી ઘુડખરની ગુજરાતમાં વસ્તી કેટલી છે?

A. 6082 B. 5082
C. 4082 D. 5582

72. Integration of Environmental Concerns in Economic and Social Development is one of the main objective of the;

A. National Environment Policy 2006
B. National Forest Policy 1988
C. Environment (Protection) Act 1986
D. National Green Tribunal Act 2010

73. An Environmental Impact Assessment (EIA) is an assessment of the possible positive or negative impact that a project may have on the environment, has to be done;

A. Before commencement of the project
B. In the mid of the project
C. After completion of the project.
D. At any time on ongoing project

74. Forest Right Act 2006 backed by the union Ministry of;

A. Ministry of Environment, Forests and Climate Change
B. Ministry of Tribal Affairs
C. Ministry of Rural Development
D. Ministry of Law and Justice

75. Forest Rights Act recognizes and vests forest rights in the scheduled tribes and other traditional forest dwellers on

A. Ensure livelihood and food security of the forest dependent communities
B. Provide for basic developmental facilities for the forest villages
C. Empower communities to protect, conserve and manage forests
D. All above

76. What is the population of Schedule-I animal Wild Ass in Gujarat?

A. 6082 B. 5082
C. 4082 D. 5582

77. ગુજરાત પર્યાવરણ પ્રબંધન સંસ્થા એ 1 ફેબ્રુઆરી, 1999 ના રોજ સ્થાપિત સ્વાયત સંસ્થા છે , જેનું મુખ્ય મથક _______ શહેરમાં આવેલું છે.

A. વડોદરા B. અહમદાબાદ
C. સુરત D. ગાંધીનગર

78. વન્યપ્રાણી (સંરક્ષણ) અધિનિયમ અંતર્ગત અનુસૂચિ I નુ પ્રાણી ઘુડખર કુદરતી રીતે કયા જિલ્લામાં જોવા મળે છે?

A. અમદાવાદ B. જામનગર
C. રાજકોટ D. બોટાદ

79. ગુજરાતમાં જૈવવિવિધતા સંરક્ષણ અને ગ્રામીણ આજીવિકા સુધારણા પ્રોજેક્ટ (BCRLIP) અંતર્ગત કયા રક્ષિત ક્ષેત્ર લાગુ કરવામાં આવી રહ્યા છે?

A. વાંસદા રાષ્ટ્રીય ઉદ્યાન
B. નારાયણ સરોવર અભ્યારણ્ય
C. નાના રણની લેન્ડસ્કેપ – ઘુડખર અભયારણ્ય
D. મરીન રાષ્ટ્રીય ઉદ્યાન

80. સુપ્રીમ કોર્ટે 4 ડિસેમ્બર, 2006 ના આદેશ અનુસાર , MoEF ને રાજ્યો / કેન્દ્રશાસિત કેન્દ્રો પાસેથી ઇકો સેન્સેટિવ ઝોન માટેની દરખાસ્તો માટે નિર્દેશ આપ્યો, અન્યથા કોઈપણ રક્ષિત વિસ્તારની આસપાસનો કેટલો વિસ્તાર ઇકો સેન્સેટિવ ઝોન વિસ્તાર હશે?

A. 5.00 કિ.મી. ની અંદર
B. 10.00 કિ.મી. ની અંદર
C. 2.00 કિ.મી. ની અંદર
D. 1.00 કિ.મી. ની અંદર

81. ઇકો સેન્સેટિવ ઝોનમાં ખાણકામ એ _____ છે

A. નિયમિત પ્રવૃત્તિ
B. પરવાનગી પ્રવૃત્તિ
C. પ્રતિબંધિત પ્રવૃત્તિ
D. આમાંથી કોઈ નહિ

77. Gujarat Environment Management Institute (GEMI) is an Autonomous Institute set up on 1st February, 1999, the HQ is at

A. Vadodara B. Ahmedabad
C. Surat D. Gandhinagar

78. Schedule I animal Wild Ass naturally found in the district

A. Ahmedabad B. Jamnagar
C. Rajkot D. Botad

79. In Gujarat, Biodiversity Conservation and Rural Livelihood Improvement Project (BCRLIP) is being implemented in the Protected Area;

A. Vansda National Park
B. Narayan Sarovar Sanctuary
C. Little Rann landscape - Wild Ass sanctuary
D. Marine National Park

80. Hon'ble Supreme Court vide order 4th December 2006, directed the MoEF to ask the proposals for ESZ from States/UTs, otherwise the ESZ area around any PA will be;

A. Within 5.00 Kms
B. Within 10.00 Kms
C. Within 2.00 Kms
D. Within 1.00 Kms

81. Commercial mining within Eco sensitive zone is a

A. Regulated activity
B. Permitted activity
C. Restricted activity
D. None of these

82. કયા સંરક્ષિત ક્ષેત્રમાં સુરક્ષાની દ્રષ્ટિએ ખુબજ કડક કાયદા હોય છે?
A. કંઝર્વેશન રિઝર્વ
B. કમ્યુનિટી રિઝર્વ
C. રાષ્ટ્રીય ઉદ્યાન
D. અભયારણ્ય

83. કયું ક્ષેત્ર વન્યપ્રાણી (સંરક્ષણ) અધિનિયમ, 1972 હેઠળ સૂચિત નથી?
A. રાષ્ટ્રીય ઉદ્યાનો
B. કમ્યુનિટી રિઝર્વ
C. કંઝર્વેશન રિઝર્વ
D. બાયોસ્ફિયર રિઝર્વ

84. વન (સંરક્ષણ) અધિનિયમ 1980 શેના પર લાગુ પડે છે?
A. અનામત જંગલો B. સંરક્ષિત જંગલો
C. ગામના જંગલો D. ઉપર્યુક્ત તમામ

85. સુપ્રીમ કોર્ટ દ્વારા ઇકો સેન્સેટિવ ઝોન કયા વિસ્તારોમાં લાગુ કરવામાં આવી રહ્યા છે?
A. અનામત જંગલો
B. રક્ષિત જંગલો
C. રક્ષિત વિસ્તારો
D. બાયોસ્ફિયર રિઝર્વ

86. 1956 ની શરૂઆતમાં ગીરને ક્યા કાયદા હેઠળ અભયારણ્ય જાહેર કરવામાં આવ્યું ?
A. વન્યપ્રાણી (સંરક્ષણ) અધિનિયમ
B. ગુજરાત જંગલી પ્રાણી અને જંગલી પક્ષી સુરક્ષા કાયદો
C. ભારતીય વન અધિનિયમ
D. વન (સંરક્ષણ) અધિનિયમ

87. વન મજૂર સહકારી મંડળીઓ મુખ્યત્વે શેના માટે કાર્યરત છે?
A. ઇમારતી કામગીરી
B. સહભાગી વન સંચાલન
C. જંગલમાં નરેગા અંતર્ગત મજૂરો
D. એક પણ નહિ

82. In which protected area, most strict as far as protection is concerned;
A. Conservation Reserve
B. Community Reserve
C. National Park
D. Sanctuary

83. Which area is not notified under Wildlife (Protection) Act, 1972;
A. National Park
B. Community Reserve
C. Conservation Reserve
D. Biosphere Reserve

84. The Forest (Conservation) Act 1980 applicable on the;
A. Reserved forests B. Protected Forests
C. Village Forests D. All of above

85. Eco sensitive zones are being enforced by Supreme Court around
A. Reserved forests
B. Protected Forests
C. Protected Areas
D. Biosphere Reserve

86. The Gir initially declared as Sanctuary in 1965 under
A. Wildlife (Protection) Act
B. Gujarat Wild animal and wild Bird Protection Act
C. Indian Forest Act
D. Forest (Conservation) Act

87. Forest labour cooperative societies are mainly working on
A. Timber operations
B. Joint forest management
C. Labour works under NREGA in forest are
D. None of them

88. સૌરાષ્ટ્ર ફેલિન્ગ એક્ટ 1951 અનુસાર વેસ્ટલેન્ડમાં અનધિકૃત વૃક્ષોને કાપવા માટે કયો વિભાગ જવાબદાર છે?

A. વન વિભાગ
B. મહેસૂલ વિભાગ
C. પોલીસ વિભાગ
D. પંચાયત વિભાગ

89. જંગલોના સંચાલન માટે વનનો વિષય બંધારણની _______ હેઠળ છે.

A. રાજ્ય સૂચિ
B. યુનિયન સૂચિ
C. સમવર્તી સૂચિ
D. કેન્દ્રની સૂચિ

90. કયા બંધારણ સુધારણામાં વન વિષયને સહવર્તી સૂચિમાં સ્થાનાંતરિત કરવામાં આવ્યો છે?

A. 43 મો બંધારણ સુધારો અધિનિયમ, 1978
B. 42 મો બંધારણ સુધારો અધિનિયમ, 1976
C. 44 મો બંધારણ સુધારો અધિનિયમ, 1978
D. 45 મો બંધારણ સુધારો અધિનિયમ, 1980

91. શા માટે વન્યપ્રાણીઓને બચાવવા જરૂરી છે?

A. આપણા વાતાવરણને સ્વસ્થ રાખવામાં મદદ કરે છે
B. ઇકોલોજીકલ સંતુલન જાળવવા માટે મહત્વપૂર્ણ
C. જૈવવિવિધતાને જાળવવામાં મદદ કરે છે
D. ઉપર્યુક્ત તમામ

92. જંગલોને કઈ રીતે સંરક્ષિત કરી શકાય છે?

A. વનીકરણ કાર્યક્રમોનો પરિચય
B. વન આગને કાબૂમાં રાખી
C. વન સંસાધનોનો યોગ્ય ઉપયોગ
D. ઉપર્યુક્ત તમામ

88. As per Saurastra Felling of Trees Act, 1951, which department is responsible for unauthorised cutting of trees in common wastelands

A. Forest department
B. Revenue department
C. Police department
D. Panchayat department

89. For the management of forests, the subject of the forest is under the _____ of constitution.

A. State List
B. Union List
C. Concurrent List
D. Central list

90. In which constitution amendment, the forest subject has been transferred to concurrent list;

A. 43rd Constitution Amendment Act, 1978
B. 42nd Constitution Amendment Act, 1976
C. 44th Constitution Amendment Act, 1978
D. 45th Constitution Amendment Act, 1980

91. Why is it necessary to save our wildlife?

A. Helps keep our environment healthy
B. Important to maintain ecological balance
C. Helps preserve biodiversity
D. All above

92. The forests can be conserved in the following manner

A. Introducing afforestation programs
B. Controlling forest fires
C. Proper utilization of forest resources
D. All above

93. ક્યુ વૃક્ષ ગુજરાતમાં રીઝર્વડ ટ્રી છે?
A. સાગ B. સીસમ
C. ખેર D. ઉપર્યુક્ત તમામ

94. કયા અધિનિયમ મુજબ કેન્દ્રની પૂર્વ મંજૂરી વિના વનીકરણ માટે વન જમીનને સ્થાનાંતરિત કરી શકાતી નથી?
A. ભારતીય વન અધિનિયમ, 1927
B. વન (સંરક્ષણ) અધિનિયમ 1980
C. વન નીતિ, 1988
D. વન્યપ્રાણી (સંરક્ષણ) અધિનિયમ 1972

95. "ડાંગ દરબાર" માં સ્થાનિક લોકોને શેના માટે એવોર્ડ આપવામાં આવે છે?
A. વન અગ્નિ નિયંત્રિત કરવા અને મદદ કરવા માટે
B. વન્યપ્રાણી શિકારને નિયંત્રિત કરવા
C. સહભાગી વન વ્યવસ્થાપન
D. અપરાધીઓથી વનનું રક્ષણ

96. સ્થાનિક લોકોને "યોગ્ય અને ન્યાયી લાભ વહેંચણી" કયા અધિનિયમ હેઠળ મૂકવામાં આવી છે?
A. રાષ્ટ્રીય વન નીતિ 1988
B. જૈવવિવિધતા અધિનિયમ 2002
C. ભારતીય વન અધિનિયમ 1927
D. વન અધિકાર અધિનિયમ2006

97. "પીપલ્સ બાયોડાઇવર્સીટી રજિસ્ટર" માં શેના પર વિસ્તૃત માહિતી શામેલ છે?
A. સ્થાનિક જૈવિક સંસાધનો
B. ઔષધીય અથવા અન્ય કોઈ ઉપયોગ
C. સંકળાયેલ પરંપરાગત જ્ઞાન
D. ઉપર્યુક્ત તમામ

98. "પીપલ્સ બાયોડાઇવર્સીટી રજિસ્ટર" તૈયાર કરવાની ફરજ કોની હોય છે?
A. સહભાગી વન વ્યવસ્થાપન સમિતિ
B. જૈવવિવિધતા વ્યવ્યસ્થાપન સમિતિ
C. ઇકો ડેવલપમેન્ટ કમિટી
D. ઇકો ટૂરિઝમ ડેવલપમેન્ટ કમિટી

93. Which is the Reserved tree in Gujarat
A. Teak B. Shisham
C. Khair D. All of above

94. Forest land cannot be transfer for non-forestry use without prior approval of central Government, as per the Act;
A. Indian Forest Act, 1927
B. Forest (Conservation) Act 1980
C. Forest Policy, 1988
D. Wildlife (Protection) Act 1972

95. In "Dangs darbar" the local villagers are being awarded for;
A. To control and help in forest fires
B. To control the wildlife poaching
C. Joint Forest Management
D. Protection of forest from offenders

96. "Fair and equitable benefit sharing" to the local peoples, is being enforced under;
A. National Forest Policy 1988
B. Biological Diversity Act 2002
C. Indian Forest Act 1927
D. Forest Right Act 2006

97. People's Biodiversity Register, contain comprehensive information on:
A. knowledge of local biological resources,
B. Medicinal or any other use
C. Traditional knowledge associated
D. All above

98. Preparation of People's Biodiversity Register is the duty of;
A. Joint Forest Management Committee
B. Biodiversity Management Committee
C. Eco Development Committee
D. Eco Tourism Development Committee

99. **પર્યાવરણનું રક્ષણ અને સુધારણા તેમજ જંગલો અને વન્યજીવોની સુરક્ષા એ બંધારણના કયા અનુચ્છેદમાં સમાવિષ્ટ છે?**

A. 48 A B. 51 A
C. 58 A D. 61 A

100. **વન વિભાગ અને આદિજાતિ સમુદાયો વચ્ચે સહજીવન વિશેની બાબત કઈ વન નીતિમાં ભાર મુકવામાં આવેલ?**

A. 1894 B. 1952
C. 1988 D. 1998

99. **Protection and improvement of environment and safeguarding of forests and wild life is in which article of constitution?**

A. 48 A B. 51 A
C. 58 A D. 61 A

100. **In which forest policy forest department and tribal's symbiotic relation was given emphasis?**

A. 1894 B. 1952
C. 1988 D. 1998

Answers | જવાબો

Que.	Ans.	Que.	Ans.	Que.	Ans.	Que.	Ans.	Que.	Ans.	Que.	Ans.	Que.	Ans.
1.	A	16.	C	31.	A	46.	D	61.	D	76.	A	91.	D
2.	C	17.	A	32.	B	47.	A	62.	D	77.	D	92.	D
3.	B	18.	C	33.	A	48.	A	63.	A	78.	A	93.	D
4.	A	19.	A	34.	C	49.	D	64.	B	79.	C	94.	B
5.	D	20.	C	35.	A	50.	C	65.	A	80.	B	95.	A
6.	B	21.	C	36.	C	51.	A	66.	C	81.	C	96.	B
7.	D	22.	C	37.	C	52.	D	67.	A	82.	C	97.	D
8.	B	23.	C	38.	C	53.	D	68.	A	83.	D	98.	B
9.	B	24.	A	39.	C	54.	B	69.	A	84.	C	99.	A
10.	C	25.	C	40.	D	55.	B	70.	D	85.	C	100.	C
11.	B	26.	A	41.	A	56.	C	71.	A	86.	B		
12.	D	27.	B	42.	C	57.	A	72.	A	87.	A		
13.	B	28.	D	43.	D	58.	B	73.	A	88.	B		
14.	A	29.	C	44.	A	59.	C	74.	B	89.	C		
15.	B	30.	A	45.	A	60.	A	75.	D	90.	B		

Chapter 15

પ્રકરણ ૧૫

Anthropogenic activities and Forests

માનવીય પ્રવૃત્તિઓ અને વનો

Branches of Forestry as per human use

Agroforestry	It is the integration of forests into agricultural systems in order to optimize the production and positive effects within the system and minimize negative side effects of farming
Analog Forestry	A management focus that seeks to establish a tree – dominated ecosystem that is similar in architectural structure and ecological function to the naturally occurring climax and sub – climax vegetation community
Boreal Forestry	It analyzes the particular challenges of forestry in the world's boreal regions
Community Forestry	Combination of forest conservation with rural development and poverty reduction objectives, accomplished through instating a legal framework that favors profitable and sustainable forest management
Dendrology	It involves the study and identification of economically useful tree species
Eco forestry or Environmental Forestry	It emphasizes tree planting practices which strive to protect and restore ecosystem
Energy Forestry	It includes specifically managing forests for the production of energy from biomass or biofuel derived from a fast-growing species of tree or woody shrub
Farm Forestry	The practice of cultivating and managing trees in compact blocks on agricultural lands
Forest Ecology	It studies the pattern and processes of a forest ecosystem
Forest Economics	It studies the impact of economics on forest management decision
Forest Hydrology	It embodies the effects of changes in forest land use on the movement, distribution and quality of water in the ecosystem
Forest Management	It comprises the overall administrative, economic, legal and social aspects of forest regulation
Forest Mensuration	It incorporates quantitative measurements of the forest stand to determine stand timber volume and productivity/ health and provides a basis of which management decision can be made
Forest Protection	It comprises both the biotic and abiotic maladies affecting the health of the forest or tree, primarily by fungus pathogens, insect, nematodes, parasitic plants and climatic factors (stress, salinity, fire etc.) and their management

Short Rotation Forestry	Managing a forest that utilizes fast growing species as a bio-based energy crop for domestic use and commercial use in power stations, alone or in combination with other fuels such as coal
Silviculture	It is the art and science of controlling the establishment, growth, composition, health and quality of forests to meet specific objectives
Social Forestry	It addresses human-forest interactions and importance of community based natural resource management
Tree Breeding	It is the method of genetically modifying or selecting forest stock for improved growth or vigor characteristics
Urban Forestry	It entails the care and management of urban tree populations for the purpose of improving the urban environment

માનવ વપરાશ મુજબ વાનિકીની શાખાઓ

કૃષિવાનિકી	સિસ્ટમમાં ઉત્પાદન અને હકારાત્મક અસરોને ઓપ્ટિમાઇઝ કરવા અને ખેતીની નકારાત્મક આડઅસર ઘટાડવા માટે તે કૃષિ સિસ્ટમોમાં જંગલોનું એકીકરણ છે
એનાલોગ ફોરેસ્ટ્રી	એક મેનેજમેન્ટ ફોકસ કે જે એક વૃક્ષ-પ્રભુત્વપૂર્ણ ઇકોસિસ્ટમ સ્થાપિત કરવા માંગે છે જે આર્કિટેક્ચરલ સ્ટ્રક્ચર અને ઇકોલોજીકલ ફંક્શનમાં કુદરતી રીતે ક્લાઈમેક્સ અને પેટા ક્લાઈમેક્સ સમુદાય માટે સમાન છે.
બોરિયલ ફોરેસ્ટ્રી	તે વિશ્વના બોરિયલ પ્રદેશોમાં વનીકરણના વિશેષ પડકારોનું વિશ્લેષણ કરે છે
કોમ્યુનિટી ફોરેસ્ટ્રી	ગ્રામીણ વિકાસ અને ગરીબી નિવારણના ઉદ્દેશો સાથે વન સંરક્ષણનું જોડાણ, નફાકારક અને ટકાઉ વન સંચાલનને અનુકૂળ એવા કાનૂની માળખાની રજૂઆત દ્વારા પરિપૂર્ણ
ડેંડ્રોલોજી	તેમાં આર્થિક રીતે ઉપયોગી વૃક્ષ પ્રજાતિઓનો અભ્યાસ અને ઓળખ શામેલ છે
ઇકો ફોરેસ્ટ્રી અથવા પર્યાવરણીય વનીકરણ	તે વૃક્ષારોપણની પદ્ધતિઓ પર ભાર મૂકે છે જે ઇકોસિસ્ટમના રક્ષણ અને પુનર્સ્થાપિત માટેના પ્રયત્નો કરે છે
એનર્જી ફોરેસ્ટ્રી	તેમાં ખાસ કરીને બાયોમાસ અથવા બાયોફ્યુઅલમાંથી ઉર્જાના ઉત્પાદન માટે જંગલોનું સંચાલન કરવાનો સમાવેશ થાય છે જે ઝડપથી વૃદ્ધિ પામેલા વૃક્ષો અથવા ઝાડમાંથી મેળવાયેલ બળતણ નો ઉપયોગ સામેલ છે
ફાર્મ ફોરેસ્ટ્રી	કૃષિ જમીનો પર કોમ્પેક્ટ બ્લોક્સમાં ઝાડની ખેતી અને તેનું વ્યવસ્થાપન
વન ઇકોલોજી	તે વન ઇકોસિસ્ટમની પદ્ધતિ અને પ્રક્રિયાઓનો અભ્યાસ કરે છે
વન અર્થશાસ્ત્ર	તે વન વ્યવસ્થાપનના નિર્ણય પર અર્થશાસ્ત્રના પ્રભાવનો અભ્યાસ કરે છે
ફોરેસ્ટ હાઇડ્રોલોજી	તે ઇકોસિસ્ટમમાં પાણીની હિલચાલ, વિતરણ અને ગુણવત્તા પર જંગલની જમીનના ઉપયોગમાં પરિવર્તનની અસરોનો સમાવેશ કરે છે
વન વ્યવસ્થાપન	તેમાં વન નિયમનના એકંદર વહીવટી, આર્થિક, કાનૂની અને સામાજિક પાસાઓનો સમાવેશ છે

ફોરેસ્ટ મેન્સ્યુરેશન	તે લાકડાના પ્રમાણ અને ઉત્પાદકતા / આરોગ્યને નિર્ધારિત કરવા માટે જંગલના સ્ટેન્ડના માત્રાત્મક માપનો સમાવેશ કરે છે અને જેના આધારે વ્યવસ્થાપનનો નિર્ણય લઈ શકાય તે આધાર પૂરો પાડે છે.
વન સંરક્ષણ	તેમાં જંગલ અથવા ઝાડના સ્વાસ્થ્યને અસર કરતી બાયોટિક અને એબાયોટિક દૂષણો બંને શામેલ છે, મુખ્યત્વે ફૂગ પેથોજેન્સ, ઉપદ્રવી જીવાત, નેમાટોડ્સ, પરોપજીવી છોડ અને આબોહવા પરિબળો (તાણ, ખારાશ, અગ્નિ વગેરે) અને તેમનુ વ્યવસ્થાપન.
શોર્ટ રોટેશન ફોરેસ્ટ્રી	એક જંગલનું વ્યવસ્થાપન જે ઝડપથી વિકસતી પ્રજાતિઓને બાયો-આધારિત ઊર્જા પાક તરીકે ઘરેલુ ઉપયોગ માટે અને પાવર સ્ટેશનોમાં એકલા અથવા કોલસા જેવા અન્ય ઇંધણ સાથે જોડાણમાં વ્યાપારી ઉપયોગ કરે છે.
સિલ્વીકલ્ચર	તે વિશિષ્ટ ઉદ્દેશોને પહોંચી વળવા માટે સ્થાપના, વિકાસ, રચના, આરોગ્ય અને જંગલોની ગુણવત્તાને નિયંત્રિત કરવાની કળા અને વિજ્ઞાન છે
સામાજિક વનીકરણ	તે માનવ-વન ક્રિયાપ્રતિક્રિયાઓ અને સમુદાય આધારિત કુદરતી સંસાધન સંચાલનનું મહત્વ સૂચવે છે
વૃક્ષ સંવર્ધન	સુધારેલ વૃદ્ધિ અથવા ઉત્તમ લાક્ષણિકતાઓ માટે વન સ્ટોકને આનુવંશિક રીતે સુધારવા કે પસંદ કરવાની તે પદ્ધતિ છે
શહેરી વનીકરણ	તે શહેરી વાતાવરણમાં સુધારો લાવવાના હેતુથી શહેરી વૃક્ષોની વસ્તીની સંભાળ અને વ્યવસ્થાપન કરે છે

Famous Books on Environment | પર્યાવરણ પર વિશ્વપ્રખ્યાત પુસ્તકો

A Blueprint for Survival	Edward Goldsmith and Robert Allen
A Hole in the Wind	David Goodrich
A Sand County Almanac	Aldo Leopald
Akoopar	Dhruv Bhatt
An Appeal to Reason	Nigel Lawson
An inconvenient truth	Al Gore
Are we smart enough to know how smart animals are?	Frans de Waal
Bee Quest	Dave Goulson
Black Faces, White Spaces	Carolyn Finney
Blackfish City	Sam J. Miller
Chesapeake Requiem	Earl Swift
Don't Even Think About It	George Marshall
Earth in the Balance	Al Gore
Elemental India	Meera Subramanian
Falter: Has the Human Game Begun to Play Itself Out?	Bill McKibben
From Soup to Superstar	Kartik Shanker
Green growth	Arun Firodiya

Green Wars	Bahar Dutt
Horizon	Barry Lopez
Indica	Pranay Lal
Irreplaceable	Julian Hoffman
Islands in Flux	Pankaj Sekhsaria
Jonathan Livingston Seagull	Richard Back
Jungle lore	Jim Corbett
Junk Raft	Marcus Eriksen
Live Sustainably Now	Karl Coplan
Losing Earth	Nathaniel Rich
Man-eaters of Kumaon	Jim Corbett
Merchants of Doubt	Dewey Decimal
My Husband and Other Animals	Janki Lenin
No one is too small to make a difference	Greta Thunberg
Pakshio ni Bhaibandhi	Lalsingh Raol
Plundering Paradise	Michael D'Orso
Rainforest	Tony Juniper
Rebirding	Benedict Macdonald
Red Sky at Morning	James Gustave Speth
Rewilding	Bahar Dutt
Rising	Elizabeth Rush
Silent Spring	Rachel Carson
Storming the Wall	Todd Miller
The Archipelago of Hope	Gleb Raygorodetsky
The Colors of Nature	Alison H. Deming and Lauret E. Savoy
The End of Nature	Bill McKibben
The End of the World	Peter Brannen
The Ends of the World	Peter Brannen
The Great Derangement	Amity Ghosh
The Greatest Show on Earth	Richard Dawkins
The Gulf	Jack E. Davis
The Himalayan Arc	Namita Gokhale
The Last Wave	Pankaj Sekhsaria
The Origin of Species	Charles Darwin
The Outlaw Ocean	Ian Urbina
The Secret Life of Tigers	Valmik Thapar
The Sixth Extinction	Elizabeth Kolbert
The Songs of Trees	David George Haskell
The Uninhabitable Earth	David Wallace-Wells

The Uninhabitable Earth	David Wallace
The Vanishing	Prerna Singh Bindra
The wall	John Lanchester
The Water Will Come	Jeff Goodell
The Wildest Sport of All	Prakash Singh
This Radical Land	Daegan Miller
Where the Water Goes	David Owen
Wolf Nation	Brenda Peterson
Working with Nature	Jeremy Purseglove

Colorful flowering trees in Urban areas

શહેરી વિસ્તારોમાં રંગબેરંગી ફૂલોના ઝાડ

Color of the flower	Tree species
Bright yellow	*Peltrophorum ferrugineum*
Flower with golden spots	*Amtherstia nobilis*
Mauve	*Pongamia glabra*
Orange crimson	*Spathodea companulata*
Orange red	*Saraca indica*
Pale pink	*Glirisidia maculate*
Pale red	*Samanea saman*
Pink	*Cassia javanica*
Pinkish white	*Careya arborea*
Purple	*Bauhinia purpurea*
Scarlet	*Callistemon lanceolatus*
Scarlet red	*Delonix regia*
Scarlet red or orange	*Butea monosperma*
Silvery white	*Millingtonia hortensis*
Violet blue	*Jacaranda minosifolia*
White	*Wrightia tinctorea*
White	*Bauhinia acumina*
Yellow	*Cassia fistula*
Yellow and crimson red	*Thespesia papulnea*

Commonly used instruments in Forestry

ફોરેસ્ટ્રીમાં સામાન્ય રીતે ઉપયોગમાં લેવામાં આવતા સાધનો

Instrument	Use
Abney's level	Measuring height and contouring
Actinometer	Heating power of sunlight
Altimeter	Altitudes
Anemometer	Wind speed
Audiometer	Hearing
Barometer	Air pressure
Bomb calorimeter	Heat of combustion
Brandis hypsometer	Measuring the height of the tree
Caliper	Measure the diameter of standing trees and logs
Calorimeter	Heat of chemical reactions
Christen hypsometer	Measuring the height of the tree
Chronometer	Time
Clinometer	Angle of a slope
Compass	Direction of North
Crescograph	Plant growth
Densimeter	Specific gravity of liquids
Dumpy Level	Horizontal levels
Evaporimeter	Rate of evaporation
Fathometer	Ocean depth
Glucometer	Blood glucose (diabetes)
GPS	To know the exact position on earth
Haga altimeter	measuring the height of the tree
Hydrometer	Specific gravity of liquids
Hygrometer	Humidity
Lactometer	Specific gravity of milk
Manometer	Pressure of gas
Mercury Barometer	Atmospheric pressure
Odometer	Distance travelled
Osmometer	Osmotic strength of solution
Pedometer	Steps
pH Meter	pH
Photometer	Light intensity
Planometer	Area

Instrument	Use
Potometer	Transpiration
Psychrometer	Humidity
Pycnometer	Fluid density
Pyranometer	Solar radiation
Rain Gauge	Measuring of rain
Relaskop	Measuring of basal area of crops, diameters over bark at any distance from the tree.
Saccharometer	Amount of sugar in a solution
Seismometer	Seismic waves (for example, earthquakes)
Spectrophotometer	Intensity of light as a function of wavelength
Sphygmomanometer	Blood pressure
Tape	To measure height, length and girth
Tensiometer	Surface tension of a liquid
Theodolite	Measuring angles in the horizontal and vertical planes
Thermometer	Temperature
Tintometer	Colour
Tong	To capture the snake
UHF/VHF Radio-telemetry	For tracking the radio collared animal
Wind Vane	Wind direction
Wooden scale	Measuring diameter of stump or end sections of logs. Also, in stump and stem analysis
Zymometer	Fermentation

Multiple Choice Questions | હેતુલક્ષી પ્રશ્નો

1. શ્રીનગરના પ્રખ્યાત દલ સરોવરને કઈ સમસ્યાઓનો સામનો કરવો પડી રહ્યો છે?
   A. માનવીય પ્રવૃત્તિઓ
   B. જળચર પ્રાણી પ્રવૃત્તિઓ
   C. ઇકોલોજીકલ અસંતુલન
   D. હાઇડ્રોલોજિકલ અસંતુલન

1. **The famous Dal lake of Srinagar is facing problems due to**
   A. Anthropogenic activities
   B. Aquatics animal activities
   C. Ecological imbalance
   D. Hydrological imbalance

2. કોરલ રીફમાં થઈ રહેલ ઘટાડાના કારણો મુખ્યત્વે શેને લીધે થાય છે (i) મહાસાગરોની વધુ પડતી ખેડ, (ii) કાંપમાં વધારો (iii) વધુ પડતી માછીમારી
   A. (i) and (iii)
   B. (i) only
   C. (ii) and (iii) only
   D. (i), (ii) and (iii)

2. **The coral reef decline is due to (i) overexploitation of seas, (ii) increased sedimentation, (iii) over fishing**
   A. (i) and (iii)
   B. (i) only
   C. (ii) and (iii) only
   D. (i), (ii) and (iii)

3. માનવ હસ્તક્ષેપ અનુસંધાને નીચેનામાંથી ________ અલગ છે.
   A. બાયોસ્ફિયર રિઝર્વ
   B. બોટનિકલ ગાર્ડન
   C. રાષ્ટ્રીય ઉદ્યાન
   D. વન્યપ્રાણી અભયારણ્ય

3. **Which one of the following is different as far as human interference is concerned.**
   A. Biosphere Reserves
   B. Botanical Garden
   C. National Park
   D. Wildlife Sanctuary

4. માણસ દ્વારા ઝાડની છાલને તેના પાયા પાસે ગોળાકાર કાઢવામાં આવે છે. આ ઝાડ ત્યાર બાદ કયા કારણોસર સુકાઈને મૃત્યુ પામે છે?
   A. જમીનમાંથી પાણી પ્રકાંડના ભાગોમાં જઈ શકતું નથી
   B. મૂળ ઉર્જા રહિત થઇ જાય છે
   C. વૃક્ષને સુક્ષ્મજીવાણુઓ દ્વારા ચેપ લાગીને
   D. મૂળ શ્વસન માટે ઓક્સિજન મેળવતા નથી

4. **Human removes the bark of a tree in a circular fashion all around near its base, it generally, dries up and dies because**
   A. Water from soil cannot rise to aerial parts.
   B. Roots are starved of energy
   C. Tree is infected by soil microbes
   D. Roots do not receive oxygen for respiration

5. ભારતના કયા રક્ષિત વિસ્તારમાંથી સ્થાનિક લોકો ને બાયોમાસ એકત્રિત અને વાપરવાની મંજૂરી નથી?
A. બાયોસ્ફિયર રીઝર્વ્સ
B. રાષ્ટ્રીય ઉદ્યાનો
C. કંઝર્વેશન રીઝર્વ્સ
D. વન્યપ્રાણી અભયારણ્ય

5. In which one among the following categories of protected areas in India are local people not allowed to collect and use the biomass?
A. Biosphere Reserves
B. National Parks
C. Conservation Reserve
D. Wildlife Sanctuaries

6. ગીધ જોખમમાં હોવાનું મુખ્ય કારણ શું છે?
A. નવી આક્રમક પ્રજાતિઓ દ્વારા તેમના માળખાના સ્થળોનો વિનાશ
B. પશુ માલિકો દ્વારા તેમના પશુઓની સારવાર માટે ઉપયોગમાં લેવામાં આવતી દવા
C. તેમને ઉપલબ્ધ ખોરાકની અછત
D. વ્યાપક, સતત અને જીવલેણ રોગ

6. Vultures are in threat, the main reason of it is;
A. the destruction of their nesting sites by new invasive species
B. a drug used by cattle owners for treating their diseased cattle.
C. scarcity of food available to them
D. a widespread, persistent and fatal disease among

7. હવામાં કાર્બન ડાયોક્સાઇડની વધતી માત્રા ને લીધે વાતાવરણ ના તાપમાન માં થતા વધારાનું મુખ્ય કારણ શું છે?
A. પાણીની વરાળ તેની ગરમી જાળવી રાખે છે
B. સૌર કિરણોત્સર્ગનો અલ્ટ્રાવાયોલેટ ભાગ
C. સૌર કિરણોત્સર્ગ
D. સૌર કિરણોત્સર્ગનો ઇન્ફ્રારેડ ભાગ

7. The increasing amount of carbon dioxide in the air is slowly raising the temperature of the atmosphere, because it absorbs
A. the water vapour of the air and retains its heat
B. the ultraviolet part of the solar radiation
C. all the solar radiations
D. the infrared part of the solar radiation

8. ભારત સરકાર સી બકથ્રોન ની ખેતીને પ્રોત્સાહિત કરે છે જેનું મુખ્ય કારણ શું છે?
A. તે જમીનના ધોવાણને નિયંત્રિત કરવામાં અને રણપ્રવાહને રોકવામાં મદદ કરે છે
B. તે બાયોડિઝલનો એક સમૃદ્ધ સ્રોત છે
C. તેમાં પોષક મૂલ્ય છે અને ઉચ્ચ ઊંચાઈવાળા ઠંડા વિસ્તારોમાં રહેવા માટે તે અનુકૂળ છે
D. તેનું લાકડું ખૂબ વ્યાપારી મૂલ્યનું છે

8. Government of India encourages the cultivation of 'sea buckthorn'. What is the importance of this plant?
A. It helps in controlling soil erosion and in preventing desertification
B. It is a rich source of biodiesel
C. It has nutritional value and is well-adapted to live in cold areas of high altitudes
D. Its timber is of great commercial value

9. સરકાર દ્વારા વન વિસ્તારમાં અનધિકૃત વાવેતરને નિયમિત કરવા કયો નિયમ અમલમાં આવ્યો છે?
A. વન (સંરક્ષણ) અધિનિયમ 1980
B. વન અધિકાર અધિનિયમ, 2006
C. ભારતીય વન અધિનિયમ 1927
D. પંચાયતો (અનુસૂચિત ક્ષેત્રોમાં વિસ્તરણ) અધિનિયમ, 1996

10. ગુજરાતમાં કરુણા અભિયાન શેના બચાવ માટે કરવામાં આવી રહ્યું છે?
A. શહેર વિસ્તારના પાશવી કૂતરા
B. શહેર વિસ્તારની ગાય
C. પક્ષીઓ
D. શહેરમાંના સાપ

11. ગુજરાતના દરિયાકાંઠાના વિસ્તારમાં પવન મિલો અને તેના માળખાકીય વિકાસને કારણે, કઈ મહત્વપૂર્ણ પક્ષી પ્રજાતિઓ જોખમમાં મુકાય છે?
A. હેરિયર્સ
B. લેસર ફ્લોરિકન
C. હૌબરા બસ્ટાર્ડ
D. ગ્રેટ ઇન્ડિયન બસ્ટાર્ડ

12. સ્પાઈની ટેલ લિઝાર્ડ શાના માટે શિકાર થાય છે?
A. માત્ર સાથે રમવા માટે
B. તેઓ ખૂબ ઝેરી છે તેથી તેમને મારી નાખવા માટે
C. તેઓ તેમના પાકનો નાશ કરે છે, તેથી પકડે છે અને દૂર કરે છે.
D. ખોરાક તરીકે અને તેમાંથી તેલ કાઢવા માટે

13. તાજેતરમાં સુપ્રીમ કોર્ટે પાવર લાઈનને ભૂગર્ભમાં કરવાનો હુકમ શેના બચાવ અર્થે જારી કર્યો છે?
A. હેરિયર્સ
B. સાઇબેરીયન ક્રેન્સ
C. હૌબરા બસ્ટાર્ડ
D. ગ્રેટ ઇન્ડિયન બસ્ટાર્ડ

9. The unauthorised cultivation in forest area is being regularized by the Government through
A. Forest (Conservation) Act 1980
B. Forest Rights Act, 2006
C. Indian Forest Act 1927
D. Panchayats (Extension to Scheduled Areas) Act, 1996

10. Karuna Abhiyan in Gujarat is being launched to save the
A. Feral dogs in city area.
B. Cows in city area
C. Birds
D. Snakes in city

11. Because of the wind mills and its infrastructural development in Coastal area of Gujarat, which endangered important avifaunal species is under threat
A. Harriers
B. Lesser Florican
C. Houbara Bustard
D. Great Indian Bustard

12. Spiny tail lizards are being caught by the humans for
A. Just to play with
B. They are most poisonous so want to kill them
C. They destroy their crops, so catch and through away
D. To eat and to extract oil

13. Recently the Supreme Court has issued order to put the power line underground to save the
A. Harriers
B. Siberian cranes
C. Houbara Bustard
D. Great Indian Bustard

14. સિમેન્ટ-કોંક્રિટ ઘરોના નિર્માણથી મુખ્યત્વે કયા પક્ષીઓને પ્રતિકૂળ અસર થઈ છે?

A. પોપટ
B. કબૂતર
C. કાબર
D. ચકલી

15. વરુના કયા મુખ્ય દુશ્મન છે કે જેમણે આ મહત્વપૂર્ણ પ્રાણીની વસ્તીમાં તીવ્ર ઘટાડો કર્યો?

A. સિંહ B. શિકારીઓ
C. ભરવાડો D. ખેડુતો

16. દીપડો સામાન્ય રીતે કોના પર હુમલો કર્યાના બનાવો થાય છે?

A. માલધારી B. શેરડીના મજૂરો
C. વન અધિકારીઓ D. પશુપાલકો

17. ગુજરાતમાં માંડવીથી કચ્છ સુધી દરિયા કાંઠે સમુદ્ર કાચબા ઇંડા મુકતા હોય છે પરંતુ એક સર્વે મુજબ, ભાગ્યે જ સેવન થઈ બચ્ચા જન્મે છે કારણ શું છે?

A. ઇંડા ક્ષતિયુક્ત છે
B. સમુદ્ર કાચબા પોતાનાજ ઇંડા નો નાશ કરે છે
C. દરિયાની જમીન ને કારણે ઇંડા જીવી શકતા નથી
D. ઇંડાનું સેવન મનુષ્યો તથા કુતરા કરે છે

18. ગીરના જંગલોની તુલનામાં ડાંગ માં ગીચ જંગલો હોવા છતાં, પક્ષી શ્રુષ્ટિ ઘણી ઓછી કેમ છે?

A. પક્ષી શ્રુષ્ટિ માટે સારી હેબીટાટ નથી
B. સ્થાનિક લોકોની ખોરાકની ટેવને કારણે
C. ડાંગ માં અવારનવાર આગના કારણે
D. જંગલી પ્રાણીઓ દ્વારા શિકાર થવાના કારણે

14. The construction of cement-concrete houses has affected the population of bird mainly

A. Parrots
B. Pigeon
C. Myna
D. Sparrow

15. The main enemy of wolf that drastically reduced the population of this important wild animal

A. Lion B. Poachers
C. Shepherds D. Farmers

16. The Leopard generally attack on

A. Maldhari
B. Sugarcane labourers
C. Forest Officers
D. Cattle Graziers

17. In Gujarat, Sea turtles lay eggs on sea shore almost from Mandvi to Kutch but as per survey, there is hardly any successful incubation and survival of younger ones, why?

A. Laying of eggs are defective
B. Sea turtles themselves destroy the eggs
C. Because of soil quality, not able to survive
D. Eggs are being eaten away by dogs and human beings

18. Despite of dense forests in Dangs in compare to Gir forests, the avifaunal number is quite less

A. not good habitat for avifauna
B. because of food habits of local people
C. Because of frequent fire in Dangs
D. wild animals pray on avifauna

19. પૂર્વી ગુજરાતમાં સ્લોથ રીંછનો સામનો મનુષ્ય સાથે ક્યારે થાય છે?
A. પાણીના સ્થળોએ
B. શેરડીનાં ખેતરોમાં
C. ખુલ્લા જંગલોમાં
D. વહેલી સવારે મહુડાના ઝાડ ઉપર ફળ એકત્રીત કરતી વખતે

20. સામાન્ય રીતે આદિવાસી સમુદાયો દ્વારા નીચેના માંથી કયા વૃક્ષની પૂજા કરવામાં આવતી નથી?
A. *Ficus religiosa*
B. *Ficus benghalensis*
C. *Adina cardifolia*
D. *Adansonia digitata*

21. કોટવાલિયા આદિવાસી સમુદાયો તેમની આજીવિકા માટે સામાન્ય રીતે શેના પર આશ્રિત હોય છે?
A. શેરડીના મજૂરો
B. જંગલી પ્રાણીઓનો શિકાર
C. વાંસની હસ્તકલા
D. MFP સંગ્રહ

22. સ્થાનિક લોકો દ્વારા ચેરને ક્યા કારણોસર કાપવામાં આવે છે?
A. લાકડા અને થાંભલા માટે
B. ઘાસચારા માટે લોપીંગ
C. માછીમારી માટે
D. દરિયામાં નૌકાવિહાર માટે

23. કઈ સંસ્થા ડાંગ અને રાજપીપળાના જંગલોમાંથી વાંસનો ઉપયોગ પલ્પ અને કાગળના ઉત્પાદન માટે કરે છે?
A. સરકારી પલ્પ મિલ, અમદાવાદ
B. સેન્ટ્રલ પલ્પ મિલ, સોનગઢ
C. મફતલાલ પલ્પ મિલ, અમદાવાદ
D. રેમન્ડ પલ્પ મિલ, મુંબઇ

19. The encounter of sloth bear in Eastern Gujarat happened with the human beings when
A. At water points.
B. In sugarcane fields
C. In open forests
D. Early morning at Mahuda trees, fruits are being collected

20. Generally the trees are being worshipped by the tribal, which one is not matching with the worshipping trees.
A. *Ficus religiosa*
B. *Ficus benghalensis*
C. *Adina cardifolia*
D. *Adansonia digitata*

21. The Kotwalia tribals are generally dependent for their livelihood on
A. Sugarcane labourers
B. Hunting of wild animals
C. Bamboo handicrafts
D. MFP collection

22. Mangroves are being damaged by locals for;
A. Timber and pole
B. Lopping for grazing fodder
C. Just cutting for fishing purpose
D. Removing mangroves for boating in sea

23. Bamboo from Dangs and Rajpipla forests are being used for manufacturing the pulp and paper by
A. Government Pulp Mill, Ahmedabad
B. Central Pulp Mill, Songadh
C. Mafat Lal Pulp Mill, Ahmdabad
D. Raymond Pulp Mill, Mumbai

24. "રાબ બર્નિંગ" એ દક્ષિણના આદિવાસી ક્ષેત્રમાં એક પ્રથા છે જેનાથી જંગલને શું અસર થાય છે?
A. વૃક્ષો મરી જાય છે
B. ઝાડનું અંકુરણ બળી બળી જાય છે
C. પૂજા માટે ઉભા ઝાડનું દહન
D. તે સ્થાયી વૃક્ષોમાંથી કાપવામાં આવેલી સામગ્રીને બાળી નાખે છે જે ઝાડની વૃદ્ધિને અસર કરે છે.

25. ગુજરાતના જંગલોમાં આગની ઘટનાના મુખ્ય કારણો કયા છે?
A. વાંસના ઘર્ષણથી
B. જંગલોમાં ભૂગર્ભ ગરમીથી
C. માનવસર્જિત આગ
D. જંગલોમાં કેટલાક પ્રકાશિત કીટકોને કારણે

26. ગુજરાત રાજ્યના જંગલોમાં ક્યા પ્રકારની આગ લાગે છે?
A. ગ્રાઉન્ડ ફાયર, સ્થાયી વૃક્ષોને નુકસાન કરતું નથી
B. ક્રાઉન ફાયર, સ્થાયી વૃક્ષોને નુકસાન
C. તે વૃક્ષો તેમજ વન્યપ્રાણીઓને ખૂબ જ નુકસાન પહોંચાડે છે
D. સ્થાયી વૃક્ષો માટે આગ ફાયદાકારક છે

27. સુપ્રીમ કોર્ટે રક્ષિત વિસ્તાર આસપાસના ઇકો સેન્સિટિવ ઝોન પર ભાર મૂક્યો હતો, આનું મુખ્યત્વે કારણ શું હતું?
A. રક્ષિત વિસ્તાર ની આસપાસ જળસંગ્રહ કરવા
B. રક્ષિત વિસ્તાર ની આસપાસના વિસ્તારોની વન્યપ્રાણી સંરક્ષણ
C. રક્ષિત વિસ્તાર ની આજુબાજુ શિકારને નિયંત્રિત કરવા
D. રક્ષિત વિસ્તાર ની આસપાસ માનવીય દબાણને નિયંત્રિત કરવા

24. "Raab burning" is a practice in southern tribal area, how does it affects the forests;
A. Because of burning fire, trees die
B. Young germination of trees burn
C. Its burning of standing tree for puja
D. It's the burning of pruned material from standing trees that affects the growth of tree

25. In Gujarat, the fire occurrence in forests, are because of
A. Rubbing of bamboos
B. Ignition of underground heats in forests
C. Manmade fires
D. Because of some lighting insects in forests

26. Fires in Forests of the Gujarat State are
A. Ground fire, does not damage standing trees
B. Crown fire, damage standing trees
C. It damages too much to trees as well as wildlife
D. Fire is beneficial to standing trees

27. Supreme Court emphasized on notification for Eco sensitive zone around protected areas, this was mainly because
A. To do water conservation around PAs
B. To conserve wildlife of the surrounding areas of PA
C. To control poaching around PAs
D. To control anthropogenic pressure around PAs

28. વન અને વન્યપ્રાણી અધિનિયમો હેઠળ, જંગલો અને પ્રાકૃતિક આવાસો જુદા જુદા દરજ્જા હેઠળ જાહેર કરવામાં આવી રહ્યા છે. કયા ક્ષેત્ર પર કોઈ માનવ દબાણને મંજૂરી અપાતી નથી?
A. રાષ્ટ્રીય ઉદ્યાન
B. અભયારણ્ય
C. સંરક્ષિત જંગલો
D. કંઝર્વેશન રિઝર્વ

29. ભૂમિગત ખારાશ પ્રદેશ શેની સંભવિત સાઇટ્સ તરીકે કાર્ય કરી શકે છે?
A. માનવસર્જિત CO_2 ઉત્સર્જન
B. કાર્બન સિકવેસ્ટ્રશન માટે માનવસર્જિત CO_2
C. માનવસર્જિત N_2O ઉત્સર્જન
D. માનવસર્જિત SO_2 ઉત્સર્જન

30. વિશ્વવ્યાપી કોરલ રીફના મરણનું મૂળભૂત કારણ શું છે?
A. માત્ર કુદરતી ઘટનાઓ
B. માનવસર્જિત કારણો
C. માનવીય અને કુદરતી ઘટનાઓ
D. ઉપર્યુક્તમાંથી કોઈ નહિ

31. કયા માનવસર્જિત સ્ત્રોતો એસિડ વરસાદમાં ફાળો આપે છે?
A. દાવાનળ
B. તેલ, કોલસો અને ગેસનું દહન
C. જ્વાળામુખી વિસ્ફોટો
D. બંને A અને B

32. ખરાબ રીતે બફર્ડ સરોવરો કેવા હોય છે?
A. કુદરતી રીતે એસિડિક તળાવો
B. એન્થ્રોપોજેનિકલી એસિડિફાઇડ તળાવો
C. કુદરતી રીતે મૂળ તળાવો
D. માનવશાસ્ત્રના મૂળભૂત તળાવો

28. Under forest & wildlife acts, forests and natural habitats are being declared under different status. Which area does not allow any anthropogenic pressure on it.
A. National Park
B. Sanctuary
C. Protected Forests
D. Conservation Reserve

29. Subterranean deep saline formations can be act as potential sites of
A. anthropogenic emissions of CO_2
B. anthropogenic emissions of CO_2 for carbon sequestration
C. anthropogenic emissions of N_2O
D. anthropogenic emissions of SO_2

30. Coral reef ecosystems world-wide have been subject to unprecedented degradation over the past few decades. Disturbances affecting coral reefs include
A. natural events only
B. anthropogenic only
C. anthropogenic and natural events
D. None of the above

31. Which of these are anthropogenic sources of N contributing to acid rain
A. Forest fires
B. Combustion of oil, coal, and gas
C. Volcanic eruptions
D. Both A and B

32. Poorly buffered lakes represent
A. Naturally acidic lakes
B. Anthropogenically acidified lakes
C. Naturally basic lakes
D. Anthropogenically basic lakes

33. નેશનલ મિશન ફોર સસ્ટેઇનિંગ ધ હિમાલયન ઈકોસીસ્ટમ (NMSHE) નો હેતુ શું છે?

A. સંભવિત પ્રભાવો પર કુદરતી અને માનવસર્જિત પર્યાવરણીય પરિવર્તન અને પ્રોજેક્ટના ભાવિ વલણો શક્ય અસરો શોધવી અને બતાવવી

B. નાણાકીય સંસાધનો અને રોકાણની જોગવાઈ પરના પગલાને સુધારવા તેમજ અનુકૂલન અને તકનીકી સહયોગની ક્રિયાને ટેકો આપવો

C. ક્લાયમેટ ફાઇનાન્સના ક્ષેત્રમાં, 'ગ્રીન ક્લાયમેટ ફંડ' ડિઝાઇન કરવાની પ્રક્રિયા કરવી

D. ઉપરોક્ત તમામ

34. શેના સંરક્ષણ માટે જંગલમાં શિકાર અને જંગલી છોડ એકત્રિત કરવા પર પ્રતિબંધ મૂકી દેવો જોઈએ?

A. મનુષ્ય જીવન B. વન્ય જીવન

C. જંગલી છોડ D. જંગલી પ્રાણીઓ

35. ભૂગર્ભજળ, પીટ, કાંકરા ક્યાંથી મળે છે?

A. પર્વતો B. જંગલો

C. વેટલેન્ડ D. ભૂગર્ભ

36. હવામાં ધૂમાડાના કણો અને હાનિકારક ગેસના મિશ્રણથી શું બને છે?

A. સ્મોક B. ફોગ

C. હેઝ D. સ્મોગ

37. અન્ટ્રીટેડ પાણી ને નદીઓ, સમુદ્ર અને તળાવોમાં સીધું ઠાલવવાથી શું પ્રતિક્રિયા આવે છે?

A. પાણીજન્ય રોગો

B. દૂષણ

C. પ્રદૂષણ

D. ઉપરોક્ત તમામ

33. National mission for sustaining the Himalayan ecosystem (NMSHE) objectives include

A. Detect and decouple natural and anthropogenic global environmental changes and project future trends on potential impacts

B. Enhanced action on the provision of financial resources and investment to support action on mitigation and adaptation and technology cooperation.

C. In the field of climate finance, a process to design a 'Green Climate Fund'

D. All of the above

34. Hunting in forest or collecting of wild plants are banned to ensure protection of

A. Human life B. Wildlife

C. Wild Plants D. Wild Animals

35. Humans attain groundwater, peat, gravels from

A. Mountains B. Forests

C. Wetlands D. Underground

36. Smoke particles and harmful gases released in air results in

A. smoke B. fog

C. haze D. smog

37. The direct spill of untreated waste water in to rivers, sea and lakes may results in

A. Water borne diseases

B. Contamination

C. Pollution

D. All of them

38. વન અધિનિયમની જોગવાઈઓના સંરક્ષણ માટેની દંડ જોગવાઈ કઈ કલમમાં છે?
A. કલમ 12A B. કલમ 8A
C. કલમ 3A D. કલમ 4A

39. જંગલની જમીનોનું એસિડિફિકેશન લાંબા સમયગાળાના ______ માનવસર્જિત સ્રોત થી થાય છે.
A. વાયુ પ્રદૂષણ
B. જળ પ્રદૂષણ
C. ભૂમિ પ્રદૂષણ
D. થર્મલ પ્રદૂષણ

40. જળચર અને ભૂગર્ભ પ્રણાલીઓમાં વધારે નાઇટ્રોજન (નાઇટ્રેટના સ્વરૂપમાં) ના સ્થાનાંતરણને સૌથી મહત્વપૂર્ણ જોખમ માનવામાં આવે છે. જેને કહે છે
A. નાઇટ્રોજન એડિશન
B. નાઇટ્રોજન સેચુરેશન
C. નાઇટ્રોજન એક્યુમ્યુલેશન
D. નાઇટ્રોજન ઈમ્પેક્ટ

41. વન ઇકોસિસ્ટમના ______ માં માનવ શોષણનાં પરિણામો બદલાય છે.
A. માળખા
B. પોષક ઇનપુટમાં
C. જૈવિક રાસાયણિક ચક્રમાં
D. ઉપર્યુક્ત તમામ

42. વર્તમાન જંગલો અને જમીનના અધોગતિનું વૈશ્વિક નુકસાનનું મુખ્ય કારણ કયુ છે?
A. વાતાવરણ મા ફેરફાર
B. પ્રદૂષણ
C. માનવ પ્રવૃત્તિઓ
D. બંને A અને C

43. HPI એટલે
A. હ્યુમન પોપ્યુલેશન ઈન્ડેક્સ
B. હ્યુમન પ્રેસર ઈન્ડેક્સ
C. હ્યુમન પોપ્યુલેશન ઇન્ટરફેરેન્સ
D. હ્યુમન પ્રેસર ઇન્ટરફેરેન્સ

38. Penalty for conservation of the provisions of the Forest Act is under
A. Section 12A B. Section 8A
C. Section 3A D. Section 4A

39. Acidification of forest soils caused by long-range ______ from anthropogenic sources
A. air pollution
B. water pollution
C. soil pollution
D. thermal pollution

40. Transfer of excess nitrogen (in the form of nitrate) to aquatic and groundwater systems is generally considered the most important risk arising from
A. nitrogen addition
B. nitrogen saturation
C. nitrogen accumulation
D. nitrogen impact

41. Human exploitation results change in ______ of forest ecosystem
A. structure
B. nutritional input
C. biogeochemical cycles
D. all of them

42. Current global loss of forests and soil degradation is majorly a result of
A. Climate change
B. Pollution
C. Human activities
D. Both A and C

43. HPI stands for
A. Human Population Index
B. Human Pressure Index
C. Human Population Interference
D. Human Pressure Interference

44. વનસ્પતિ અને પ્રાણીસૃષ્ટિના અવક્ષય ની બાબતમાં શું સાચું નથી?
A. કૃષિ વિસ્તરણ
B. ખાણકામ પ્રવૃત્તિઓ
C. રહેવા માટે જરૂરી જમીન
D. કૃષિ સ્થળાંતર

45. કઈ સંરક્ષણ વ્યૂહરચનામાં સામુદાયિક ભાગીદારી નો સમાવેશ થતો નથી?
A. બીજ બચાવો આંદોલન
B. ચિપકો આંદોલન
C. સહભાગી વન વ્યવસ્થાપન
D. વન્યપ્રાણી અભયારણ્યોનું સીમાંકન

46. વિશ્વભરમાં જમીનના ડીગ્રેડેશનના મુખ્ય કારણો કયા છે?
A. વનનાબૂદી
B. ઓવરગ્રેઝિંગ
C. કૃષિ ગેરવહીવટ
D. ઉપર્યુક્ત તમામ

47. ઉત્તર પૂર્વીય ભારતના ભાગોમાં આદિવાસી પટ્ટાઓના જંગલોનો કાપ શેના થકી થયો?
A. માળખાગત વિકાસ
B. શિફટિંગ કલ્ટીવેશન
C. ખાણકામ
D. બંને A અને C

48. નીચેનામાંથી કયા રીન્યુએબલ એક્સોસ્ટીવ નેચરલ એનર્જી રિસોર્સ છે?
A. કોલસો B. તેલ
C. બાયોમાસ D. પેટ્રોલિયમ

49. બક્સર ટાઇગર રિઝર્વ શેના કારણે જોખમ માં છે?
A. વનનાબૂદી
B. શિકારની પ્રજાતિઓની અછત
C. ડોલોમાઇટ માઇનિંગ
D. જ્વાળામુખી

44. Which one of the following statements is not true with regard to depletion of flora and fauna?
A. Agricultural expansion
B. Mining activities
C. Land required for housing
D. Shifting agriculture

45. The conservation strategy that do not directly involve community participation is
A. Beej Bachao Andolan
B. Chipko Movement
C. Joint forest management
D. Demarcation of Wildlife Sanctuaries

46. The major causes of soil degradation worldwide are
A. deforestation
B. overgrazing
C. agricultural mismanagement
D. All of them

47. Substantial parts of the tribal belts in North-East India have been deforested by
A. Infrastructure development
B. Shifting cultivation
C. Mining
D. Both A and C

48. Which is the renewable exhaustible natural energy resource?
A. Coal B. Oil
C. Biomass D. Pêtroleum

49. The Buxar Tiger Reserve is seriously threatened by
A. Deforestation
B. Scarcity of prey species
C. Dolomite mining
D. Volcanic eruption

50. ભારતના કયા વિસ્તારમાં સાગ મોનોકલ્ચરને લીધે કુદરતી જંગલોને નુકસાન થયું છે?
A. ઉત્તર ભારત
B. દક્ષિણ ભારત
C. બ્રહ્મપુત્રા મેદાન
D. પૂર્વ ભારત

51. ભારતનો પ્રથમ સ્લોથ રીંછ કોરિડોર ક્યાં સ્થાપિત થયેલ છે?
A. માઉન્ટ આબુ અને જેસોર
B. રતનમહાલ અને જેસોર
C. રતનમહાલ અને જાંબુઘોડા
D. રતનમહાલ અને બલારામ અંબાજી

52. ડિસેમ્બર 2020 માં, મદ્રાસ હાઇકોર્ટે ભારતના કયા વિસ્તારમાં વિદેશી ઇન્વેસિવ જાતોના વાવેતર પર પ્રતિબંધ મૂક્યો હતો?
A. ગિન્ડી નેશનલ પાર્ક
B. મુદુમલાઈ વન્યપ્રાણી અભયારણ્ય
C. નીલગિરી હિલ્સ
D. પશ્ચિમ ઘાટ

53. હાલની સ્લોથ રીંછ ની 90% રેન્જ ક્યાં જોવા મળે છે?
A. નેપાળ B. શ્રીલંકા
C. બાંગ્લાદેશ D. ભારત

54. ભારત વિશ્વમાં _____ વિશાળ જૈવવિવિધતા દેશોમાંનો એક છે.
A. 12 B. 11
C. 17 D. 15

55. તાન્ઝાનિયનમાં તાજેતરમાં જે કાળો ગેંડો 'ફાઉસ્તા' નું અવસાન થયું. તે ગેંડાનું IUCN સ્ટેટસ શું છે?
A. વિનાશને આરે
B. ભયગ્રસ્ત
C. ભયને આરે
D. ભયમુક્ત

50. Teak monoculture has damaged the natural forests in
A. North India
B. South India
C. Brahmaputra Plain
D. East India

51. The first sloth bear corridor of India is set up between
A. Mount Abu and Jessore
B. Ratanmahal and Jessore
C. Ratanmahal and Jambughoda
D. Ratanmahal and Balaram Ambaji

52. Recently in December 2020, the Madras High Court banned planting of exotic invasive species in
A. Guindy National Park
B. Mudumalai Wildlife Sanctuary
C. Nilgiri Hills
D. Western Ghats

53. At least 90% of the present Sloth Bear range occurs in
A. Nepal B. Shri Lanka
C. Bangladesh D. India

54. India is one of the _____ mega biodiversity countries of the world.
A. 12 B. 11
C. 17 D. 15

55. A female black rhino 'Fausta' recently died in Tanzanian. The IUCN status of black rhino is
A. Critically Endangered
B. Endangered
C. Near threatened
D. Least concern

56. કઈ મહિલાએ 80 વર્ષમાં 8,000 થી વધુ વૃક્ષો વાવ્યા અને 2019 માં પદ્મશ્રી આપવામાં આવ્યો?

A. સાલુમરાદા થિમ્મક્કા

B. મૌસમ કાઝી

C. ઇન્દિરા બોરા

D. મીનાક્ષી જૈન

57. કર્ણાટકના હોન્નાલી ગામની એક વૃદ્ધ મહિલાએ 30,000 થી વધુ રોપાઓ વાવ્યા અને 2020 માં પદ્મશ્રી એનાયત થયો હતો?

A. ઇન્દિરા બોરા

B. તુલસી ગૌડા

C. મૌસમ કાઝી

D. સાલુમરાદા થિમ્મક્કા

58. હર્પેટોલોજીમાં તેમના નોંધપાત્ર કામ બદલ વન્યપ્રાણી સંરક્ષણમાં પદ્મશ્રી કોને મળ્યો છે?

A. જેસી ડેનિયલ

B. અભિજિત દાસ

C. રોમ્યુલસ વ્હાઇટેકર

D. અજય દેસાઇ

59. ચેન્નાઇમાં મદ્રાસ ક્રોકોડાઇલ બેંક ટ્રસ્ટ અને સેન્ટર ફોર હર્પેટોલોજીની સ્થાપનામાં કોનો ફાળો છે?

A. વિવેક દેસાઈ

B. ઇન્દ્રનીલ દાસ

C. રિચાર્ડ ડોકિન્સ

D. રોમ્યુલસ વ્હાઇટેકર

60. યુ.એસ. એર ક્વોલિટી ઈન્ડેક્સના 2020 ના અહેવાલમાં બહાર પાડવામાં આવેલું કયું શહેર વિશ્વનું સૌથી પ્રદુષિત શહેર છે જેનો PM 423 છે?

A. નવી દિલ્હી

B. કાઠમંડુ

C. ન્યુ યોર્ક

D. લાહોર

56. A lady planted over 8,000 trees in 80 years and awarded Padma Shri in 2019

A. Saalumarada Thimmakka

B. Mausam Kazi

C. Indira Bora

D. Meenakshi Jain

57. An old lady from Honnali village in Karnataka planted more than 30,000 saplings and awarded Padma Shri in 2020

A. Indira Bora

B. Tulsi Gowda

C. Mausam Kazi

D. Saalumarada Thimmakka

58. Who has been awarded Padma Shri in wildlife conservation for his notworthy work in herpetology?

A. J. C. Daniel

B. Abhijeet Das

C. Romulus Whitaker

D. Ajay Desai

59. Who has been instrumental in establishing Madras Crocodile Bank Trust and Centre for Herpetology in Chennai?

A. Vivek Desai

B. Indraneil Das

C. Richard Dawkins

D. Romulus Whitaker

60. According US Air Quality Index report released in 2020, Which is the world's most polluted city with the particulate matter 423

A. New Delhi

B. Kathmandu

C. New York

D. Lahore

61. તાજેતરના માં મહારાષ્ટ્ર માં થયેલા બ્લ્યુ ટાઇડ થવાનું મુખ્ય કારણ શું છે ?
A. ફાયટોપ્લાંકટોન્સ દ્વારા પ્રકાશનું ઉત્સર્જન
B. તેલને કારણે સમુદ્રનું પ્રદૂષણ
C. શેવાળની વધુ વૃદ્ધિ
D. પ્લાસ્ટિકના કારણે સમુદ્રનું પ્રદૂષણ

61. In the recent past, the blue tide phenomenon was observed in many beaches of Maharashtra, this phenomenon is due to
A. Emission of light by phytoplanktons
B. Pollution of sea due to oil
C. Over growth of algae
D. Pollution of sea due to plastic

62. 2020 માં, ઉત્તરપ્રદેશના કયા ટાઇગર રિઝર્વ ને ચાર વર્ષમાં બમણી સંખ્યા બદલ આંતરરાષ્ટ્રીય એવોર્ડ TX2 મળ્યો?
A. દુધવા B. અમનગઢ
C. પીલીભીત D. કોર્બેટ

62. In 2020, the tiger reserve of Uttar Pradesh received international award TX2 for doubling number of tigers in four years.
A. Dudhwa B. Amangarh
C. Pilibhit D. Corbett

63. નેશનલ એક્શન પ્લાન _____ કન્ઝર્વેશન 2020-25 માટે નેશનલ બોર્ડ ફોર વાઇલ્ડલાઇફ દ્વારા નવેમ્બર 2020માં મંજૂરી આપવામાં આવી છે .
A. ડોલ્ફિન B. સ્વેમ્પ ડીયર
C. મોર D. ગીધ

63. The National Action Plan for _____ Conservation 2020-25 has been approved by National Board for Wildlife in November 2020.
A. Dolphin B. Swamp Deer
C. Peafowl D. Vulture

64 . નવેમ્બર 2020 માં મહારાષ્ટ્રમાં કઈ રામસર સાઇટ ઉમેરાઈ?
A. લોનાર સરોવર
B. સુર સરોવર સરોવર
C. નંદુર માધમેશ્વર
D. સરસાઈ નવર સરોવર

64. In November 2020, _____was added Ramsar site in Maharashtra.
A. Lonar Lake
B. Sur Sarovar lake
C. Nandur Madhameshwar
D. Sarsai Nawar Jheel

65. નવેમ્બર 2020 માં ઉત્તરપ્રદેશમાં કઈ રામસર સાઇટ ઉમેરાઈ?
A. લોનાર તળાવ
B. સુર સરોવર તળાવ
C. નંદુર માધમેશ્વર
D. સરસાઈ નવર ઝીલ

65. In November 2020, _____was added Ramsar site in Uttar Pradesh.
A. Lonar Lake
B. Sur Sarovar lake
C. Nandur Madhameshwar
D. Sarsai Nawar Jheel

66. તાજેતરમાં સ્થાપિત ઓસ્ટ્રેલિયા-ઇન્ડિયા વોટર સેન્ટર નું નેતૃત્વ યુનિવર્સિટી વેસ્ટર્ન સિડની દ્વારા અને ભારતથી સંચાલન _____ સંસ્થા કરે છે.
A. IIT, કાનપુર B. IIT, ગુવાહાટી
C. IIT, દિલ્હી D. IIT, મુંબઇ

66. The recently established Australia-India Water Centre (AIWC) is led by University of Western Sydney from the Australian side and _____from Indian side.
A. IIT, Kanpur B. IIT, Guwahati
C. IIT, Delhi D. IIT, Mumbai

67. ભારતમાં સૌપ્રથમ કયા રાજ્યએ IUCN સાથે ઓર્કિડ રેડ લિસ્ટ તૈયાર કરવા માટે કરાર કર્યા છે?

A. ગુજરાત B. મહારાષ્ટ્ર
C. અશમ D. અરુણાચલ પ્રદેશ

68. તાજેતરમાં કયા ભારતીયને પર્યાવરણીય સિદ્ધિ માટે "ટાઈલર પ્રાઈઝ" એનાયત કરવામાં આવ્યો??

A. પાવન સુખદેવ
B. કાર્તિકેય સારાભાઇ
C. ઇન્દિરા બોરા
D. સુનિતા નારાયણ

69. "ભારતમાં વાઘની સ્થિતિ" પર MoEFCC નો તાજેતરનો અહેવાલ કયા વર્ષ માં બહાર પાડવામાં આવ્યા?

A. 2020 B. 2019
C. 2018 D. 2017

70. MoEFCC "બ્લુ ફ્લેગ" ઇન્ટરનેશનલ ઇકો-લેબલ માટે 8 ભારતીય સમુદ્રતટની ભલામણ કરી છે, જેમાં ગુજરાતના કયા સમુદ્રતટને બ્લુ ફ્લેગ કેટેગરીમાં શામેલ છે?

A. શિવરાજપુર B. માધવપુર
C. દ્વારકા D. તિથલ

71. તાજેતરમાં ભારતનાં કયા રાષ્ટ્રીય ઉદ્યાનમાં તેની હદમાં 3053 હેક્ટરનો ઉમેરો થયો છે ?

A. ગીર રાષ્ટ્રીય ઉદ્યાન
B. કાજીરંગા રાષ્ટ્રીય ઉદ્યાન
C. કોર્બેટ રાષ્ટ્રીય ઉદ્યાન
D. સાયલન્ટ વેલી રાષ્ટ્રીય ઉદ્યાન

72. IUCN ના વર્લ્ડ હેરિટેજના અહેવાલ મુજબ ભારતનું કયુ ક્ષેત્ર મહત્વનું છે?

A. હિમાલય
B. સુંદરવન
C. પશ્ચિમ ઘાટ
D. ઉત્તર પૂર્વ ક્ષેત્ર

67. India's first ever state to sign up MoU with IUCN to prepare a red-list of Orchids

A. Gujarat B. Maharashtra
C. Asham D. Arunachal Pradesh

68. Recently, the prestigious Tyler Prize winner Indian for Environmental achievement.

A. Pavan Sukhdev
B. Karthikey Sarabhai
C. Indira Bora
D. Sunita Narain

69. The latest report of MoEFCC on "Status of Tigers in India" was released in

A. 2020 B. 2019
C. 2018 D. 2017

70. MoEFCC has recommended 8 Indian beaches for the "Blue Flag" International eco-label, which of the Gujarat's beach is included in blue flag category

A. Shivrajpur B. Madhavpur
C. Dwarka D. Tithal

71. Which of India's national park has recently added 3053 hectares into its extent?

A. Gir National Park
B. Kaziranga National Park
C. Corbett National Park
D. Silent Velley National Park

72. Which region of India is of great concern as per IUCN's latest report on World Heritage?

A. Himalayas
B. Sundarbans
C. Western Ghats
D. North East Region

73. ભારતનાં કયા રાજ્યએ તાજેતરમાં અમુર ફાલ્કનનાં શિકાર સામે સરકારી ચેતવણી આપી છે?
A. અરુણાચલ પ્રદેશ
B. ત્રિપુરા
C. પશ્ચિમ બંગાળ
D. ગુજરાત

74. કઈ સંસ્થાએ જાહેર કર્યું કે ઓગસ્ટ 2020 માં 1926 થી લઇ આજદિન સુધીનો સૌથી વધુ વરસાદ નોંધાયો?
A. BISAG B. ISRO
C. IMD D. WMD

75. ભારતમાં કઈ સંસ્થા દ્વારા વન આવરણના દ્વિવાર્ષિક આકારણી કરવામાં આવે છે?
A. ભારતીય રિમોટ સેન્સિંગ સંસ્થા, દહેરાદૂન
B. વન સંશોધન સંસ્થા, દહેરાદૂન
C. ભારતીય વન વ્યવસ્થાપન સંસ્થા, ભોપાલ
D. ભારતીય વન સર્વે, દહેરાદૂન

76. ભારતમાં વન સંશોધન સંસ્થા, દહેરાદૂનની સ્થાપના કયા વર્ષ માં કરવામાં આવી હતી?
A. 1986 B. 1896
C. 1906 D. 926

77. અમૃતા દેવી એવોર્ડ દર વર્ષે શાના માટે આપવામાં આવે છે?
A. શિક્ષણ
B. બાળ આરોગ્ય
C. જળ સંરક્ષણ
D. વૃક્ષ અને વન્યપ્રાણી સંરક્ષણ

78. અમૃતા દેવી બિસ્નોઇ શાના માટે જાણીતા છે?
A. જળસંગ્રહ
B. બાળકી શિક્ષણ
C. અનાથાશ્રમ
D. ખીજરીના વૃક્ષનું રક્ષણ

73. Which Indian state has recently issued official warning against hunting of Amur falcons?
A. Arunachal Pradesh
B. Tripura
C. West Bengal
D. Gujarat

74. Which organization revealed that India received the highest rainfall in August 2020 since 1926?
A. BISAG B. ISRO
C. IMD D. WMD

75. Biennial assessment of forest cover in India is done by
A. Indian Institute of Remote Sensing, Dehradun
B. Forest research Institute, Dehradun
C. Indian Institute of Forest Management, Bhopal
D. Forest Survey of India, Dehradun

76. In India systematic forest research started after establishment of Forest Research Institute at Dehradun in the year
A. 1986 B. 1896
C. 1906 D. 1926

77. Amrita Devi award is given every year for
A. Education
B. Child health
C. Water conservation
D. Tree and wildlife conservation

78. Amrita Devi Bisnoi Known for?
A. Water conservation
B. Girl child education
C. Orphanage
D. Protection of Khejri tree

79. પર્યાવરણને બચાવવા માટે કયા ત્રણ સામાન્ય "R" વપરાય છે.

A. Reproduce, Reduce, Recycle
B. Reduce, Reuse, Recreate
C. Reduce, Reuse, Recycle
D. Recreate, Reserve, Reuse

80. બિશ્નોઇ સમુદાયની સ્થાપના કોને કરી હતી?

A. ભગવાન બિશ્નેશ્વર
B. ભગવાન જાડેશ્વર
C. ભગવાન જંભેશ્વર
D. ભગવાન કરુણેશ્વર

81. રાજસ્થાનનો કયો સમુદાય તેમની ધાર્મિક માન્યતાઓ પર વન અને વન્યપ્રાણીઓ નો બચાવ કરે છે?

A. ગુર્જર B. ઠાકુર
C. રાજપૂત D. બિશ્નોઇ

82. જંગલો અને પર્યાવરણ સંરક્ષણ માટે કયું આંદોલન થયું હતું?

A. નવનિર્માણ આંદોલન
B. વૃક્ષ આંદોલન
C. ચિપકો આંદોલન
D. સંરક્ષણ આંદોલન

83. શંકુદ્રુપ વન ક્યાં મળી આવે છે?

A. ઘાસના મેદાનો B. શુષ્ક વિસ્તારો
C. નદીના પટ્ટાઓ D. હિમાલય

84. ભારતમાં કયા રાજ્યમાં ક્ષેત્રની દ્રષ્ટિએ મહત્તમ જંગલો આવેલા છે?

A. અરુણાચલ પ્રદેશ
B. મધ્યપ્રદેશ
C. મિઝોરમ
D. નાગાલેન્ડ

85. સામાજિક વનીકરણ, ગ્રામીણ વસ્તીને કઈ દ્રષ્ટિએ લાભ પ્રદાન કરી શકે છે?

A. બળતણ B. ઘાસચારો
C. ઇમારતી લાકડું D. ઉપરોક્ત તમામ

79. The three common R's used to save the environment

A. Reproduce, Reduce, Recycle
B. Reduce, Reuse, Recreate
C. Reduce, Reuse, Recycle
D. Recreate, Reserve, Reuse

80. Bishnoi community was founded by

A. Lord Bishneshwar
B. Lord Jadeshwar
C. Lord Jambheshwar
D. Lord Karuneshwar

81. The community of Rajasthan saving forest and wildlife on their religious beliefs

A. Gurjar B. Thakur
C. Rajput D. Bishnoi

82. Which movement was for the forests and environment conservation?

A. Navnirman Andolan
B. Vruksh Andolan
C. Chipko Movement
D. Sanrakshan Movement

83. Coniferous forest found in

A. Grassland B. Arid areas
C. River belts D. Himalayas

84. Which state in India has the maximum percentage of its area covered by forests?

A. Arunachal Pradesh
B. Madhya Pradesh
C. Mizoram
D. Nagaland

85. Social forestry is providing benefits to rural population in terms of

A. fuelwood B. fodder
C. Timber D. All above

86. ડેઝર્ટીફીકેશન માટેનાં મુખ્ય કારણો કયા છે?
A. વધુ પડતું ચારણ
B. વધારે ખેતી
C. ડીફોરેસ્ટેશન
D. ઉપરોક્ત તમામ

87. માનવની કઈ ક્રિયા દ્વારા જૈવવિવિધતામાં કુદરતી વૈશ્વિક પેટર્ન બદલાય છે?
A. મોટા પાયે જમીન પરિવર્તન
B. બળતણ લાકડું અને કોલસા ઉત્પાદન
C. સ્લેષ અને બર્ન કૃષિ
D. ઉપરોક્ત તમામ

88. જંગલોના વિસ્તારમાં ફાયર લાઇનોને ક્યા કારણે તૈયાર કરવામાં આવે છે?
A. આગની ઘટનાઓ અને ફેલાવાને અંકુશમાં લેવા
B. વન્યપ્રાણી જીવન નિરીક્ષણ કરવા
C. વન્યપ્રાણીસૃષ્ટિને ખુલ્લી જગ્યા પુરી પાડવા
D. ટ્રેકિંગ પાથ વિકસાવવા

89. ઘાસના મેદાનોના સંચાલનમાં રેન્ડમ ફાયર ફાયદાકારક છે, કારણ કે
A. ઘાસના મેદાનોમાં નીંદણને નિયંત્રિત કરવા
B. ઘાસના વિકાસને નિયંત્રિત કરવા
C. ઘાસના નવા ફણગાઓ માટે ખુલ્લી જગ્યા પુરી પાડવા
D. જમીનના માળાને બાળી નાખવા

90. ડાંગમાં ધરુંના ઉત્પાદન માટે પરંપરાગત આદર પદ્ધતિ કરવામાં આવે છે જેની કઈ નકારાત્મક અસરો થાય છે ?
A. વૃક્ષોના વિકાસને અસર કરે છે
B. જમીનના ધોવાણમાં વધારો
C. જમીનમાં પાણીની જાળવણી ક્ષમતામાં ઘટાડો
D. નિંદામણમાં વધારો

86. The Primary reasons for desertification are
A. overgrazing
B. over cultivation
C. deforestation
D. All above

87. Natural global patterns in biodiversity are also altered by human action
A. large scale land conversion
B. fuelwood and charcoal production
C. slash and burn agriculture
D. All above

88. Fire lines in forests area are prepared because of
A. To control fire incidences and spread.
B. To observe wildlife
C. To create openings to wildlife
D. To develop trekking path

89. In grasslands management, randomly fires are beneficial to area, because of
A. To control weeds in grasslands
B. To control growth of grasses
C. To create openings for new sprouts of grass
D. To Burn the ground nests

90. Raab burnig in Dangs is a traditional practice for raising Dharu, but negative impacts of this practice
A. affects the growth of trees
B. increase soil erosion
C. decrease water retention capacity of soils
D. enhanced weeds germination

91. વન વિસ્તારોમાં સ્થળાંતર ખેતી ને લીધે શું થાય છે?

A. કુદરતી સંસાધનોને નુકસાન પહોંચાડે છે
B. વનસ્પતિ અંકુરણ વધારો
C. વિસ્તારની જૈવવિવિધતામાં વધારો
D. વૃક્ષોના કુદરતી વિકાસમાં મદદ કરે છે

92. સુંદરવનના જંગલોમાં સામાન્ય રીતે વાઘ માણસો પર ક્યારે હુમલો કરે છે?

A. માછીમારી
B. પ્રવાસન
C. ઢોર ચરાવવા
D. મધ સંગ્રહ

93. કચ્છના નાના રણમાં પાણીનાં પક્ષીઓની વસ્તી કયા કારણે અસરગ્રસ્ત છે?

A. પશુ ચરાવવા
B. સોલ્ટ પાન્સ
C. માછીમારી
D. લોકોની અવરજવર

94. જંગલોના વિસ્તારમાં જમીનમાં સંકોચન શા માટે થાય છે?

A. સતત વરસાદ B. ફાયર લાઇનો
C. ઘાસ કાપવા D. પશુ ચરાવવા

95. જંગલમાં ઝાડનું ડિફોર્મેશન ક્યારે થાય છે?

A. લોકો ડાળીઓ આડેધડ કાપે તો
B. જંગલની આગને કારણે
C. કીટકો અને જીવાતોને લીધે
D. પશુ ચરાવવાને લીધે

96. માનવની કઈ પ્રવૃત્તિને લીધે ઉભા લીલા ઝાડ સૂકાઈ જાય છે?

A. શાખાઓ કાપણી
B. ઘાસચારા માટે પાંદડા
C. બધા ફળોનો સંગ્રહ
D. ગોળ આકારમાં છાલ કાઢવી

91. Shifting cultivation in forest areas

A. damages the natural resources
B. increase vegetation germination
C. increase biodiversity of the area
D. help in natural growth of the trees

92. In Sundervan forests, human deaths by Bengal Tiger caused when, people go for

A. Fishing
B. Tourism
C. Cattle grazing
D. Honey collection

93. The population of water birds in Little Rann of Kutch is affected by

A. Cattle Grazing
B. Salt pans
C. Fishing
D. People movements

94. Compaction of soils occurs in forests area because of

A. Continuous rains B. Fire lines
C. Cutting of Grasses D. Cattle grazing

95. In forest, the deformation of trees occurs when

A. People cut the branches haphazardly
B. Because of forest fire
C. Because of insects and pests
D. Because of grazing

96. Standing green trees dry, because of the human activity of

A. Pruning the branches
B. Removing leaves for fodder
C. Collection of all the fruits
D. Removal of bark in circular shape

97. સામાન્ય રીતે ખેડૂતો ખેતી જમીન ની અંદર ના ઝાડો કાપી નાખે છે પરંતુ રણ પ્રદેશના ખીજરીના ઝાડને કયા કારણોસર કાઢવામાં આવતા નથી?

A. ધાર્મિક વૃક્ષ

B. કાપવા માટે ખૂબ સખત

C. ઘાસચારા માટે

D. સૂર્યની ગરમીથી કૃષિ પાકને બચાવે છે

97. Generally, farmers remove the trees inside the agriculture fields, but the desert region, Khejri (Prosopis cineraria) trees are sporadically retained in agriculture field because of

A. Religious tree

B. Very hardy to cut

C. For fodder purpose

D. Save the agriculture crops from heat of the Sun

98. કયા કારણોથી ગીરમાં માલધારીઓ દીપડા અથવા સિંહ ઉપર ભાગ્યે જ હુમલો કરે છે?

A. જંગલી પ્રાણીઓ માટે જાણીતા છે

B. શાકાહારીઓ અને દૂધના ઉત્પાદનોનો ઉપયોગ કરે છે

C. પ્રાણીઓ સફેદ કપડાથી ભયભીત છે

D. પ્રાણીઓ માલધારીઓથી ડરે છે

98. The Cattle Graziers (Maldharis) rarely attack leopard or lion in Gir, because of

A. Known to wild animals

B. Vegetarians and use milk products

C. Afraid of white clothes

D. Afraid of Maldharis

99. કેટલીકવાર મોરના કૃષિ ક્ષેત્રોમાં મૃત્યુ થાય છે, કારણ કે

A. ખેડુતોએ ઇરાદાપૂર્વક અનાજમાં ઝેર ભેળવે છે

B. સરિસૃપોને મારવા માટે ખેડૂતો ઝેર ભેળવે છે

C. મગફળીના દાણા એન્ટી ટર્મિટેટ જંતુનાશકો સાથે કોટેડ કરવામાં આવે છે

D. લીલા પાક પર જંતુનાશક દવા છંટકાવ

99. Sometimes peacocks are caused death in agriculture fields, because of

A. Farmers intentionally mixed poison in grains

B. Farmers mixed poison to kill reptiles

C. Groundnut seeds are being coated with anti-termite pesticides

D. Spraying of insecticide on green crops

100. અળસિયા ખેડૂતોના મિત્ર છે, કારણ કે

A. કૃષિ ક્ષેત્રોમાં જમીન છિદ્રાળુ કરે છે

B. કૃમિ ખાતરમાં વપરાય છે

C. અળસિયા કીટકોને ખાય છે

D. બંને A અને B

100. Earthworms is friend of the farmers because of

A. Loosen the earth in agriculture fields

B. Used in vermicomposting

C. Earthworms eat the insects

D. Both A & B

Answers | જવાબો

Que.	Ans.	Que.	Ans.	Que.	Ans.	Que.	Ans.	Que.	Ans.	Que.	Ans.	Que.	Ans.
1.	A	16.	B	31.	D	46.	D	61.	A	76.	C	91.	A
2.	D	17.	D	32.	B	47.	B	62.	C	77.	D	92.	D
3.	B	18.	B	33.	A	48.	C	63.	D	78.	D	93.	C
4.	A	19.	D	34.	B	49.	C	64.	A	79.	C	94.	D
5.	B	20.	C	35.	C	50.	B	65.	B	80.	C	95.	A
6.	B	21.	C	36.	C	51.	A	66.	B	81.	D	96.	D
7.	D	22.	B	37.	D	52.	D	67.	D	82.	C	97.	D
8.	A	23.	B	38.	C	53.	D	68.	A	83.	D	98.	B
9.	B	24.	D	39.	A	54.	C	69.	C	84.	C	99.	C
10.	C	25.	C	40.	B	55.	A	70.	A	85.	D	100.	D
11.	D	26.	A	41.	D	56.	A	71.	B	86.	D		
12.	D	27.	D	42.	C	57.	B	72.	C	87.	D		
13.	D	28.	A	43.	B	58.	C	73.	B	88.	A		
14.	D	29.	B	44.	C	59.	D	74.	C	89.	A		
15.	C	30.	C	45.	D	60.	D	75.	D	90.	A		

Chapter 16

પ્રકરણ ૧૬

Environmental Science and Climate Change

પર્યાવરણ વિજ્ઞાન અને ક્લાઈમેટ ચેન્જ

Missions under National Action Plan for Climate Change (NAPCC)

Released in	2008
Missions	8

Mission	Key points
Green India Mission	• Aims at afforestation of 6 million ha of degraded forest lands and expanding forest cover from 23 to 33% of India's territory • Mission to increase forest and trees cover in 5 million ha and to improve ecosystem services in 10 million ha of land • Increase flow of forest-based livelihood services and income of about 3 million forest dependent household
National Mission for Enhanced Energy Efficiency	• Trade of energy-saving certificates, financing for public-private partnership to reduce energy consumption through demand-side management programs in the municipal, buildings and agricultural sectors • Energy incentives including reduced taxes on energy-efficient appliances
National Mission for Sustainable Agriculture	• Aims to support climate adaptation in agriculture through the development of climate-resilient crops, expansion of weather insurance mechanisms, and agricultural practices
National Mission for Sustaining Himalayan Ecosystem	• Sets the goal to prevent melting of the Himalayan glaciers and to protect biodiversity in the Himalayan region
National Mission on Strategic Knowledge for Climate Change	• Aims to gain better understanding of climate science, impacts, and challenges, the plan envisions a new Climate Science Research Fund, improved climate modeling, and increased international collaboration. • To encourages private sector initiatives to develop adaptation and mitigation technologies through venture capital funds
National Mission on Sustainable Habitat	• Promoting energy efficiency as a core component of urban planning by extending the existing Energy Conservation Building Code

Mission	Key points
	• Strengthening the enforcement of automotive fuel economy standards of public transport • Emphasizes on waste management and recycling
National Solar Mission	• To promote the development and use of solar energy for power generation and other uses • Making solar competitive with fossil fuels energy options • Establishment of solar research centers, increased international collaboration on technology development. • Strengthening of domestic manufacturing capacity and increased govt. funding and international support
National Water Mission	• Sets a goal of 20% improvement in water use efficiency through pricing and other measures to deal with water scarcity as a result of climate change

ક્લાઈમેટ ચેન્જના રાષ્ટ્રીય એક્શન પ્લાન અંતર્ગતના મિશન

મિશન	મહત્વના મુદ્દા
ગ્રીન ઇન્ડિયા મિશન	• ભારતના 23 થી 33 ટકા ક્ષેત્રના ડિગ્રેડેડ જંગલની જમીનોના 6 મિલિયન હેક્ટરના વનીકરણ અને વન વિસ્તારને વિસ્તૃત કરવાના લક્ષ્ય • 5 મિલિયન હેક્ટરમાં જંગલ અને ઝાડને આવરી લેવાનું અને 10 મિલિયન હેક્ટર જમીનમાં ઇકોસિસ્ટમ સેવાઓ સુધારવાનું મિશન • વન આધારિત આજીવિકા સેવાઓનો પ્રવાહ અને આશરે 30 મિલિયન વન આધારિત પરિવારોની આવક વધારવી
ઉન્નત ઊર્જા કાર્યક્ષમતા માટે રાષ્ટ્રીય મિશન	• ઊર્જા બચત પ્રમાણપત્રો, નગરપાલિકા, મકાનો અને કૃષિ ક્ષેત્રમાં માંગ-બાજુના સંચાલન કાર્યક્રમો દ્વારા ઊર્જા વપરાશ ઘટાડવા માટે જાહેર-ખાનગી ભાગીદારી માટે ધિરાણ • ઉર્જા-કાર્યક્ષમ ઉપકરણો પર ઓછા કર સહિત ઊર્જા પ્રોત્સાહન
ટકાઉ કૃષિ માટે રાષ્ટ્રીય મિશન	• પાકના ક્લાઈમેટ એડેપ્ટેશન થકી વિકાસ, હવામાન વીમા મિકેનિઝમના વિસ્તરણ અને કૃષિ પદ્ધતિઓ દ્વારા કૃષિ ક્ષેત્રમાં આબોહવા અનુકૂલનને ટેકો આપવાના હેતુ
હિમાલય ઇકોસિસ્ટમને ટકાવી રાખવા માટેનું રાષ્ટ્રીય મિશન	• હિમાલયના હિમનદીઓનું ગલન અટકાવવા અને હિમાલયના ક્ષેત્રમાં જૈવવિવિધતાને સુરક્ષિત રાખવા માટેનું લક્ષ્ય નક્કી કરે છે
ક્લાઈમેટ ચેન્જ માટે વ્યૂહાત્મક જ્ઞાન પર રાષ્ટ્રીય મિશન	• આબોહવા વિજ્ઞાન, પ્રભાવો અને પડકારોની વધુ સારી સમજણ મેળવવાના હેતુથી, આ યોજનામાં એક નવું ક્લાઈમેટ વિજ્ઞાન સંશોધન ભંડોળ, સુધારેલ હવામાન મોડેલિંગ અને આંતરરાષ્ટ્રીય સહયોગમાં વધારો કરવાની કલ્પના છે.

મિશન	મહત્વના મુદ્દા
	• સાહસ મૂડી ભંડોળ દ્વારા અનુકૂલન અને શમન ટેકનોલોજી વિકસાવવા માટે ખાનગી ક્ષેત્રની પહેલને પ્રોત્સાહન આપવું
સસ્ટેનેબલ હેબીટાટ પર રાષ્ટ્રીય મિશન	• હાલની ઊર્જા સંરક્ષણ બિલ્ડિંગ કોડનો વિસ્તાર કરીને શહેરી આયોજનના મુખ્ય ઘટક તરીકે ઊર્જા કાર્યક્ષમતા ને પ્રોત્સાહન આપવું • જાહેર પરિવહનના ઓટોમોટિવ ફ્યુઅલ ઇકોનોમી ધોરણોના અમલીકરણને મજબૂત બનાવવું • કચરો વ્યવસ્થાપન અને રિસાયક્લિંગ પર ભાર મુકવો
રાષ્ટ્રીય સોલર મિશન	• વીજ ઉત્પાદન અને અન્ય ઉપયોગ માટે સૌર ઊર્જાના વિકાસ અને ઉપયોગને પ્રોત્સાહન આપવું • અશ્મિભૂત ઇંધણ ઊર્જા વિકલ્પો સાથે સૌર સ્પર્ધાત્મક બનાવવી • સૌર સંશોધન કેન્દ્રોની સ્થાપના, તકનીકી વિકાસ પર આંતરરાષ્ટ્રીય સહયોગમાં વધારો. • ઘરેલું ઉત્પાદન ક્ષમતાને મજબૂત બનાવવી અને સરકારી ભંડોળમાં વધારો કરવો અને આંતરરાષ્ટ્રીય મદદ મેળવવી
રાષ્ટ્રીય જળ મિશન	• ક્લાઈમેટ ચેન્જને પરિણામે પાણીની અછતને પહોંચી વળવા ભાવો અને અન્ય પગલાં દ્વારા પાણીના વપરાશ કાર્યક્ષમતામાં 20% સુધારણાના લક્ષ્યાંકને સુયોજિત કરવો.

UN's Climate Change Conferences | યુએનની ક્લાઈમેટ ચેન્જ પરિષદો

COP 1	1995	Berlin, Germany (First UNFCCC Conference of Parties)
COP 2	1996	Geneva, Switzerland
COP 3	1997	Kyoto, Japan (The Kyoto Protocol on Climate Change)
COP 4	1998	Buenos Aires, Argentina
COP 5	1999	The Hague, Netherlands
COP 6	2000	Bonn, Germany
COP 7	2001	Marrakech, Morocco
COP 8	**2002**	**New Delhi, India**
COP 9	2003	Milan, Italy
COP 10	2004	Buenos Aires, Argentina
COP 11/ CMP 1	2005	Montreal, Canada
COP 12/ CMP 2	2006	Nairobi, Kenya
COP 13/ CMP 3	2007	Bali, Indonesia
COP 14/ CMP 4	2008	Poznan, Poland
COP 15/ CMP 5	2009	Copenhagen, Denmark
COP 16/ CMP 6	2010	Cancun, Mexico

COP 17/ CMP 7	2011	Durban, South Africa
COP 18/ CMP 8	2012	Doha, Qatar
COP 19/CMP 9	2013	Warsaw, Poland
COP 20/CMP 10	1014	Lima, Peru
COP 21/CMP 11	2015	Paris, France
COP 22/CMP 12/ CMA 1	2016	Marrakech, Morocco
COP 23/CMP 13/ CMA 1-2	2017	Bonn, Germany
COP 24/CMP 14/ CMA 1-3	2018	Katowice, Poland
COP 25/CMP 15/ CMA 2	2019	Madrid, Spain
COP 26/CMP 16/CMA 3	2021	Glasgow, UK

COP - the Conference of the Parties.
CMP - the Conference of the Parties serving as the meeting of the Parties to the Kyoto Protocol.
CMA - the Conference of the Parties serving as the meeting of the Parties to the Paris Agreement.

UNFCCC 2020

Financial Architecture of Global Climate | વૈશ્વિક ક્લાઇમેટનું નાણાકીય આર્કિટેક્ચર

Program	Operational date	Administered by	Focus
GEF Trust Fund - Climate Change focal area	1991	The Global Environment Facility (GEF)	Adaptation, Mitigation-general
Special Climate Change Fund	2002	The Global Environment Facility (GEF)	Adaptation
Least Developed Countries Fund	2002	The Global Environment Facility (GEF)	Adaptation
Biocarbon Fund	2004	The World Bank	Adaptation, Mitigation-general, Mitigation-REDD
Strategic Priority on Adaptation	2004	The Global Environment Facility (GEF)	Adaptation
MDG Achievement Fund-Environment and Climate Change thematic window	2007	UNDP	Adaptation, Mitigation-general
Strategic Climate Fund	2008	The World Bank	Adaptation, Mitigation-general, Mitigation-REDD
Pilot Program for Climate Resilience	2008	The World Bank	Adaptation
Clean Technology Fund	2008	The World Bank	Mitigation-general
Forest Carbon Partnership Facility	2008	The World Bank	Mitigation-REDD
Global Climate Change Alliance	2008	The European Commission	Adaptation, Mitigation-general, Mitigation-REDD

Program	Operational date	Administered by	Focus
Global Energy Efficiency and Renewable Energy Fund	2008	The European Commission	Mitigation-general
UN-REDD Programme	2008	UNDP	Mitigation-REDD
Congo Basin Forest Fund	2008	African Development Bank	Mitigation-REDD
Forest Investment Program	2009	The World Bank	Mitigation-REDD
Scaling-Up Renewable Energy Program for Low Income Countries	2009	The World Bank	Mitigation-general
Adaptation Fund	2009	Adaptation Fund Board	Adaptation
Amazon Fund (Fundo Amazônia)	2009	Brazilian Development Bank (BNDES)	Mitigation-REDD
Indonesia Climate Change Trust Fund	2010	Indonesia's National Development Planning Agency	Adaptation, Mitigation-general, Mitigation-REDD
Partnership for Market Readiness	2011	The World Bank	Mitigation-general
Adaptation for Smallholder Agriculture Program	2012	The International Fund for Agricultural Development	Adaptation
Green Climate Fund	2015	UNFCC	Adaptation, Mitigation-general, Mitigation-REDD

Multiple Choice Questions | હેતુલક્ષી પ્રશ્નો

1. વહીવટની સરળતા અને અસરકારક સંરક્ષણ માટે MoEFCC એ ભારતને ______ ઝોનમાં વહેંચ્યું છે.

A. 10 B. 8
C. 6 D. 4

2. ભારતની નેશનલ એક્શન પ્લાન ફોર ક્લાઇમેટ ચેન્જ ની શરૂઆત ક્યારે કરવામાં આવી હતી

A. 2015 B. 2018
C. 2010 D. 2008

3. પર્યાવરણીય કાર્યકર જાદવ પાયેંગ ક્યા નામે ઓળખાય છે?

A. જળ માણસ
B. વન માણસ
C. પર્વતનો માણસ
D. જમીનનો માણસ

4. વિશ્વ પર્વત દિવસ ક્યારે ઉજવવામાં આવે છે?

A. 12 જૂન B. 15 ઓગસ્ટ
C. 11 ડિસેમ્બર D. 29 ડિસેમ્બર

5. 40% એમેઝોન વન સવાનામાં બદલાવાનું જોખમ ધરાવે છે. આ મુખ્યત્વે શા કારણે છે?

A. ગ્રીનહાઉસ ગેસ ઉત્સર્જનમાં વધારો
B. ઓછો વરસાદ
C. વનનાબૂદી
D. બંને A અને B

6. UN એન્વાયર્નમેન્ટ પ્રોગ્રામ એ ______ નામનું એક નવું પ્રકાશન શરૂ કર્યું.

A. The Little Book of Green Nudges
B. Environment friendly habits and greener lifestyles
C. The Little Book of environment friendly habits
D. The book of greener lifestyles

1. For the ease of administration and effective conservation, MoEFCC has divided India into ______ zones.

A. 10 B. 8
C. 6 D. 4

2. India's National Action Plan for Climate Change (NAPCC) was launched in

A. 2015 B. 2018
C. 2010 D. 2008

3. Environmental activist Jadav Payeng is known as the

A. Water man
B. Forest man
C. Mountain man
D. Soil man

4. World mountain day is celebrated on

A. 12 June B. 15 August
C. 11 December D. 29 December

5. 40% of Amazon forest is at the risk of changing into Savannah like environment. This is mainly because of

A. increased Greenhouse gas emissions
B. reduced rainfall
C. deforestation
D. Both A and B

6. The UN Environment Programme (UNEP) launched a new publication named

A. The Little Book of Green Nudges
B. Environment friendly habits and greener lifestyles
C. The Little Book of environment friendly habits
D. The book of greener lifestyles

7. વિશ્વનો સૌથી મોટું સોલર ટ્રી ______ દ્વારા વિકસિત કરવામાં આવ્યું છે.
A. CSIR-IHBT B. CSIR-AMPRI
C. CSIR-CMERI D. CSIR-CDRI

8. 26મી સપ્ટેમ્બર શા માટે ઉજવવામાં આવે છે?
A. વિશ્વ પ્રદૂષણ દિવસ
B. વિશ્વ વન દિવસ
C. વિશ્વ સંરક્ષણ દિવસ
D. વિશ્વ પર્યાવરણીય આરોગ્ય દિવસ

9. સંયુક્ત રાષ્ટ્રએ સસ્ટેનેબલ ડેવલપમેન્ટ ગોલ માટે 2020 વર્ગના યુવા નેતાઓની સૂચિમાં એક ભારતીય નું નામ લીધું છે, તે કોણ છે?
A. ઉદિત સિંઘલ B. વિશાલ ગર્ગ
C. ઉદય સિંગલ D. રોહિત શર્મા

10. છઠ્ઠા બ્રિક્સ પર્યાવરણની બેઠક ક્યાં મળી હતી?
A. જર્મની B. રશિયા
C. ઇંગ્લેન્ડ D. ચીન

11. ______ ક્લાઈમેટ ચેન્જનું સર્વવ્યાપક પ્રતીક માનવામાં આવે છે.
A. વ્હેલ B. સિંહો
C. ધ્રુવીય રીંછ D. પેંગ્વીન

12. ______ વર્ષને જૈવવિવિધતા માટેના સુપર વર્ષ તરીકે ઓળખવામાં આવે છે.
A. 2015 B. 2017
C. 2019 D. 2020

13. કયા વર્ષમાં રાષ્ટ્રીય પર્યાવરણ સંરક્ષણ અધિનિયમ આપવામાં આવ્યો હતો?
A. 1967 B. 1973
C. 1982 D. 1986

14. પર્યાવરણ સાથે વિવિધ પ્રજાતિઓના જૂથના સંબંધનો અભ્યાસને શું કહે છે?
A. ઓટેકોલોજી B. સિનેકોલોજી
C. ઇકોલોજી D. બાયોલોજી

7. **World's largest solar tree has been developed by**
A. CSIR-IHBT B. CSIR-AMPRI
C. CSIR-CMERI D. CSIR-CDRI

8. **26th September is observed as**
A. World Pollution Day
B. World Forest Day
C. World Conservation Day
D. World Environmental Health Day

9. **The United Nations has named an Indian teenager in the list of 2020 Class of Young leaders for Sustainable Development Goals (SDG)**
A. Udit Singhal B. Vishal Garg
C. Uday Singal D. Rohit Sharma

10. **6th BRICS Environment meeting was held in**
A. Germany B. Russia
C. England D. China

11. **______ are considered as ubiquitous symbol of climate change**
A. Whales B. Lions
C. Polar beers D. Penguins

12. **______ year is called as the Super Year for Biodiversity**
A. 2015 B. 2017
C. 2019 D. 2020

13. **National Environment Protection Act was given in the year-**
A. 1967 B. 1973
C. 1982 D. 1986

14. **The study of relationship of group of different species with their environment is called**
A. Autecology B. Synecology
C. Ecology D. Biology

15. ઇકોસિસ્ટમ એ પર્યાવરણનું સૌથી નાનું ______ એકમ છે.

A. માળખાકીય

B. કાર્યાત્મક

C. જૈવિક

D. બંને A અને B

16. દર વર્ષે સપ્ટેમ્બરનો ત્રીજો શનિવાર ______ તરીકે ઉજવાય છે.

A. આંતરરાષ્ટ્રીય સમુદ્રતટ સફાઇ દિવસ

B. આંતરરાષ્ટ્રીય આરોગ્ય દિવસ

C. આંતરરાષ્ટ્રીય પર્યાવરણ દિવસ

D. આંતરરાષ્ટ્રીય ક્લાઈમેટ ચેન્જ દિવસ

17. ગ્રીનપીસ ઇન્ડિયા દ્વારા તાજેતરમાં જારી કરાયેલા અહેવાલ મુજબ, ભારત વિશ્વનો સૌથી મોટો______ ઉત્સર્જક દેશ છે.

A. નાઇટ્રોજન ઓક્સાઇડ

B. કાર્બન મોનોક્સાઈડ

C. સલ્ફર ડાયોક્સાઇડ

D. બંને A અને C

18. યુએન ક્લાઇમેટ ચેન્જ સમીટ, 2019 નું આયોજન ક્યાં કરવામાં આવ્યું હતું.

A. લંડન B. ન્યુ યોર્ક

C. ટોક્યો D. દિલ્હી

19. ભારતમાં સૌથી ઓછું જંગલ અને ઝાડનું આવરણ કયા રાજ્યમાં છે.

A. પંજાબ

B. રાજસ્થાન

C. હરિયાણા

D. ઉત્તરપ્રદેશ

20. MOSAIC એ ______ ના અભ્યાસ માટે મલ્ટિડિસિપ્લિનરી ડ્રિફ્ટિંગ વેધશાળા છે.

A. એન્ટાર્કટિકાનો ક્લાઈમેટ

B. આર્કટિકનો ક્લાઈમેટ

C. હિમાલયનો ક્લાઈમેટ

D. શુષ્ક રણનો ક્લાઈમેટ

15. Ecosystem is the smallest ______ unit of environment

A. Structural

B. Functional

C. Biotic

D. Both A and B

16. Every year the third Saturday of September is marked as the

A. International Coastal Cleanup Day

B. International Health Day

C. International Environment Day

D. International Climate Change Day

17. According to recently released report by Greenpeace India, India is the largest emitter of ______ in the world

A. Nitrogen oxides

B. Carbon monoxide

C. Sulphur dioxide

D. Both A and C

18. The UN Climate Change Summit, 2019 was held in

A. London B. New York

C. Tokyo D. Delhi

19. In India lowest forest and tree cover is in which state

A. Punjab

B. Rajasthan

C. Haryana

D. Uttar Pradesh

20. MOSAIC is Multidisciplinary drifting Observatory for the Study of

A. Antarctica Climate

B. Arctic Climate

C. Himalayan Climate

D. Arid Desert Climate

21. સસ્ટેનેબલ ડેવલપમેન્ટ ગોલ 13 મુજબ, ________ વિશ્વના વિકાસ તરફનો સૌથી મોટો ખતરો છે

A. પ્રદૂષણ
B. વસ્તી વધારો
C. ક્લાઈમેટ ચેન્જ
D. બંને A અને B

22. ગ્લોબલ વોર્મિંગના સંદર્ભમાં "ગ્રીન હાઉસ ઇફેક્ટ" શું છે?

A. ઠંડક અને ભેજવાળી સ્થિતિ
B. વરસાદ અને લીલોતરીનો વધારો
C. ડેઝર્ટિફિકેશન
D. વોર્મિંગની અસર

23. વાર્ષિક રિંગ્સના આધારે મુખ્યત્વે ઝાડમાંથી ભૂતકાળનું વાતાવરણ નક્કી કરવાનું વિજ્ઞાન ________ કહેવામાં આવે છે.

A. ડેંડ્રોલોજી
B. ડેંડ્રોક્રોનોલોજી
C. ડેંડ્રોક્લીમેટોલોજી
D. ડેંડ્રોઇકોલોજી

24. વૈશ્વિક તાપમાન વૃદ્ધિની દ્રષ્ટિએ સરકારો ગ્લોબલ વોર્મિંગને કેટલું મર્યાદિત કરવા માંગે છે?

A. 2° સેલ્સિયસ
B. 5° સેલ્સિયસ
C. 7° સેલ્સિયસ
D. 10° સેલ્સિયસ

25. કયા દેશોમાં 80% બાળકો કાં તો સપાટીનાં પાણીનો ઉપયોગ કરે છે અથવા સંરક્ષિત જળ સ્રોત શોધવા માટે 15 મિનિટથી વધુ ચાલવું પડે છે?

A. ઉત્તર કોરિયા, રશિયા, યુક્રેન
B. ઇથોપિયા, રવાન્ડા અને યુગાન્ડા
C. દક્ષિણ કોરિયા, ઝિમ્બાબ્વે, એન્ગોલા
D. કોલમ્બિયા, ચીન, ન્યુઝીલેન્ડ

21. According to the Sustainable Development Goal 13, the biggest threat towards world development is

A. Pollution
B. Population growth
C. Climate change
D. Both A and B

22. "Greenhouse effect" with respect to global warming refers to

A. Cooling & moist condition
B. Increased rainfall & greenery
C. Desertification
D. Warming effect

23. Science of determining past climate from trees primarily on the basis of annual rings is called

A. Dendrology
B. Dendrochronology
C. Dendroclimatology
D. Dendroecology

24. How much governments want to limit global warming to in terms of a global temperate increase?

A. 2° Celsius
B. 5° Celsius
C. 7° Celsius
D. 10° Celsius

25. In which countries do 80% of children either using surface water or have to walk more than 15 minutes to find a protected water source?

A. North Korea, Russia, Ukraine
B. Ethiopia, Rwanda and Uganda
C. South Korea, Zimbabwe, Angola
D. Colombia, China, New Zealand

26. ઘણી જીવંત પ્રજાતિઓ લુપ્ત થવાનું મુખ્ય કારણ શું છે?

A. કુદરતી આપત્તિ
B. પ્રદૂષણ
C. વાતાવરણમાં પરિવર્તન
D. માનવ પ્રવૃત્તિઓ

27. પર્યાવરણ શબ્દ એ એક પ્રાચીન ફ્રેન્ચ શબ્દ "એન્વિરો" નો પરથી આવ્યો છે, જેનો અર્થ શું થાય છે?

A. અજૈવિક ઘટકો
B. આસપાસની પરિસ્થિતિ
C. પ્રકૃતિ
D. જૈવિક ઘટકો

28. જંગલોની કાપણીથી શું ઓછું થાય છે?

A. વરસાદ B. માટીનું ધોવાણ
C. દુકાળ D. ગ્લોબલ વોર્મિંગ

29. ક્લાઈમેટ ચેન્જ માટેનું મુખ્ય પરિબળ

A. પ્રદૂષણ વધી રહ્યું છે
B. પૃથ્વીનું વૈશ્વિક તાપમાન વધી રહ્યું છે
C. માનવ વસ્તી વધી રહી છે
D. બંને A અને C

30. ક્લાઈમેટ ચેન્જએ ________ માં બદલાવ છે.

A. તાપમાન B. વરસાદ
C. પવન D. ઉપરોક્ત તમામ

31. INDCs નો અર્થ છે

A. Intended Nationally Determined Contributions
B. Intended Nationally Domestic Contributions
C. Intended National Determined Contributions
D. International Noise Determined Contributions

26. What is the main cause of the extinction of many living species?

A. Natural catastrophes
B. Pollution
C. Changes in climate
D. Human activities

27. The term Environment is derived from an old French word "enviro" means

A. Abiotic components
B. Surroundings
C. Nature
D. Biotic components

28. Deforestation decreases

A. Rainfall B. Soil erosion
C. Draught D. Global warming

29. The major factor for climate change is

A. Pollution is increasing
B. Earth's global temperature is increasing
C. Human population is increasing
D. Both A and C

30. Climate change is the changes in

A. Temperature
B. Rain
C. Wind
D. All of the above

31. INDCs stands for

A. Intended Nationally Determined Contributions
B. Intended Nationally Domestic Contributions
C. Intended National Determined Contributions
D. International Noise Determined Contributions

32. મોટા ગ્રીન હાઉસ ગેસ ઉત્સર્જકો મુખ્યત્વે કોણ છે?
A. ઉચ્ચ આવકવાળા દેશો
B. મધ્યમ આવકવાળા દેશો
C. ઓછી આવકવાળા દેશો
D. બંને A અને C

33. 2° Cની નીચે તાપમાન જાળવી રાખવા માટે ન્યૂનતમ કેટલા કાર્બન કેપ્ચર અને સ્ટોરેજ પ્લાન્ટ્સની આવશ્યકતા છે?
A. 500 જેટલા CCS પ્લાન્ટ
B. 800 જેટલા CCS પ્લાન્ટ
C. 1000 થી વધુ CCS પ્લાન્ટ
D. 2000 થી વધુ CCS પ્લાન્ટ

34. પર્યાવરણીય લેપ્સ દર કેટલો હોય છે?
A. -6.5ºC / કિ.મી. B. 7.5ºC / કિ.મી.
C. - 8.5ºC / કિ.મી. D. 5.5ºC / કિ.મી.

35. તમારા કાર્બન ફૂટપ્રિન્ટમાં નીચેનામાંથી શેનો ફાળો છે?
A. મોટર સાયકલ ચલાવવી
B. કાર ચલાવવી
C. પગપાળા ચાલવું
D. વિમાનમાં જવું

36. INDCના અમલીકરણ માટે સમયમર્યાદા કેટલી છે?
A. 2015-2025 B. 2020-2025
C. 2020-2030 D. 2015-2020

37. પૃથ્વી કેવા પ્રકારની પ્રણાલી છે?
A. બંધ પ્રણાલી
B. ખુલ્લી પ્રણાલી
C. નિર્બળ પ્રણાલી
D. સક્રિય પ્રણાલી

38. પ્રકૃતિમાં પૃથ્વીની આસપાસના વાયુઓને નિયંત્રિત કરવા અને સંતુલિત કરવા માટે ________ ચક્રો છે.
A. ફીઝીકલ B. કેમિકલ
C. મિકેનિકલ D. ઇલેક્ટ્રોનિક

32. The greatest GHG emitters are
A. High income countries
B. Middle income countries
C. Low-income countries
D. Both A and C

33. Minimum how many carbon capture and storage plants will be needed to hold temperature increase below 2°C?
A. Around 500 CCS plants
B. Around 800 CCS plants
C. More than 1000 CCS plants
D. More than 2000 CCS plants

34. Environmental lapse rate is found to be
A. -6.5°C/km B. 7.5°C/km
C. -8.5°C/km D. 5.5°C/km

35. Which of the following contributes least to your carbon footprint?
A. Riding a bike
B. Driving a car
C. Walking on foot
D. Taking a plane

36. The timeframe for the implementation of the INDCs is
A. 2015-2025 B. 2020-2025
C. 2020-2030 D. 2015-2020

37. Earth is known as
A. Closed system
B. Open system
C. Vulnerable system
D. Active system

38. In nature there are ________ cycles to control and balance the gases that surround the Earth
A. Physical B. Chemical
C. Mechanical D. Electronic

39. ____________ દર વર્ષે હવામાંથી લગભગ 10 અબજ ટન કાર્બન કાઢે છે, જે સમુદ્રોમાં થતાં કુલ કાર્બન ફિક્સેશનના લગભગ બે તૃતીયાંશ છે.

A. *પ્રોક્લોરોકસ* B. *સિનેકોકોકસ*
C. *રાઇઝોબિયમ* D. બંને A અને B

40. મેથેનોજેન્સ, આર્કેયાનું એક જૂથ જે રૂમેન માં રહે છે અને પ્રાણીના ખોરાકને ક્યા ગેસમાં રૂપાંતરિત કરે છે?

A. CO_2 B. CH_4
C. CO D. NO_2

41. જો બાયફ્યુઅલ વિશાળ માત્રામાં ઊર્જાનો ઉપયોગ કર્યા વિના બનાવી શકાય છે, તો તે કેવું હશે?

A. કાર્બન ઊર્જા બળતણ
B. કાર્બન ચેનલિંગ બળતણ
C. કાર્બન તટસ્થ બળતણ
D. કાર્બનવિનાનું બળતણ

42. IPCC એટલે

A. International Panel on Climate Change
B. International Platform on Climate Change
C. Intergovernmental Panel on Climate Change
D. Intergovernmental Platform on Climate Change

43. IPCC ના અહેવાલ મુજબ, ક્લાઈમેટ ચેન્જથી

A. મનુષ્ય અને પ્રાણીઓમાં ચેપી રોગ ફાટી નીકળવાની કોઈ અસર નહી કરે
B. મનુષ્ય અને પ્રાણીઓમાં ચેપી રોગ ફાટી નીકળવાના દાખલામાં ફેરફાર થશે
C. જમીનના પેથોજેનેસિસમાં વધારો થશે
D. જમીનના પેથોજેનેસિસ પર કોઈ અસર નહીં કરે

39. ____________ remove about 10 billion tons of carbon from the air each year; this is about two-thirds of the total carbon fixation that occurs in the oceans.

A. *Prochlorococcus* B. *Synechococcus*
C. *Rhizobium* D. Both A and B

40. The methanogens, a group of archaea that live in the rumen, specialize in breaking down the animal's food into which gas?

A. CO_2 B. CH_4
C. CO D. NO_2

41. If biofuels can be made without using huge amounts of energy then they would be

A. Carbon energy fuel
B. Carbon channeling fuel
C. Carbon neutral fuel
D. Carbonless fuel

42. IPCC stands for

A. International Panel on Climate Change
B. International Platform on Climate Change
C. Intergovernmental Panel on Climate Change
D. Intergovernmental Platform on Climate Change

43. According to the IPCC report, climate change will

A. produce no effect on infectious disease outbreaks in humans and animals
B. alter patterns of infectious disease outbreaks in humans and animals
C. increase soil pathogenesis
D. produce no effect on soil pathogenesis

44. જમીનમાં સંગ્રહિત કાર્બનની માત્રા શાના સંતુલન પર આધારિત છે?

A. જમીન નીચેના માઇક્રોબાયલ શ્વસનમાંથી કાર્બન ઇનપુટ્સ અને પાંદડાના કચરા અને રૂટ ડેટ્રિટસમાંથી કાર્બન આઉટપુટ

B. પાંદડાંના કચરા અને રૂટ ડેટ્રિટસ કાર્બન ઇનપુટ્સ, જમીન નીચેના માઇક્રોબાયલ શ્વસન કાર્બન આઉટપુટ

C. ફક્ત પાંદડાના કચરામાંથી કાર્બન ઇનપુટ્સ અને સપાટીના માઇક્રોબાયલ શ્વસનમાંથી કાર્બન આઉટપુટ્સ

D. સપાટીના માઇક્રોબાયલ શ્વસનમાંથી કાર્બન ઇનપુટ્સ અને ફક્ત પાંદડાના કચરામાંથી કાર્બન આઉટપુટ

45. કાંપવાળો બાયોજીયોકેમિકલ ________ ચક્ર છે.

A. નાઇટ્રોજન B. સલ્ફર

C. કાર્બન D. ફોસ્ફરસ

46. ઇકોસિસ્ટમમાં ઊર્જા પ્રવાહ કઈ દિશામાં હોય છે?

A. બંને દિશામાં

B. એક દિશામાં

C. નીચેની દિશામાં

D. ઉપરની દિશામાં

47. નીચેનામાંથી કયા સ્થાનને ભારતનું 'કોલ્ડ ડેઝર્ટ' કહેવામાં આવે છે ?

A. લદાખ B. લાહૌલ

C. લેહ D. મનાલી

48. નોન-રિન્યુએબલ સ્રોતો ________ તરીકે પણ ઓળખાય છે

A. અમર્યાદિત સંસાધનો

B. અનંત સંસાધનો

C. મર્યાદિત સંસાધનો

D. વિશાળ સંસાધનો

44. The amount of carbon stored in the soil is dependent on the balance between

A. carbon inputs from microbial respiration underground and carbon outputs from leaf litter and root detritus

B. carbon inputs from leaf litter and root detritus and carbon outputs from microbial respiration underground

C. carbon inputs from leaf litter only and carbon outputs from surface microbial respiration

D. carbon inputs from surface microbial respiration and carbon outputs from leaf litter only

45. The sedimentary biogeochemical cycle is

A. Nitrogen B. Sulphur

C. Carbon D. Phosphorus

46. The energy flow in an ecosystem is

A. Bidirectional

B. Unidirectional

C. Downward direction

D. Upward direction

47. Which of the following places is termed as 'Cold Desert' of India?

A. Ladakh B. Lahaul

C. Leh D. Manali

48. Non-renewable resources are also known as

A. Unlimited resources

B. Infinite resources

C. Finite resources

D. Vast resources

49. કીસ્ટોન પ્રજાતિ એ છે કે જેની વિપુલ પ્રમાણમાં તેના પર્યાવરણ પર અસંગત _____ અસર હોય છે

A. નાની B. મોટી
C. સ્થિર D. મામૂલી

50. ઇકોલોજીકલ સમુદાયના પ્રજાતિના બંધારણમાં સમય જતાં પરિવર્તનની પ્રક્રિયા શું છે?

A. ઇકોલોજીકલ પ્રોગ્રેશન
B. એન્વાયરોન્મેન્ટલ સક્સેસન
C. એન્વાયરોન્મેન્ટલ પ્રોગ્રેશન
D. ઇકોલોજીકલ સક્સેસન

51. જ્યારે હવામાનના આંકડાકીય વિતરણમાં ફેરફાર થાય અને તે ફેરફાર લાંબા સમય સુધી ચાલે તો તેને શું કહેવાય છે?

A. ક્લાઈમેટ ચેન્જ
B. સ્ટેટિસ્ટિકલ ચેન્જ
C. વેધર ચેન્જ
D. એન્વાયરમેન્ટલ ચેન્જ

52. 1986માં અમલમાં આવેલ પર્યાવરણ (સંરક્ષણ) અધિનિયમનો ઉદ્દેશ્ય શું હતો?

A. પર્યાવરણ રક્ષણ
B. પર્યાવરણમાં સુધારો
C. રક્ષણ અને પર્યાવરણ સુધારણા
D. ક્લાઈમેટ ચેન્જથી રક્ષણ

53. CVP index એટલે

A. Climate, Vegetation and Precipitation index
B. Climate, Vegetation and Production index
C. Climate, Vegetation and Productivity index
D. Climate, Vegetation and Potential index

49. A keystone species is a species that has a disproportionately _________ effect on its environment relative to its abundance.

A. Small B. Large
C. Static D. Insignificant

50. ___________ is the observed process of change in the species structure of an ecological community over time.

A. Ecological Progression
B. Environmental Succession
C. Environmental Progression
D. Ecological Succession

51. The change in the statistical distribution of weather patterns when that change lasts for an extended period of time is known as

A. Climate change
B. Statistical change
C. Weather change
D. Environmental change

52. The Environment (Protection) Act was enacted in 1986 with the objective of providing for the

A. protection of the environment
B. improvement of the environment
C. protection and improvement of the environment
D. protection for the climate change

53. CVP index means

A. Climate, Vegetation and Precipitation index
B. Climate, Vegetation and Production index
C. Climate, Vegetation and Productivity index
D. Climate, Vegetation and Potential index

54. પોષણ કડીમાં ઉપભોક્તાઓ કોણ હોય છે?
A. શાકાહારી B. માંસભક્ષક
C. સર્વભક્ષી D. ઉપરોક્ત તમામ

55. નીચેનામાંથી કયુ પુનઃ અપ્રાપ્ય સંસાધન છે?
A. ભૂસ્તર ઊર્જા B. મહાસાગર ઊર્જા
C. સૌર ઊર્જા D. વન સંસાધનો

56. નીચેનામાંથી કયું ઇકોસિસ્ટમનું કાર્યાત્મક પાસું નથી?
A. આહાર શૃંખલાઓ
B. ઉત્ક્રાંતિ
C. ટોપોગ્રાફી
D. પોષણ ચક્ર

57. ગ્લોબલ વોર્મિંગના હાલના તબક્કામાં મુખ્ય ફાળો આપનાર ____________ છે.
A. કુદરતી ગ્રીનહાઉસ અસર
B. વનનાબૂદી
C. માનવ વસ્તી
D. અવક્ષયિત ઓઝોન

58. વન્યપ્રાણી (સંરક્ષણ) અધિનિયમ 1972 પહેલા, ભારતમાં ફક્ત ________ રાષ્ટ્રીય ઉદ્યાન હતા.
A. ત્રણ B. પાંચ
C. સાત D. નવ

59. ક્લાઈમેટને અસર કરતુ એન્થ્રોપોજેનિક રેડિયેટિવ નીચેનામાંથી કયું છે?
A. એરોસોલ્સ B. સિમેન્ટ
C. ગ્લાસ D. રેતી

60. કુદરતી સંસાધન સંચાલન માટે લાગુ અભિગમો કયા છે?
A. ટોપ-ડાઉન (આદેશ અને નિયંત્રણ)
B. સાવચેતી અભિગમ
C. સખ્ત નીતિઓ
D. ઉપરોક્ત તમામ

54. Consumers in food chain include
A. Herbivores B. Carnivores
C. Omnivores D. All of the above

55. Which of the following is non-renewable resource?
A. Geothermal energy B. Ocean energy
C. Solar energy D. Forest resources

56. Which of the following is not a functional aspect of ecosystem?
A. Food chains
B. Evolution
C. Topography
D. Nutrient cycles

57. The principal contributor to the present phase of global warming is considered to be the enhancement of the
A. natural greenhouse effect
B. deforestation
C. human population
D. Ozone depletion

58. Before Wildlife Protection Act 1972, India had only ____________ designated national parks.
A. Three B. Five
C. Seven D. Nine

59. Which one of the following is the anthropogenic radiative forcing of climate?
A. Aerosols B. Cement
C. Glass D. Sand

60. The approaches applied to natural resource management are
A. Top-down (command and control)
B. Precautionary approach
C. Concrete policy tools
D. All of the above

61. *In-situ* સંરક્ષણમાં પ્રજાતિઓનું સંરક્ષણ ક્યાં થાય છે?
A. તેમના કુદરતી હેબીટાટમાં
B. તેમના કુદરતી હેબીટાટની બહાર
C. કોઈપણ જગ્યાએ સંરક્ષણ
D. તેમના કુદરતી હેબીટમાં

62. ગ્લોબલ વોર્મિંગ રેટ કોણ માપે છે?
A. જીવવૈજ્ઞાનિક
B. ભૌતિકશાસ્ત્રી
C. ભૂસ્તરશાસ્ત્રી
D. હવામાનશાસ્ત્રી

63. પૃથ્વી દિવસ ક્યારે આવે છે ?
A. 21 ડિસેમ્બર
B. 12 સપ્ટેમ્બર
C. 22 એપ્રિલ
D. 31 ઓક્ટોબર

64. બરફની ચાદરો ઓગળવાનું મુખ્ય કારણ શું છે ?
A. જંગલોની કાપણીમાં વધારો
B. ગ્લોબલ વોર્મિંગ
C. કાર્બન ડાયોક્સાઇડની માત્રામાં ઘટાડો
D. અવાજ પ્રદૂષણ

65. ક્લાઈમેટ ચેન્જ સામે લડવામાં શું કરી શકાય?
A. અવશેષ ઇંધણ કંપનીઓને જાકારો આપવો
B. તમારી જાતને ક્લાઈમેટ ચેન્જ અંગેના વિજ્ઞાનમાં રસ લેવો
C. હવામાન સંબંધિત કાયદા અને નીતિ સુધારણાના હિમાયતીઓને મત આપો
D. ઉપરોક્ત તમામ

66. સૂટ (Soot) પૃથ્વીની આબોહવા ને શું કરે છે ?
A. ઠંડી કરે છે
B. ગરમ કરે છે
C. ઠંડુ અથવા ગરમ કરે છે
D. નિયમિત કરે છે

61. ***In-situ* conservation, the conservation of species in**
A. their natural habitat
B. outside their natural habitat
C. conservation in any place
D. their natural habit

62. **Who measures the global warming rate?**
A. Biologist
B. Physicist
C. Geologist
D. Climatologist

63. **Which day is Earth Day?**
A. 21 December
B. 12 September
C. 22 April
D. 31 October

64. **What is the main reason of melting of ice sheets?**
A. Increase in the deforestation
B. Global warming
C. Decrease in carbon dioxide content
D. Noise pollution

65. **What can one do to help fight climate change?**
A. Deny from fossil fuel companies
B. Engage yourself in the science behind climate change
C. Vote for advocates of climate-related legislation and policy improvements
D. All of the above

66. **Soot __________ Earth's climate system**
A. Cools
B. Warms
C. Either cools or warms
D. Controls

67. વરસાદના પાણીના નમૂના લેવા માટે સામાન્ય રીતે ગ્લાસ કન્ટેનર પસંદ કરવામાં આવતા નથી. કેમ?
A. ગ્લાસ કન્ટેનર પાણીના BODને બદલી નાખે છે
B. ગ્લાસ કન્ટેનર સરળતાથી તૂટી જાય છે
C. ગ્લાસ કન્ટેનર વરસાદના પાણીના pHને અસર કરે છે
D. ઉપરોક્ત તમામ

68. સમાજને પર્યાવરણ વધુ વાકેફ કરાવવા માટેનું પ્રથમ પુસ્તક કયું હતું?
A. Silent Spring by Rachel Carson
B. The Lorax by Dr. Seuss
C. The Omnivore's Dilemma by Michael Pollan
D. Population Bomb by Paul Ehrlich

69. પુન: પ્રાપ્ય ઊર્જાનો ગેરફાયદો શું છે ??
A. ઉચ્ચ પ્રદૂષણ
B. ફક્ત થોડા સ્થળોએ જ ઉપલબ્ધ છે
C. ઉચ્ચ ખર્ચ
D. અવિશ્વસનીય પુરવઠો

70. વાતાવરણની રચનામાં પરિવર્તન એ ________ નું ઉદાહરણ છે
A. આંતરિક દબાણ
B. ન્યુટન ફોર્સિંગ
C. બાહ્ય દબાણ
D. ગુરુત્વાકર્ષણ દબાણ

71. વૈશ્વિક સ્તરે, ________ ક્ષેત્ર ગ્રીનહાઉસ ગેસ ઉત્સર્જનમાં સૌથી મોટી ટકાવારીનો ફાળો આપે છે.
A. પરિવહન
B. મકાનો
C. ઉદ્યોગ
D. વીજળી અને ગરમીનું ઉત્પાદન

67. Glass containers are generally not preferred for sampling rain water. Why?
A. Glass containers change BOD of the water
B. Glass containers can break easily
C. Glass containers affect the pH of the rain water
D. All of the above

68. Which was the first book let the society to become more environmentally aware?
A. Silent Spring by Rachel Carson
B. The Lorax by Dr. Seuss
C. The Omnivore's Dilemma by Michael Pollan
D. Population Bomb by Paul Ehrlich

69. Which of the following is a disadvantage of renewable energy?
A. High pollution
B. Available only in few places
C. High running cost
D. Unreliable supply

70. Changes in the composition of the atmosphere is an example of ________.
A. Internal forcing
B. Newton forcing
C. External forcing
D. Gravitational forcing

71. Globally, ________ economic sector emits the largest percentage of greenhouse gas emissions
A. Transportation
B. Buildings
C. Industry
D. Electricity and heat production

72. આમાંથી કયું વિધાન ખોટું છે?

A. પર્યાવરણ શિક્ષણ લોકોમાં જાગૃતિ અને જ્ઞાન વધારે છે

B. પર્યાવરણ શિક્ષણ કોઈ ખાસ દ્રષ્ટિકોણની હિમાયત કરે છે

C. પર્યાવરણ શિક્ષણ વ્યક્તિગત આલોચનાત્મક વિચારશીલતા કૌશલ્યો શીખવે છે

D. પર્યાવરણ શિક્ષણ વ્યક્તિની સમસ્યાનું નિરાકરણ અને નિર્ણય લેવાની કુશળતાને વધારે છે

73. ________ પુન: પ્રાપ્ય ઉર્જા સ્ત્રોતોના સંશોધન અને વિકાસ માટે મંત્રાલય જવાબદાર છે

A. નવા અને પુન: પ્રાપ્ય ઉર્જા મંત્રાલય

B. માનવ સંસાધન વિકાસ

C. આરોગ્ય અને પરિવાર કલ્યાણ

D. કૃષિ અને ખેડૂત કલ્યાણ

74. વિશ્વનું પ્રથમ 100% સૌર સંચાલિત એરપોર્ટ ક્યાં છે?

A. બેંગલુરૂ, કર્ણાટક

B. કોચિન, કેરળ

C. અમદાવાદ, ગુજરાત

D. મુંબઈ, મહારાષ્ટ્ર

75. 1990 માં રાષ્ટ્રીય પર્યાવરણીય શિક્ષણ અધિનિયમનો હેતુ શું હતો?

A. યુનાઇટેડ સ્ટેટ્સ એન્વાયર્નમેન્ટલ પ્રોટેકશન એજન્સી પર્યાવરણીય સાક્ષરતાને પ્રોત્સાહન આપવામાં અગ્રેસર બનવાની જરૂર છે

B. પર્યાવરણીય શિક્ષણ કચેરીની સ્થાપના કરવી

C. હવામાન પરિવર્તન વિશે વિશ્વને જાગૃત કરવી

D. બંને A અને B

72. Which of these statements is false?

A. Environment education increases public awareness and knowledge

B. Environment education advocates a particular viewpoint

C. Environment education teaches individual's critical thinking skills

D. Environment education enhances an individual's problem solving and decision-making skills

73. ________ Ministry is responsible for research and development in renewable energy sources

A. Ministry of New and Renewable Energy

B. Human Resource Development

C. Health and Family Welfare

D. Agriculture and Farmers Welfare

74. The world's first 100% solar powered airport is located at ________

A. Bengaluru, Karnataka

B. Cochin, Kerala

C. Ahmedabad, Gujarat

D. Mumbai, Maharashtra

75. What was the purpose of the National Environmental Education Act in 1990?

A. To require that the United States Environmental Protection Agency be a leader in promoting environmental literacy

B. To establish the Office of Environmental Education

C. To aware world about climate change

D. Both A and B

76. ભારતમાં ક્રોમાઇટ ઓર(Ore)નું ઉત્પાદન સૌથી વધુ કયા રાજ્યમાં થાય છે?
A. ઉત્તરપ્રદેશ B. તામિલનાડુ
C. ઓરિસ્સા D. આંધ્રપ્રદેશ

77. દાવાનળ નું મુખ્ય કારણ ________ છે.
A. જ્વાળામુખી
B. ગ્લોબલ વૉર્મિંગ
C. પ્રદૂષણ
D. માનવ પ્રવૃત્તિ

78. તાજા પાણીમાં રંબદલાવ નું કારણ શું છે?
A. ફૂગ બ્લુમ
B. અલ્ગલ બ્લુમ
C. બેક્ટેરિયલ બ્લુમ
D. પ્રદૂષણ

79. જંગલોની કાપણીનું સૌથી મોટું કારણ ________ છે.
A. દાવાનળ B. જ્વાળામુખી વિસ્ફોટ
C. કૃષિ D. માટીનું ધોવાણ

80. ભારતમાં થોરિયમની સૌથી વધુ થાપણ ધરાવતું રાજ્ય ________ છે.
A. તામિલનાડુ B. ઝારખંડ
C. આંધ્રપ્રદેશ D. બિહાર

81. 'બ્લુ બોડી સિન્ડ્રોમ' શાને કારણે થાય છે?
A. નાઇટ્રેટ પ્રદૂષણ B. કાર્બન પ્રદૂષણ
C. ઝીંક પ્રદૂષણ D. માટી પ્રદૂષણ

82. શા માટે ઓઝોન સ્તર દક્ષિણ ધ્રુવ પર વધુ છે ?
A. CFCની અસમાન ખલેલને કારણે
B. એન્ટાર્કટિક વાતાવરણમાં ભારે હવામાનની સ્થિતિને કારણે
C. દક્ષિણ ધ્રુવમાં ચક્રવાતના કારણે
D. ભારે દબાણને કારણે

76. Which state is largest producer of chromite ore in India?
A. Uttar Pradesh B. Tamil Nadu
C. Orissa D. Andhra Pradesh

77. The main cause of forest fire is
A. Volcanic activity
B. Global warming
C. Pollution
D. Human activity

78. What is the cause of discoloration in fresh water?
A. Fungal bloom
B. Algal bloom
C. Bacterial bloom
D. Pollution

79. The biggest cause of deforestation is
A. Forest fire B. Volcanic eruption
C. Agriculture D. Soil erosion

80. The state which is having largest deposit of thorium in India is
A. Tamil Nadu B. Jharkhand
C. Andhra Pradesh D. Bihar

81. 'Blue Body Syndrome' is caused due to
A. Nitrate pollution B. Carbon pollution C. Zinc pollution
D. Soil pollution

82. Why ozone depletion is more over the South Pole?
A. Due to uneven disturbance of CFCs
B. Due to extreme weather conditions in the Antarctic atmosphere
C. Due to cyclonic depression in South Pole
D. Due to extreme pressure

83. ________ પ્રોટોકોલનો ઉદ્દેશ ઓઝોન ઘટાડતા પદાર્થોના ઉત્પાદન અને વપરાશને નિયંત્રિત કરવાનો છે.

A. ક્યોટો પ્રોટોકોલ
B. મોન્ટ્રીયલ પ્રોટોકોલ
C. ગોથેનબર્ગ પ્રોટોકોલ
D. ઉપર્યુક્તમાંથી કોઈ નહિ

84. પર્યાવરણની ચિંતા સાથે ઉદ્યોગોનું સ્થાન કેમ મહત્વપૂર્ણ ભૂમિકા ભજવે છે?

A. ઉદ્યોગો માટે વધુ ઇનપુટ મેળવવા માટે
B. ઉદ્યોગો બનાવવા માટે જંગલો સાફ કરવા
C. વધુ વ્યવસાય પેદા કરવા
D. ઉદ્યોગો દ્વારા થતા પ્રદૂષણને રોકવા

85. પર્યાવરણીય હાનિથી માનવ સ્વાસ્થ્યને બચાવવા માટે સમર્પિત કઇ સંસ્થા છે ?

A. Environment and Human Health
B. Environmental and Health Science
C. Ecological Protection Organization
D. Ecological Science and Health

86. કઈ અસર કિરણોત્સર્ગના સંપર્કથી આવે છે?

A. હોર્મોનલ અસંતુલન
B. શ્વેત રક્તકણોની ખોટ
C. વાળ ખરવા
D. દાંતનો સડો

87. ભારતે મોન્ટ્રીયલ પ્રોટોકોલ પર ક્યારે સહી કરી હતી?

A. 1988
B. 1990
C. 1992
D. 1994

83. ________ protocol's aim is to control the production and consumption of ozone depleting substances.

A. Kyoto Protocol
B. Montreal Protocol
C. The Gothenburg Protocol
D. None of the above

84. Why industries location plays an important role with concern to the environment?

A. To get more input for industries
B. To clear the forests in order to build the industries
C. To generate more occupation
D. To prevent pollution causing by industries

85. Which one of the following organizations dedicated to protecting human health from environmental harms?

A. Environment and Human Health
B. Environmental and Health Science
C. Ecological Protection Organization
D. Ecological Science and Health

86. Which one of the following is a result of high exposure of radiation?

A. Hormonal imbalance
B. Loss of white blood cells
C. Hair loss
D. Tooth decay

87. India signed the Montreal Protocol in

A. 1988
B. 1990
C. 1992
D. 1994

88. પાણીનો રંગ માપવા માટે કઈ પદ્ધતિનો ઉપયોગ કરવામાં આવે છે?

A. ગ્રેવીમેટ્રિક વિશ્લેષણ
B. ક્રોમેટોગ્રાફી
C. હાઇડ્રોમીટર વિશ્લેષણ
D. ટિન્ટોમીટર પદ્ધતિ

89. ગ્રીનહાઉસ વાયુઓ વિના પૃથ્વીનું સરેરાશ તાપમાન શું હોત?

A. 0°C B. -17°C
C. -19°C D. -29°C

90. ગ્રહને ગરમ કરવા માટે ગ્રીનહાઉસ વાયુઓ દ્વારા સૂર્યની કિરણોત્સર્ગ શક્તિ કેટલી ગ્રહણ કરે છે?

A. 75 પીડબ્લ્યુ B. 1550GW
C. 1000MW D. 850TW

91. કયા પ્રકારનાં વાદળો પૃથ્વીના વાતાવરણની સૌથી વધુ ઊંચાઈએ જોવા મળે છે?

A. સ્ટ્રેટસ B. નોકટીક્યુસન્ટ
C. સિરોસ્ટ્રેટસ D. સિરસ

92. ટ્રોફોસ્ફીયરમાં જથ્થા મુજબ લગભગ ________ % વાતાવરણ હોય છે?

A. 20% B. 50%
C. 65% D. 80%

93. વોટરપેપર અને વાયુના ભરના ગુણોત્તરને ________

A. એબ્સોલ્યૂટ હ્યુમિડિટી
B. સ્પેસિફિક હ્યુમિડિટી
C. રિલેટિવ હ્યુમિડિટી
D. એપ્રોક્ષીમેટ હ્યુમિડિટી

94. ITCZ વૈશ્વિક પવનના દાખલાને અનુરૂપ શું છે?

A. Inner Tropospheric Convergence Zone
B. Inter Tropical Convergence Zone
C. Inner Tropical Continental Zone
D. Inter Tropical Continental Zone

88. Which method is used to measure the color of water?

A. Gravimetric analysis
B. Chromatography
C. Hydrometer analysis
D. Tintometer method

89. What would have been the average temperature of Earth without greenhouse gases?

A. 0°C B. -17°C
C. -19°C D. -29°C

90. How much of the sun's radiation energy is absorbed by the greenhouse gases to warm the planet?

A. 75PW B. 1550GW
C. 1000MW D. 850TW

91. Which type of clouds is found in the highest altitude of the Earth's atmosphere?

A. Stratus B. Noctilucent
C. Cirrostratus D. Cirrus

92. Troposphere contains nearly ________ of the atmosphere by mass.

A. 20 % B. 50 %
C. 65 % D. 80 %

93. The ratio of mass of water vapor to mass of air is known as ________

A. Absolute humidity
B. Specific humidity
C. Relative humidity
D. Approximate humidity

94. What does ITCZ stand for corresponding to global wind patterns?

A. Inner Tropospheric Convergence Zone
B. Inter Tropical Convergence Zone
C. Inner Tropical Continental Zone
D. Inter Tropical Continental Zone

95. સમુદ્ર સપાટી પર વાતાવરણીય દબાણ શું છે?

A. 101325 Pa B. 14.696 psi
C. 760 Torr D. ઉપરોક્ત તમામ

96. "ઓવરકાસ્ટ" શબ્દ શું વ્યાખ્યાયિત કરે છે?

A. હવામાનનો જ્યારે ક્લાઉડ કવર 8 ઓક્ટસ બરાબર છે
B. છૂટાછવાયા પરોક્ષ કિરણોત્સર્ગના સંપર્કમાં આવવાની ઘટના
C. A અને B બંનેનો સમાવેશ
D. ઉપર્યુક્તમાંથી કોઈ નહિ

97. વાતાવરણના વિવિધ સ્તરો દ્વારા અવલોકન કરાયેલ દૂરની લાઈટ ઈમિટિંગ બોડીની સ્પષ્ટ સ્થિતિ અથવા તેજની અસર શું છે?

A. સિંટિલેશન ઇફેક્ટ
B. પેંડ્યુલમ ઇફેક્ટ
C. જીઓથર્મલ ઇફેક્ટ
D. ગેલિલિઓ ઇફેક્ટ

98. નીચેનામાંથી કયુ હવામાનનું ઉદાહરણ છે?

A. ધુમ્મસવાળો દિવસ
B. ગરમ દિવસ
C. છેલ્લાં 50 વર્ષથી એમેઝોનના જંગલોનું સરેરાશ તાપમાન
D. બંને A અને B

99. ક્લાઈમેટ ચેન્જ ઘટાડવા માટે જંગલો કેમ મહત્વપૂર્ણ છે?

A. જંગલો કાર્બન સિંક કરે છે
B. જંગલો ઓક્સિજન પ્રદાન કરે છે
C. જંગલો એ કાર્બનિક પદાર્થોનો મહત્વપૂર્ણ સ્રોત છે
D. વન વૃક્ષોના પાંદડા સૂર્યપ્રકાશને પ્રતિબિંબિત કરે છે

95. What is the atmospheric pressure at sea level?

A. 101325 Pa B. 14.696 psi
C. 760 Torr D. All of the above

96. What does the term "overcast" define?

A. Weather when cloud cover is equal to 8 oktas
B. Phenomenon of indirect radiation exposure due to scattering
C. Comprises of both A and B
D. None of the above

97. The effect of apparent position or brightness of a distant light emitting body, observed through various layers of the atmosphere is known as

A. Scintillation effect
B. Pendulum effect
C. Geothermal effect
D. Galileo effect

98. Which of the following is an example of climate?

A. A foggy day
B. A hot day
C. The average temperature in Amazon forests over the past 50 years
D. Both A and B

99. Why are forests important for mitigating climate change?

A. Forests serve as a sink in the carbon cycle
B. Forests provide oxygen
C. Forests are an important source of organic matter
D. Leaves of forest trees reflect the sunlight away from Earth

100. નીચેનામાંથી શું પૃથ્વી પરના કાર્બન ચક્રમાં કાર્બન ડાયોક્સાઇડ ઉમેરે છે?

A. શ્વસન

B. કાર્બનિક પદાર્થોનો સડો

C. જ્વાળામુખી

D. ઉપરોક્ત તમામ

101. જો કોઈ સમુદ્રમાં ફાયટોપ્લાંકટોન સંપૂર્ણ નાશ પામે છે તો

A. કાર્બન સિંક તરીકે સમુદ્રને વિપરીત અસર થશે

B. સમુદ્રમાં ફૂડ ચેઇન પર વિપરીત અસર થશે

C. સમુદ્રના પાણીની ઘનતામાં તીવ્ર ઘટાડો થશે

D. બંને A અને B

102. મહાસાગરોનું એસિડિફિકેશન ________ છે.

A. વધતો જાય છે

B. ઘટતો જાય છે

C. સતત રહે છે

D. સમુદ્રના પ્રકાર પર આધારિત છે

103. ભારત સરકાર 'સમુદ્ર બકથ્રોન' ના વાવેતરને પ્રોત્સાહિત કરે છે કારણ કે

A. તે બાયોડિઝલનો એક સમૃદ્ધ સ્રોત છે

B. તે જમીનના ધોવાણને નિયંત્રિત કરવામાં અને રણને રોકવામાં મદદ કરે છે

C. તેમાં ઉચ્ચ પોષક મૂલ્ય છે

D. તે ઊંચાઇવાળા ઠંડા વિસ્તારોમાં રહેવા માટે સારી રીતે અનુકૂળ છે

104. થર્મલ પાવર પ્લાન્ટોમાં કોલસાના દહનમાંથી ઉત્સર્જન નીચેનામાંથી કયુ છે?

A. CO_2

B. નાઇટ્રોજનનો ઓક્સાઇડ

C. સલ્ફરનો ઓક્સાઇડ

D. ઉપરોક્ત તમામ

100. Which of the followings add carbon dioxide to the carbon cycle on Earth?

A. Respiration

B. Decay of organic matter

C. Volcanic action

D. All of the above

101. If phytoplanktons in an ocean are completely destroyed then

A. The ocean as a carbon sink would be adversely affected

B. The food chains in the ocean would be adversely affected

C. The density of ocean water would drastically decrease

D. Both A and B

102. The acidification of oceans is

A. Increasing

B. Decreasing

C. Constant

D. Depends on type of ocean

103. Government of India encourages the cultivation of 'sea buckthorn' because

A. It is a rich source of biodiesel

B. It helps in controlling soil erosion and in preventing desertification

C. It has high nutritional value

D. It is well-adapted to live in cold areas of high altitudes

104. Which of the following is/are the emission/s from coal combustion at thermal power plants?

A. CO_2

B. Oxides of nitrogen

C. Oxides of Sulphur

D. All of the above

105. નીચેના ઉદ્યોગો પૈકી કયુ ભારતમાં પાણીનો મહત્તમ વપરાશકાર છે?

A. ખોરાક

B. કાગળ અને પલ્પ

C. થર્મલ પાવર

D. કાપડ

106. ____________ એ ઇકોસિસ્ટમ્સમાં ફૂડ ચેનના સંદર્ભમાં વિઘટનકર્તા તરીકે ગણવામાં આવે છે.

a. બેક્ટેરિયા

A. ફૂગ

B. એક્ટિનોમિસેટ્સ

C. વાયરસ

107. "આબોહવા આત્યંતિક, વરસાદ ઓછો અને વિચરતા લોકો." નીચેના કયા ક્ષેત્રમાં આવું જોવા મળે છે?

A. આફ્રિકન સવાન્નાહ

B. ઉત્તર અમેરિકન પ્રેરી

C. સેન્ટ્રલ એશિયન સ્ટેપ્પી

D. સાઇબેરીયન ટુન્ડ્રા

108. નીચેની શરતોમાંથી કઈ તેના સમુદાયમાં જીવતંત્રની કાર્યાત્મક ભૂમિકાનું વર્ણન કરે છે?

A. ઇકોલોજીકલ નિશ

B. ઇકોટોન

C. હોમ રેન્જ

D. હેબીટાટ

109. ઉત્પાદકતામાં ઘટાડો થવાના ક્રમમાં ઇકોસિસ્ટમ્સનો સાચો ક્રમ કયો છે?

A. મહાસાગરો, તળાવો, ઘાસના મેદાનો, ચેર

B. સરોવરો, ચેર, મહાસાગરો, ઘાસના મેદાનો

C. ઘાસના મેદાનો, મહાસાગરો, તળાવો, ચેર

D. ચેર, ઘાસના મેદાનો, તળાવો, મહાસાગરો

105. Which one among the following industries is the maximum consumer of water in India?

A. Food

B. Paper and pulp

C. Thermal power

D. Textiles

106. ____________ are considered as decomposers with reference to food chains in ecosystems

A. Bacteria

B. Fungi

C. Actinomycetes

D. Viruses

107. "Climate is extreme, rainfall is scanty and the people used to be nomadic herders". In which of the following regions

A. African Savannah

B. North American Prairie

C. Central Asian Steppe

D. Siberian Tundra

108. Which one of the following terms describes the functional role of organism in its community?

A. Ecological niche

B. Ecotone

C. Home range

D. Habitat

109. The correct sequence of ecosystems in the order of decreasing productivity is

A. Oceans, lakes, grasslands, mangroves

B. Lakes, Mangroves, oceans, grasslands

C. Grasslands, Oceans, lakes, mangroves

D. Mangroves, grasslands, lakes, oceans

110. માઇક્રોરિઝાનો ઉપયોગ ડિગ્રેડેડ સાઇટ્સના પુનર્વસન માટે થાય છે કારણ કે તે છોડને

A. દુષ્કાળનો પ્રતિકાર અને શોષક વિસ્તારમા વધારો કરે છે.

B. pHની ચરમસીમા ટોલરેટ કરે છે.

C. રોગના ઉપદ્રવને પ્રતિકાર કરે છે.

D. ઉપરોક્ત તમામ

110. Mycorrhizae are used to rehabilitate degraded sites as it enables the plants to

A. Resist drought and increase absorptive area

B. Tolerate extremes of pH

C. Resist disease infestation

D. All of the above

111. 'ઇકો સેન્સિટિવ ઝોન' ના સંદર્ભમાં, નીચેનામાંથી કયું વિધાન યોગ્ય છે?

A. ઇકો સેન્સિટિવ ઝોનની ઘોષણા કરવાનો હેતુ એ છે કે તે ઝોનમાં કૃષિ સિવાય તમામ પ્રકારની માનવ પ્રવૃત્તિઓને પ્રતિબંધિત કરવાનો છે

B. ઇકો સેન્સિટિવ ઝોન એ એવા ક્ષેત્ર છે કે જેને વન્યપ્રાણી (સંરક્ષણ) અધિનિયમ, 1972 હેઠળ જાહેર કરવામાં આવ્યા છે

C. બંને A અને B

D. તે સુરક્ષિત વિસ્તાર છે

111. With reference to 'Eco-Sensitive Zones', which of the following statements is correct?

A. The purpose of the declaration of Eco-Sensitive Zones is to prohibit all kinds of human activities in those zones except agriculture

B. Eco-Sensitive Zones are the areas that are declared under the Wildlife (Protection) Act, 1972

C. Both A and B

D. It is a secure zone

112. દર વર્ષે, એક મહિના લાંબો ઇકોલોજીકલ મહત્વનો ઉત્સવ યોજાય છે, જે દરમિયાન અમુક સમુદાયો/જાતિઓ ફળ આપતા ઝાડના રોપા રોપતા હોય છે. આ કઈ જાતિઓ છે

A. ગોંડ અને કોર્કુ

B. ભુતીયા અને લેપ્ચા

C. ઇરુલા અને તોડા

D. સહારિયા અને અગરિયા

112. Every year, a month long ecologically important festival is held during which certain communities/tribes plant saplings of fruit-bearing trees. These are

A. Gond and Korku

B. Bhutia and Lepcha

C. Irula and Toda

D. Sahariya and Agariya

113. ઓક અને રોડોડેન્ડ્રોન ક્યાં જોવા મળે છે?

A. પશ્ચિમ ઘાટ

B. ડેક્કન પ્લેટુ

C. હિમાલયનો પ્રદેશ

D. બંને A અને C

113. Oak and Rhododendron are the naturally occurring plants of

A. Western Ghats

B. Deccan Plateau

C. Himalayan Region

D. Both A and C

114. જ્યારે ઇકોલોજીકલ કેરેક્ટરમાં પરિવર્તન આવ્યું છે, બન્યું છે અથવા માનવ દખલને પરિણામે વેટલેન્ડમાં થવાની સંભાવના છે, તો તે ________ હેઠળ આવે છે

A. વેટલેન્ડ રેકોર્ડ
B. મોન્ટ્રેક્સ રેકોર્ડ
C. જિયોફિઝીકલ રેકોર્ડ્સ
D. ડિજિટલ લેન્ડ રેકોર્ડ

115. પર્યાવરણીય માહિતી સિસ્ટમ (ENVIS), મંત્રાલયની એક સેન્ટ્રલ સેક્ટર યોજના ક્યારથી લાગુ કરવામાં આવી છે?

A. 1982
B. 1984
C. 1987
D. 1990

116. રાષ્ટ્રીય ઉદ્યાન જેની આબોહવા ટ્રોપિકલથી સબટ્રોપિકલ, ટેમ્પરેટ અને આર્કટિકમાં બદલાય છે.

A. નંદાદેવી રાષ્ટ્રીય ઉદ્યાન
B. કાંચનજંગા રાષ્ટ્રીય ઉદ્યાન
C. નામદાફા રાષ્ટ્રીય ઉદ્યાન
D. નિયોરા વેલી રાષ્ટ્રીય ઉદ્યાન

117. ________ "ઇકોસિસ્ટમ" નું શ્રેષ્ઠ વર્ણન છે.

A. સજીવનો સમુદાય તેમના રહેવાસી વાતાવરણ સાથે
B. ભૌગોલિક ક્ષેત્રની વનસ્પતિ અને પ્રાણીસૃષ્ટિ
C. સજીવનો સમુદાય એક બીજા સાથે ક્રિયાપ્રતિક્રિયા કરે છે
D. ભૌગોલિક ક્ષેત્રની વનસ્પતિ અને પ્રાણીસૃષ્ટિ એક બીજા સાથે ક્રિયાપ્રતિક્રિયા કરે છે

114. When changes in ecological character have occurred, are occurring or are likely to occur in the wetland as a result of human interference, then it is brought under

A. Wetland Record
B. Montreux Record
C. Geophysical records
D. Digital land records

115. Environmental Information System (ENVIS), a Central Sector Scheme of the Ministry has been implemented since

A. 1982
B. 1984
C. 1987
D. 1990

116. The National Park whose climate varies from tropical to subtropical, temperate and arctic is

A. Nandadevi National Park
B. Khangchendzonga National park
C. Namdapha National park
D. Neora Valley National Park

117. The best description of the term "ecosystem" is

A. A community of organisms together with their living environment
B. The flora and fauna of a geographical area
C. A community of organisms interacting with one another
D. The flora and fauna of a geographical area interacting with one another

118. 'BioCarbon Fund Initiative for Sustainable Forest Landscapes' નું સંચાલન કોણ કરે છે?

A. અંતરરાષ્ટ્રીય નાણાંકીય ભંડોળ

B. એશિયન વિકાસ બેંક

C. સંયુક્ત રાષ્ટ્ર પર્યાવરણ કાર્યક્રમ

D. વર્લ્ડ બેંક

119. કોવિડ -19 રોગચાળો

A. પ્રકૃતિના સ્વાસ્થ્ય સુધારવામાં મદદ કરી

B. પ્રદૂષણ ઓછું કર્યું

C. વન્યપ્રાણીઓને તેમની ખૂબ જરૂરી જગ્યા મેળવવા માટે મદદ કરી

D. ઉપરોક્ત તમામ

120. રિયો +20 કોન્ફરન્સ શું છે?

A. તે સસ્ટેનેબલ ડેવલપમેન્ટ પર યુનાઇટેડ નેશન્સની કોન્ફરન્સ છે

B. તે વિશ્વ વેપાર સંગઠનની મંત્રી સભા છે

C. તે જૈવિક વિવિધતા પરના સંમેલનના સભ્ય દેશોનું એક સંમેલન છે

D. તે હવામાન પરિવર્તન પર આંતર-સરકારી પેનલની એક પરિષદ છે

121. 'ગ્રીન ક્લાયમેટ ફંડ' સંબંધિત નીચેનામાંથી કયું નિવેદન યોગ્ય છે?

A. તે UNEP, OECD, એશિયન ડેવલપમેન્ટ બેન્ક અને વર્લ્ડ બેંકની આગેવાની હેઠળ સ્થપાયેલ છે

B. વિકાસશીલ દેશોને ક્લાઈમેટ ચેન્જ નો સામનો કરવા માટે અનુકૂલન અને શમન પદ્ધતિઓમાં સહાય કરવાનો હેતુ છે

C. બંને A અને B

D. તે IMD હેઠજ સ્થપાયેલ છે

118. 'BioCarbon Fund Initiative for Sustainable Forest Landscapes' is managed by the

A. International Monetary Fund

B. Asian Development Bank

C. United Nations Environment Program

D. World Bank

119. COVID-19 pandemic

A. Helped the nature to heal

B. Reduced the pollution

C. Helped the wildlife to get their much-needed space

D. All of the above

120. What is Rio+20 Conference?

A. It is the United nations Conference on Sustainable Development

B. It is a Ministerial Meeting of the World Trade Organization

C. It is a Conference of the Member Countries of the Convention on Biological Diversity

D. It is a Conference of the Inter-governmental Panel on Climate Change

121. Which of the following statement regarding 'Green Climate Fund' is correct?

A. It is founded under the aegis of UNEP, OECD, Asian Development Bank and World Bank

B. It is intended to assist the developing countries in adaptation and mitigation practices to counter climate change

C. Both A and B

D. It is founded by IMD

122. શબ્દ 'ઉદ્દેશિત રાષ્ટ્રીય નિર્ધારિત ફાળો' શેના સંદર્ભમાં છે?

A. યુરોપિયન દેશો દ્વારા યુદ્ધ અસરગ્રસ્ત મધ્ય પૂર્વથી શરણાર્થીઓનું પુનર્વસન કરવા માટે કરેલા વચનો

B. ટકાઉ કૃષિની સ્થાપનામાં સભ્ય દેશો દ્વારા ફાળો આપેલ મૂડી

C. વિશ્વના દેશો દ્વારા હવામાન પલટા સામે લડવાની કાર્યવાહીની યોજના

D. વિશ્વના દેશો દ્વારા વૈશ્વિક કક્ષાએ વનો અંગેની રૂપરેખાની યોજના

123. પૃથ્વીની __________ આ ગ્રહ પર સ્વસ્થ બાયોસ્ફિયરના અસ્તિત્વ માટે યોગ્ય પરિસ્થિતિઓ બનાવવા માટે જવાબદાર છે.

A. ઇકોસિસ્ટમ B. વાતાવરણ

C. માટી D. પાણી

124. ઇકોલોજીકલ દ્રષ્ટિકોણથી, પૂર્વીય ઘાટ અને પશ્ચિમ ઘાટ વચ્ચે સારી કડી હોવા માટે __________નું મહત્વ છે.

A. સત્યમંગલમ્ ટાઇગર રિઝર્વ

B. નાગરહોલ નેશનલ પાર્ક

C. શેષાચલમ બાયોસ્ફિયર રિઝર્વ

D. નલ્લમલા વન

125. CCAC શેના પર ધ્યાન કેન્દ્રિત કરે છે?

A. મિથેન

B. બ્લેક કાર્બન

C. હાઇડ્રોફ્લોરોકાર્બન્સ

D. ઉપરોક્ત તમામ યોગ્ય છે

126. TRAFFICનું લક્ષ્ય એ સુનિશ્ચિત કરવાનું છે કે જંગલી છોડ અને પ્રાણીઓનો વેપાર પ્રકૃતિના __________ માટે ખતરો નથી.

A. સુધારણા B. સંરક્ષણ

C. ટકાઉપણા D. સુંદરતા

122. The term 'Intended Nationally Determined Contributions' is in the context of

A. pledges made by the European countries to rehabilitate refugees from the war-affected Middle East

B. capital contributed by the member countries in the establishment of sustainable agriculture

C. plan of action outlined by the countries of the world to combat climate change

D. plan of action outlined by the countries of the world regarding forestation at global level

123. The earth's __________ is responsible for creating conditions suitable for the existence of a healthy biosphere on this planet

A. Ecosystem B. Atmosphere

C. Soil D. Water

124. From the ecological point of view, __________ importance in being a good link between the Eastern Ghats and the Western Ghats

A. Sathyamangalam Tiger Reserve

B. Nagarhole National Park

C. Seshachalam Biosphere Reserve

D. Nallamala Forest

125. The CCAC focuses on

A. Methane

B. Black carbon

C. Hydrofluorocarbons

D. All of the above

126. The mission of TRAFFIC is to ensure that trade in wild plants and animals is not a threat to the __________ of nature

A. Healing B. Conservation

C. Sustainability D. Beauty

127. United Nations Convention to Combat Desertification'નું શું મહત્વ છે?

A. તેનો ઉદ્દેશ નવીન રાષ્ટ્રીય કાર્યક્રમો અને સહાયક આંતરરાષ્ટ્રીય ભાગીદારી દ્વારા અસરકારક પગલાને પ્રોત્સાહન આપવાનો છે.

B. દક્ષિણ એશિયા અને ઉત્તર આફ્રિકાના પ્રદેશો પર તેનું વિશેષ ધ્યાન છે અને તેનું સચિવાલય આ પ્રદેશોમાં નાણાકીય સંસાધનોના મોટા ભાગની ફાળવણીની સુવિધા આપે છે.

C. તે રણની લડતમાં લડવામાં સ્થાનિક લોકોની ભાગીદારીને પ્રોત્સાહિત કરીને સૈધ્ધાંતિક અભિગમ માટે પ્રતિબદ્ધ છે.

D. બંને A અને C

128. સક્શેશનએ ફેરફારોની પ્રગતિશીલ શ્રેણી છે જે __________તરફ દોરી જાય છે જે સ્થિર સ્થિતિ છે.

A. પાયોનિયર કોમ્યુનિટી

B. સેરેસ

C. ગૌણ કોમ્યુનિટી

D. ક્લાઇમેક્ષ કોમ્યુનિટી

129. જો પતંગિયાની પ્રજાતિઓની વસ્તીમાં મોટો ઘટાડો થાય છે, તો તેનું પરિણામ શું હોઈ શકે છે?

A. કેટલાક છોડના પરાગનયનને વિપરીત અસર થઈ શકે છે

B. તેનાથી ભમરી, કરોળિયા અને પક્ષીઓની કેટલીક પ્રજાતિઓની વસતીમાં ઘટાડો થઈ શકે છે

C. કેટલાક વાવેલા છોડના ફંગલ ઇન્ફેક્શનમાં તીવ્ર વધારો થઈ શકે છે

D. બંને A અને B

127. What is/are the importance/s of the 'United Nations Convention to Combat Desertification'

A. It aims to promote effective action through innovative national programmes and supportive international partnerships.

B. It has a special/particular focus on South Asia and North Africa regions, and its Secretariat facilitates the allocation of major portion of financial resources to these regions.

C. It is committed to bottom-up approach, encouraging the participation of local people in combating the desertification.

D. Both A and C

128. Succession is a progressive series of changes which leads to the establishment of a relatively stable

A. Pioneer community

B. Seres

C. Secondary community

D. Climax community

129. If there is a huge fall in the population of species of butterflies, what could be its likely consequences?

A. Pollination of some plants could be adversely affected

B. It could lead to a fall in the population of some species of wasps, spiders and birds

C. There could be a drastic increase in the fungal infections of some cultivated plants

D. Both A and B

130. 'નેશનલ ગ્રીન ટ્રિબ્યુનલ (NGT) ______ ન્યાય પૂરો પાડે છે અને ઉચ્ચ અદાલતોમાં દાવાઓનો ભાર ઘટાડવામાં મદદ કરે છે.

A. પર્યાવરણીય B. પ્રદૂષણ
C. વન્યજીવ D. પાણી

131. 'ગ્લોબલ એલાયન્સ ફોર ક્લાઇમેટ સ્માર્ટ એગ્રિકલ્ચર (GACSA)' વિશે, નીચેના નિવેદનો યોગ્ય છે.

A. GACSA એ 2015 માં પેરિસમાં આયોજિત ક્લાઈમેટ સમિટનું પરિણામ છે
B. GACSA નું સભ્યપદ કોઈપણ બંધનકર્તા જવાબદારીઓનું નિર્માણ કરતું નથી
C. GACSA ના નિર્માણમાં ભારતનો મહત્વ નું યોગદાન હતું
D. બંને A અને C

132. The Partnership for Action on Green Economy (PAGE), કે જે UN સંચાલિત છે અને તે હરિયાળી અને વધુ સારા અર્થશાસ્ત્ર તરફના પ્રયાસોમાં દેશોને મદદ કરે છે. આ PAGE ક્યાં બન્યું હતું ?.

A. વર્લ્ડ સસ્ટેનેબલ ડેવલપમેન્ટ સમિટ 2016, નવી દિલ્હી
B. અર્થ સમિટ ઓન સસ્ટેનેબલ ડેવલપમેન્ટ 2002, જોહાનિસબર્ગ
C. સસ્ટેનેબલ ડેવલપમેન્ટ કોન્ફરન્સ 2012, રિયો ડી જનેરો
D. યુનાઇટેડ નેશન્સ ફ્રેમવર્ક કન્વેશન ઓન ક્લાઇમેટ ચેન્જ 2015, પેરિસ

133. વાતાવરણમાં કાર્બન ડાયોક્સાઇડની વધેલી સાંદ્રતાને કારણે છોડની વૃદ્ધિમાં વધારો ______ તરીકે ઓળખાય છે

A. કાર્બન સિક્વેસ્ટરેશન
B. કાર્બન કવીંન્ચિંગ
C. પ્રકાશસંશ્લેષણ
D. કાર્બન ફર્ટિલાઇઝેશન

130. The 'National Green Tribunal (NGT) provides ______ justice and helps to reduce the burden of litigation in the higher courts

A. Environmental B. Pollution
C. Wildlife D. Water

131. The following statement/s about 'Global Alliance for Climate Smart Agriculture (GACSA)', is/are correct

A. GACSA is an outcome of the Climate Summit held in Paris in 2015
B. Membership of GACSA does not create any binding obligations
C. India was instrumental in the creation of GACSA
D. Both A and C

132. The Partnership for Action on Green Economy (PAGE), a UN mechanism to assist countries transition towards greener and more inclusive economies, emerged at

A. The World Sustainable Development Summit 2016, New Delhi
B. The Earth Summit on Sustainable Development 2002, Johannesburg
C. The United Nations Conference on Sustainable Development 2012, Rio de Janeiro
D. The United Nations Framework Convention on Climate Change 2015, Paris

133. Increased plant growth due to increased concentration of carbon dioxide in the atmosphere is termed as

A. Carbon sequestration
B. Carbon quenching
C. Photosynthesis
D. Carbon fertilization

134. "Momentum for Change: Climate Neutral Now" એ________દ્વારા શરૂ કરાયેલ એક પહેલ છે.

A. UNFCCC સચિવાલય
B. વિશ્વ હવામાન સંસ્થા
C. IPCC
D. UNEP સચિવાલય

135. પ્રખ્યાત કહેવત 'પૃથ્વી દરેક માણસની જરૂરિયાત પૂરી કરવા માટે પૂરતી છે પરંતુ એક માણસના લોભ માટે પણ પૂરતી નથી' કોણે કહ્યું છે?

A. જવાહરલાલ નહેરુ
B. સુંદરલાલ બહુગુણા
C. મોહનદાસ કરમચંદ ગાંધી
D. સરદાર વલ્લભભાઈ પટેલ

136. FAO પરંપરાગત કૃષિ પ્રણાલીઓને GIAHS નું સ્ટેટસ આપે છે જેનો ધ્યેય ________ છે.

A. આધુનિક ટેક્નોલોજી, આધુનિક ખેતી પદ્ધતિઓની તાલીમ અને ઓળખાયેલ GIAHSના સ્થાનિક સમુદાયોને આર્થિક સહાય પ્રદાન કરવી જેથી તેમની કૃષિ ઉત્પાદકતામાં મોટા પ્રમાણમાં વૃદ્ધિ થાય
B. ઇકો-ફ્રેન્ડલી ખેતી પદ્ધતિઓ અને તેનાથી સંકળાયેલ લેન્ડસ્કેપર્સ, કૃષિ જૈવવિવિધતા અને સ્થાનિક સમુદાયોની જ્ઞાન પ્રણાલીને ઓળખવી અને સુરક્ષિત કરવી
C. આવી ઓળખાયેલ GIASH માં કૃષિ પેદાશની તમામ જાતોને જિયોગ્રાફિક ઈન્ડિકેશન સ્ટેટસ પ્રદાન કરવા
D. બંને A અને C

137. પર્યાવરણના જૈવિક ઘટકોમાં કયા જીવોનો સમાવેશ થાય છે?

A. છોડ B. પ્રાણીઓ
C. સુક્ષ્મજીવાણુઓ D. ઉપરોક્ત તમામ

134. "Momentum for Change: Climate Neutral Now" is an initiative launched by

A. The UNFCCC Secretariat
B. The World Meteorological Organization
C. The Intergovernmental Panel on Climate Change
D. The UNEP Secretariat

135. The famous saying 'Earth provides enough to satisfy every man's need but not one man's greed' is of

A. Jawaharlal Nehru
B. Sunderlal Bahuguna
C. Mohandas Karamchand Gandhi
D. Sardar vallabhbhai patel

136. The FAO accords the status of GIAHS to traditional agricultural systems and the overall goal of this initiative is

A. To provide modern technology, training in modern farming methods and financial support to local communities of identified GIAHS so as to greatly enhance their agricultural productivity
B. To identify and safeguard eco-friendly traditional farm practices and their associated landscapers, agricultural biodiversity and knowledge systems of the local communities
C. To provide Geographical Indication status to all the varieties of agricultural produce in such identified GIAHS
D. Both A and C

137. Biotic components of environment include living organisms comprising

A. Plants B. Animals
C. Microbes D. Above all

138. __________ એ બે અથવા વધુ વૈવિધ્યસભર ઇકોસિસ્ટમ્સ વચ્ચેનો જંકશનનો એક ઝોન છે

A. ઇકોલોજીકલ નિશ B. ઇકોટોન
C. ઇકોસિસ્ટમ D. હેબીટાટ

139. નગર વન ઉદ્યાન યોજનાનો હેતુ શું છે ?

A. ઓછામાં ઓછા 25 હેક્ટરના વિસ્તારનું દરેક શહેરમાં ઓછામાં ઓછું શહેરનું એક જંગલ બનાવવું
B. દેશમાં 200 સિટી ફોરેસ્ટ્સ બનાવવા
C. પ્રકૃતિ સાથે વિદ્યાર્થીઓ કાયમી બોન્ડ બનાવવો
D. બંને A અને B

140. બરફ સાથે જોડાયેલા ધ્રુવોને સંલગ્ન વિશ્વના ઉત્તરી અને દક્ષિણના મોટાભાગના ક્ષેત્રનો સમાવેશ કરતું બાયોમ __________ તરીકે ઓળખાય છે

A. તાઈગા B. ટુંડ્ર
C. સવાન્નાહ D. ટેમ્પરેટ

141. 'ગ્લોબલ એન્વાયર્નમેન્ટ ફેસિલિટી' સંદર્ભે, નીચેનામાંથી કયું વિધાન યોગ્ય છે?

A. તે CBD અને UNFCC' માટે નાણાકીય મિકેનિઝમ તરીકે કામ કરે છે.
B. તે વૈશ્વિક સ્તરે પર્યાવરણીય મુદ્દાઓ અને ક્લાઈમેટ ચેન્જ પર વૈજ્ઞાનિક સંશોધન કરે છે
C. તે પર્યાવરણને સુરક્ષિત રાખવા માટેના વિશિષ્ટ ઉદ્દેશ સાથે અવિકસિત દેશોમાં તકનીકી અને ભંડોળના સ્થાનાંતરણની સુવિધા આપવા માટેની એક એજન્સી છે
D. બંને A અને C

142. સેડિમેન્ટરી ચક્ર __________ તરીકે ગણવામાં આવે છે.

A. સંપૂર્ણ ચક્ર B. જૈવિક ચક્ર
C. અપૂર્ણ ચક્ર D. આદર્શ ચક્ર

138. __________ is a zone of junction between two or more diverse ecosystems

A. Ecological Niche B. Ecotone
C. Ecosystem D. Habitat

139. Nagar Vana Udyan Yojana aims to

A. create at least one city forest in each city with a minimum area of 25 ha
B. create 200 City Forests in the country
C. build a lasting bond of students with nature
D. Both A and B

140. The biome comprising of Northern and Southern most region of world adjoining the ice bound poles is known as

A. Taiga B. Tundra
C. Savannah D. Temperate

141. With reference to 'Global Environment Facility', which of the following statements is/are correct?

A. It serves as financial mechanism for 'Convention on Biological Diversity' and 'United Nations Framework Convention on Climate Change'
B. It undertakes scientific research on environmental issues and climate change at global level
C. It is an agency to facilitate the transfer of technology and funds to under-developed countries with specific aim to protect their environment
D. Both A and C

142. Sedimentary cycles are considered as

A. Perfect cycles B. Biological cycles
C. Imperfect cycles D. Ideal cycles

143. ઇકોસિસ્ટમમાં ટ્રોફિક સ્તરની ક્રિયાપ્રતિક્રિયા ના અભ્યાસથી ઇકોસિસ્ટમ દ્વારા ઊર્જા પ્રવાહ વિશેનો ખ્યાલ આવે છે. ટ્રોફિક સ્તરની ક્રિયાપ્રતિક્રિયામાં ________ સામેલ છે

A. આહાર શૃંખલા
B. આહાર જાળ
C. ઇકોલોજીકલ પિરામિડ્સ
D. ઉપરોક્ત તમામ

144. પ્રદૂષકોનું ધ્યાન કેન્દ્રિત કરવાનુ વલણ જ્યારે તેઓ એક ટ્રોફિક સ્તરથી બીજા સ્થાને જાય છે, તે ________ તરીકે ઓળખાય છે.

A. બાયોએક્યુમ્યુલેશન
B. બાયોમેગ્નિફિકેશન
C. બાયોકોનસ્નટ્રેશન
D. પોલ્યુટેન્ટ એક્યુમ્યુલેશન

145. વધુ વરસાદ વાળા ઠંડા પ્રદેશો, લાંબો શિયાળો અને ટૂંકા ઉનાળા સાથે મોસમી હવામાન ધરાવતો વિસ્તાર કયા નામે ઓળખાય છે?

A. બોરિયલ શંકુદ્રુપ વન
B. ઉષ્ણકટિબંધીય વરસાદનું વન
C. સમશીતોષ્ણ પાનખર જંગલ
D. ઘાસના મેદાનો

146. ટુન્ડ્રા એટલે

A. ફળદ્રુપ જમીન B. રણ જમીન
C. ઉજ્જડ જમીન D. પર્વતની જમીન

147. ઘાસ ________ ને કારણે નીચા ટ્રોફિક સ્તરમાં હોય છે.

A. કદ B. વિપુલતા
C. વજન D. બંને A અને B

148. જ્યારે તે સમુદાયના જીવંત રહેવાસીઓ દ્વારા સક્સેશન લાવવામાં આવે છે, ત્યારે તે પ્રક્રિયા ________ કહેવામાં આવે છે

A. ઓટોજેનિક સક્સેશન
B. પ્રાયમરી સક્સેશન
C. એલોજેનિક સક્સેશન
D. સેકન્ડરી સક્સેશન

143. The study of trophic level interaction in an ecosystem gives an idea about the energy flow through the ecosystem. The trophic level interaction involves

A. Food Chain
B. Food Web
C. Ecological Pyramids
D. All of the above

144. The tendency of pollutants to concentrate as they move from one trophic level to the next is known as

A. Bioaccumulation
B. Biomagnification
C. Bioconcentration
D. Pollutant accumulation

145. Cold regions with high rainfall, strong seasonal climates with long winters and short summers are characterized by

A. Boreal coniferous forest
B. Tropical rain forest
C. Temperate deciduous forest
D. Grasslands

146. Tundra refers to

A. Fertile land B. Desert land
C. Barren land D. Mountain land

147. The grasses occupy the lowest trophic level because of their

A. Size B. Abundance
C. Weight D. Both A and B

148. When succession is brought about by living inhabitants of that community itself, the process is called

A. Autogenic succession
B. Primary succession
C. Allogenic succession
D. Secondary succession

149. Global Climate Change Alliance'ના સંદર્ભમાં, નીચેનામાંથી કયું વિધાન ખોટું છે?

A. તે યુરોપિયન યુનિયનની પહેલ છે

B. તે લક્ષિત વિકાસશીલ દેશોને તેમની વિકાસ નીતિઓ અને બજેટમાં ક્લાઈમેટ ચેન્જ ને એકીકૃત કરવા તકનીકી અને નાણાકીય સહાય પૂરી પાડે છે

C. તે World Resources Institute (WRI) અને World Business Council for sustainable Development (WBCSD) દ્વારા સંકલન કરવામાં આવે છે.

D. ઉપર્યુક્તમાંથી કોઈ નહિ

150. લિટોરલ અને સ્વેમ્પ જંગલો ક્યાં મળી આવે છે?

A. આંદામાન અને નિકોબાર આઇલેન્ડ્સ

B. ગંગાના ડેલ્ટા વિસ્તાર

C. બ્રહ્મપુત્રનો ડેલ્ટા વિસ્તાર

D. ઉપરોક્ત તમામ

151. ઘાસના મેદાનોત્યાં જોવા મળે છે, જ્યાં વરસાદ ________ જેટલો હોય છે

A. દર વર્ષે 10-15 સે.મી.

B. દર વર્ષે 25-75 સે.મી.

C. દર વર્ષે 85-95 સે.મી.

D. > દર વર્ષે 100 સે.મી.

152. ક્યોટો પ્રોટોકોલના વિકાસશીલ દેશની પાર્ટીઓમાં કોંક્રીટ એડેપ્ટિવ પ્રોજેક્ટ્સ અને કાર્યક્રમોના નાણાં માટે શું સ્થાપના કરવામાં આવ્યું હતું, જે ખાસ કરીને તે પાર્ટીઓ (દેશો) માટે જે ક્લાઈમેટ ચેન્જની પ્રતિકૂળ અસરોથી વધુ અસરગ્રસ્ત હોય છે?

A. એડેપ્ટેશન ફંડ

B. અલાયન્સ ઓફ સ્મોલ આઇલેન્ડ સ્ટેટ્સ

C. બાલી એક્શન પ્લાન

D. ક્યોટો ફાઇનાન્સ પ્રોગ્રામ

149. With reference to 'Global Climate Change Alliance', which of the following statement is correct?

A. It is an initiative of the European Union

B. It provides technical and financial support to targeted developing countries to integrate climate change into their development policies and budgets

C. It is coordinated by World Resources Institute (WRI) and World Business Council for sustainable Development (WBCSD)

D. None of the above

150. Littoral and swamp forests are found in

A. Andaman and Nicobar Islands

B. Delta area of the Ganga

C. Delta area of the Brahmaputra

D. All of the above

151. The grasslands are found where rainfall is about

A. 10-15 cm per year

B. 25-75 cm per year

C. 85-95 cm per year

D. >100 cm per year

152. ________ was established to finance concrete adaptation projects and programmes in developing country Parties to the Kyoto Protocol that are particularly vulnerable to the adverse effects of climate change

A. The Adaptation Fund

B. Alliance of Small Island States

C. Bali Action Plan

D. Kyoto finance program

153. ક્યોટો પ્રોટોકોલ હેઠળ અથવા અન્ય કરારો હેઠળ, જેમ કે યુરોપિયન યુનિયનના સભ્ય દેશો વચ્ચે, દેશના ઉત્સર્જન પરની તેમની રાષ્ટ્રીય મર્યાદાને પહોંચી વળવા માટેના પ્રયત્નોમાં ગ્રીનહાઉસ-ગેસ ઉત્સર્જનના એકમો ખરીદી અથવા વેચી શકે તેવી વેપાર સિસ્ટમ ને શું કહે છે ?

A. યુરોપિયન યુનિયનનું બજાર
B. કાર્બન માર્કેટ
C. GHG માર્કેટ
D. ઉત્સર્જન વેપાર

153. The trading system through which countries may buy or sell units of greenhouse-gas emissions in an effort to meet their national limits on emissions, either under the Kyoto Protocol or under other agreements, such as that among member states of the European Union. What is this system called?

A. European Union market
B. Carbon Market
C. GHG Market
D. Emissions trading

154. UNFCCC પ્રક્રિયા અને લો કાર્બન ડોમેસ્ટિક આઉટપુટથી મહત્વાકાંક્ષી પરિણામો મેળવવા માંગતા 27 દેશોનો સમૂહ ________ તરીકે ઓળખાય છે

A. ક્યોટો ગ્રુપ
B. અલાયન્સ ઓફ સ્મોલ આઇલેન્ડ સ્ટેટ્સ
C. કારટેજેના ગ્રુપ
D. અમ્બ્રેલા ગ્રુપ

154. A collection of 27 countries seeking ambitious outcomes from the UNFCC process and low carbon domestic output is known as

A. Kyoto Group
B. Alliance of Small Island States
C. Cartagena Group
D. Umbrella group

155. બાય-પ્રોડક્ટ્સ તરીકે ગ્રીનહાઉસ-ગેસ ઉત્સર્જન અથવા બળતણ પ્રોડક્શન, સ્ટોરેજ અથવા પરિવહનની પ્રક્રિયામાં કચરો અથવા નુકસાન કયા નામે ઓળખાય છે ?

A. સ્પીલ-ઓવર ઇફેક્ટ્સ
B. ફ્યુજીટીવ ફ્યુઅલ એમિશન
C. ગ્લોબલ વોર્મિંગ પોટેન્શિયલ
D. એમિશન પોટેન્શિયલ

155. Greenhouse-gas emissions as by-products or waste or loss in the process of fuel production, storage, or transport is known as

A. Spill-over effects
B. Fugitive fuel emissions
C. Global warming potential
D. Emission potential

156. અતિશય પોષક તત્વોને લીધે ઉચ્ચ પ્રાથમિક ઉત્પાદકતા સાથેનો તળાવ અને પાણીની ગુણવત્તા અલ્ગાલ બ્લુમ આધીન છે, આવા સરોવરને શું કહે છે?

A. યુટ્રોફિક સરોવર
B. ઓલિગોટ્રોફિક સરોવર
C. મેસોટ્રોફિક સરોવર
D. ફ્રેશવોટર સરોવર

156. Lake with high primary productivity due to excessive nutrients and is subject to algal blooms resulting in poor water quality

A. Eutrophic lake
B. Oligotrophic lake
C. Mesotrophic lake
D. Fresh water lake

157. રિબાઉન્ડ ઇફેક્સ્ટ્સ શું છે ?

A. પ્રોટોકોલ્સ સામાન્ય રીતે નવી, વધુ વિગતવાર પ્રતિબદ્ધતાઓ ઉમેરીને સંમેલનને મજબૂત બનાવે છે

B. સીધા માનવ-પ્રેરિત જમીન વપરાશ, જમીનના ઉપયોગમાં ફેરફાર અને વનીકરણની પ્રવૃત્તિઓના પરિણામે ગ્રીનહાઉસ વાયુઓના ઉત્સર્જન અને દૂર કરવા

C. વિકાસશીલ દેશોમાં ગ્રીનહાઉસ ગેસના ઉત્સર્જનને ઘટાડવા માટે વિકસિત દેશોએ લીધેલી પગલાંને લીધે થતાં બદલાવ

D. વાતાવરણમાં કાર્બન ડાયોક્સાઇડની સમકક્ષતાની સાંદ્રતા, જે વિશ્વને 2 ડિગ્રી સેલ્સિયસ લક્ષ્યને સાબિત કરવા માટે અથવા તેનાથી નીચે રહેવું જોઈએ તે મિલિયન દીઠ 450મો ભાગ છે

158. પર્યાવરણીય ડિગ્રેડેશનને કારણે થતા રોગોમાં ________ સામેલ છે.

A. મીનામાતા રોગ

B. યોક્કાઇચી અસ્થમા

C. બ્લુ બેબી સિન્ડ્રોમ

D. ઉપરોક્ત તમામ

159. પર્યાવરણ અને તેના પરના બાયોટા પર પ્રકાશિત પ્રદૂષક તત્વોના પ્રભાવનો શેનો અભ્યાસ છે?

A. ઇકો-ટોક્સિકોલોજી

B. બાયો-ટોક્સિકોલોજી

C. ઇકો-એપીડેમિઓલોજી

D. બાયો-એપીડેમિઓલોજી

160. ________ પર થયેલ મીનામાતા સંમેલન માનવ આરોગ્ય અને પર્યાવરણને એન્થ્રોપોજેનિક ઉત્સર્જનથી બચાવવા વૈશ્વિકસંધિ છે.

A. કોબાલ્ટ B. કોપર

C. મરક્યુરી D. આર્સેનિક

157. Rebound effects are

A. Protocols typically strengthen a convention by adding new, more detailed commitments

B. Emissions and removals of greenhouse gases resulting from direct human-induced land use, land-use change and forestry activities

C. Reverberations in developing countries caused by actions taken by developed countries to cut greenhouse gas emissions

D. The concentration of carbon dioxide equivalent in the atmosphere that the world must stay at or under to stay true to the 2 degrees Celsius goal is 450 parts per million

158. Diseases caused by environmental degradation include/s

A. Minamata disease

B. Yokkaichi asthma

C. Blue baby syndrome

D. All of the above

159. ________ is a study of the effects of released pollutants on the environment and on the biota that inhabit it

A. Eco-toxicology

B. Bio-toxicology

C. Eco-epidemiology

D. Bio-epidemiology

160. The Minamata Convention on ________ is a global treaty to protect human health and the environment from anthropogenic emissions

A. Cobalt B. Copper

C. Mercury D. Arsenic

161. કયો આંતરરાષ્ટ્રીય કરાર કાયમી ધોરણે પર્યાવરણ અને વિકાસને સસ્ટેનેબલ લેન્ડ મેનેજમેન્ટ સાથે જોડે છે?

A. UNCCD

B. IPCC

C. UNFCCC

D. બેસલ કન્વેનશન

162. ________ તે વિશ્વની સૌથી મોટી વન્યજીવન વેપાર પર દેખરેખ કાર્યક્રમ અને વન્યજીવન વેપાર મુદ્દાઓ પર એક વૈશ્વિક નિષ્ણાત સંસ્થા છે.

A. CITES B. TRAFFIC

C. CAWT D. WWF

163. નાગોયા પ્રોટોકોલ શું છે ?

A. લાભ-વહેંચણી અને પાલન માટેની જોગવાઈઓ સાથે આનુવંશિક સંસાધનો સાથે સંકળાયેલ પરંપરાગત જ્ઞાન

B. આનુવંશિક સંસાધનો જ્યાં સ્વદેશી અને સ્થાનિક સમુદાયોને તેમને પ્રવેશ આપવા માટે સ્થાપિત અધિકાર છે

C. બંને A અને B

D. ઉપર્યુક્તમાંથી કોઈ નહિ

164. The National Biodiversity Authority (NBA) ની સ્થાપના ________ માં ભારતના જૈવિક વિવિધતા અધિનિયમ (2002) ના અમલીકરણ માટે કરવામાં આવી હતી.

A. 2003 B. 2004

C. 2007 D. 2009

165. નોન- આયનાઇઝિંગ રેડિએશન ________ ધરાવે છે

A. ઉચ્ચ પ્રવેશ શક્તિ

B. ઓછી પ્રવેશ શક્તિ

C. એક્સ-રે

D. બંને A અને C

161. The legally binding international agreement linking environment and development to sustainable land management is

A. UNCCD

B. IPCC

C. UNFCCC

D. Basel Convention

162. ________ is the world's largest wildlife trade monitoring programme, and a global expert on wildlife trade issues.

A. CITES B. TRAFFIC

C. CAWT D. WWF

163. The Nagoya Protocol addresses

A. traditional knowledge associated with genetic resources with provisions on access, benefit-sharing and compliance

B. genetic resources where indigenous and local communities have the established right to grant access to them

C. Both A and B

D. None of the above

164. The National Biodiversity Authority (NBA) was established in ______ to implement India's Biological Diversity Act (2002)

A. 2003 B. 2004

C. 2007 D. 2009

165. Non-ionising radiations have

A. High penetration power

B. Low penetration power

C. X-rays

D. Both A and C

166. "સાયન્સ એક્સપ્રેસ" એ એક નવીન મોબાઇલ સાયન્સ પ્રદર્શન છે જે 16 કોચની એસી ટ્રેનમાં લગાવેલી છે, જેને ભારતીય રેલ્વે દ્વારા કસ્ટમ બિલ્ટ કરી કોના માટે બનાવવામાં આવી છે?

A. બાયોટેકનોલોજી વિભાગ

B. વિજ્ઞાન અને ટેકનોલોજી વિભાગ

C. WWF

D. WII

167. બાયરોમીડીએશન જેમાં ફૂગનો ઉપયોગ આ ક્ષેત્રને ફરીથી કોન્ટામીનેટ કરવા માટે થાય છે તે વિસ્તાર _____ તરીકે ઓળખવામાં આવે છે.

A. ફાયટોરેમીએશન

B. માયકોમેડીએશન

C. *In-situ* બાયરોમેડિએશન

D. *Ex-situ* બાયરોમેડિએશન

168. ભારતમાં એસિડ વરસાદનો પહેલો અહેવાલ ક્યારે આવ્યો?

A. 1968 માં દિલ્હી

B. 1982 માં કોલકાતા

C. 1971 માં બેંગ્લોર

D. 1974 માં બોમ્બે

169. International Solar Alliance (ISA) એ પેરિસ ખાતે યોજાયેલ CoP21 ક્લાયમેટ કોન્ફરન્સ માં શરૂ કરવામાં આવેલ જે _____________ ને સમર્પિત છે.

A. ISA સભ્ય દેશોની વિશેષ ઊર્જા જરૂરિયાતોને સંબોધવા

B. વિકાસશીલ દેશોની વિશેષ ઉર્જા જરૂરિયાતોને સંબોધવા

C. વિકસિત દેશોની વિશેષ ઉર્જા જરૂરિયાતોને સંબોધવા

D. સમશીતોષ્ણ દેશોની વિશેષ ઉર્જા જરૂરિયાતોને સંબોધવા

166. "Science Express" is an innovative mobile science exhibition mounted on a 16-coach AC train, which has been custom-built for _____________ by Indian Railway

A. Department of Biotechnology (DBT)

B. Department of Science & Technology (DST)

C. WWF

D. WII

167. Bioremediation in which fungi are used to decontaminate the area is known as

A. Phytoremediation

B. Mycoremediation

C. *In-situ* bioremediation

D. *Ex-situ* bioremediation

168. In India, the first report of acid rain came from

A. Delhi in 1968

B. Kolkata in 1982

C. Bangalore in 1971

D. Bombay in 1974

169. International Solar Alliance (ISA) was launched at the CoP21 Climate Conference in Paris and is dedicated to

A. address special energy needs of ISA member countries

B. address special energy needs of developing countries

C. address special energy needs of developed countries

D. address special energy needs of temperate countries

170. પાણી કેવી રીતે વહી જાય ત્યારે હાઇડ્રોલિક પાવર કબજે કરી શકાય છે?
A. ઉચ્ચ સ્તરથી નીચલા સ્તર પર
B. વધુ ઝડપે
C. ધીમી ગતિએ
D. સપાટ વિસ્તારમાં

171. બાયોમેથેનેશન એક પ્રક્રિયા છે જેમાં એનારોબિક સુક્ષ્મસજીવો બાયોડિગ્રેડેબલ મેટર ________ નિર્માણને વિઘટન કરે છે.
A. CO_2 B. CO
C. CH_4 D. C_2H_6

172. ભારત પાસે ____________
A. પાંચ બાયોમ છે
B. દસ બાયો-જિયોગ્રાફિક ક્ષેત્ર છે
C. પચીસ બાયો-જિયોગ્રાફિક પ્રાંત છે
D. ઉપરોક્ત તમામ

173. The Indian Himalayan Region (IHR) વહીવટી રીતે કેટલા રાજ્યોને શમાવેશ કરે છે?
A. 6 B. 8
C. 10 D. 12

174. દલ સરોવર કોના દ્વારા સંરક્ષણ કાર્યક્રમ સમાવેશ કરવામાં આવ્યો છે?
A. MoEF, GOI B. LAWDA
C. CEE D. NLCP

175. નીચેનીમાંથી કઈ નદી કેરળમાંથી વહેતી નથી
A. ભરતપુઝા B. કુતિયાદિ
C. કાવેરી D. અંચનકોવિલ

176. ____________ એ રેતી અને કાંકરીને દૂર કરવાની પ્રક્રિયા છે જે પર્યાવરણીય મુદ્દો બની રહી છે.
A. કાંકરી ખાણકામ
B. રેતી ખનન
C. કાંકરી ખોદકામ
D. રેતી ખોદકામ

170. Hydraulic power can be captured when water flows
A. downward from a higher level to a lower level
B. at high speed
C. at slow speed
D. in plain area

171. Biomethanation is a process in which anaerobic microorganisms decompose biodegradable matter producing
A. CO_2 B. CO
C. CH_4 D. C_2H_6

172. India has ____________
A. Five Biomes
B. Ten Bio-geographic Zones
C. Twenty-five Bio-geographic provinces
D. All of the above

173. The Indian Himalayan Region (IHR) administratively covers how many states in their entirety?
A. 6 B. 8
C. 10 D. 12

174. The Dal lake has been included in the lake conservation programme of the
A. MoEF, GOI B. LAWDA
C. CEE D. NLCP

175. Which of the following river do not flow through Kerala
A. Bharatapuzha B. Kuttiyadi
C. Cauvery D. Achankovil

176. ____________is the process of removal of sand and gravel which is becoming an environmental issue
A. Gravel mining
B. Sand mining
C. Gravel excavating
D. Sand excavating

177. પામ તેલના ઉત્પાદન પર્યાવરણીય અસરો

A. વનનાબૂદી
B. જૈવવિવિધતાનું નુકસાન
C. ક્લાઈમેટ ચેન્જ
D. ઉપરોક્ત તમામ

178. પર્યાવરણ વ્યવસ્થાપન યોજનામાં શું સામેલ હોવું જોઈએ

A. બધી ઓળખાતી નોંધપાત્ર અસરો માટે શમન અને વળતરનાં પગલાંનું નિર્દેશન
B. અનિશ્ચિત અસરોનું ચિત્રણ
C. કાર્યક્રમ, સમયનું શેડ્યૂલ અને શમન અને વળતર પ્રણાલીને સ્થાને મૂકવા માટેના સ્થળો સહિતના આયોજન
D. ઉપરોક્ત તમામ

179. કેન્દ્રિય પ્રાયોજિત યોજનાઓ હેઠળ, 2015- 16 ના સુધારેલા ભંડોળના દાખલા મુજબ, ભારત સરકારનો હિસ્સો બાકીના ભારત માટે 50 ટકા અને ________ માટે 80 ટકા છે

A. ઉત્તર પૂર્વી રાજ્યો અને ત્રણ હિમાલયન રાજ્યો
B. ઉત્તર પૂર્વી રાજ્યો
C. હિમાલયના રાજ્યો
D. જમ્મુ કાશ્મીર

180. નિયોનિકોટિનોઇડ્સ એ એક ________ નો નવો વર્ગ છે નિકોટિનથી રાસાયણિક રીતે સંબંધિત છે.

A. જંતુનાશકો B. ફૂગનાશકો
C. જંતુનાશકો D. નીંદણનાશક દવાઓ

181. EMR માઇક્રોવેવ કિરણોત્સર્ગના શોષણને કારણે પેદા થતી ________ અસરોને કારણે માનવમાં સેલ્યુલર અને માનસિક ફેરફારો લાવી શકે છે.

A. ન્યુક્લિઅર B. ઇલેક્ટ્રોનિક
C. મેગ્નેટિક D. થર્મલ

177. Environmental impacts of palm oil production

A. Deforestation
B. Loss of biodiversity
C. Climate change
D. All of the above

178. Environment Management Plan should include

A. Delineation of mitigation and compensation measures for all the identified significant impacts
B. Delineation of unmitigated impacts
C. Physical planning including work programme, time schedule and locations for putting mitigation and compensation systems in place
D. All of the above

179. Under the Centrally Sponsored Schemes, as per the revised funding pattern from 2015- 16 onwards, the Government of India's share is 50 percent for rest of India and 80 percent for the

A. North Eastern States and three Himalayan States
B. North Eastern States
C. Himalayan states
D. Jammu and Kashmir

180. Neonicotinoids are a new class of - ________ chemically related to nicotine

A. Pesticides B. Fungicides
C. Insecticides D. Weedicides

181. EMR may cause cellular and psychological changes in human beings due to ________ effects that are generated due to absorption of microwave radiation

A. Nuclear B. Electronic
C. Magnetic D. Thermal

182. આનુવંશિક રીતે ઇજનેરીવાળા ________ ઝાડ સાથેનો ભારતમાં પહેલો પ્રયોગ હતો.

A. રબરનું ઝાડ B. પામું ઝાડ
C. નાળિયેરનું ઝાડ D. કેરીનું ઝાડ

183. ભારત સરકાર દ્વારા વાયુ પ્રદૂષણને નિયંત્રણમાં લેવાતા પગલાં ક્યા છે ?

A. સુધારેલ ઓટો-ફ્યુઅલ
B. વિશુદ્ધ તકનીકોનો પ્રમોશન
C. ઉલ્લેખિત ઉદ્યોગો માટે પર્યાવરણીય મંજૂરી ફરજિયાત
D. ઉપરોક્ત તમામ

184. ભારતની પ્રથમ UNESCO's Mixed World Heritage Site કઈ છે ?

A. ગુગમલ રાષ્ટ્રીય ઉદ્યાન
B. બાલપક્રમ રાષ્ટ્રીય ઉદ્યાન
C. કાંચનજંઘા રાષ્ટ્રીય ઉદ્યાન
D. નાહારહોલ રાષ્ટ્રીય ઉદ્યાન અને ટાઇગર રિઝર્વ

185. ________ એ પર્યાવરણના મહત્વના પરિણામો અને તેના વિકલ્પોના મહત્વના પાસાઓના પરિણામોને માપવાની એક રીત છે.

A. સ્ક્રીનીંગ B. સ્કોપિંગ
C. બેઝલાઇન ડેટા D. ઈમ્પેક્ટ પ્રેડિક્શન

186. ભારતમાં ઓઇલ પામ ડેવલપમેન્ટ પ્રોગ્રામની શરૂઆત ક્યારે કરવામાં આવી હતી?

A. 1989-1990 B. 1991-1992
C. 1992-1993 D. 1993-1994

187. 2006 ના સુધારા પછી EIA ચક્રમાં ક્યા ચાર તબક્કાઓનો સમાવેશ થાય છે?

A. સ્ક્રીનીંગ, સ્કોપિંગ, હીયરીંગ, અપ્રેઈઝલ
B. સ્ક્રીનિંગ, સ્કોપિંગ, હીયરીંગ, ઇન્ડેક્સિંગ
C. પ્રાથમિક અને માધ્યમિક સ્ક્રીનીંગ, હીયરીંગ, અપ્રેઈઝલ
D. સ્ક્રીનીંગ, સ્કોપિંગ, ઇન્ડેક્સિંગ , અપ્રેઈઝલ

182. The first experiment in India with genetically engineered tree was with

A. Rubber tree B. Palm tree
C. Coconut tree D. Mango tree

183. Steps taken by the Indian Government for control air pollution

A. Supply of improved auto-fuel
B. promotion of cleaner technologies
C. mandatory environmental clearance for specified industries
D. All of the above

184. ________ has been inscribed as India's first "Mixed World Heritage Site" on UNESCO World Heritage List

A. Gugamal National Park
B. Balpakram National Park
C. Khangchendzonga National Park
D. Naharhole Nat. Park and Tiger Reserve

185. ________ is a way of mapping the environmental consequences of the significant aspects of the project and its alternatives

A. Screening B. Scoping
C. Baseline data D. Impact prediction

186. Oil Palm Development Programme in India was launched during

A. 1989-1990 B. 1991-1992
C. 1992-1993 D. 1993-1994

187. After 2006 Amendment the EIA cycle comprises of four stages

A. Screening, Scoping, Public hearing, Appraisal
B. Screening, Scoping, Public hearing, Indexing
C. Primary and Secondary Screening, Public hearing, Appraisal
D. Screening, Scoping, Indexing, Appraisal

188. National Clean Development Mechanism Authority (NCDMA)ની સ્થાપના ________ માં CDM પ્રોજેક્ટ્સ માટે હોસ્ટ કન્ટ્રી એપ્રુવલ (HCA) માટે કરવામાં આવી હતી.

A. ફેબ્રુઆરી 2005 B. મે 2007
C. ડિસેમ્બર 2003 D. ઓગસ્ટ 2001

189. નીચેનામાંથી કયા પર્યાવરણીય સંવેદનશીલ સ્થળો છે?

A. કોરલ્સ, મેંગ્રોવ્સ, વિશિષ્ટ જાતિઓના સંવર્ધન મેદાનથી સમૃદ્ધ કાંઠાના વિસ્તારો
B. મેંગ્રોવથી સમૃદ્ધ એસ્ચ્યુઅરીઝ, પ્રજાતિઓનું સંવર્ધન સ્થળ
C. ગલ્ફ એરિયા
D. ઉપરોક્ત તમામ

190. કોઈ ચોક્કસ ક્ષેત્ર અથવા ઇકોસિસ્ટમની વિવિધતા, અને સામાન્ય રીતે તે ઇકોસિસ્ટમની પ્રજાતિઓની સંખ્યા દ્વારા વ્યક્ત કરવામાં આવે છે, જેને શું કહેવાય છે?

A. આલ્ફા વિવિધતા
B. બીટા વિવિધતા
C. ગામા વિવિધતા
D. ડેલ્ટા વિવિધતા

191. ઝાડના પાંદડા સૂટ (soot) અને ધૂળ (dust) ના નાના કણોને ફસાવવામાં અને દૂર કરવામાં મદદ કરે છે જે અન્યથા __ ને નુકસાન કરે છે

A. માનવ ફેફસાં B. માનવ લીવર
C. માનવ કિડની D. માનવ હૃદય

192. ____________ એ વૈશ્વિક પુનઃ પ્રાપ્ય ઊર્જા નીતિ મલ્ટી-સ્ટેકહોલ્ડર નેટવર્ક છે.

A. REN27 B. REN21
C. REN24 D. REN29

193. ભારતનું બાયો જિયોગ્રાફિક વર્ગીકરણ કોના દ્વારા કરવામાં આવ્યું હતું

A. રોજર્સ અને પનવર (1988)
B. એડવર્ડ અને પનવર (1985)
C. રોજર્સ અને હેરિસ (1989)
D. રોઝન અને થોમસ (1987)

188. National Clean Development Mechanism Authority (NCDMA) was established in ________for according Host Country Approval (HCA) to the CDM projects

A. February 2005 B. May 2007
C. December 2003 D. August 2001

189. Which of the following are environmentally sensitive places?

A. Coastal areas rich in corals, mangroves, breeding grounds of specific species
B. Estuaries rich in mangroves, breeding ground of specific species
C. Gulf areas
D. All of the above

190. The diversity within a particular area or ecosystem, and is usually expressed by the number of species in that ecosystem is known as

A. Alpha diversity
B. Beta diversity
C. Gamma diversity
D. Delta diversity

191. Tree leaves help trap and remove tiny particles of soot and dust which otherwise damages

A. human lungs B. human liver
C. human kidneys D. human heart

192. ____________ is the global renewable energy policy multi-stakeholder network

A. REN27 B. REN21
C. REN24 D. REN29

193. Biogeographic classification of India was done by

A. Rodgers and Panwar (1988)
B. Edward and Panwar (1985)
C. Rodgers and Harris (1989)
D. Rosen and Thomas (1987)

194. ________ છોડના શરીરને નાના દાંડી અને સરળ પાંદડામાં અલગ પાડવામાં આવે છે, પરંતુ સાચા મૂળ ગેરહાજર છે.

A. ટિરીડોફાઇટ્સ B. શેવાળ

C. બ્રાયોફાઇટ્સ D. જિમ્નોસ્પર્મ્સ

195. ગ્લોબલ વોર્મિંગ ઇફેક્ટ્સમાં શું સામેલ છે?

A. રોગનો ફેલાવો (મેલેરિયાની જેમ)

B. કોરલ રીફ્સનું બ્લીચિંગ

C. વરસાદમાં ફેરફાર

D. ઉપરોક્ત તમામ

196. ક્લાઈમેટ ચેન્જ એ સતત ________ વલણની માપી શકાય તેવી અસરો છે

A. ઠંડક

B. ગરમી

C. ઠંડક અને ગરમ

D. બદલાવ

197. વૈજ્ઞાનિક રૂપે વિશ્વના વિવિધ ભાગોમાંથી જીવંત વૃક્ષો, ઝાડીઓ, ઔષધિઓ, લતા અને અન્ય છોડ સંગ્રહ કરવાની યોજના ને ________ કહે છે.

A. રાષ્ટ્રીય ઉદ્યાન

B. વનસ્પતિ બગીચો

C. અભયારણ્ય

D. જૈવિક ઉદ્યાન

198. ક્લાઇમેટ ________ એ ક્લાઇમેટ પ્રણાલીના પરિબળો છે જે હવામાન પ્રણાલીમાં અસરને વધારે છે અથવા ઘટાડે છે.

A. ફોરસિંગ્સ B. મેઝર્સ

C. ઈવેન્ટ્સ D. ડિલિંગ્સ

199. પર્યાવરણ (સંરક્ષણ) અધિનિયમ, 1986 ના ક્યા નિયમને છૂટછાટ આપી છે?

A. જીવનનો અધિકાર

B. અનિવાર્ય હકો

C. લોકસ સ્ટેન્ડી

D. આર્ટિકલ્સ ઓફ કોન્ફીડરેશન

194. The plant body is differentiated into a small stem and simple leaves, but true roots are absent

A. Pteridophytes B. Algae

C. Bryophytes D. Gymnosperms

195. Global warming impacts include

A. Spread of disease (like malaria)

B. Bleaching of Coral Reefs

C. Changes in rainfall patterns

D. All of the above

196. Climate change is the measurable effects of the continual ________ trend

A. Cooling

B. Warming

C. Cooling and warming

D. Changing

197. The scientifically planned collection of living trees, shrubs, herbs, climbers and other plants from various parts of the world

A. National park

B. Botanical garden

C. Sanctuary

D. Biological park

198. Climate ________ are factors in the climate system that either increase or decrease the effects to the climate system.

A. Forcings B. Measures

C. Events D. Dealings

199. The Environment (Protection) Act, 1986 has relaxed the rule of

A. Right to life

B. The inalienable rights

C. Locus Standi

D. The articles of confederation

200. કઈ COP ભારતમાં યોજાઈ હતી?

A. COP 6 B. COP 8

C. COP 10 D. COP 12

200. Which COP was held in India?

A. COP 6 B. COP 8

C. COP 10 D. COP 12

Answers | જવાબો

Que.	Ans.	Que.	Ans.	Que.	Ans.	Que.	Ans.	Que.	Ans.	Que.	Ans.	Que.	Ans.
1.	A	31.	A	61.	A	91.	B	121.	B	151.	B	181.	D
2.	D	32.	B	62.	D	92.	D	122.	C	152.	A	182.	A
3.	B	33.	C	63.	C	93.	B	123.	B	153.	B	183.	D
4.	C	34.	A	64.	B	94.	B	124.	A	154.	C	184.	C
5.	D	35.	C	65.	D	95.	D	125.	D	155.	B	185.	D
6.	A	36.	C	66.	C	96.	C	126.	B	156.	A	186.	B
7.	C	37.	A	67.	C	97.	A	127.	D	157.	C	187.	A
8.	D	38.	B	68.	A	98.	C	128.	D	158.	D	188.	C
9.	A	39.	D	69.	D	99.	A	129.	D	159.	A	189.	D
10.	B	40.	B	70.	C	100.	D	130.	A	160.	C	190.	A
11.	C	41.	C	71.	D	101.	D	131.	B	161.	A	191.	A
12.	D	42.	C	72.	B	102.	A	132.	C	162.	B	192.	B
13.	D	43.	B	73.	A	103.	B	133.	D	163.	C	193.	A
14.	B	44.	B	74.	B	104.	D	134.	A	164.	A	194.	C
15.	D	45.	D	75.	A	105.	C	135.	C	165.	B	195.	D
16.	A	46.	B	76.	C	106.	B	136.	B	166.	B	196.	B
17.	C	47.	A	77.	D	107.	C	137.	D	167.	B	197.	B
18.	B	48.	C	78.	B	108.	A	138.	B	168.	D	198.	A
19.	C	49.	B	79.	C	109.	D	139.	D	169.	A	199.	C
20.	B	50.	D	80.	C	110.	D	140.	B	170.	A	200.	B
21.	C	51.	A	81.	A	111.	D	141.	A	171.	C		
22.	D	52.	C	82.	B	112.	A	142.	D	172.	D		
23.	C	53.	C	83.	B	113.	C	143.	B	173.	C		
24.	A	54.	D	84.	D	114.	B	144.	C	174.	A		
25.	B	55.	D	85.	A	115.	A	145.	A	175.	C		
26.	C	56.	C	86.	B	116.	C	146.	C	176.	B		
27.	B	57.	A	87.	C	117.	A	147.	B	177.	D		
28.	A	58.	B	88.	D	118.	D	148.	A	178.	D		
29.	B	59.	A	89.	C	119.	D	149.	C	179.	A		
30.	D	60.	D	90.	A	120.	A	150.	D	180.	C		

Chapter 17

પ્રકરણ ૧૭

Global Conventions related to Forests and Environment

વન અને પર્યાવરણને લગતા વૈશ્વિક સંમેલનો

Important International Environment Organizations

મહત્વપૂર્ણ આંતરરાષ્ટ્રીય પર્યાવરણ સંસ્થાઓ

Organization	HQ	Started in	Head
Convention on International Trade in Endangered Species of Flora and Fauna (CITES)	Washington D.C.	1975	Ivonne Higuero
Global Environment Facility (GEF)	Washington DC, USA	1991	Carlos Manuel Rodriguez
Global Green Growth Institute (GGGI)	Seoul, Korea	2012	Frank Rijsberman
Global Tiger Forum (GTF)	New Delhi	1993	Rajesh Gopal
Intergovernmental Panel on Climate Change (IPCC)	Geneva, Switzerland	1988	Hoesung Lee
International Tropical Timber Organization (ITTO)	Japan	1985	Gerhard Dieterle
International Whaling Commission (IWC)	Cambridge, United Kingdom	1946	Rebecca Lent
The International Centre for Research in Agroforestry (ICRAF)	Nairobi, Kenya	1977	Tony Simmons
The World Conservation Union- IUCN	Morgis, Switzerland	1948	Dr Bruno Oberle
TRAFFIC- Wildlife Trade Monitoring Network	Cambridge, UK	1976	Mark Helle
United Nations Educational, Scientific and Cultural Organization (UNESCO)	Paris, France	1945	Audrey Azoulay, Director General
United Nations Environment Programme (UNEP)	Nairobi, Kenya	1972	Inger Andersen
United Nations World Tourism Organization (UNWTO)	Madrid, Spain	1974	Zurab Pololikashvili
World Food Programme	Rome, Italy	1961	David Beasley

Organization	HQ	Started in	Head
World Intellectual Property Organization (WIPO)	Geneva, Switzerland	1967	Daren Tang
World Trade Organization (WTO)	Geneva, Switzerland	1995	Roberto Azevêdo
World Wide Fund for Nature (WWF)	Gland, Switzerland	1961	Carter Roberts

Important Protocols | મહત્ત્વપૂર્ણ પ્રોટોકોલ્સ

Protocol	Remarks/ Theme
Kyoto protocol	Climate change
Geneva protocol	Prohibition of the Use in War of Asphyxiating, Poisonous or other Gases, and of Bacteriological Methods of Warfare
Montreal Protocol	Ozone protection
Cartagena Protocol	Biosafety to the Convention on Biological Diversity
Nagoya Protocol	Access to Genetic Resources and the Fair and Equitable Sharing of Benefits Arising from their Utilization (ABS)
Greenhouse Gas Protocol	to understand, quantify and manage greenhouse gas emissions

Environment Conventions / પર્યાવરણ સંમેલનો

Convention	Year	Remarks
Basel Convention	1989	Control of Transboundary Movements of Hazardous Wastes and Their Disposal
Bonn Convention	1979	Aims to conserve migratory species within their migratory ranges. Also known as- The Convention on the Conservation of Migratory Species of Wild Animals (CMS).
Convention on International Trade in Endangered Species of Wild Fauna and Flora (CITES)	1963	A multilateral treaty to protect endangered plants and animals to ensure that international trade in specimens of wild animals and plants does not threaten their survival.
Convention on Biological Diversity (CBD)	1992	For the conservation of biological diversity, the sustainable use of its components and the fair and equitable sharing of benefits arising from genetic resources.
Minamata Convention	2013	An international treaty designed to protect human health and the environment from anthropogenic emissions and releases of mercury and mercury compounds.

Convention	Year	Remarks
Ramsar Convention	1971	An international treaty for the conservation and sustainable use of wetlands. Also known as the Convention on Wetlands.
Rotterdam Convention	1998	A multilateral treaty to promote shared responsibilities in relation to importation of hazardous chemicals.
Stockholm Convention	2001	An international environmental treaty on Persistent Organic Pollutants, that aims to eliminate or restrict the production and use of persistent organic pollutants.
United Nations Programme on Reducing Emissions from Deforestation and Forest Degradation (UN-REDD)	2008	To reduce the emissions from deforestation and Forest degradation
United Nations Convention to Combat Desertification (UNCCD)	1994	An international agreement linking environment and development to sustainable land management. To Combat Desertification in those Countries experiencing serious drought and/or desertification, particularly in Africa.
United Nations Framework Convention on Climate Change (UNFCCC)	1992	An international environmental treaty addressing climate change, and to stabilize greenhouse gas concentrations in the atmosphere at a level.
Vienna Convention	1985	Convention for the protection of the Ozone layer.

Important International Reports | મહત્વપૂર્ણ આંતરરાષ્ટ્રીય અહેવાલો

Report	Frequency	Started in	Releasing body
Actions on Air Quality	—	—	United Nations Environment Programme (UNEP)
Climate Change and Land	5 years	—	Intergovernmental Panel on Climate Change (IPCC)
Emission Gap Report	Annual		UN Environment Programme (UNEP)
Global Agricultural Information Network (GAIN)	Annual	1995	
Global Assessment Report on Biodiversity and Ecosystem Services	Annual	—	UN's Intergovernmental Science—Policy Platform on Biodiversity and Ecosystem Services (IPBES)
Global Climate Risk report	Annual	2007	Germanwatch
Global Environment Outlook (GEO)	5 years	1997	UN Environment Programme (UNEP)

Report	Frequency	Started in	Releasing body
Global Environment Performance Index	Annual	2002	World Economic Forum (WEF)
Global Forest Resource Assessment (GFRA)	5 years	1946	Food and Agriculture Organization (FAO), Rome, Itly
Global Futures Report	4 Years	—	World Wide Fund for Nature, Global Trade Analysis Project and the Natural Capital Project.
India State Forest Report (ISFR)	2 year	1987	Forest Survey of India (FSI), Dehradun
India's National Wildlife Action Plan (NWAP)	15 year	1983	MoEFCC
National Air Quality Indices		2014	Central Pollution Control Board (CPCB)
Nature risk rising report	Annual	2020	World Economic Forum (WEF)
Red Data Book	Red List is updated at least twice each year.	1964	The World Conservation Union-IUCN
State of the Global Climate Report	Annual	1996	World Meteorological Organization (WMO)
The Energy Report & Living Planet Report	2 Years	1998	World Wide Fund for Nature (WWF)
The Rise of Environmental Crime Report	Annual	—	UNEP & INTERPOL
World Air Quality Report 2019	Annual	—	Pollution tracker IQ Air and Greenpeace
World Climate and Security Report	Annual	—	International Military Council on Climate and Security (IMCCS)
World Disaster Risk Index	Annual	2011	United Nations University Institute for Environment and Human Security (UNUEHS)
World Energy Outlook	Annual	1977	International Energy Agency
World Water Development Report	Annual	2003	UNESCO World Water Assessment Programme (WWAP)
World Wildlife Crime Report	Annual	2017	United Nations Office on Drugs and Crime (UNODC)

Important Environment Conferences | મહત્વપૂર્ણ પર્યાવરણ પરિષદો

Conference	Theme	Year	Held at
Human Environment	Environment conservation	1972	Stockholm
Montreal Protocol	Ozone protection	1987	Montreal, Canada
Earth Summit	Biodiversity conservation & Reducing GHGs	1992	Rio-de-Janerio
Kyoto Protocol	Climate change	1997	Kyoto, Japan
United Nation	Desertification	1997	Nairobi, Kenya
World Summit	Sustainable development	2002	Johncsberg

Chapter 18

પ્રકરણ ૧૮

National and International Organizations related to Forests and Environment

વન અને પર્યાવરણ સંબંધિત રાષ્ટ્રીય અને આંતરરાષ્ટ્રીય સંસ્થાઓ

No	Organization	Full form	HQ	Year of establishment
1.	AFED	Arab Forum for Environment and Development	Beirut, Lebanon	2006
2.	AFRI	Arid Forestry Research Institute	Jodhpur	1985
3.	AKRSP	Aga Khan Rural Support Programme	Gilgit	1982
4.	ARCBR	Advanced Research Centre for Bamboo and Rattan	Aizawl, Mizoram	2004
5.	ATREE	Ashoka Trust for Research in Ecology And The Environment	Bangalore	1996
6.	AWBI	Animal Welfare Board of India	Faridabad, Haryana	1962
7.	BAIF	Bhartiya Agro-industries Foundation	Pune	1967
8.	BCSG	Bird Conservation Society Gujarat	Ahmedabad	
9.	BNHS	Bombay Natural History Society	Mumbai	1883
10.	BSI	Botanical Survey of India	Kolkata	1890
11.	BWC	Beauty Without Cruelty (India)	Pune	1974
12.	CAZRI	Central Arid Zone Research Institute	Jodhpur	1959
13.	CBMTRI	Crocodile Breeding and Management Training Research Institute	Chennai	1980
14.	CBRI	Central Bee Research Institute	Pune	1962
15.	CDRI	Central Drug Research Institute	Lucknow	1951
16.	CEE	Centre for Environment Education	Ahmedabad	1984
17.	CEMDC	Centre for Environmental Management of Degraded Ecosystem	Delhi	1990
18.	CES	Centre for Ecological Sciences	Bengaluru	1983
19.	CFLE	Centre for Forest-based Livelihoods and Extension	Agartala	2012

No	Organization	Full form	HQ	Year of establishment
20.	**CFRHRD**	Centre for Forestry Research and Human Resource Development	Chhindwara, MP	1995
21.	**CIFOR**	Centre for International Forestry Research	Indonesia	1993
22.	**CIMAP**	Central Institute of Medicinal and Aromatic Plants	Lucknow	1959
23.	**CITES**	Convention on International Trade in Endangered Species of Wild Fauna and Flora	Geneva, Switzerland	1975.
24.	**CME**	Centre for Mining Environment	Dhanbad, Jharkhand	1987
25.	**CNH**	Central National Herbarium	Kolkata	1787
26.	**CPCB**	Central Pollution Control Board	New Delhi	1974
27.	**CSE**	Centre for Science and Environment	New Delhi	1980
28.	**CSFER**	Centre for Social Forestry and Eco-Rehabilitation	Allahabad	1992
29.	**CSIR**	Council of Scientific and Industrial Research	New Delhi	1942
30.	**CSMCRI**	Central Salt and Marine Chemical Research Institute	Bhavnagar	1954
31.	**CSSRI**	Central Soil Salinity Research Institute	Karnal, Haryana	1969
32.	**CTFS**	Centre for Tropical Forest Science	Panama City, Panama	1980
33.	**CZA**	Central Zoo Authority	New Delhi	1992
34.	**EEA**	European Environment Agency	Copenhagen, Denmark	1993
35.	**EFI**	European Forest Institute	Joensuu, Finland	1993
36.	**F&B**	Forest and Bird	Wellington, New Zealand	1923
37.	**FAO**	Food and Agriculture Organization	Rome, Italy	1945
38.	**FRI**	Forest Research Institute	Dehradun	1906
39.	**FSI**	Forest Survey of India	Dehradun	1981
40.	**GBPIHED**	Govind Ballabh Pant Institute of Himalayan Environment & Development	Almora, Uttarakhand	1988
41.	**GEC**	Gujarat Ecology Commission	Gandhinagar	1992
42.	**GEDA**	Gujarat Energy Development Agency	Vadodara	1979

No	Organization	Full form	HQ	Year of establishment
43.	GEER	Gujarat Ecological Education and Research Foundation	Gandhinagar	1982
44.	GEF	Global Environment Facility	Washington, D.C	1991
45.	GEMI	Gujarat Environment Management Institute	Gandhinagar	1999
46.	GES	Gujarat Ecology Society	Vadodara	1995
47.	GFRC	Gujarat Forest Rangers College	Rajpipla	1987
48.	GFRI	Gujarat Forestry Research Institute	Gandhinagar	2008
49.	GGGI	Global Green Growth Institute	Seoul, Republic of Korea	2010
50.	GPCB	Gujarat Pollution Control Board	Gandhinagar	1974
51.	GSBTM	Gujarat State Biotechnology Mission	Gandhinagar	2004
52.	GSFDC	Gujarat State Forest Development Corporation	Vadodara	1976
53.	GSI	Geological Survey of India	Kolkata, West Bengal	1851
54.	GSRDC	Gujarat State Rural Development Corporation	Gandhinagar	1977
55.	GTF	Global Tiger Forum	New Delhi	1993
56.	HFRI	Himalayan Forest Research Institute	Shimla	1977
57.	ICAR	Indian Council of Agricultural Research	New Delhi	1929
58.	ICBP	International Centre for Birds of Prey	Gloucestershire, UK	1967
59.	ICFRE	Indian Council of Forestry Research and Education	Dehradun	1986
60.	ICIPE	International Centre of Insect Physiology and Ecology	Nairobi, Kenya	1970
61.	ICRAF	International Centre for Research in Agroforestry	Nairobi, Kenya	1977
62.	ICRISAT	International Crop Research Institute for Semi-arid Tropics	Hyderabad	1972
63.	ICSU	International Council for Science	Paris, France	1931
64.	IDRC	International Development Research Centre	Ottawa, Canada	1970
65.	IEP	Institute for Environmental Policy in Albania	Tirana, Albania	2008
66.	IFAD	International Fund for Agricultural Development	Rome, Italy	1977

No	Organization	Full form	HQ	Year of establishment
67.	**IFAW**	International Fund for Animal Welfare	Yarmouth Port, USA	1969
68.	**IFB**	Institute of Forest Biodiversity	Hyderabad	1997
69.	**IFGTB**	Institute of Forest Genetics and Tree Breeding	Coimbatore	1988
70.	**IFP**	Institute of Forest Productivity	Ranchi	1993
71.	**IFPRI**	International Food Policy Research Institute	Washington	1975
72.	**IGFRI**	Indian Grassland and Fodder Research Institute	Jhansi	1962
73.	**IGNFA**	Indira Gandhi National Forest Academy	Dehradun	1938
74.	**IIED**	International Institute for Environment and Development	London, UK	1973
75.	**IIFM**	Indian Institute of Forest Management	Bhopal	1982
76.	**IIHR**	Indian Institute of Horticultural Research	Bangalore	1967
77.	**IINRG**	Indian Institute of Natural Resins and Gums	Ranchi	1924
78.	**IIRS**	Indian Institute of Remote Sensing	Dehradun	1966
79.	**IISR**	Indian Institute of Spices Research	Kozhikode	1975
80.	**ILRAD**	International Laboratory for Research on Animal Diseases	Nairobi, Kenya	1973
81.	**ILRI**	Indian Lac Research Institute	Ranchi	1924
82.	**INSA**	Indian National Science Academy	New Delhi	1935
83.	**IPCC**	Intergovernmental Panel on Climate Change	Geneva, Switzerland	1988
84.	**IPIRTI**	Indian Plywood Industries Research & Training Institute	Bangalore	1970
85.	**IRMDFR**	Institute of Rain and Moist Deciduous Forests Research	Jorhat	
86.	**ITCI**	International Tree Crops Institute	Braintree, UK	1977
87.	**ITE**	Institute of Terrestrial Ecology	Cambridge, UK	1973
88.	**IUBS**	International Union of Biological Sciences	Paris, France	1919
89.	**IUCN**	International Union for Conservation of Nature and Natural Resources	Gland, Switzerland	1948
90.	**IUFRO**	International Union of Forestry Research Organization	Vienna, Austria	1892

No	Organization	Full form	HQ	Year of establishment
91.	**IVRI**	Indian Veterinary Research Institute	Izatnagar, UP	1889
92.	**IWST**	Institute of Wood Science and Technology	Bangalore	1938
93.	**JNTBGRI**	Jawaharlal Nehru Tropical Botanic Garden and Research Institute	Thiruvananthapuram	1979
94.	**KFRI**	Kerala Forest Research Institute	Peechi, Kerala	1975
95.	**NAARM**	National Academy of Agricultural Research and Management	Hyderabad	1976
96.	**NABARD**	National Bank for Agriculture and Rural Development	Mumbai	1982
97.	**NBA**	National Biodiversity Authority	Chennai	2003
98.	**NBB**	National Bee Board	New Delhi	2006
99.	**NBPGR**	National Bureau of Plant Genetic Resources	New Delhi	1976
100.	**NBRI**	National Botanical Research Institute	Lucknow	1953
101.	**NBWL**	National Board for Wildlife	New Delhi	2003
103.	**NCF**	Nature Conservation Foundation	Mysore	1996
103.	**NCL**	National Chemical Laboratory	Pune	1950
104.	**NDRI**	National Dairy Research Institute	Kernal, Haryana	1923
105.	**NEERI**	National Environmental Engineering and Research Institute	Nagpur	1958
106.	**NGRBA**	National Ganga River Basin Authority	New Delhi	2009
107.	**NIAW**	National Institute of Animal Welfare	Faridabad, Haryana	1999
108.	**NIRD**	National Institute of Rural Development	Hyderabad	2006
109.	**NRCMAP**	National Research Centre of Medicinal and Aromatic Plants	Boriyavi, Anand	1992
110.	**NRSA**	National Remote Sensing Agency	Hyderabad	1974
111.	**NSC**	National Seed Corporation	Gandhinagar	1963
112.	**NTCA**	National Tiger Conservation Authority	New Delhi	2005
113.	**NWDB**	National Wasteland Development Board	New Delhi	1985
114.	**OECD**	Organisation for Economic Cooperation and Development	Paris, France	1961
115.	**PEMSEA**	Partnerships in Environmental Management for the Seas of East Asia	Quezon City, Philippines	1994
116.	**RFRI**	Rain Forest Research Institute	Jorhat	1988
117.	**SAARC**	South Asian Association for Regional Cooperation	Kathmandu, Nepal	1985

No	Organization	Full form	HQ	Year of establishment
118.	**SACON**	Salim Ali Centre for Ornithology and Natural history	Coimbatore	1990
119.	**SAZARC**	South Asian Zoo Association for Regional Cooperation	Nepal	2000
120.	**TBGRI**	Tropical Botanical Garden and Research Institute	Thiruvananthapuram, Kerala	1979
121.	**TERI**	The Energy Research Institute	New Delhi	1974
122.	**TES**	The Ecotourism Society	USA	1990
123.	**TFRI**	Tropical Forestry Research Institute	Jabalpur	1988
124.	**TRAFFIC**	Trade Record Analysis for Flora and Fauna in Commerce	Cambridge, United Kingdom	1976
125.	**UNDP**	United Nations Development Programme	New York, United States	1965
126.	**UNEP**	United Nations Environment Programme	Nairobi, Kenya	1972
127.	**UNESCO**	United Nations Educational, Scientific and Cultural Organization	Paris, France	1945
128.	**UNO**	United Nations Organization	New York, United States	1945
129.	**USEPA**	United States Environmental Protection Agency	Washington D.C	1970
130.	**VIKSAT**	Vikram Sarabhai Centre for Development Interaction	Ahmedabad	1977
131.	**WASMO**	Water and Sanitation Management Organization	Gandhinagar	2002
132.	**WAZA**	World Association of Zoos and Aquariums	Barcelona, Spain	1935
133.	**WCCB**	Wildlife Crime Control Bureau	New Delhi	2006
134.	**WCMC**	World Conservation Monitoring Centre	Cambridge, UK	2000
135.	**WFP**	World Food Programme	Rome, Italy	1961
136.	**WII**	Wildlife Institute of India	Dehradun	1982
137.	**WMO**	World Meteorological Organization	Geneva, Switzerland	1950
138.	**WTO**	World Trade Organization	Geneva, Switzerland	1995
139.	**WWF**	World Wide Fund for Nature	Gland, Switzerland	1961
140.	**ZOO**	Zoo Outreach Organization	Coimbatore	1999
141.	**ZSI**	Zoological Survey of India	Kolkata	1916

Chapter 19

પ્રકરણ ૧૯

Forestry Terms

વાનિકી શબ્દાવલી

Agriculture — Latin word — "ager" meaning 'soil' and "cultura" meaning 'cultivation'.

Communication — Latin word — "communis" meaning 'common'.

Dendrology — Greek word — "dendron" meaning 'tree' and "logia" meaning 'study'.

Ecology — Greek word — "oikos" meaning 'habitat' and "logos" meaning 'study'.

Economics — Greek word — "oikonomia" meaning 'household management'.

Entomology — Greek word — "entomo" meaning 'insect' and "logos" meaning 'study'.

Ergonomics — Greek word — "ergon" meaning 'work' and "nomoi" meaning 'natural law'.

Extension — Latin word — "ex" meaning 'out' and "tensio" meaning 'stretching'.

Forest mensuration — Latin word — "Mensura" meaning 'measure'.

Forest Pathology — Greek word — "pathos" meaning 'ailments' and "logos" meaning 'study'.

Meteorology — Greek word — "meteoro" meaning 'above earth surface' and "logy" meaning 'indicating science'.

Monsoon — Arabic word — "mausim" meaning 'season'.

Silviculture — Latin word — "silvi" meaning 'forest and 'culture" meaning 'as in growing'.

Soil — Latin word — "solum" meaning 'ground'

Soil Science — Latin word — "catena" meaning 'chain'.

Adaptation: Modifications in living organisms over time that allow them to function better in their environment.

Aestivation: A state of animal dormancy, characterized by inactivity and a lowered metabolic rate that is entered in response to high temperatures and arid conditions. It takes place during times of heat and dryness, the hot dry season, which is often but not necessarily the summer months. Both terrestrial and aquatic animals undergo aestivation.

Afforestation: The establishment of a stand of trees or forest in an area where there was no previous tree cover.

Agroforestry: Is an integrated approach of using the interactive benefits from combining trees and shrubs with crops and livestock. It combines agricultural and forestry technologies.

Algal bloom: is a rapid increase or accumulation in the population of algae in an aquatic system.

All-Aged or Uneven-Aged Management: The practice of managing a forest by periodically selecting and harvesting individual trees or groups of trees from the stand while preserving its natural appearance. Most common in hardwood forests.

Angiosperms: Flowering plants which produce seeds enclosed in an ovary.

Annuals: A plant that lives or grows for only one year or one growing season.

Aspect: The compass direction towards which a slope faces, for example south-west.

Backburning: A bushfire preventative technique that involves lighting a controlled fire at a safe point to burn back towards the front of the bushfire.

Basal Area: The cross-sectional area of a tree at breast height (1.3m) expressed in square metres (m^2). It's a tool used to calculate sustainable timber yield.

Bedding: Land prepared before planting in the form of small mounds. The prepared land concentrates topsoil and elevates the root zone of seedlings above temporary standing water. Fertilizer is often incorporated into the bedding.

Bio-link zones: A land use category developed for biodiversity conservation and landscape adaptation under changing climates

Biobank: A cryogenic storage facility used to archive biological samples for use in research and experiments

Biocapacity: The capacity of an area to provide resources and absorb wastes. When the area's ecological footprint exceeds its biocapacity, unsustainability occurs.

Biodiversity: The genetic variety of life forms and their ecosystems. Comprises genetic diversity (within species), species diversity (between species) and ecosystem diversity.

Bioenergy: Renewable energy produced from organic matter that is used directly as fuel or has been converted into liquids or gasses for use as a fuel.

Biological pump: The sum of a suite of biologically mediated processes that transport carbon from the surface euphotic zone to the ocean's interior.

Biomass: Any organic matter, including forest and mill residues. Forest biomass is usually obtained through in-woods chipping of all or some portion of trees, to include tops and limbs, for energy production.

Biopiracy: The theft of genetic materials especially plants and other biological materials by the patent process. Biopiracy is a situation where indigenous knowledge of nature, originating with indigenous people, is exploited for commercial gain without permission from and with no compensation to the indigenous people themselves.

Bioprospecting: A term describing the discovery of new and useful biological samples and mechanisms, typically in less-developed countries, either with or without the help of indigenous knowledge, and with or without compensation.

Bioterrorism: The intentional re-lease or dissemination of biological agents. These agents are (bacteria, viruses, or toxins), and may be in a naturally occurring or a human-modified form.

Block: An area of land or timber that has been defined for management purposes. One block may be composed of stands of different species or ages.

Board Foot: A unit of wood measuring 144 cubic inches. A 1-inch by 12-inch shelving board that is 1 foot long is equal to 1 board foot. Board foot volume is determined by: [(length (feet) x width (inches) x thickness (inches)) ÷ 12]

Bole Wood: Section of the bole that is utilizable for a commercial product and is cut square at both ends to be made ready for delivery for processing.

Bole: The tree trunk from the ground to the crown break. The bole doesn't include major branches supporting the tree crown, and may/not be straight.

Brackish water: A water that has more salinity than fresh water, but not as much as seawater. It may result from mixing of seawater with fresh water.

Breast height: This is the standard height (1.37 m above the ground) at which a tree diameter is measured.

Broadleaf forests: Forests composed principally of trees and shrubs of the botanical group of flowering species which is in contrast to conifers or pines.

Bucking: A harvesting function of cutting a felled tree into shorter lengths.

Bushfire: A wildfire burning through bushland such as a State forest.

Cable logging: Logging method based on a skyline wire cable system, used where slopes are too steep for conventional logging machinery.

Calcicole or **Calciphyte:** Types of plant that does not tolerate acidic soil.

Calcifuge: Types of plant that does not tolerate alkaline soil.

Canopy: The uppermost level of foliage formed by the branches and leaves of a tree.

Capacity: The ability to apply knowledge in problem solving processes and innovate to create change.

Captive carbon: Once CO_2 has been used by a plant the carbon will stay captive (or stored) in the plant until it decomposes, is burnt or breaks down.

Carbon Credit: A component of national and international attempts to mitigate the growth in concentrations of greenhouse gases (GHGs). One carbon credit is equal to one metric ton of carbon dioxide, or in some markets, carbon dioxide equivalent gases. Carbon trading is an application of an emissions trading approach.

Carbon footprint: A measurement of all greenhouse gases we individually produce and has units of tons (or kg) of carbon dioxide equivalent. A carbon footprint is made up of the sum of two parts, the primary foot-print and the secondary footprint.

Carbon Sequestration: Process that removes carbon dioxide from the atmosphere — trees, grasses, and other plants uptake carbon dioxide during photosynthesis and store it in biomass.

Carrying capacity: The maximum population size of the species that the environment can sustain indefinitely, given the food, habitat, water and other necessities available in the environment

Catchment Area: A drainage area in the landscape that is the source of water for a river or reservoir. drainage line.

Cellulose: A large molecule made up of glucose units produced naturally by plants. It provides strength to plant cell walls.

Chemotrophs: are organisms that obtain energy by the oxidation of electron donors in their environments

Clear-felling: The lopping and removal of all or most of the tree cover and associated understorey from a forested or vegetated area.

Clearing: An area where all or most of the tree cover & associated understorey have been removed & the land use changed eg. for agriculture or urban & industrial development.

Climate change: A change of climate which is attributed directly or indirectly to human activity that alters the composition of the global atmosphere and which is in addition to natural climate variability observed over comparable time periods.

Closed forest: A forest with a canopy cover of greater than 70 per cent.

Cold-blooded: A term often used to refer to animals that do not use their metabolism to maintain body temperature. Examples: reptiles, insects, arachnids, amphibians and fish

Community (ecological): An ecological unit composed of a group of organisms or a population of different species occupying a particular area usually interacting with each other and their environment.

Competition: The struggle between trees to obtain sunlight, nutrients, water, and growing space.

Cone: The fruit of coniferous trees that contains the seeds of the plant.

Conifers: Evergreen trees and shrubs of the botanical group gymnosperms. These plants produce naked seeds usually in cones.

Conservation biology: The scientific study of the nature and status of Earth's biodiversity with the aim of protecting species, their habitats, and ecosystems from excessive rates of extinction.

Conservation Easement: A voluntary legal agreement between a landowner and a land trust, or government agency, that limits the use of the property in order to protect the conservation values of the land.

Conservation: The protection, improvement, and wise use of natural resources for present and future generations.

Coppice: Regrowth that grows from dormant buds under the bark of tree stumps after the tree has been felled.

Coupe: A small area of forest within a compartment that is harvested in a single operation.

Crop Tree: Any tree selected to grow to final harvest or to a selected size. Crop trees are selected for quality, species, size, timber potential or wildlife value.

Crown cover: The area covered by the crowns of trees growing closely together, often expressed as a percentage for the combined crown cover of trees in a defined area.

Crown-Class: A tree classification system based on the tree's relative height, foliage density, and ability to intercept light.

Crown: The branches and foliage at the top of a tree.

Cutting Cycle: The planned time interval between major harvesting operations within the same stand usually within uneven-aged stands. For example, on a 10-year cutting cycle in a hardwood stand, trees are harvested every 10 years.

Deciduous: Plants that shed their leaves annually.

Decomposer: Name for organisms that break down dead and decaying organisms in the process of decomposition.

Deforestation Is the removal of a forest or stand of trees where the land is thereafter converted to a non-forest use.

Detritivores: Heterotrophs that obtain nutrients by consuming detritus (decomposing organic matter).

Dicotyledons (dicots): One of the two sub classes of flowering plants. There are two (di) cotyledons, or seed leaves, present in the seeds of these plants.

Dieback: The eventual death of trees from environmental stresses, such as pest attack, exposure to fungal disease, increasing soil salinity & human-induced changes.

Dominant Trees: Average-to-fairly large trees with medium-sized crowns that form the forest canopy. These trees receive full light from above but are crowded on the sides.

Drainage line: A discernible natural depression along which surface water runoff concentrates and flows towards a stream, drainage plain or swamp.

Dry sclerophyll forests: Open forests that are found in drier regions, well adapted to fire and low moisture. Leaves are hard textured with fewer pores to minimise water loss.

Dryland salinity: Accumulation of salt in soil and water of non-irrigated areas; caused by clearing trees and vegetation on outflow zones for saline water tables.

Ecological footprint: A measure of human demand on the Earth's ecosystems. It is a standardized measure of demand for natural capital that may be contrasted with the planet's ecological capacity to regenerate

Ecology: The science or study of the relationships between organisms and their environment.

Ecosystem Services: Benefits people obtain from the goods and services produced by an ecosystem.

Ecosystem: A biological community of living organisms and nonliving components such as air, water and mineral soil that interact. An ecosystem can be of any size — e.g., log, field, pond, woodlot, or Earth's biosphere.

Ecotopes: The smallest ecologically distinct landscape features in a landscape mapping and classification system.

Ecotourism: Nature-based tourism which is ecologically sustainable.

Ecozones: Delineate large areas of the Earth's surface within which organisms have been evolving in relative isolation over long periods of time, separated from one another by geographic features, such as oceans, broad deserts, or high mountain ranges, that constitute barriers to migration

Edge: The transition between two different types or ages of vegetation.

Endolith: An organism that lives inside rock, coral, animal shells, or in the pores between mineral grains of a rock.

Environment: The interaction of climate, soil, topography, and other plants and animals in any given area.

Ephemeral stream: A watercourse which flows infrequently.

Erosion: The washing away of land or soil by the action of wind, water, or ice.

Eutrophic lake: A lake has high primary productivity due to excessive nutrients and is subject to algal blooms resulting in poor water quality. The bottom waters of such bodies are commonly deficient in oxygen, ranging from hypoxic to anoxic.

Evapotranspiration: The evaporation of water from the soil and the transpiration of water from the plants that live in that soil. Approximately one-quarter of a forest's annual rainfall returns to the air through evapotranspiration.

Farm forestry: Commercial tree production on farmland.

Fauna: The animal life of a region.

Filter strip: A strip of permanent retained vegetation, 10-50 metres wide, along each side of a drainage line or bank of stream or river for retarding the flow of runoff water.

Fire regime: The fire history or strategy of particular place including the frequency, intensity and season of burning.

Firebreak: Any nonflammable barrier used to slow or stop fires. Several types of firebreaks are mineral soil barriers; barriers of green, slow-burning vegetation; and mechanically cleared areas.

Flora : The plant life of a region.

Forage: Vegetation such as leaves, stems, buds, and some types of bark that can be eaten for food and energy.

Forb : Any herb other than grass.

Forest cover: The presence of trees on lands more than one hectare in area with a tree canopy of more than 10 percent, irrespective of ownership and legal status.

Forest Management: The practical application of scientific, economic, and social principles to forest property.

Forestry: The science, art, and practice of managing and using trees, forests, and their associated resources for human benefit.

Gene banks: It help preserve genetic material, be it plant or animal. In plants, this could be by freezing cuts from the plant, or stocking the seeds.

Gene pool: It is the complete set of unique alleles in a species or population.

Genetic erosion: It is a process whereby an already limited gene pool of an endangered species of plant or animal diminishes even more when individuals from the surviving population die off without getting a chance to meet and breed with others in their endangered low population.

Genus: Scientific class of living things, uniquely named to group species that are very similar to one another and closely related. A sub-division of a family or subfamily.

Girdling: A physical cutting or disruption of the cambial sap flow within a tree. Girdling by humans, animals, or insects can often kill a tree.

Global warming: The expected consequence of an increase in the concentration of greenhouse gases by human activity

Greenhouse effect: Warming of the Earth's surface due to increases in radiated energy trapped by higher levels of atmospheric gases, including carbon dioxide, due primarily to the combustion of fossil fuels.

Gymnosperms: Non-flowering plants. Seeds not enclosed in an ovary, enclosed in a cone.

Habitat: An area in which a specific plant or animal can naturally live, grow, and reproduce.

Hardwoods (Deciduous Trees): Trees with broad, flat leaves as opposed to coniferous or needled trees. Wood hardness varies among the hardwood species, and some are actually softer than some softwoods.

Harvesting: The felling of trees, either as a group selection operation or a thinning or a clear-felling operation.

Hemotrophs: Organisms that obtain energy by the oxidation of electron donors in their environments.

Hibernation: A state of inactivity and metabolic depression in animals, characterized by lower body temperature, slower breathing, and lower metabolic rate. Hibernating animals conserve food, especially during winter when food supplies are limited, tapping energy reserves, body fat, at a slow rate. It is the animal's slowed metabolic rate which leads to a reduction in body temperature and not the other way around.

Intermediate Trees: Medium-sized trees with small crowns below the general level of the canopy. Intermediate trees receive little direct light, are poor crop trees, and should be removed during thinning operations.

Inventory: The process and product of a survey that assesses a resource. May also be termed forest inventory, stand inventory or plantation inventory.

Land Use: The human use of land for economic, social, environmental, or other purposes. Land use is often associated with urban development.

Legumes: Plants that produce organic nitrogen from nitrogen gas in the air. These plants, which typically form seeds in pods, include soybeans, peas, alfalfa, lespedeza, and locust.

Lignin: An organic carbon compound found in cell walls of plants. It helps conduct water in living trees and provides structural strength.

Lignotuber: Swollen underground root structure developed by most eucalypts, capable of sprouting new shoots if the tree is damaged or coppiced.

Lithophiles: Micro-organisms that can live within the pore interstices of sedimentary and even igneous rocks to depths of several kilometers.

Lithotroph: An organism that uses an inorganic substrate to obtain reducing equivalents for use in biosynthesis or energy conservation via aerobic or anaerobic respiration

Log: To cut trees into logs; to cut down trees for timber; a portion of cut tree trunk or branch.

Logging: The process of cutting down trees for wood.

Management Plan: Written guidelines for current and future management practices recommended to meet objectives.

Microclimate: A local atmospheric zone where the climate differs from the surrounding area. The term may refer to areas as small as a few square feet or as large as many square miles.

Mixed Stand: A timber stand in which less than 80 percent of the trees in the main canopy are of a single species.

Monocotyledons (monocots): One of the two sub classes of flowering plants. There is one (mono) cotyledon or seed leaf present in the seeds of these plants.

Monoculture: One type or species of plant cultivated in an area, for example plantations of a single species of tree.

Mor soil: It is characterized by smaller soil animals, the buildup of a litter layer on top and decomposition mainly laid by fungi.

Natural Stand: A stand of trees grown from natural seed fall or sprouting.

Non-Timber Forest Products: Any product derived from the forest that is not timber:e.g., medicinal herbs, resins, mushrooms, berries, ferns, cones.

Nutrients: Elements necessary for growth and reproduction. Primary plant nutrients are nitrogen, phosphorus, and potassium.

Oil spill: A release of a liquid petroleum hydrocarbon into the environment due to human activity, and is a form of pollution. The term often refers to marine oil spills, where oil is released into the ocean or coastal waters.

Old growth forest: Forests in which the upper stratum is ecologically mature and has been subjected to negligible unnatural disturbance such as logging, road building and clearing.

Oligotrophic Lake: A lake with low primary productivity, the result of low nutrient content. These lakes have low algal production, and consequently, often have very clear waters, with high drinking-water quality

Open forest: A forest with a canopy cover of between 30 and 70 per cent.

Parasites: Organisms which obtain their nutrients by feeding off other organisms.

Perennial: Plants that live or grow for more than one year. Some resprout from a root system or reseed themselves every year.

Pest: Any organism that is out of place or causes stress to a desired organism.

Phloem: Part of the plant vascular system where sugars are transported around the parts of a plant.

Photosynthesis: The chemical reaction in green plants that uses carbon dioxide, water and sunlight to produce food-glucose.

Poaching: The illegal taking of wild plants or animals contrary to local and international conservation and wildlife management laws. Violations of hunting laws and regulations are normally punishable by law and, collectively, such violations are known as poaching.

Poles or Pole-timber: Trees from 5 to 7 inches in diameter at breast height.

Prey: An animal which is being prayed by the predator.

Predator: An animal that preys on and devours other animals.

Pruning: The technique of removing dead, damaged or excess branches from trees. It is usually done from the base of the tree, and encourage trees to grow straight.

Pulp: Wood fibre processed to make paper. 'Pulp' logs are processed into wood chips or pulp for wood-based panels, paper and paper products.

Pure forest: A forest in which at least 80 percent of the trees in the main crown canopy are of a single species.

Rainforest: Type of forest that grows in higher rainfall regions and in fertile soils.

Raptor: A bird of prey such as an owl, hawk, osprey, or eagle.

Reforestation: Reestablishing a forest by planting or seeding an area from which forest vegetation has been removed.

Regeneration: The process of growing new trees on areas previously forested, whether naturally or planted.

Restoration- Practice of restoring a forest to its previous condition in terms of ecological function and structure as it was prior to disturbance.

Rotation: The number of years in a production cycle from planting trees till harvest.

Runoff: The proportion of rain falling in a catchment which flows across the surface rather than infiltrating the soil.

Salvage logging: Logging carried out to use trees that have been damaged by fire or storm.

Sapling:A small tree, usually between 2 and 4-inches diameter at breast height.

Sedimentation: The deposition or settling of soil particles suspended in water.

Seed orchard- Collection of seed-producing trees; trees cultivated for the production of seeds used to grow and improve tree qualities.

Seed Year: A year in which a given species produces a large seed crop over a sizable area.

Seedling: A nursery-grown tree that has not been transplanted in the field.

Selective Cutting: The periodic removal of individual trees or groups of trees to improve or regenerate a stand.

Selective logging: Selected trees within an area nominated for logging are logged while other trees are retained to maintain the forest structure & wildlife habitat.

Senescent trees: Similar to over-mature trees.

Shade-Intolerant Trees: Trees that cannot thrive in the shade of larger trees.

Shrub: A woody plant smaller than a tree, usually divided into separate stems near the ground.

Silviculture: The branch of forestry that involves the use of science and art to grow and maintain trees. It involves tree establishment, growth, health and quality.

Sink: Any process, activity or mechanism which removes a greenhouse gas, an aerosol or a pre- cursor of a greenhouse gas from the atmosphere

Site Index: A relative measure of forest site quality based on the height (in feet) of the dominant trees at a specific age (usually 25 or 50 years, depending on rotation length). Site index information helps estimate future returns and land productivity for timber and wildlife.

Slash: Tree tops, branches, bark, or other residue left on the ground after logging or other forestry operations.

Softwood: Tree species that have soft wood that is useful in making newspapers, toilet paper and tissues as well as house frames and furniture.

Soil Texture: The feel or composition of the soil (sand, silt, or clay) as determined by the size of the soil particles.

Species: A group of related organisms having common characteristics and capable of interbreeding.

Stand: An easily defined area of the forest that is relatively uniform in species composition or age and can be managed as a single unit.

Standing crop: The quantity or total weight or energy content of the organisms which are in a particular location at a particular time.

Stem- Part of a plant. A stalk which supports a leaf, flower or fruit.

Subtropical: A type of rainforest occurring between the tropics and temperate geographic regions.

Succession: The natural sequence of plant community replacement beginning with bare ground and resulting in a final, stable community in which a climax forest is reached.

Suppressed Trees: Small trees that grow below the tree canopy and receive no direct sunlight from any direction.

Sustained Yield: Management of forestland to produce a relatively constant amount of wood products, revenue, or wildlife.

Thinning: A technique to remove thin or weak trees to leave more nutrients and space access for sunlight to the remaining trees to maximise their growth potential and wood properties.

Timber: The general term used to describe sawn wood suitable for furniture.

Tolerant Species: A species of tree that has the ability to grow in the shade of other trees and in competition with them.

Topping: A harvesting function of cutting off the top of a tree at predetermined, minimum diameter of the upper stem of a tree.

Tree: A perennial plant having a self-supporting woody stem or trunk which usually develops woody branches.

Understorey: The layer of forest vegetation between the overstorey or canopy and the ground layer.

Urban forestry: The cultivation and management of trees and forests within cities for their present and potential contribution to the physiological, sociological and economic wellbeing of urban society.

Warm-blooded: A term to describe animal species which have a relatively higher blood temperature, and maintain thermal homeostasis primarily through internal metabolic processes. Examples: Mammals and birds.

Wildlife corridor: A strip of vegetation retained between areas of used and unused land to allow wildlife to move from sites of activity to other habitat sites.

Wood: The hard, fibrous inner part of tree trunks, branches and stems.

Woodland: Plant communities dominated by trees whose crowns shade on average 10 to 30 per cent of the ground.

Xerosere: A plant succession which is limited by water availability.

Xylem: The part of a plants vascular system that transports water and minerals within the plant, from the roots to the shoots.

Chapter 20

પ્રકરણ ૨૦

Forest and Environment Abbreviations

વન અને પર્યાવરણ સંક્ષેપાક્ષર

ABA: Abscisic Acid

ACF: Assistant Conservator of Forests

ACS: Additional Chief Secretary

ADP: Annual Development Plan

AFM: Advanced Forest Management

AFOLU: Agriculture, Forestry and Other Land Use

AFRI: Arid Forest Research Institute

AGB: Above Ground Biomass

AI: Aridity Index

AKRSP: Aga Khan Rural Support Programme

AOFFPS: Area-Oriented Fuelwood & Fodder Project Schemes

AOSA: Association of Official Seed Analysts

AOSIS: Alliance of Small Island States

APCCF: Additional Principal Chief Conservator of Forests

APEDA: Agricultural and Processed Food Products Export Development Authority

APFC: Asia-Pacific Forestry Commission

APL: Above Poverty Line

APMC: Agricultural Produce Market Committee

ASCOBANS: Agreement on the Conservation of Small Cetaceans of the Baltic and North Sea

ASEAN: Association of South East Asian Nations

ASTRP: Association of STs & Rural Poor in Regeneration of Degraded Forests on Usufruct Basis

ATIC: Agriculture Technology Information Centre

ATMA: Agricultural Technology Management Agency

ATO: African Timber Organization

AWiFS: Advanced Wide Field Sensor

AYUSH: Department of Ayurveda, Yoga & Naturopathy, Unani, Siddha & Homeopathy

BAF: Basal Area Factor

BAIF: Bharatiya Agro Industries Foundation

BC: Buffering Capacity

BDA: Biological Diversity Act

BEF: Biomass Expansion Factor

BGB: Below Ground Biomass

BISAG-N: Bhaskaracharya National Institute for Space Applications and Geo-informatics

BMC: Biodiversity Management Committee

BPL: Below Poverty Line

BR: Biosphere Reserve

BSH: Biodiversity Heritage Site

BSI: Botanical Survey of India

BUR: Biennial Update Reports

C&I: Criteria and Indicators

C : N Ratio: Carbon Nitrogen Ratio

CAI: Current Annual Increment

CAZRI: Central Arid Zone Research Institute

CBD: Convention on Biological Diversity

CBF: Central Board of Forestry

CBP: Conservation Breeding Programme

CBT: Conservation Bench Terraces

CCF: Chief Conservator of Forests

CCS: Carbon Capture and Sequestration

CD: Causeway-cum-Checkdam

CDM: Clean Development Mechanism

CEC: Cation Exchange Capacity

CER: Certified Emission Reduction

CF: Conservator of Forests

CFR & HRD: Centre for Forestry Research & Human Resource Development

CGIAR: Consultative Group on International Agricultural Research

CIDA: Canadian International Development Agency

CIFOR: Centre for International Forestry Research

CITES: Convention on International Trade in Endangered Species of Wild Fauna and Flora

CMA: the Conference of the Parties serving as the meeting of the Parties to the Paris Agreement

CMP: the Conference of the Parties serving as the meeting of the Parties to the Kyoto Protocol

CMS: Convention on the Conservation of Migratory Species of Wild Animals

CNFA: Culturable Non-Forest Area

CO_2: Carbon Dioxide

CO_2 eq: Carbon Dioxide Equivalent

COC: Chain of Custody

COFO: Committee on Forestry

COP: Conference of the Parties

CPF: Collaborative Partnership on Forests

CS: Conservation & Survey

CSD: United Nations Commission on Sustainable Development

CSIRO: Commonwealth Scientific and Industrial Research Organization

CSO: Clonal Seed Orchard

CSO: Central Statistical Organization

CSS: Centrally-sponsored Scheme

CWLW: Chief Wild Life Warden

CZA: Central Zoo Authority

DANIDA: Danish International Developmental Agency

DBH: Diameter at Breast Height

DBHOB: Diameter at Breast Height Over Bark

DCF: Deputy Conservator of Forest

DEM: Digital Elevation Model

DF: Dense Forest

DFE: Directorate of Forest Education

DFID: Department for International Development

DFO: Divisional Forest Officer

DIP: Digital Image Processing

DoLR: Department of Land Resources

DOM: Dead Organic Matter

DRDA: District Rural Development Agency

EAP: Externally-Aided Project

ECOSOC: United Nations Economic and Social Council

EDC: Eco Development Committee

EEC: European Economic Commission

EIA: Environmental Impact Assessment

EMC: Equilibrium Moisture Content

ENSO: El Nino -Southern Oscillation

EPA: Environment Protection Act

ERUs: Emission Reduction Units

ESP: Exchangeable Sodium Percent

ESZ: Eco Sensitive Zone

ETDC: Eco Tourism Development Committee

FAO: Food and Agricultural Organization of the United Nations

FC: Forest Conservation

FCA: Forest Conservation Act

FCC: False Color Composite

FCS: Forest Carbon Stock

FD: Forest Department

FDA: Forest Development Agency

FDC: Forest Development Corporation

FFCM: Forest Fire Control & Management

FFPS: Fuelwood and Fodder Projects Scheme

FFV: Forest Fringe Villages

FLCP: Forest Labour Cooperative Society

FMU: Forest Management Unit

FORWORKNET: Forestry Workforce Network

FPC: Forest Protection Committee

FPO: Farmer Producer Organization

FRA: Global Forest Resources Assessment

FRA: Forest Right Act

FRC: Forest Research Centre

FRI: Forest Research Institute

FRL: Forest Reference Level

FRM: Forest Reproductive Material

FRO: Forest Range Officer

FSC: Forest Stewardship Council

FSI: Forest Survey of India

FSO: Forest Settlement Officer

FSP: Fiber Saturation Point

FWI: Fire Weather Index

FYM: Farm Yard Manure

FYP: Five Year Plan

GA: Geographic Area

GATT: General Agreement on Tariffs and Trade

GBB: Gujarat Biodiversity Board

GCA: General Combining Ability

GCP: Ground Control Point

GDP: Gross Domestic Product

GEDA: Gujarat Energy Development Agency

GEER: Gujarat Ecological Education and Research

GEF: Global Environment Facility

GFD: Gujarat Forest Department

GFDP: Gujarat Forestry Development Project

GFMC: Global Fire Monitoring Center

GFRA: Global Forest Resource Assessment

GHGs: Green House Gases

GIB: Great Indian Bustard

GIM: Green India Mission

GIS: Geographical Information System

GLASOD: Global Assessment of Soil Degradation

GM: Genetically Modified

GOI: Government of India

GPG: Good Practices Guidance

GPS: Global Positioning System

GSFDC: Gujarat State Forest Development Corporation

GSIDS: Gujarat Social Infrastructure Development Society

GtC: Gigatons of Carbon

GTZ: Deutsche Gesellschaft Fur Technische Zusammenarbeit

GW: Green Wash

GWP: Global Warming Potential

GWP: Global warming potential

HDEP: High Density Energy Plantation

HFRI: Himalayan Forest Research Institute

HoFF: Head of the Forest Force

HRD: Human Resource Development

HYV: High Yielding Variety

IAA: Indole Acetic Acid

IAEPS: Integrated Afforestation & Eco-Development Projects Schemes

IBA: Indole Butyric Acid

IBWL: Indian Board for Wildlife

IC & CES: Indian Customs & Central Excise Service

ICAR: Indian Council of Agricultural Research

ICBN: International Code of Botanical Nomenclature

ICFRE: Indian Council of Forestry Research and Education

ICIMOD: International Centre for Integrated Mountain Development

ICRAF: International Centre for Research in Agroforestry (the World Agroforestry Centre)

IDRC: International Development Research Centre, Canada

IEG: Institute of Economic Growth

IFAD: International Fund for Agricultural Development

IFF: Intergovernmental Forum on Forests

IFP: Institute of Forest Productivity

IFPS: Integrated Forest Protection Scheme

IFS: Indian Forest Service

IGNFA: Indira Gandhi National Forest Academy

IIFM: Indian Institute of Forest Management

IIRS: Indian Institute of Remote Sensing

INBAR: International Network for Bamboo and Rattan

INC: Initial National Communication

IPCC: Intergovernmental Panel on Climate Change

IPF: Intergovernmental Panel on Forests

IPIRTI: Indian Plywood Industries Research & Training Institute

IPM: Integrated Pest Management

IPPU: Industrial Processes and Product Use

IRDP: Integrated Rural Development Programme

IRS: Indian Remote Sensing Satellite

ISFR: India State of Forest Report

ISRO: Indian Space Research Organization

ISTA: International Seed Testing Association

ITTA: International Tropical Timber Agreement

ITTO: International Tropical Timber Organization

IUCN: International Union for Conservation of Nature

IUDZG: International Union of Directors of Zoological Gardens

IUFRO: International Union of Forest Research Organizations

IVI: Importance Value Index

IWDP: Integrated Watershed Development Programme

IWRM: Integrated Water Resources Management

IWST: Institute of Wood Science & Technology

JBIC: Japanese Bank for International Cooperation

JFM: Joint Forest Management

JFMC: Joint Forest Management Committees

JICA: Japan International Cooperation Agency

JRY: Jawahar Rojgar Yojana

KFRI: Kerala Forest Research Institute

KVK: Krishi Vigyan Kendra

LaCONES: Laboratory for Conservation of Endangered Species

LADA: Land Degradation Assessment in Drylands

LISS: Linear Imaging and Self-scanning Sensor

LULUCF: Land use, land-use change, and forestry

M ha: Million Hectare

MAB: Man and Biosphere Programme

MAI: Mean Annual Increment

MAP: Medicinal and Aromatic Plants

MC: Moisture Content

MDF: Moderately Dense Forest

MDGs: Millennium Development Goals

MDP: Management Development Programme

MEA: Millennium Ecosystem Assessment

MFFCM: Modern Forest fire Control Methods

MFP: Minor Forest Produce

MGNREGA: Mahatma Gandhi National Rural Employment Guarantee Act

MIKE: Monitoring of Illegal Killing of Elephants

MMT: Million Metric Tonnes

MOA: Ministry of Agriculture

MODIS: Moderate Resolution Imaging Spectroradiometer

MoEF: Ministry of Environment & Forests

MoEFCC: Ministry of Environment, Forests & Climate change

MoRD: Ministry of Rural Development

MPT: Multipurpose Tree

MSS: Multi Spectral Scanner

MUSLE: Modified Universal Soil Loss Equation

MW: Megawatt

MXD: Maximum Latewood Density

NABL: National Accreditation Board for Testing and Calibration Laboratories.

NAEB: National Afforestation and Eco-Development Board

NAP: National Afforestation Programme

NAPCC: National Action Plan for Climate Change

NAPCD: National Action Programme to Combat Desertification

NASA: National Aeronautics and Space Administration

NATCOM: National Communication. India's Initial National Communication (NATCOM) to United Nations Framework Convention on Climate Change and the Forestry Sector.

NBA: National Biodiversity Authority

NBRP: National Biosphere Reserve Programme

NBSS&LUP: National Bureau of Soil Survey and Land Use Planning

NCA: National Commission on Agriculture

NCAER: National Council of Applied Economic Research

NCDMA: National CDM Authority

NDC: Nationally Determined Contributions

NDMA: National Disaster Management Authority

NDVI: Normalized Difference Vegetation Index

NE: North-East

NF: Non-Forest

NFAP: National Forestry Action Programme

NFMS: National Forest Monitoring System

NFTS: Nitrogen Fixing Tree Species

NGO: Non-Governmental Organization

NHAI: National Highway Authority of India

NLBI: Non-Legally Binding Instrument on All Types of Forests

NMPB: National Medicinal Plants Board

NMSA: National Mission for Sustainable Agriculture

NP: National Park

NPV: Nuclear Polyhedrosis Virus/ Net Present Value

NRCAF: National Research Centre for Agroforestry

NRSA: National Remote Sensing Agency

NSO: National Statistical Office

NSSO: National Sample Survey Organisation

NTFP: Non-Timber Forest Produce

NWAP : National Wildlife Action Plan

NWDB: National Wasteland Development Board

NWFP: Non-Wood Forest Products

NWLD: National Wildlife Database

NZP: National Zoological Park

ODA: Overseas Development Agency

OECD: Organization for Economic Co-operation and Development

OECE: Overseas Economic Cooperation Fund of Japan.

OF: Open Forest

OM: Organic Matter

PA: Protected Area

PAN: Pan-Chromatic

PAs: Protected Areas

PBR: People's Biodiversity Register

PC: Plan Co-ordination

PCCF: Principal Chief Conservator of Forests

PCCF & HoFF: Principal Chief Conservator of Forests & Head of the Forest Force

PCR: Polymerase Chain Reaction

PF: Protected Forest

PHC: Primary Health Centre

PHVA: Population and Habitat Viability Assessment

PIP: Pugmark Impression Pads

PISFR: Pre-investment Survey of Forest Resources

POPs: Convention on Persistent Organic Pollutants

PPME: Project, Planning, Monitoring and Evaluation

PPVFRA: Protection of Plant Varieties and Farmer's Right Authority

PRA: Participatory Rural Appraisal

PROFOR: Programme on Forests

PT: Project Tiger

PTGs: Primitive Tribal Groups

PVTGs: Particularly Vulnerable Tribal Groups

PWD: Public Works Department

PWPR: Preliminary Working Plan Report

QELROs: Quantified Emissions Limitation and Reduction Commitments

RAPD: Random Amplified Polymorphic DNA

REDD: Reducing Emissions from Deforestation and forest Degradation

REDD+: Reducing Emissions from Deforestation and Forest Degradation Plus

REL: National Forest Reference Emission Level

RF: Reserved Forest

RFA: Recorded Forest Area

RFLP: Restriction Fragment Length Polymorphism

RFO: Range Forest Officer

RFPP: Rural Fuel wood Plantation Programme

RFRI: Rain Forest Research Institute

RGI: Registrar General of India

RIL: Reduced Impact Logging

RKVY: Rashtiya Krishi Vikas Yojana

RO HQs: Regional Offices Headquarters

RRA: Rapid Rural Appraisal

RSG: Re-introduction Specialist Group

RUPFOR: Resource Unit for Participatory Forestry

RWE: Raw Wood Estimates

RWE: Round Wood Equivalent

SAC: Space Application Centre

SBA: Stand Basal Area

SBB: State Biodiversity Board

SC: Scheduled Caste

SCA: Specific Combining Ability

SCCF: Special Climate Change Fund

SDG: Sustainable Development Goals

SES: Socio-economic Status

SFD: State Forest Department

SFM: Sustainable Forest Management

SFR: State of Forest Report

SFS: State Forest Service

SHG: Self Help Group

SIA: Social Impact Assessment

SIDA: Swedish International Development Agency

SLEM: Sustainable Land and Ecosystem Management

SMAF: Sub-Mission on Agroforestry

SMPB: State Medicinal Plants Board

SNPP-VIIRS: Suomi National Polar orbiting Partnership - Visible Infrared Imaging Radiometer Suite

SOC: Soil Organic Carbon

SOFO: State of the World's Forests

SOI: Survey of India

SOM: Soil Organic Matter

SOS: Save our Species

SPA: Seed Production Area

SRTM: Shuttle Radar Topography Mission

SSO: Seedling Seed Orchard

SSP: Species Survival Plan

SSR: Simple Sequence Repeat

ST: Schedule Tribe

SU: Survey & Utilization

SWOT: Strength Weakness Opportunity and Threat

TAC: Technical Advisory Committee

TASP: Tribal Area Sub Plan

TBA: Tree Basal Area

TBOs: Tree Borne Oilseeds

TC: Tree Cover

TEV: Total Economic Value

TFRI: Tropical Forest Research Institute

TIES: The International Ecotourism Society

TM: Thematic Mapper

TOF: Trees Outside Forests

UF: Unclassed Forest

UFS: Urban Frame Survey

UNCCD: United Nations Convention to Combat Desertification

UNCED: United Nations Conference on Environment and Development

UNCHE: United Nations Conference on the Human Environment

UNCOD: United Nations Conference on Desertification

UNCTAD: United Nations Conference on Trade and Development

UNDESA: United Nations Department of Economic and Social Affairs

UNDP: United Nations Development Programme

UNEP: United Nations Environment Programme

UNESCO: United Nations Educational, Scientific and Cultural Organization

UNFCCC: United Nations Framework Convention on Climate Change

UNFF: United Nations Forum on Forests

UNISDR: United Nations International Strategy for Disaster Reduction

UNPFII: United Nations Permanent Forum on Indigenous Issues

UN-REDD: United Nation Reduced Emissions from Degradation and Deforestation

UNWTO: United Nations World Tourism Organization

USAID: United States Agency for International Development.

USEPA: United States Environmental Protection Agency

USLE: Universal Soil Loss Equation

UT: Union Territory

VA: Village Associations

VDF: Very Dense Forest

VERs: Verified Emission Reductions

VFC: Village Forest Committee

VFPC: Village Forest Protection Committee

VIIRS: Visible Infrared Imaging Radiometer Suite

WAZA: World Association of Zoos and Aquariums

WB: World Bank

WBG: World Bank Group

WCFSD: World Commission on Forests and Sustainable Development

WF: Weightage Factor

WHC: World Heritage Convention

WII: Wildlife Institute of India

WLPA: Wildlife Protection Act

WLPSI: Wildlife Preservation Society of India

WMO: World Meteorological Organization

WMS: Web Map Service

WP: Working Plan

WPO: Working Plan Officer

WSSD: World Summit on Sustainable Development

WTO: World Trade Organization

WWF: World Wide Fund for Nature

ZSI: Zoological Survey of India

ZSL: Zoological Society of London

Chapter 21

પ્રકરણ ૨૧

GPSC Questions

જી.પી.એસ.સી.ના પ્રશ્નપત્રો

Gujarat Administrative Service, Class-1, Gujarat Civil Services, Class-1 & Class-2 & Gujarat Municipal Chief Officer Service, Class-2 10/2019-20 (13/10/2019)

1. Which of the following statements is/are correct regarding GRAMIN AGRICULTURAL MARKETS (GrAMS)?

1. It aims to develop and upgrade existing 40,000 rural haats into Gramin Agricultural Markets.
2. It will be linked to e-NAMs.
3. It would be outside APMC Act regulation.

(A) 1 and 3 only (B) 1 and 2 only
(C) 2 and 3 only (D) 1, 2 and 3

2. Which of the following statements is/are correct regarding "Kisan Urja Suraksha Evem Utthan Mahabhiyan" (KUSUM) Scheme?

1. This scheme has planned for setting up biogas electricity plants for farmers
2. This scheme will enable farmers to earn extra income and help in their irrigation facility

(A) 1 only (B) 2 only
(C) Both 1 and 2 (D) Neither 1 nor 2

3. Which of the following statements is/are correct regarding UNESCO's Heritage sites in India?

1. Old Delhi became the 2nd Heritage city in India followed by Jaipur in the year 2019
2. Ahmedabad became the first Heritage City in India in the year 2018
3. With the inclusion of Jaipur, the number of heritage sites across India that are on UNESCO World heritage list, has grown to 38.

(A) 1 and 2 only (B) 3 only
(C) 2 and 3 only (D) 1 and 3 only

4. Ministry of Food Processing Industries, New Delhi (MOFPI) has a number of schemes for capital investment and other subsidies in the food processing sector. Which is the nodal agency in Gujarat to implement schemes of MOFPI?

(A) Gujarat Agro Industries Corporation Ltd.
(B) Gujarat Industrial Development Corporation
(C) Directorate of Agriculture, Gujarat
(D) Food Corporation of India, Gujarat Regional Office

5. An ecological footprint is measured in the unit_______________.

(A) Global Hectares (B) Nanometre
(C) Hoppus cubic foot (D) Cubic ton

6. Consider the following statements:

1. Agriculture crop year in India is July to June.
2. Kharif season is from July to October.
3. Rabi season is October to March.
4. Jowar, Bajra and Maize are main Rabi crops in India.

(A) 1, 2 and 4 only (B) 2, 3 and 4 only
(C) 1, 2 and 3 only (D) 1, 2, 3 and 4

7. Which of the following statements is/are correct regarding e-Shakti which is a pilot project of National Bank for Agriculture and Rural Development (NABARD) for digitization of Self Help Groups (SHGs) ?

(A) It was initiated to improve the quality of bookkeeping of SHGs and to enable banks to take informed credit decisions about the group through a Management Information System (MIS).
(B) Integrating SHG members with the National Financial Inclusion Agenda
(C) Both A and B
(D) Neither A nor B

8. Which of the following statements is/are correct regarding Mission Organic Value Chain Development for North Eastern Region (MOVCDNER)?

1. It is a Sub-Mission under National Mission for Sustainable Agriculture (NMSA)
2. Developing crop specific organic production clusters with necessary infrastructural, technical and financial support
3. This scheme is implemented in the States of Arunachal Pradesh, Assam, Manipur, Meghalaya, Mizoram, Nagaland, Sikkim and Tripura

(A) 1 and 3 only (B) 2 and 3 only
(C) 1 and 2 only (D) 1, 2 and 3

9. If a ship sailing from Tokyo (Japan) on Sunday at 8 AM for San Francisco (USA) takes 8 hours to reach there. On which day it will reach there?

(A) On Sunday
(B) On Saturday
(C) On Monday
(D) On Sunday midnight

10. Which among the following air pressure belt is called "Horse Latitude"?

(A) Equatorial low pressure belt
(B) Subtropical high pressure belt
(C) Sub polar low pressure belt
(D) none of the above

11. Which of the following statements are correct regarding coral reefs?

1. Corallite is the house of the corals
2. Coral are living organisms
3. Reefs are uncemented and soft structure of calcium carbonate
4. Reefs are cemented and hard structure of calcium carbonate

(A) 1, 2, 3 and 4 (B) 2 and 3 only
(C) 2 and 4 only (D) 1, 2 and 4 only

12. Which of the following statements is/are correct regarding landslide vulnerability zones in India?

1. Steep and rainy slopes of Western Ghats and Nilgiris fall in very high landslide vulnerability zone
2. States from the north-eastern region except the plains of Assam are covered under moderate to low landslide vulnerability zone
3. Coastal regions of the southern states fall under high landslide vulnerability zone

(A) 1 only (B) 1 and 3 only
(C) 1, 2 and 3 (D) 2 and 3 only

13. Which of the following statements are true with refrence to Economic Survey 2018-19?

1. The annual population growth rate is set to increase from 1.1% in 2011-21 to 1.5% in 2031-41.
2. The service sector (excluding construction) has a share of 54.3% to India's Gross Value Added (GVA) in 2018-19.
3. Share of employment of service sector is at 34%.
4. India accounts for 18% of the world's population but uses only 6% of world's primary energy.

(A) 1, 2 and 3 only (B) 1, 3 and 4 only
(C) 2, 3 and 4 only (D) 1, 2, 3 and 4

14. The duration of Dawn and Twilight is of maximum period at ________.

(A) The Equator
(B) The Arctic/Antarctic circle
(C) The Tropics
(D) The Poles

15. Which one of the following water bodies contains highest salinity?

(A) Dead Sea (B) Mono Lake
(C) Great Salt Lake (D) Red Sea

16. Which of the following pairs are correctly matched?

1. Rotterdam Convention - Prior consent of trade for hazardous chemicals and pesticides
2. Basel Convention - Control of transboundary movements of hazardous wastes
3. Minamata convention - control adverse effects of lead on human health
4. Stockholm convention - restrict the production and use of persistent organic pollutants

(A) 1 and 3 only (B) 1, 2, 3 and 4
(C) 1, 2 and 4 only (D) 2, 3 and 4 only

17. Which of the following river does not flow in the Dangs?

(A) Purna (B) Ambica
(C) Gira (D) Par

18. Which of the following pairs of places and products are correctly matched?

1. Coorg - Coffee Plantation
2. Pulwama - Saffron Cultivation
3. Chickmagalur - Tea plantation
4. Godavari - Cocoa plantation

(A) 1 and 3 only
(B) 2 and 4 only
(C) 1, 2 and 4 only
(D) 1, 2, 3 and 4

19. What is the notational system of salinity of the oceans?

(A) Parts per hundred (%)
(B) Parts per gram (g)
(C) Parts per cubic meter (m^3)
(D) Parts per thousand (%0)

20. Who does tiger census in India?

(A) Ministry of environment, forest and climate change
(B) National Tiger conservation authorities
(C) State Governments
(D) Worldwide Fund for Nature-India

21. Which among the following is the longest continuous mountain chain of the world?

(A) The Rockies
(B) The Andes
(C) Great Dividing Range
(D) The Himalayas

22. Wallace line is a ________

(A) An International divide between India and Nepal
(B) Line separating coastal plains and south plateau
(C) Line separating the distinct flora and fauna of South East Asia and Australia
(D) Line dividing Atlantic and Pacific Oceans

23. Which one of the following ocean currents belongs to the Indian Ocean?

(A) Aleutian current
(B) Canaries Current
(C) Agulhas current
(D) Benguela current

24. The Chilka Lake lies at ________.

(A) North Circar Coast
(B) Coromandal Coast
(C) Malabar Coast
(D) None of the above

25. Soils of the Humid Tropical and Equatorial Zones are called as________.

(A) Chernozem Soil (B) Chestnut Soil
(C) Halomorphic Soil (D) Lateritic Soil

26. Which one of the following gases is in the lowest percentage of the total gaseous composition of the atmosphere?

(A) Helium (B) Hydrogen
(C) Ozone (D) Neon

27. Prairie Region is famous for ________.

(A) Horticulture
(B) Dairy Industry
(C) Commercial grain Farming
(D) Agro Forestry

28. Which of the following are true?

1. Epiphytes are plants that derive their water and nutrients from air
2. Hydrophytes are flowering plants adapted to living submerged water
3. Xerophytes are plants which are adapted to seasonal or perennial droughts

(A) 1, 2 and 3 (B) 1 and 3 only
(C) 2 and 3 only (D) 1 and 2 only

29. Which of the following is/are true for coniferous forests?

1. Coniferous forests cover a broad belt of land in North America and Eurasia.
2. Coniferous forests have tall, straight, evergreen trees
3. The trees of these forests shed their leaves in autumn

(A) 2 only (B) 1, 2 and 3
(C) 1 and 2 only (D) 3 only

30. Which of the following statements is/are INCORRECT regarding Geography of Gujarat?

1. Gujarat shares boundaries with Pakistan in North West.
2. Rajasthan is in North as well as North East and Maharashtra is in South East
3. The northern portion of Rann is called Greater Rann and eastern portion is referred to as Little Rann.
4. Madhya Pradesh is in West

(A) 1 only (B) 2 and 3 only
(C) 3 and 4 only (D) 4 only

31. Which of the following statements is/are correct?

1. Tropical savannah type of climate is found in the north eastern states of India
2. Semi-arid steppe climate is found in Central Rajasthan
3. Cold humid winter type of climate is found in Jammu and Kashmir

(A) 1 and 3 only (B) 2 only
(C) 1, 2 and 3 (D) 2 and 3 only

32. Which of the following have volcanic origin?

1. Andaman and Nicobar Islands
2. Lakshadweep 3. Deccan Trap
4. Aravalli Ranges

(A) 1, 2 and 3 only (B) 2 and 4 only
(C) 1 and 3 only (D) 1, 2, 3 and 4

33. For Short-term climate predictions, which one of the following events is associated with weak monsoon rains in the Indian Subcontinent?

(A) La Nina
(B) Movement of Jet Streams
(C) El Nino (D) Global Warming

34. Monkey fever or Kyasanur disease recently seen in the news is caused by which of the following?

(A) Bacteria (B) Virus
(C) Fungi (D) None of the above

35. In the context of environmental pollution, what does the term 'dirty dozen' refer to?

(A) The twelve ozone depleting substances.
(B) The twelve gases emanating from vehicular exhaust.
(C) The twelve persistent organic pollutants listed under Stockholm Convention.
(D) The twelve hazardous chemicals listed under the Air Pollution Act, 1981.

36. With reference to World Water Development Report, which of the following statements is/are correct?

1. It is published annually by the World Economic Forum.
2. The latest report was released at the 8th World Water Forum in Brasilia.

(A) 1 only (B) 2 only
(C) Both 1 and 2 (D) Neither 1 nor 2

37. Which of the following statements is/are correct with reference to blue waves?

1. It is a rare natural phenomenon also referred to as bio-luminescence.
2. It is caused by the light emitted by fluorescent corals present in ocean.

(A) 1 only (B) 2 only
(C) Both 1 and 2 (D) neither 1 nor 2

38. With reference to One Planet City Challenge (OPCC), which of the following statements is/are correct?

1. It is an initiative of United Nations Environment Programme.
2. It invites cities from the participating countries to report innovative climate actions.

(A) 1 only (B) 2 only
(C) Both 1 and 2 (D) Neither 1 nor 2

39. In the context of environmental agreements, which of the following statements is/are correct?

1. Doha Amendment is the amendment to Montreal Protocol.
2. Kigali Agreement is an amendment to Kyoto Protocol to limit the carbon emissions.
3. India recently ratified the Doha Amendment.

(A) 1 and 2 only (B) 2 and 3 only
(C) 3 only (D) 1, 2 and 3

40. With reference to the National Biodiversity Authority, which of following statements are true?

1. It was formed under the Environment Protection Act, 1986.
2. It protects the intellectual property rights of local communities related to biodiversity.
3. It is headquartered in Chennai.

(A) 1, 2 and 3 (B) 2 and 3 only
(C) 1 and 3 only (D) 1 and 2 only

41. With the help of given statements, identify the animal?

1. It is India's most critically endangered reptile species
2. Chambal wildlife sanctuary located on river Chambal is the home for them
3. According to IUCN nearly 250 of them are present in the world

(A) Olive Ridley Turtle
(B) Gharial Crocodiles
(C) Salt water crocodile
(D) Sea Cow (Dugong)

42. Which of the following Indian cricketers has started campaign for conservation of Rhinos in India?

(A) Virat Kohli (B) Rohit Sharma
(C) Mithali Raj (D) MS Dhoni

Answers | જવાબ

Que.	Ans.	Que.	Ans.	Que.	Ans.	Que.	Ans.	Que.	Ans.	Que.	Ans.	Que.	Ans.
1.	C	7.	C	13.	C	19.	D	25.	D	31.	B	37.	A
2.	B	8.	D	14.	D	20.	B	26.	B	32.	C	38.	B
3.	B	9.	B	15.	A	21.	B	27.	C	33.	C	39.	C
4.	A	10.	B	16.	C	22.	B	28.	A	34.	D	40.	B
5.	A	11.	D	17.	D	23.	C	29.	C	35.	C	41.	B
6.	D	12.	A	18.	C	24.	A	30.	D	36.	B	42.	B

Gujarat Administrative Service Class-I & Gujarat Civil Service Class I & II Advertisement No. 121/2016-17 Preliminary Test on - 04-06-2017

1. Under the Green Highways (Plantation and Maintenance) Policy, 2015 the percentage of total civil cost of National Highway Development Projects is to be set aside as plantation of trees.

(A) 1 percent (B) 3 percent
(C) 5 percent (D) 10 percent

2. The average height of the Troposphere over the Equator is about

(A) 5 Kms (B) 16 Kms
(C) 20 Kms (D) 50 Kms

3. The Most of the hot deserts of the world lie between the latitude of

(A) 0° = 5° (B) 5 - 15°
(C) 15° - 30° (D) 30°-50°

4. Gutenberg discontinuity is found between

(A) Crust and Mantle
(B) Mantle and Core
(C) Upper Mantle and Lower Mantle
(D) Outer Core and Inner Core

5. The Mediterranean type of climate is characterised by

(A) Rainfall throughout the year
(B) Wet winter and dry summer
(C) Maximum rain during summer
(D) Scanty rainfall throughout the year

6. Which one of the following drainage pattern is found in the Kathiawar region of Gujarat?

(A) Rectangular (B) Dendritic
(C) Trellis (D) Radial

7. The layer of the atmosphere where temperature decreases to a maximum of about -90°C is known as

(A) Stratosphere (B) Mesosphere
(C) Troposphere (D) Troposphere

8. Arrange the following gases in descending order of percentage in the atmosphere.

1. Oxygen
2. Nitrogen
3. Carbon Dioxide
4. Argon

Codes:

(A) 1, 2, 3 and 4 (B) 2, 1, 4 and 3
(C) 3, 4, 2 and 1 (D) 2, 1, 3 and 4

9. Which of the following pairs is mismatched?

(A) Blue Revolution – Fish and Sea food
(B) Rainbow Revolution – Agriculture and Horticulture
(C) White Revolution – Milk and Milk production
(D) Yellow Revolution – Fruits and Flowers

10. According to the Indian Noise Pollution Regulation and Control Rules, 2000 the Leq levels should not increase more than __________at night time in residential area.

(A) 40 dB (A)
(B) 45 dB (A)
(C) 50 dB (A)
(D) 55 dB (A)

11. The Pennines (Europe), Appalachians (America) and Aravallies (India) are examples of

(A) Block Mountains
(C) Young Mountains
(B) Old Mountains
(D) Fold Mountains

12. The Largest percentage of forest area in India is in which of the following regions
(A) Peninsular Plateau
(B) Western Ghats
(C) Eastern Ghats
(D) Himalayan Region

13. Select the correct option showing the following districts in the descending order in reference to the distribution of forest area.
(A) Kutch, Junagadh, Narmada, Dang
(B) Dang, Narmada, Junagadh, Kutch
(C) Narmada, Dang, Kutch, Junagadh
(D) Junagadh, Dang, Narmada, Kutch

14. The Indian Council of Agricultural Research (ICAR) has classified the Indian soils into how many categories?
(A) 6 (B) 7 (C) 8 (D) 9

15. Which of the following is UN Decade on Biodiversity?
(A) 2011 - 2020 (B) 2016 - 2025
(C) 2014 - 2024 (D) 2005 - 2014

16. What is bio-diversity hotspot?
(A) A biogeographic on characterized both by exceptional level of plant endemism and by serious levels of habitat loss
(B) A smaller area of forested area with high temperature
(C) An area of volcanic, having few flora and fauna species
(D) None of the above

17. Which of the following Article in Indian constitution is related to protection and improvement of the environment and safeguarding of forests and wildlife of the country?
(A) Article 47A
(B) Article 48A
(C) Article 49A
(D) Article 50A

18. 'Bhoor Singh the Barasingha' is the official mascot of
(A) Jim Corbett National Park
(B) Kanha Tiger Reserve
(C) Kaziranga National Park
(D) Dudhwa Tiger Reserve

19. The natural cycle that circulates elements between the earth and the environment (particularly between living and non-living matter) is called
(A) Biogeochemical cycle
(B) Chemical cycle
(C) Biological cycle
(D) Bioorganic cycle

20. Where is the regional station of Central Soil Salinity Research Institute located in Gujarat?
(A) Bhavnagar
(B) Bharuch
(C) Verawal
(D) Mundra

21. Amazon Basin of Brazil has the largest expanse of which forest?
(A) Tropical Forest
(C) Monsoon Forest
(B) Temperate Forest
(D) Coniferous Forest

22. China has launched a satellite to monitor global carbon emissions, the name of the satellite is
(A) Tan Sat
(B) Con Sat
(C) CC Sat
(D) None of the above

23. Under the National Mission for Green India outlined under the National Action Plan on Climate Change (NAPCC), India aims to increase its forest cover by_________ million hectares.
(A) 2 (B) 3
(C) 4 (D) 5

24. Indian fisheries and aquaculture sector contribute________ of the GDP and ________ of the agriculture GDP respectively

(A) 6.5%,11.5% (B) 3.5%, 7.5%
(C) 2.4%, 6.5% (D) 1.1%, 5.15%

Answers | જવાબ

Que.	Ans.	Que.	Ans.	Que.	Ans.	Que.	Ans.	Que.	Ans.	Que.	Ans.
1.	A	5.	B	9.	D	13.	A	17.	B	21.	A
2.	B	6.	D	10.	B	14.	C	18.	B	22.	A
3.	C	7.	B	11.	B	15.	A	19.	A	23.	D
4.	B	8.	B	12.	A	16.	A	20.	B	24.	D

Gujarat Administrative Service, Class-1, Gujarat Civil Services, Class-1 and Class-2 and Gujarat Municipal Chief Officer Service, Class-2 40/2018-19 (21/10/2018)

1. The main aim of National Mission for Green India (GIM) is to -

(A) Protect, restore and enhance India's diminishing forest cover and responding to climate change

(B) Reduce pollution in 10 most polluted cities of India

(C) Increase the use of non-commercial sources of energy

(D) Encourage the use of alternative sources of energy

2. Among the following types of forest, which has the highest species diversity?

(A)Temperate forests

(B) Tropical forests

(C) Temperate deciduous forests

(D) Temperate grass lands

3. Which of the following statement(s) is (are) correct with reference to Marine National Parks?

1. Marine National Park in Gulf of Kutch is situated on the southern shore of the Gulf in the Devbhumi Dwaraka District of Gujarat and it is the first National Marine Park of India.
2. Gahirmatha is the first and the only Marine Sanctuary of Orissa, where Olive Ridely sea turtles migrate in huge numbers.

(A) Only 1 (B) Only 2

(C) Both 1 and 2 (D) Neither 1 nor 2

4. Which of the following statement(s) is (are) correct in relation to Western Ghats?

1. The Western Ghats are the main reason for the monsoon rain in the western coast of India.
2. Western Ghats are locally known as Sahyadri in Tamil Nadu and Nilgiri hills in Telangana.
3. Heights of the Western Ghats decrease from north to south.

(A) Only 1 (B) Only 1 and 2

(C) Only 2 and 3 (D) Only 2

5. The tribal population in Andaman island largely belongs to the ....

(A) Australoid race (B) Caucasoid race

(C) Mongoloid race (D) Negrito race

6. Which of the following statement(s) (is) are correct about the Forests in India?

A. Major regions of Tropical Evergreen Forests in India are Western Ghats, Plateau of Shillong, Andaman and Nicobar Islands and Lakshadweep.

B. Major trees in the Tropical Deciduous Forests are Teak, Sal, Mango, Bamboo and Sandal.

C. Tropical Evergreen Forests are also called Monsoon Forest.

7. The National Forest Policy of 1988 proposed 33% of the total geographical area should be under forest/tree cover.

(A) Only 1, 2 and 3 (B) Only 1, 2 and 4

(C) Only 2, 3 and 4 (D) Only 2 and 3

8. Which of the following statements are correct with reference to the Coral Reefs in India?

1. Coral reefs are the counterpart to the tropical rain forest in terms of species diversity and biological productivity in the Ocean.
2. The major reef formations in India are restricted to the Gulf of Mannar, Pal bay, Gulf of Kutch, Andaman and

Nicobar Islands and the Lakshadweep Islands.

3. All the coral reefs in India are fringing reefs.
4. The coral formations of the Gulf of Kutch represent one of the extreme northern limits of corals in the Indian Ocean.

(A) Only 1, 2 and 3 (B) Only 1, 2 and 4
(C) Only 2, 3 and 4 (D) Only 2 and 3

9. Which of the following statement(s) is (are) correct with reference to the Biodiversity Hotspot?

1. A biodiversity hotspot is a biogeography region with significant levels of biodiversity that is threatened with destruction.
2. To qualify as a biodiversity hotspot, a region must meet two strict criteria: it must contain at least 500 species of vascular plants as endemic and it has to have lost at least 50% of its primary vegetation.
3. Western Ghats is a biodiversity hotspot in India.

(A) Only 1 and 3 (B) Only 2 and 3
(C) Only 2 (D) 1, 2 and 3

10. Singapore lays in ________ zone.

(A) Equatorial rain
(B) Tropical summer rain
(C) Sub-tropical dry
(D) None of the above

11. The term Green Bonus is related with __.

(A) Indian space Programme
(B) Environment and Forest
(C) Agriculture
(D) Indian Military

12. Which of the following statements is correct about the Datum Line?

(A) It is a divide between Rann of Kutch & Kathiyawar Plateau.
(B) It is a natural divide between the Himalayas & the great northern plain.
(C) It is an imaginary line which passes through the magnetic poles of the Earth.
(D) It is an imaginary horizontal line used as a reference for measuring heights & depths.

13. The name Hurricane is given to tropical cyclone in ________ .

(A) The North Pacific Ocean
(B) Australia
(C) The North Atlantic Ocean
(D) The Bay of Bengal

14. Which of the following pairs are correctly matched?

1. Nainital lake – Tectonic wetland
2. Loktak lake – Oxbow wetland
3. Lonar lake – Crater wetland
4. Chilka lake – Lagoon wetland

(A) Only 1 and 3 (B)Only 2 and 4
(C) Only 1, 2 and 4 (D) 1, 2, 3 and 4

15. Which of the following statements is (are) correct?

1. Onam shifting cultivation is practised in Kerala.
2. Podu shifting cultivation is practised in Andhra Pradesh and Odisha.
3. Jhum shifting cultivation is practised in Madhya Pradesh.
4. Penda shifting cultivation is practised in Assam.

(A) Only 2 (B) Only 3 and 4
(C) Only 1 and 2 (D) Only 1

16. Polar Stratospheric Clouds are associated with which of the following?

(A) Ozone layer depletion
(B)Greenhouse effect
(C)Acid Rain
(D) Artificial Rain

17. Which of the following statements isn't correct?

(A) Increase in carbon dioxide levels in the atmosphere will lead to increase in the rate of photosynthesis.

(B) Increase in carbon dioxide levels in the atmosphere will lead to decrease in the rate of photosynthesis.

(C) Increase in carbon dioxide levels in the atmosphere will cause a net transfer of carbon dioxide from the atmosphere to the oceans.

(D) Increase in the carbon dioxide levels will lead to a rise in the global average temperature of earth.

18. The Gujarat Livelihood Promotion Company (GLPC) has joined hands with Central Salt Marine and Chemical Research Institute (CSMCRI) to encourage seaweed cultivation along the state coast. Which of the following statements are correct with reference to the same?

1. In Gujarat, Seaweed culture can be best developed along the coast lines of Amreli and Kutch districts.
2. Seaweeds provide valuable source of raw material for industries like health, food, medicine, pharmaceuticals, textiles, fertilizers and animal feed.
3. Chemicals from red seaweeds such as alginicacid, mannitol, laminarin, fucoidinand iodine are extracted on a commercial basis.
4. Seaweeds are rich in minerals, vitamins, trace elements and bioactive substances and called as medical food of the 21st century.

(A) Only 1 and 2 (B) Only 2 and 3

(C) Only 1, 2 and 4 (D) Only 1, 2, and 3

19. The Montreux Record is a register of ________________.

1. Wetland sites under threat of anthropogenic activities.
2. Endangered species of tropical and subtropical Fauna.
3. Coastal cities under direct threat of consequences of global warming.

(A) Only 1 (B) Only 2

(C) Only 2 and 3 (D) 1, 2 and 3

20. Under Indian Forest Act (1927) which of the following is the correct classification of forest?

(A) Grasslands, tropical forest, wetlands

(B) Wildlife Sanctuary, National park, Biosphere reserves

(C) Grassland, Wildlife Sanctuary, National Parks

(D) Protected Forests, Reserved Forests, Village Forests

21. Ozone layer prevents entry of which of the following rays in the atmosphere?

(A) UV-B only

(B) UV-A only

(C) UV-C only

(D) Both UV-B and UV-C

22. Scientists from the Botanical Survey of India have identified new plant species and named it as *Drypetes Kalamii* after the name of Dr. A.P.J. Abdul Kalam. From which State was this specie discovered?

(A) Andhra Pradesh

(B) West Bengal

(C) Mizoram

(D) Assam

23. In which one of the following categories of protected areas in India local people aren't allowed to own private lands?

(A) Biosphere Reserves

(B) National Parks

(C) World Heritage Sites

(D) All of the above

24. Which of the following places has maximum biodiversity and endemic animals?

(A) Eastern Ghats and West Bengal

(B) Western Ghats and Eastern Himalayas

(C) Thar Desert and Punjab

(D) Nicobar and Kerala

25. Which of the following cities of Gujarat is placed in the list of top 10 of Overall Emissions and Energy Consumption Rankings released by Centre for Science and Environment?

(A) Surat

(B) Ahmedabad

(C) Vadodara

(D) None of the above

26. Which of the following statement(s) is (are) incorrect about India's first Biofuel flight?

1. It travelled from Shimla to Delhi.
2. The flight was operated by Air India.
3. It used completely Biofuel during its travel.
4. This Biofuel was developed by IIT Delhi

(A) 1, 2, 3 and 4 (B) Only 2, 3 and 4

(C) Only 3 and 4 (D) Only 4

27. Bombay Natural History Society opened its first regional centre at ______.

(A) Chilka Lake

(B) Nal Sarovar

(C) Kolleru Lake

(D) All of the above

Answers | જવાબ

Que.	Ans.	Que.	Ans.	Que.	Ans.	Que.	Ans.	Que.	Ans.	Que.	Ans.	Que.	Ans.
1.	A	5.	D	9.	A	13.	C	17.	B	21.	D	25.	A
2.	B	6.	A	10.	A	14.	D	18.	C	22.	B	26.	A
3.	C	7.	B	11.	B	15.	C	19.	A	23.	B	27.	A
4.	A	8.	B	12.	D	16.	A	20.	D	24.	B		

Name of The Post - Assistant Manager (Environment), (GMDC), Class-2 Advertisement No. 138/2019-20 Preliminary Test Held On - 23-12 -2020

1. The provisions of environmental protection in the constitution were made under:

(A) Article 5-A

(B) Article 21-B

(C) Article 48-A and Article 51-A (g)

(D) Article 27-B (h)

2. The first of the major environmental protection act to be promulgated in India was:

(A) Air Act

(B) Noise pollution Act

(C) Environmental Act

(D) Water Act

3. When environmental Lapse Rate (ELR) is less is than Adiabatic Lapse Rate (ALR), then which of the following occurs?

(A) Neutral lapse rate

(B) Super adiabatic lapse rate

(C) Neutral lapse rate

(D) Sub adiabatic lapse rate

4. Which of the following statements are true?

1. Aerobic digestion is not as sensitive to environment as that of anaerobic digestion.
2. Production of sludge in aerobic digestion is relatively inert than anaerobic digestion.
3. Aerobic process is more energy consumptive than anaerobic digestion.
4. Digested sludge dewaters poorly in aerobic digestion.

Select the correct answer using the codes given below:

(A) 1, 2 and 4

(B) 1, 2, 3 and 4

(C) 2, 3, and 4

(D) None of the above

5. If the Arctic ice was somehow replaced with dense forest, which of the following situation may arise:

(A) It will accelerate Global Warming

(B) It will have no affect on Global warming

(C) It will decelerate Global Warming

(D) It may or may not affect the Global Warming

6. The U.S. Environmental Protection Agency was formed in _______.

(A) 1970 (B) 1963

(C) 1967 (D) 1955

7. The Wildlife (Protection) Act was enacted in the year:

(A) 1972 (B) 1974

(C) 1994 (D) 1986

8. The humus in the soil is the:

(A) Organic portion of the soil remaining after prolonged microbial decomposition

(B) Heterogeneous mixture of products resulting from the microbial and chemical transformation of organic materials

(C) Complex of organic compounds, living organisms, water and includes gases

(D) All of these

9. Which of the following statements is/are correct:

1. To obtain the lowest hydraulic conductivity for a compacted clay liner you must compact the soil with the water content higher than the optimum

water content (wet side of liner of optima) in order to obtain "dispersed structure".

2. In general as you decrease the hydraulic conductivity of clay liner the main mechanism of a contaminant migration changes from advection-dispersion to constraint molecular diffusion.
3. A material is considered hazardous when it contains material listed as hazardous by the USEPA displays hazardous characteristics (i.e., ignitability, corrosivity, reactivity or toxicity) and/or causes environmental or health damage.

Answer from the codes given below:

(A) 1 and 3
(B) 2 and 3
(C) 3 only
(D) All of the above

10. The compression of wet waste in the mixture releases _______ (a black foul-smelling liquid) that is difficult to dispose of. It percolates into the soil and contaminates groundwater.

(A) Leachate
(B) Pulp
(C) Drainage
(D) Mother liquor

11. Major criteria for considering an area as "hotspot"

(A) Ecological fragile
(B) Richness in endemic species and impact by human activities
(C) Richness in rare and endangered species
(D) All the these

12. Which of the following statement is correct about Intergovernmental Panel on Climate Change (IPCC)?

(A) IPCC published its first report in 1989
(B) IPCC does not evaluate the risk of climate change on human
(C) IPCC was established in 1988
(D) All of the above

13. FAME scheme was launched in 2015 under NEMMP. Its main aim is to promote _______.

(A) Climate change
(B) CNG based vehicles
(C) Improved road facility
(D) Incentive for electric vehicles

Answers | જવાબ

Que.	Ans.	Que.	Ans.	Que.	Ans.	Que.	Ans.	Que.	Ans.	Que.	Ans.	Que.	Ans.
1.	C	3.	D	5.	A	7.	A	9.	D	11.	B	13.	D
2.	C	4.	B	6.	A	8.	D	10.	A	12.	C		

Chapter 22

પ્રકરણ ૨૨

Practice sets

પ્રેક્ટીસ સેટ

Practice Paper-1 | પ્રેક્ટિસ પેપર-૧

1. રાષ્ટ્રીય જૈવવિવિધતા સત્તામંડળ નું મુખ્ય મથક ક્યાં છે ?

A. મુંબઈ B. ચેન્નાઈ

C. હૈદરાબાદ D. નવી દિલ્હી

1. The headquarter of National Biodiversity Authority (NBA) is

A. Mumbai B. Chennai

C. Hyderabad D. New Delhi

2. સજીવની ક્રિયાપ્રતિક્રિયા અને જીવસૃષ્ટિના આંતર સંબંધ ના અભ્યાસ ને શું કહે છે ?

A. પર્યાવરણીય વિજ્ઞાન

B. ઇકોલોજી

C. ાણીશાસ્ત્ર

D. A અને B બન્ને

2. The study of interaction and inter-relationship of organisms with their environment

A. Environmental Sciences

B. Ecology

C. Zoology

D. Both A and B

3. ____________ જંગલો દરિયાઇ અને જમીની ઇકોસિસ્ટમ વચ્ચેના ઇકોટોનનું પ્રતિનિધિત્વ કરે છે

A. સમશીતોષ્ણ

B. ઉષ્ણકટિબંધીય

C. બોરિયલ

D. મેંગ્રોવ (ચેરના જંગલો)

3. ________________ forests represent an ecotone between marine and terrestrial ecosystem.

A. Temperate

B. Tropical

C. Boreal

D. Mangrove

4. પરંપરાગત માનવ જીવન સાથે જૈવવિવિધતાના સંરક્ષણ માટેની સૌથી મહત્વપૂર્ણ વ્યૂહરચના ____________ની સ્થાપના છે.

A. વનસ્પતિ ઉદ્યાનો

B. રાષ્ટ્રીય ઉદ્યાનો

C. બાયોસ્ફિયર રિઝર્વ

D. વન્યપ્રાણી અભયારણ્ય

4. The most important strategy for the conservation of biodiversity together with traditional human life is the establishment of

A. Botanical gardens

B. National parks

C. Biosphere reserves

D. Wildlife sanctuaries

5. આપણા દેશમાં સુરખાબની એક માત્ર જાણીતી બ્રીડિંગ કોલોની ક્યાં મળી આવે છે?

A. થાર રણ
B. લદ્દાખ રણ
C. કચ્છનો મોટો રણ
D. શીત રણ

5. The only known breeding colony of Greater Flamingos in our country is found in

A. The Thar desert
B. Ladakh desert
C. The Great Rann of Kutch
D. Cold desert

6. સસ્તન પ્રાણીઓની પૂંછડી, કાન, અંગો, મોઢું એ ઠંડા પ્રદેશમાં નાના હોય છે અને ગરમ પ્રદેશમાં મોટા હોય છે. આ નિયમને શું કહે છે ?

A. બર્ગરનો નિયમ B. એલેન્સનો નિયમ
C. બર્ગમેન નિયમ D. જોર્ડન નિયમ

6. The tails, ears, limbs, snout of mammals are smaller in colder region and larger in warmer region. This rule is known as

A. Berger's rule B. Allen's rule
C. Bergman's rule D. Jordan's rule

7. ભેજવાળા પ્રદેશના શંકુદ્રુપ જંગલો હેઠળ બરછટ ટેક્ષ્ચરને શું કહેવાય છે ?

A. ઇનસેપ્ટિસોલ્સ
B. વર્ટિસોલ્સ
C. ઓક્સિસોલ્સ
D. સ્પોડોસોલ્સ

7. Coarse textured deposits under coniferous forests of humid region consists of

A. Inceptisols
B. Vertisols
C. Oxisols
D. Spodosols

8. તાજેતરમાં પ્રકાશિત વર્લ્ડ વાઇલ્ડલાઇફ ક્રાઇમ રિપોર્ટ 2020 મુજબ વિશ્વનું સૌથી વધુ શિકાર થતું સસ્તન પ્રાણી કયું છે?

A. વાઘ B. પેંગોલિન
C. આગરા D. હાથી

8. The recently released World Wildlife Crime Report 2020 revealed that the world's most trafficked mammal is

A. Tiger B. Pangolin
C. Leopard D. Elephant

9. પ્રિસેન્ટિવ પ્રજાતિઓ ________ તરીકે પણ ઓળખાય છે.

A. સર્વદેશીય પ્રજાતિઓ
B. ભયગ્રસ્ત જાતિઓ
C. દુર્લભ પ્રજાતિઓ
D. સ્થાનિક જાતિઓ

9. Precentive species is also known as

A. Cosmopolitan species
B. Endangered species
C. Rare species
D. Endemic species

10. લાઇકેન અને માઇકોરિઝા ________નાં ઉદાહરણો છે

A. પ્રોટોકોપરેશન
B. સહભોજિતા
C. સિમ્બાયોટિક એસોસિએશન
D. અલ્ટ્રુઈઝમ

10. Lichens and Mycorrhizae are the examples of

A. Protocooperation
B. Commensalism
C. Symbiotic associations
D. Altruism

11. રામસાર કન્વેન્શન શાને લગતું છે ?
A. પર્વતો
B. નદીઓ
C. જલપ્લાવિત વિસ્તારો
D. રણ

12. ભારતમાં વન્યપ્રાણી સંરક્ષણ ચળવળનો ____________ દ્વારા પ્રારંભ કરાયો હતો
E. BNHS F. IUCN
G. WWF H. CITES

13. પર્યાવરણનો મુખ્ય ઊર્જા સ્ત્રોત કયો છે?
A. સૌર ઊર્જા B. પરમાણુ ઊર્જા
C. હાઈડ્રો ઊર્જા D. રાસાયણિક ઊર્જા

14. વાતાવરણને સ્વચ્છ અને લીલું રાખવા માટેની રીતોમાં ____________ શામેલ છે.
A. ઓછી ઊર્જાનો ઉપયોગ
B. પર્યાવરણ મૈત્રીપૂર્ણ ટેકનોલોજી
C. વન્યજીવન રક્ષણ
D. ઉપરોક્ત તમામ

15. સપ્રોફાઇટ્સ શું છે ?
A. ઝાડની છાલમાંથી સેપ ખાનાર
B. અન્ય સજીવો ખાનાર
C. મૃત અથવા સડેલા કાર્બનિક પદાર્થો ખાનાર
D. ઉપર્યુક્તમાંથી કોઈ નહિ

16. કાળી ક્રાંતિ ____________ સાથે સંબંધિત છે
A. દૂધ ક્રાંતિ
B. કૃષિ પાકની ક્રાંતિ
C. બાયોફ્યુઅલ / જાટ્રોપા ક્રાંતિ
D. કાલા બીજ ક્રાંતિ

17. કયા જંગલોને પૃથ્વીના ફેફસાં કહે છે?
A. એમેઝોન
B. ઓસ્ટ્રેલિયન જંગલો
C. અર્ધ પાનખર જંગલો
D. સદાબહાર જંગલો

11. Ramsar Convention refers to the conservation of
A. Mountains B. Rivers
C. Wetlands D. Deserts

12. First wildlife conservation movement in India was started by
A. BNHS B. IUCN
C. WWF D. CITES

13. The chief energy source of environment is
A. Solar energy B. Nuclear energy
C. Hydro energy D. Chemical energy

14. Methods to keep environment clean and green include
A. Use of less energy
B. environment friendly technology
C. protecting wildlife
D. All of the above

15. Saprophytes are
A. feed on the sap from tree bark
B. feed on other living beings
C. feed on dead or decayed organic matter
D. None of the above

16. Black revolution is related with
A. Milk revolution
B. Agriculture crop revolution
C. Biofuel/Jatropa revolution
D. Black seed revolution

17. The forests which are called the lungs of the earth?
A. Amazon
B. Australian forests
C. Semi deciduous forests
D. Evergreen forests

18. વનોની જમીનમાં શાનું પ્રભુત્વ હોય છે?

A. બેક્ટેરિયા
B. ફૂગ
C. એક્ટિનોમિસેટ્સ
D. A અને B બન્ને

19. વન જમીનના ભૌતિક ગુણધર્મો મોટા પ્રમાણમાં શાના દ્વારા નિયંત્રિત થાય છે?

A. માટીના કણોનું કદ
B. માટીના કણોનું વિતરણ
C. માટીના કણોની ગોઠવણ
D. ઉપરોક્ત તમામ

20. વન જમીનો કૃષિ જમીનો કરતા કુદરતી રીતે વધુ ________ હોય છે

A. બેઈઝીક
B. ન્યુટ્રલ
C. એસિડિક
D. સંતુલિત

21. ક્લાઈમેટ ચેન્જના લગભગ તમામ મોડેલો ગ્લોબલ વોર્મિન્ગના પરિણામે માટીમાંથી _____ ની આગાહી કરે છે

A. કાર્બનનો લાભ
B. કાર્બનનું નુકસાન
C. નાઇટ્રોજનનો લાભ
D. A અને C બન્ને

22. Indian Institute of Soil Science ક્યાં આવેલ છે?

A. ભોપાલ B. મુંબઈ
C. અમદાવાદ D. દિલ્હી

23. મેગ્નેટાઇટ ભારે ખનીજ છે વિદારણ માટે પ્રતિરોધક છે અને ઘણી વખત જમીનમાં ક્યાં પ્રકારે થાય છે?

A. મજબૂત ચુંબકત્વ સાથે લાલ રેતી
B. મજબૂત ચુંબકત્વ સાથે પીળી રેતી
C. મજબૂત ચુંબકત્વ સાથે કાળી રેતી
D. મજબૂત ચુંબકત્વ સાથે સફેદ રેતી

18. The microbial community in forest soil is dominated by

A. Bacteria
B. Fungi
C. Actinomycetes
D. Both A and B

19. Forest soil physical properties are largely controlled by

A. size of soil particles
B. distribution of soil particles
C. arrangement of soil particles
D. all of the above

20. Forest soils tend to be naturally more ________ than agricultural soils

A. Basic
B. Neutral
C. Acidic
D. Balanced

21. Nearly all models of global climate change predict a ______ from soils as a result of global warming

A. Gain of carbon
B. Loss of carbon
C. Gain of nitrogen
D. Both A and C

22. Indian Institute of Soil Science is situated in

A. Bhopal B. Mumbai
C. Ahmedabad D. Delhi

23. Magnetite is a heavy mineral resistant to weathering and often occurs in soils as a

A. red sand with strong magnetism
B. yellow sand with strong magnetism
C. black sand with strong magnetism
D. white sand with strong magnetism

24. ભારતમાં National Air Quality Index ની શરૂઆત ક્યારે કરવામાં આવી હતી?

A. 2010 B. 2012
C. 2014 D. 2015

25. WCED નું આખું નામ શું છે?

A. ઊર્જા અને વિકાસ પર વિશ્વ પંચ
B. પર્યાવરણ અને વિકાસ પર વિશ્વ પંચ
C. વિશ્વ સ્વચ્છ પર્યાવરણ દિવસ
D. વિશ્વ સ્વચ્છ ઊર્જા દિવસ

26. ____________ પાણીની કાયમી કઠિનતાનું કારણ બને છે.

A. મેગ્નેશિયમ બાયકાર્બોનેટ
B. મેગ્નેશિયમ કાર્બોનેટ
C. મેગ્નેશિયમ સલ્ફેટ
D. ઉપરોક્ત તમામ

27. આગરા ખાતે તાજમહેલને મુખ્યત્વે શાનાથી નુકસાન થાય છે?

A. હાઇડ્રોજન પેરોક્સાઇડ
B. સલ્ફર ડાયોક્સાઇડ
C. ક્લોરિન
D. ઉપર્યુક્ત તમામ

28. BOD અને COD એ શેના માપદંડ છે?

A. જળચર ઇકોસિસ્ટમ્સમાં પ્રદૂષણનું પૃથક્કરણ
B. લોહીમાં ઓક્સિજનનું સ્તર માપવા
C. વન ઇકોસિસ્ટમ્સમાં ઓક્સિજન સ્તરની ગણતરી
D. ઉંચાઇવાળા પ્રદેશોમાં ઓક્સિજનના સ્તરોનું મૂલ્યાંકન

29. ઝાડની વૃદ્ધિ અને સ્વરૂપમાં સુધારો લાવવાના હેતુથી અપરિપક્વ સ્ટેન્ડમાં કરવામાં આવતી ફેલીન્ગને ____________ તરીકે ઓળખવામાં આવે છે

A. હેડિંગ બેક B. પ્રુનિંગ
C. થિનિંગ D. કોપિસિન

24. National Air quality index was launched in India in

A. 2010 B. 2012
C. 2014 D. 2015

25. WCED stands for

A. World Commission on Energy and Development
B. World Commission on Environment and Development
C. World Clean Environment Day
D. World Clean Energy Day

26. Which salt is the main cause of permanent hardness of water?

A. Magnesium bicarbonate
B. Magnesium carbonate
C. Magnesium sulphate
D. All of the above

27. The Taj Mahal at Agra is damaged mainly by which compound?

A. Hydrogen Peroxide
B. Sulphur dioxide
C. Chlorine
D. All of the above

28. BOD and COD are standard criterion for

A. Pollution assay in aquatic ecosystems
B. Measuring oxygen levels in blood
C. Computing oxygen levels in forest ecosystems
D. Assessing oxygen levels in high altitude regions

29. The felling made in an immature stand for the purpose of improving the growth and form of tree is termed as

A. Heading back B. Pruning
C. Thinning D. Coppicing

30. વડા પ્રધાન દ્વારા એપ્રિલ, 2015 માં હવામાન ગુણવત્તાની માહિતીના પ્રસાર માટે 14 શહેરોથી ____________ શરૂ કરવામાં આવી હતી.

A. રાષ્ટ્રીય હવા ગુણવત્તા સૂચકાંક
B. રાષ્ટ્રીય જળ ગુણવત્તા સૂચકાંક
C. રાષ્ટ્રીય માટી ગુણવત્તા સૂચકાંક
D. રાષ્ટ્રીય પર્યાવરણ ગુણવત્તા સૂચકાંક

30. ____________ was launched by the Prime Minister in April, 2015 starting with 14 cities to disseminate air quality information

A. National Air Quality Index
B. National Water Quality Index
C. National Soil Quality Index
D. National Environment Quality Index

31. ન્યુમેટોફોર્સ સામાન્ય રીતે કઈ પ્રજાતિ માં જોવા મળે છે?

A. સુકા પાનખર જંગલ
B. ચેરના વન
C. ખુલ્લા વન
D. ઉષ્ણકટિબંધીય વન

31. Pneumatophores are commonly seen in

A. Dry deciduous forest
B. Mangrove forest
C. Open forest
D. Sub-tropical forest

32. એન્ડેમિક છોડની પ્રજાતિઓ એટલે શું

A. છૂટાછવાયા વિતરણ
B. તમામ સ્થળોએ વિતરણ
C. વિતરણ ચોક્કસ ક્ષેત્ર સુધી મર્યાદિત છે
D. સમાન વિતરણ

32. Endemic means plant species which has

A. Scattered distribution
B. Distribution all over places
C. Distribution is limited to a specific area
D. Distribution evenly

33. ઝાડના થડ પર ફૂલ અને ફળ આવવાને વૈજ્ઞાનિક રીતે શું કહેવામાં આવે છે

A. પાર્થેનોકાર્પી
B. વિવિપરી
C. કોલિફ્લોરી
D. ડોર્મેનસી

33. Stem of trees bear flowers and fruits habit called

A. Parthenocarpy
B. Vivipary
C. Cauliflory
D. Dormancy

34. જંગલને ઉછેરવા અને વૃદ્ધિ કરવાની કળાને શું કહેવામાં આવે છે?

A. સિલ્વીકલ્ચર
B. મેન્સ્યુરેશન
C. ઇકોલોજી
D. વાનિકી

34. The art of growing and reproducing a forest is called

A. Silviculture
B. Mensuration
C. Ecology
D. Forestry

35. ઝુમિંગ નું બીજું નામ શું છે ?

A. શિફ્ટિંગ કલ્ટિવેશન
B. ટોન્ગયા
C. A અને B
D. હોમ ગાર્ડન

35. Jhuming is the other name for

A. Shifting cultivation
B. Taungya
C. A and B
D. Home Garden

36. શુદ્ધ વન નો અર્થ શું છે?
A. પાકનો 30% ભાગ એક પ્રજાતિનો હોય
B. પાકનો 50% ભાગ એક પ્રજાતિનો હોય
C. પાકનો 70% ભાગ એક પ્રજાતિનો હોય
D. પાકનો 80% ભાગ એક પ્રજાતિનો હોય

37. ભારતમાં વન મોહોત્સવની શરૂઆત ક્યા વર્ષ માં કરવામાં આવી હતી ?
A. 1847 B. 1947
C. 1850 D. 1950

38. વુડલોટનો બેઝલ એરિયા શું દર્શાવે છે?
A. ઝાડનો વિસ્તાર ચોરસ ફૂટમાં
B. ક્રાઉન ડેન્સીટી
C. સ્ટેન્ડ ડેન્સીટી
D. ટ્રંકનો ટેપર

39. "પરંપરાગત જંગલોની બહારનું વનીકરણ જેનો હેતુ મુખ્યત્વે લોકોના ફાયદા માટે સામાન અને સેવાઓનો સતત પ્રવાહ પૂરો પાડવાનો છે" જે__________ તરીકે ઓળખાય છે?
A. સામાજિક વનીકરણ
B. ફાર્મ ફોરેસ્ટ્રી
C. એક્સ્ટેંશન ફોરેસ્ટ્રી
D. રૂરલ ફોરેસ્સ્ટ્રી

40. ભારતમાં, સહભાગી વન આયોજન અને વ્યવસ્થાપન પ્રથમ ક્યા રાજ્ય માં શરૂ કરવામાં આવ્યું હતું ?
A. ગુજરાત B. કેરળ
C. હિમાચલ પ્રદેશ D. પશ્ચિમ બંગાળ

41. સાગ વાવેતર કેવી રીતે કરવામાં આવે છે?
A. એક વર્ષ રોપાઓ વાવીને
B. ખાડાઓમાં રોપાઓ વાવીને
C. પ્રથમ વરસાદ પછી સ્ટમ્પ વાવેતર
D. પરિપક્વ વૃક્ષની શાખાઓનું વાવેતર

36. What do you mean by Pure Forest?
A. 30% of the crop consist of a single species
B. 50% of the crop consist of a single species
C. 70% of the crop consist of a single species
D. 80% of the crop consist of a single species

37. In India, Van Mohotsav was started in
A. 1847 B. 1947
C. 1850 D. 1950

38. Basal area of a woodlot represents
A. Area occupied by the tree in square feet
B. Crown density
C. Stand density
D. Taper of trunk

39. "Forestry outside the conventional forests which primarily aims at providing continuous flow of goods and services for the benefit of people". What this forestry is known as __________?
A. Social Forestry
B. Farm Forestry
C. Extension Forestry
D. Rural Forestry

40. In India, the concept of joint forest planning and management was first started in _______.
A. Gujarat B. Kerala
C. Himachal Pradesh D. West Bengal

41. How does the Teak plantation is being carried?
A. Planting of one-year seedlings
B. Planting in big size of pits
C. Stump planting just after first rain
D. Planting of branches of mature tree

42. ગુજરાતમાં રાજ્ય કક્ષાના વન-મહોસ્તવની ઉજવણી કેવી રીતે કરવામાં આવે છે?
A. વિધાનસભા ખાતે વિધિપૂર્વકનું વાવેતર કરીને
B. દરેક સરકારી કચેરી ખાતે વૃક્ષારોપણ કરીને
C. શૈક્ષણિક સંસ્થાઓમાં વૃક્ષારોપણ કરીને
D. દર વર્ષે અમુક થીમ પર સાંસ્કૃતિક વન વિકસાવીને

43. ગુજરાત વન વિભાગની સામાજિક વનીકરણ યોજના અંતર્ગત રસ્તાની બાજુના વાવેતર ઉભા કરવામાં આવી રહ્યા છે, પરિપક્વતા પછી વૃક્ષોની હરાજી કરવામાં આવે છે અને ચોખ્ખી આવકનો કેટલા % વિકાસના કામોમાં તાલુકા પંચાયતને સોંપવામાં આવે છે ?
A. 75% B. 50%
C. 25% D. 100%

44. ISFR 2019 મુજબ, કયા વૃક્ષની પ્રજાતિઓ ગુજરાતમાં જંગલની બહાર સૌથી વધુ સંખ્યા ધરાવે છે?
A. દેશી બાવળ
B. આંબો
C. લીમડો
D. નીલગીરી

45. નીચેનામાંથી કઈ જાતનો ઉપયોગ મોટાભાગે પાંદડાના તેલ માટે થાય છે?
A. *Santalum album*
B. *Cedrus deodara*
C. *Eucalyptus* spp.
D. *Azadirecta indica*

46. ડાળીઓનું વિશ્લેષણ __________ નક્કી કરવા માટે કરવામાં આવે છે.
A. ઝાડની ઉંમર
B. ઝાડની ઉંચાઈ
C. ઝાડનું વજન
D. ઝાડની ઘનતા

42. In Gujarat, the State level . Vanmahostava is being celebrated as
A. Ceremonial plantation at Vidhan sabha
B. Plantations at every Govt office
C. Plantations in Educational institutions
D. Development of Cultural forest (Sanskritic Van) on certain theme every year

43. Under Social Forestry scheme of Gujarat forest department, the road side plantations are being raised, after maturity, the trees are auctioned and how much % of the net income is handed over to Taluka Panchayat for use in development works?
A. 75% B. 50%
C. 25% D. 100%

44. As per last ISFR 2019, the dominant tree species is having the highest number outside the forest in Gujarat, is;
A. Desi Baval
B. Mango
C. Neem
D. Nilgiri

45. Which of the following species is mostly use for leaf oil?
A. *Santalum album*
B. *Cedrus deodara*
C. *Eucalyptus* spp.
D. *Azadirecta indica*

46. Stem analysis is done to determine
A. Age of a tree
B. Height of a tree
C. Weight of a tree
D. Density of a tree

47. આમાંથી કયો છોડ બાયોડિઝલ પ્લાન્ટ છે?
A. જાટ્રોપા B. મહુડો
C. લીમડો D. કુંવારપાઠું

48. કયા લાકડાનો કોલસો વધુ કેલરીફિક મૂલ્ય ધરાવે છે?
A. *Prosopis juliflora*
B. *Bombex ceiba*
C. *Ailenthus excela*
D. *Butea monosperma*

49. વન્યજીવોના શિકાર પર સખત પ્રતિબંધ છે, વન્યજીવ (સંરક્ષણ) અધિનિયમ, 1972 મુજબ ગુનેગારને કેટલા વર્ષ કેદની સજા ફટકરી શકાય ?
A. 2 વર્ષ B. 7 વર્ષ
C. 5 વર્ષ D. 9 વર્ષ

50. BNHSનું મુખ્ય મથક કયા નામ તરીકે ઓળખાય છે?
A. સ્પેરો હાઉસ B. હોર્નબિલ હાઉસ
C. સારસ હાઉસ D. પિફોલ હાઉસ

51. જમ્મુ અને કાશ્મીરના ડાચીગામ નેશનલ પાર્કનું મુખ્ય પ્રાણી ક્યું છે
A. કસ્તુરી હરણ
B. હંગુલ
C. માર્ખોર
D. કાળું રીંછ

52. એશિયાઇ સિંહ નું ટ્રાન્સલોકેશન સાસણ ગીર થી ______, મધ્ય પ્રદેશ માટે સૂચન કરવામાં આવે છે.
A. કાન્હા B. ધોની
C. કુનો D. કારેરા

53. ગુજરાતનું એકમાત્ર રામસાર સ્થળ કયું છે ?
A. પોરબંદર અભયારણ્ય
B. પાનીયા અભયારણ્ય
C. નલસરોવર વન્યપ્રાણી અભયારણ્ય
D. લાલા પરજન અભયારણ્ય

47. Biodiesel plant is
A. Jatropa B. Madhuca
C. Azadirachta D. Aloe vera

48. Which of the wooden coal is having higher calorific value?
A. *Prosopis juliflora*
B. *Bombex ceiba*
C. *Ailenthus excela*
D. *Butea monosperma*

49. Hunting is strictly prohibited for wild animals, how much imprisonment can be imposed to the culprit as per Wildlife (Protection) Act, 1972.
A. 2 years B. 7 years
C. 5 years D. 9 years

50. The Headquarter of BNHS is known as
A. Sparrow House B. Hornbill House
C. Sarus House D. Peafowl House

51. The main animal of Dachigam National Park, Jammu and Kashmir is
A. Musk deer (*Moschus chrysogaster*)
B. Hangul (*Cervus canadensis*)
C. Markhor (*Capra falconeri*)
D. Black Bear (*Ursus thibetanus*)

52. The translocation of Asiatic lion (*Panthera leo*) from Sasan Gir to ______ in Madhya Predesh is suggested.
A. Kanha B. Kheoni
C. Kuno D. Karera

53. The only Ramsar Site of Gujarat is
A. Porbandar Bird Sanctuary
B. Paniya Wildlife sanctuary
C. Nal Sarovar Wildlife Sanctuary
D. Lala parjan Sanctuary

54. વન્યપ્રાણી (સંરક્ષણ) અધિનિયમ 1972 મુજબ ______ રક્ષિત વિસ્તાર નથી.

A. અભયારણ્ય

B. રાષ્ટ્રીય ઉદ્યાન

C. કનઝર્વેશન રિઝર્વ

D. બાયોસ્ફિયર રિઝર્વ

55. કોમ્યુનિટી રિઝર્વ અને કંઝર્વેશન રિઝર્વ વચ્ચેનો મુખ્ય તફાવત શું છે?

A. સમુદાય અને વન વિભાગ દ્વારા સંચાલન

B. ખાનગી અથવા સમુદાયની જમીન અને સરકારી પડતર જમીન પર

C. સમુદાયની જમીન અને જાહેર જંગલની જમીન પર

D. JMF વિસ્તાર અને Non-JMF વિસ્તાર પર

56. વન અધિકાર કાયદો, 2006 ક્યાં લાગુ થશે નહી?

A. રાષ્ટ્રીય ઉદ્યાન

B. વન્યપ્રાણી અભયારણ્ય

C. ક્રિટિકલ એરિયા

D. અનામત વનો

57. કયા કાયદા હેઠળ, સુરક્ષિત વિસ્તારોની આસપાસના ઇકો સેન્સિટિવ ઝોનને સૂચિત કરવામાં આવી રહ્યું છે?

A. વન (સંરક્ષણ) અધિનિયમ 1980

B. ભારતીય વન અધિનિયમ 1927

C. વન અધિકાર અધિનિયમ 2006

D. પર્યાવરણ (સંરક્ષણ) અધિનિયમ 1986

58. વન્યપ્રાણી (સંરક્ષણ) અધિનિયમ હેઠળ કોને ચીફ વાઇલ્ડલાઇફ વોર્ડન તરીકે નિમણુંક આપવામાં આવે છે ?

A. રાજ્યના મુખ્યમંત્રી

B. વન મંત્રી

C. વરિષ્ઠ વન અધિકારી

D. સચિવ, વન

54. Which is not the Protected area as per Wildlife (Protection) Act 1972

A. Sanctuary

B. National Park

C. Conservation reserve

D. Biosphere Reserve

55. The main difference between Community Reserve and Conservation Reserve is

A. Management by Community and by Forest department

B. Private or Community land and Government wasteland

C. Community land and declared forest land

D. JFM area and Non JFM area.

56. Forest right Act, 2006 will not be applicable in

A. National Park

B. Wildlife Sanctuary

C. Critical Areas

D. Reserved Forests

57. Under which Act, the Eco sensitive zone around protected areas are being notified?

A. Forest (Conservation) Act 1980

B. Indian Forest Act 1927

C. Forest Right Act 2006

D. Environment (Protection) Act, 1986

58. The Chief Wildlife Warden (CWLW) is the statutory authority, under the Wildlife Protection Act of any State is

A. Chief Minister of the State

B. Forest Minister

C. Senior Rank Forest Officer

D. Secretary, Forests

59. ગુજરાતમાં જૈવવિવિધતા સંરક્ષણ અને ગ્રામીણ આજીવિકા સુધારણા પ્રોજેક્ટ (BCRLIP) અંતર્ગત કયા રક્ષિત ક્ષેત્રે લાગુ કરવામાં આવી રહ્યા છે?

A. વાંસદા રાષ્ટ્રીય ઉદ્યાન
B. નારાયણ સરોવર અભ્યારણ્ય
C. નાના રણનો લેન્ડસ્કેપ– ઘુડખર અભયારણ્ય
D. મરીન રાષ્ટ્રીય ઉદ્યાન

60. "ડાંગ દરબાર" માં સ્થાનિક લોકોને શેના માટે એવોર્ડ આપવામાં આવે છે?

A. વન અગ્નિ નિયંત્રિત કરવા અને મદદ કરવા માટે
B. વન્યપ્રાણી શિકારને નિયંત્રિત કરવા
C. સહભાગી વન વ્યવસ્થાપન
D. અપરાધીઓથી વનનું રક્ષણ

61. ભારતના કયા રક્ષિત વિસ્તારમાંથી સ્થાનિક લોકો ને બાયોમાસ એકત્રિત અને વાપરવાની મંજૂરી નથી?

A. રાષ્ટ્રીય ઉદ્યાનો
B. વન્યપ્રાણી અભયારણ્ય
C. બાયોસ્ફિયર રીઝર્વ્સ
D. કંઝર્વેશન રીઝર્વ્સ

62. તાજેતરમાં સુપ્રીમ કોર્ટે પાવર લાઈનને ભૂગર્ભમાં કરવાનો હુકમ શેના બચાવ અર્થે જારી કર્યો છે?

A. હેરિયર્સ
B. સાઇબેરીયન ક્રેન્સ
C. હૌબરા બસ્ટાર્ડ
D. ગ્રેટ ઇન્ડિયન બસ્ટાર્ડ

63. કોટવાલિયા આદિવાસી સમુદાયો તેમની આજીવિકા માટે સામાન્ય રીતે શેના પર આશ્રિત હોય છે?

A. શેરડીના મજૂરો
B. જંગલી પ્રાણીઓનો શિકાર
C. વાંસની હસ્તકલા
D. માઇનોર ફોરેસ્ટ પ્રોડક્ટ્સનો સંગ્રહ

59. In Gujarat, Biodiversity Conservation and Rural Livelihood Improvement Project (BCRLIP) is being implemented in the Protected Area

A. Vansda National Park
B. Narayan Sarovar Sanctuary
C. Little Rann landscape - Wild Ass sanctuary
D. Marine National Park

60. In "Dangs darbar" the local villagers are being awarded for

A. To control and help in forest fires
B. To control the wildlife poaching
C. Joint Forest Management
D. Protection of forest from offenders

61. In which one among the following categories of protected areas in India are local people not allowed to collect and use the biomass?

A. National Parks
B. Wildlife Sanctuaries
C. Biosphere Reserves
D. Conservation Reserve

62. Recently the Supreme Court has issued order to put the power line underground to save the;

A. Harriers
B. Siberian cranes
C. Houbara Bustard
D. Great Indian Bustard

63. The Kotwalia tribals are generally dependent for their livelihood on

A. Sugarcane labourers
B. Hunting of wild animals
C. Bamboo handicrafts
D. MFP collection

64. નેશનલ એક્શન પ્લાન ______ કન્સર્વેશન 2020-25 માટે નેશનલ બોર્ડ ફોર વાઇલ્ડલાઇફ દ્વારા નવેમ્બર 2020 માં મંજૂરી આપવામાં આવી છે.
A. ડોલ્ફિન
B. સ્વેમ્પ ડિયર
C. મોર
D. ગીધ

65. અળસિયા ખેડૂતોના મિત્ર છે, કારણ કે
A. કૃષિ ક્ષેત્રોમાં જમીન છિદ્રાળુ કરે છે
B. કૃમિ ખાતરમાં વપરાય છે
C. અળસિયા કીટકોને ખાય છે
D. બંને A અને B

66. પર્યાવરણ સાથે વિવિધ પ્રજાતિઓના જૂથના સંબંધનો અભ્યાસને શું કહે છે?
A. ઓટેકોલોજી
B. સિનેકોલોજી
C. ઇકોલોજી
D. ઉપર્યુક્તમાંથી કોઈ નહિ

67. CVP index એટલે
A. Climate, Vegetation and Precipitation index
B. Climate, Vegetation and Production index
C. Climate, Vegetation and Productivity index
D. Climate, Vegetation and Potential index

68. ભારતમાં પહેલો એસિડ વરસાદ ક્યારે આવ્યો?
A. 1968 માં દિલ્હી
B. 1982 માં કોલકાતા
C. 1971 માં બેંગ્લોર
D. 1974 માં બોમ્બે

64. The National Action Plan for ______ Conservation 2020-25 has been approved by National Board for Wildlife in November 2020.
A. Dolphin
B. Swamp Deer
C. Peafowl
D. Vulture

65. Earthworms is friend of the farmers because of
A. Loosen the earth in agriculture fields
B. Used in vermicomposting
C. Earthworms eat the insects
D. Both A & B

66. The study of relationship of group of different species with their environment is called
A. Autecology
B. Synecology
C. Ecology
D. None of the above

67. CVP index means
A. Climate, Vegetation and Precipitation index
B. Climate, Vegetation and Production index
C. Climate, Vegetation and Productivity index
D. Climate, Vegetation and Potential index

68. In India, the first report of acid rain came from
A. Delhi in 1968
B. Kolkata in 1982
C. Bangalore in 1971
D. Bombay in 1974

69. ભારતનું બાયોજીયોગ્રાફિક વર્ગીકરણ કોના દ્વારા કરવામાં આવ્યું હતું?
A. રોજર્સ અને પનવર (1988)
B. એડવર્ડ અને પનવર (1985)
C. રોજર્સ અને હેરિસ (1989)
D. રોઝન અને થોમસ (1987)

70. સૌથી વિનાશક આગ કઇ છે?
A. ગ્રાઉન્ડ
B. ક્રાઉન
C. સપાટી
D. પેટા સપાટી

71. ડીબીએચ એ સ્થાયી ઝાડના વ્યાસને માપવાની એક માનક પદ્ધતિ છે જે સામાન્ય રીતે કેટલી ઊંચાઈએ લેવામાં આવે છે?
A. 2 મી B. 2.37 મી
C. 1 મી D. 1.37 મી

72. ફૂલો, પર્ણસમૂહ અને ફળના વિકાસમાં મોસમી ફેરફારને ________ તરીકે ઓળખવામાં આવે છે.
A. મોર્ફોલોજી B. ફિનોલોજી
C. ગ્રોથ D. ડેવલપમેન્ટ

73. રાજસ્થાનના સુકા વિસ્તારનું કયું વૃક્ષ કલ્પવૃક્ષ તરીકે ઓળખાય છે.?
A. *Acacia nilotica*
B. *Calligonum polygonoides*
C. *Prosopsis cineraria*
D. *Euphorbia tirucalli*

74. વૃક્ષની જે પ્રજાતિઓ 5Fs આપે છે, એટલે કે fuel, fodder, fiber, food and fertilizers તેને શું કહેવામાં આવે છે?
A. ઘાસચારા ના વૃક્ષો
B. ખાતરના વૃક્ષો
C. લેગ્યુંમિનશ વૃક્ષો
D. બહુહેતુક વૃક્ષો

69. Biogeographic classification of India was done by
A. Rodgers and Panwar (1988)
B. Edward and Panwar (1985)
C. Rodgers and Harris (1989)
D. Rosen and Thomas (1987)

70. Which is the most destructive fire?
A. Ground
B. Crown
C. Surface
D. Sub surface

71. Diameter at breast height- DBH is a standard method of measuring the diameter of the trunk a standing tree, it is generally taken at
A. 2 m B. 2.37 m
C. 1 m D. 1.37 m

72. The seasonal changes in the development of flowering, foliage and fruiting is termed as ________.
A. Morphology B. Phenology
C. Growth D. Development

73. Kalpavriksha of dry part of Rajasthan?
A. *Acacia nilotica*
B. *Calligonum polygonoides*
C. *Prosopsis cineraria*
D. *Euphorbia tirucalli*

74. Tree species which gives 5Fs i.e. fuel, fodder, fiber, food and fertilizers are called:
A. Fodder trees
B. Fertilizer trees
C. Leguminous tress
D. Multipurpose trees

75. રાષ્ટ્રીય એગ્રોફોરેસ્ટ્રી નીતિ________દ્વારા ઘડવામાં આવેલ હતી.
A. કૃષિ અને ખેડૂત કલ્યાણ મંત્રાલય
B. પર્યાવરણ, વન અને જલવાયુ પરિવર્તન મંત્રાલય
C. વિજ્ઞાન અને ટેકનોલોજી વિભાગ
D. આર્થિક બાબતોના વિભાગ

75. National Agroforestry Policy formulated by
A. Ministry of Agriculture & Farmers Welfare
B. Ministry of Environment, Forests and Climate Change
C. Department of Science and Technology
D. Department of Economic Affairs

76. માટીમાંથી છોડ અને જમીનમાં પોષક તત્વોના સાયકલિંગની પ્રક્રિયાને________કહેવામાં આવે છે.
A. ન્યુટ્રીઅન્ટ સાયક્લિંગ
B. ન્યુટ્રીઅન્ટ અપટેક
C. ન્યુટ્રીઅન્ટ લિન્ચ
D. ન્યુટ્રીલાઇઝેશન

76. The process of cycling of nutrients from soil to the plants and back to the soil is called as________.
A. Nutrient cycling
B. Nutrient uptake
C. Nutrient lynch
D. Nutrilization

77. કઇ ઝાડની પ્રજાતિને તેના પોષક મૂલ્યને કારણે શાળાઓમાં વાવેતર કરવા પ્રોત્સાહન આપવામાં આવે છે?
A. જામફળ B. દાડમ
C. લીંબુ D. સરગવો

77. Which of tree species is promoted to be planted in schools because of its nutritional value
A. Guava B. Anar
C. Lemon D. Drumstick

78. ભારત સરકાર દ્વારા ક્યાં વર્ષમાં રાષ્ટ્રીય એગ્રોફોરેસ્ટ્રી નીતિ શરૂ કરવામાં આવી?
A. 2014 B. 2016
C. 1992 D. 2010

78. National Agroforestry Policy has been launched by Govt. of India in
A. 2014 B. 2016
C. 1992 D. 2010

79. ગુજરાત માં કોના દ્વારા વિકેન્દ્રિત નર્સરી ઉભી કરવામાં આવી રહી છે ?
A. શાળાઓ B. ખેડુતો
C. સ્ત્રીઓ D. ઉપરોક્ત તમામ

79. In Gujarat, the decentralized nurseries are being raised in social forestry by
A. Schools B. Farmers
C. Women D. All of above

80. ભારતના ક્યાં રાજ્ય એ સામાજીક વનીકરણ પ્રથમ વખત સફળતાપૂર્વક અપનાવ્યું?
A. ગુજરાત
B. કેરળ
C. હિમાચલ પ્રદેશ
D. પશ્ચિમ બંગાળ

80. The concept of Social forestry was first successfully adopted in ______ State of India.
A. Gujarat
B. Kerala
C. Himachal Pradesh
D. West Bengal

81. "વન પંડિત એવોર્ડ " ગુજરાત વન વિભાગ દ્વારા કોને આપવામાં આવે છે?
A. જંગલો વિશે જાણકાર વ્યક્તિને
B. જંગલો વિશે ઉચ્ચ શિક્ષણ પ્રાપ્ત કરેલ વ્યક્તિને
C. ખાનગી અને સામાન્ય જમીન પર મોટા વાવેતરમાં ફાળો આપનાર વ્યક્તિને
D. વન સંરક્ષણ માં ફાળો આપનાર વ્યક્તિને

82. નીચેની ઝાડમાંથી કઈ પ્રજાતિઓનો ઉપયોગ એલે ક્રોપીંગ માટે સૌથી વધુ રીતે થાય છે?
A. લેગ્યુંમિનશ વૃક્ષોની પ્રજાતિઓ
B. નોન- લેગ્યુંમિનશ વૃક્ષોની પ્રજાતિઓ
C. A અને B બંને
D. બહુહેતુક પ્રજાતિઓ

83. અમદાવાદના કાલિદાસ ભાઈ પટેલ શેના માટે જાણીતા હતાં ?
A. દૂધ ઉત્પાદનમાં તેમનું યોગદાન
B. કોટન મિલોમાં તેમનું યોગદાન
C. નીલગિરીના વાવેતર વધારવામાં તેમનું યોગદાન
D. ગાંધીવાદી ફિલસૂફીના જાણીતા ઉપદેશક

84. તાજેતરમાં ગુજરાત રાજ્ય વન મહોત્સવ કાર્યક્રમ અંતર્ગત, કચ્છ ખાતેના સાંસ્કૃતિક વનનો વિકાસ કરવામાં આવ્યો છે, તેનું શું નામ છે ?
A. પુનિત વન B. જૈવવિવિધતા વન
C. રક્ષક વન D. શક્તિ વન

85. ગુજરાત વન વિભાગ દ્વારા 2015 માં એગ્રોફોરેસ્ટ્રીને પ્રોત્સાહન આપવા માટે સૌરાષ્ટ્ર ફેલિંગ એક્ટ હેઠળ વૃક્ષની કેટલી જાતોને પરવાનગી માંથી મુક્તિ કરવામાં આવી છે?
A. 67 B. 77
C. 87 D. 97

81. "Van Pandit Award" is being given by the Gujarat forest department to the person, who
A. having deep knowledge about the forests
B. having higher education about the forests
C. contributed in mass plantations on private and common lands
D. contributed in natural forest protection

82. Which of the following tree species is most commonly used for alley cropping.
A. Legumes tree species
B. Non-leguminous tree species
C. Both A and B
D. Multipurpose species

83. Ahmedabad's Kalidas Bhai Patel was known for?
A. His contribution in Milk production
B. His contribution in Cotton mills
C. His contribution in raising Eucalyptus spp plantations
D. Well known preacher on Gandhian philosophy

84. Recently under Gujarat State Van mahostava programme, the Cultural forest at Kutch has been developed, known as
A. Punit Van B. Biodiversity Van
C. Rakshak Van D. Shakti van

85. In 2015, how many tree species have been freed for permission under Saurastra Felling of Tree Act 1951 to promote the Agroforestry by the Gujarat forest department.
A. 67 B. 77
C. 87 D. 97

86. કેન્ડિડેટ પ્લસ વૃક્ષો શાને માટે કરવામાં આવે છે?

A. સારા લાકડા મેળવવા માટે
B. નર્સરી માટે બીજ એકત્રિત કરવા
C. ગમ એકત્રિત કરવા માટે
D. તે વૃક્ષનું જતન કરવા

87. "તમારા છોડને અવલોકન કરી ઉગાડી અને તેનું પાલન કરવું" એ શાનો મુખ્ય ઉદ્દેશ છે?

A. શાળા નર્સરી
B. કિસાન નર્સરી
C. મહિલા નર્સરી
D. વિભાગ નર્સરી

88. કોમ્પ્રિગ્નેટેડ વૂડ્સનો મુખ્ય ઉપયોગ શું છે ?

A. એરોપ્લેન પ્રોપેલર બ્લેડ
B. કોર બોર્ડ
C. બોર્ડ
D. રેલ્વે સ્લીપર્સ

89. ઉચ્ચ ગુણવત્તાના બીડી પાંદડા મેળવવા માટે નીચેની કઈ જાતિનો ઉપયોગ થાય છે?

A. *Bauhini vahlii*
B. *Diospyros melanxylon*
C. *Borasus flabellifer*
D. ઉપરોક્ત તમામ

90. કયો દેશ વિશ્વભરમાં કડાયા ગુંદરનું અગ્રણી ઉત્પાદક છે?

A. જાપાન B. ભારત
C. સુદાન D. ઓસ્ટ્રેલિયા

91. એક લોકપ્રિય ફૂલ રંગ જે *Butea monosperma* માંથી મેળવવામાં આવે છે તે ___________ તરીકે ઓળખાય છે.

A. ધક, પલાશ
B. આર્નોટો
C. કમેલા
D. રોહીડો

86. Candidate Plus trees are selected for

A. To get good timber
B. To collect seeds for nursery
C. To collect gum
D. To conserve that tree

87. "To observe, learn, grow and nurture your plant" is main objective of

A. School Nursery
B. Kisan Nursery
C. Mahila Nursery
D. Department Nursery

88. What is the main use of Compregnated woods?

A. Aeroplanes propeller blades
B. Core boards
C. Particle boards
D. Railway sleepers

89. Which of the following species is use to obtain premium bidi leaves?

A. *Bauhini vahlii*
B. *Diospyros melanxylon*
C. *Borasus flabellifer*
D. All of the above

90. Which Country is the leading producer of Karaya Gums worldwide?

A. Japan B. India
C. Sudan D. Australia

91. A popular flower dye which is obtained from *Butea monosperma* is known as

A. Dhak, Palash
B. Arnotto
C. Kamela
D. Rohido

92. કયા ઔષધીય ઉપયોગ માટે *Commiphora wightii* નું ગુંદર જાણીતું છે?

A. મુખ્યત્વે એલિવેટેડ રક્ત કોલેસ્ટરોલનું સ્તર ઘટાડવા માટે વપરાય છે.

B. ઘાને મટાડવા માટે

C. તાવને કાબૂમાં રાખવા

D. તીવ્ર ઠંડીમાં વપરાય છે

93. બાયોસ્ફિયરમાં સૌથી મોટો હેબીટાટ કયો છે?

A. ઇશ્ચ્યુરાઇન હેબીટાટ

B. ફ્રેશ વોટર હેબીટાટ

C. મરીન હેબીટાટ

D. ટેરેસ્ટ્રિયલ હેબીટાટ

94. ______________એ એક વિસ્તાર છે જે તેની દૈનિક પ્રવૃત્તિમાં વન્ય પ્રાણી દ્વારા આવરી લેવામાં આવે છે

A. ટેરીટરી

B. હોમ રેન્જ

C. કોર એરિયા

D. ઉપર્યુક્તમાંથી કોઈ નહિ

95. સોલ્ટ લીક બનાવવા માટે ______________ માટી સાથે ભેળવવામાં આવે છે.

A. HCL B. H_2SO_4

C. NaCl D. NH_4

96. NBWLના વાઇસ ચેરમેન કોણ હોય છે?

A. વન મહાનિદેશક

B. સચિવ, MoEFCC

C. મંત્રી, MoEFCC

D. રાજ્ય કક્ષાના મંત્રી, MoEFCC

97. નીચેના માંથી ક્યું દરિયાઇ સસ્તન પ્રાણી નથી?

A. બ્લુ વ્હેલ

B. ડોલ્ફિન

C. ડુગોંગ

D. વ્હેલ શાર્ક

92. For which medicinal value the gum of *Commiphora wightii* is known?

A. used primarily to reduce elevated blood cholesterol levels.

B. To heal up wounds

C. To control fever

D. Used in severe cold

93. Which is the largest habitat in biosphere?

A. Estuarine habitat

B. Freshwater habitat

C. Marine habitat

D. Terrestrial habitat

94. ______________ is an area covered by a wild animal in its day to day activity

A. Territory

B. Home range

C. Core area

D. None of above

95. For making artificial salt licks, ______________ is mixed with soil.

A. HCL B. H_2SO_4

C. NaCl D. NH_4

96. Who is the Vice Chairman of NBWL?

A. DGF

B. Secretory, MoEFCC

C. Minister, MoEFCC

D. State Minister, MoEFCC

97. Which of the following is not a marine mammal

A. Blue whale

B. Dolphin

C. Dugong

D. Whale shark

98. માર્ચ 2020 માં ગુજરાત સરકાર દ્વારા સાપ સંશોધન સંસ્થા ની સ્થાપના ક્યાં કરવામાં આવી ?

A. ગાંધીનગર B. રાજપીપળા
C. આહવા D. ધરમપુર

98. Govt of Gujarat established Snake Research Institute in March 2020 at

A. Gandhinagar B. Rajpipla
C. Ahwa D. Dharampur

99. આ નિવેદન કોના દ્વારા આપવામાં આવ્યું - "એશિયાટિક સિંહ ત્યાં ન હોત તો, ગીર રાષ્ટ્રીય ઉદ્યાન દેશ નુ સૌથી વધુ રસપ્રદ પક્ષી અભયારણ્યમાંનુ એક હોત"

A. જિમ કોર્બેટ
B. સાલીમ અલી
C. રિચાર્ડ ડોકિન
D. લવકુમાર ખાચર

99. Who dictated the statement – "had the Asiatic lion not been there, the area of Gir national park would have been one of the most fascinating bird sanctuaries of the country"

A. Jim Corbett
B. Salim Ali
C. Richard Dawkin
D. Lavkumar Khachar

100. ઇકો સેન્સેટિવ ઝોન માટે ઝોનલ માસ્ટર પ્લાનમાં _________ શામેલ હોવું જોઈએ

A. તમામ સંબંધિત રાજ્ય વિભાગોની ભાગીદારી
B. ડીન્યુડેડ વિસ્તારોની ફેરસુધારણા
C. જમીનના ઉપયોગમાં કોઈ ફેરફાર નહીં
D. ઉપરના તમામ

100. Zonal master Plan for Eco sensitive zone should include the

A. Involvement of all concerned State Departments
B. Restoration of denuded areas,
C. No change of land use from green uses
D. All above

Refer concern chapters for answers
જવાબો માટે સંબંધિત પ્રકરણો નો અભ્યાસ કરો

Practice Paper-2 | પ્રેક્ટિસ પેપર-૨

1. ____________ ભારતમાં બાયોડાયવર્સીટી હોટસ્પોટ તરીકે જાણીતું છે.
A. સુંદરવન ડેલટા
B. પશ્ચિમ ઘાટ
C. પૂર્વી ઘાટ
D. ગેન્ગેટીક પ્લેન

2. ઇકોસિસ્ટમના ઓટોટ્રોપ એટલે
A. પ્રોડ્યુસર B. ટ્રાંસડ્યુસર
C. કન્વર્ટર D. ઉપરોક્ત તમામ

3. ઇકોસિસ્ટમમાં ચોક્કસ સમયે ચોક્કસ ક્ષેત્રમાં જીવંત કાર્બનિક પદાર્થોની કુલ માત્રાને શું કહે છે?
A. સ્ટેન્ડિંગ ફોર્મ B. સ્થાયી સ્ટેટ
C. સ્ટેન્ડિંગ ક્રોપ D. સ્ટેન્ડિંગ મેટર

4. પ્રાથમિક ઉત્પાદકતાને કેવી રીતે વિભાજિત કરી શકાય છે?
A. GPP B. NPP
C. TPP D. A અને B બન્ને

5. બાયોસ્ફીયર બીજા ક્યાં નામે ઓળખાય છે ?
A. હાઇડ્રોસ્ફિયર B. એટમોસ્ફિઅર
C. ઇકોસ્ફિયર D. B અને C બન્ને

6. છાયાવાળા નિવાસસ્થાન ને પસંદ કરનાર વનસ્પતિને શું કહે છે ?
A. મેસોફાઇટ્સ B. સાયોફાઇટ્સ
C. એપિફાઇટ્સ D. હેલિઓફાઇટ્સ

7. 'K' સિલેક્ટેડ પ્રજાતિ કઈ છે?
A. *એસ્પરગિલસ* B. ઘાસ
C. માનવ D. વરુ

8. સૌથી નાનું કૃત્રિમ ઇકોસિસ્ટમ પસંદ કરો કે જે લાંબા સમય સુધી ટકી રહ્યું છે.
A. ફોલ્સમ બોટલ
B. ફોલ્સમ તળાવ
C. ફોલ્સમ સરોવર
D. B અને C બન્ને

1. **____________ are known as hotspot of biodiversity in India.**
A. Sunderbans Deltas
B. Western Ghats
C. Eastern Ghats
D. Gangetic Plain

2. **Autotrophs of an ecosystem are called**
A. Producer B. Transducer
C. Converter D. All of the above

3. **Total amount of living organic matter present in particular area at particular time in an ecosystem is**
A. Standing form B. Standing state
C. Standing crop D. Standing matter

4. **Primary productivity can be divided into**
A. GPP B. NPP
C. TPP D. Both A and B

5. **Biosphere is also known as**
A. Hydrosphere B. Atmosphere
C. Ecosphere D. Both B and C

6. **Plants that prefer Shady habitats are known as**
A. Mesophytes B. Sciophytes
C. Epiphytes D. Heliophytes

7. **Which one is a 'K' selected species?**
A. *Aspergillus* B. Grass
C. Human D. Wolf

8. **Select the smallest artificial ecosystem that has sustained for a long period**
A. Folsom bottle
B. Folsom pond
C. Folsom lake
D. Both B and C

9. બાયોમની લાક્ષણિકતા શું છે ?
A. વનસ્પતિ અને પ્રાણી જીવન
B. આબોહવા
C. માટીના પ્રકારો
D. ઉપરોક્ત તમામ

10. જૈવવિવિધતા હેરિટેજ સાઇટ અમિનપુર તળાવને જૈવવિવિધતા અધિનિયમ હેઠળ દેશમાં પ્રથમ બાયોડિવર્સીટી હેરિટેજ સાઇટનો દરજ્જો ક્યારે મળ્યો?
A. 2002 B. 2004
C. 2006 D. 2008

11. તાજેતરમાં પ્રકાશિત વર્લ્ડ વાઇલ્ડલાઇફ ક્રાઇમ રિપોર્ટ 2020 મુજબ વિશ્વનું સૌથી વધુ શિકાર થતું સસ્તન પ્રાણી કયું છે?
A. વાઘ B. પેંગોલિન
C. ચિત્તો D. હાથી

12. બેંથોનિક સજીવો ________ છે .
A. સમુદ્રના સેડેન્ટરી સજીવો
B. સમુદ્ર સપાટી પરના સજીવ
C. જમીનના સેડેન્ટરી સજીવ
D. જમીનની સપાટી પરના સજીવ

13. રામસાર કન્વેન્શન શાને લાગતું છે ?
A. પર્વતો
B. નદીઓ
C. જલપ્લાવિત વિસ્તારો
D. રણ

14. દરેક જીવતંત્રમાં પર્યાવરણીય પરિબળોના સંદર્ભમાં સહનશીલતાની ન્યૂનતમ અને મહત્તમ મર્યાદા હોય છે અને તે સાથે કેન્દ્રીય મહત્તમ શ્રેણી જેમાં સજીવો વિપુલ પ્રમાણમાં જોવા મળે છે; આ નિયમ ________ના નિયમ તરીકે ઓળખાય છે.
A. ડાર્વિનનો નિયમ
B. શેલ્ફોર્ડનો નિયમ
C. એલનનો નિયમ
D. અલીનો નિયમ

9. Biomes are characterized by
A. Vegetation and animal life
B. Climate
C. General soil type
D. All of the above

10. First Biodiversity Heritage Site Ameenpur Lake gets the status of the first Biodiversity Heritage Site in the country under the biodiversity act
A. 2002 B. 2004
C. 2006 D. 2008

11. The recently released World Wildlife Crime Report 2020 revealed that the world's most trafficked mammal is
A. Tiger B. Pangolin
B. Leopard D. Elephant

12. Benthonic organisms are
A. Sedentary organisms of sea
B. Surface organisms of sea
C. Sedentary organisms of land
D. Surface organisms of land

13. Ramsar Convention refers to the conservation of
A. Mountains B. Rivers
C. Wetlands D. Deserts

14. Every organism has minimum and maximum limit of tolerance with respect to environmental factors along with central optimum range in which organisms are found abundant; this law is known as
A. Darwin law
B. Shelford law
C. Allen law
D. Ali law

15. __________ નો ઉપયોગ જીવસૃષ્ટિની ગુણવત્તાની તુલના કરવા માટે થાય છે.
A. વિઘટનકર્તા
B. શિકારી
C. બાયોઇન્ડિકેટર
D. જૈવ-ઉપચારક

16. યોગ્ય રીતે મેળ ખાતી જોડીને ઓળખો
A. બેઝલ કન્વેનશન - જૈવવિવિધતા સંરક્ષણ
B. મોન્ટ્રીયલ પ્રોટોકોલ - ગ્લોબલ વોર્મિંગ
C. ક્યોટો પ્રોટોકોલ - ક્લાઈમેટ ચેન્જ
D. રામસર કન્વેનશન - ભૂગર્ભ જળ પ્રદૂષણ

17. પાણી ઇકોસિસ્ટમમાં BODના સૂચક __________ છે.
A. ટ્રાઉટ B. ડાફનીયા
C. ગાંબુસિયા D. લાઇકેન

18. એબીસોપેલેજિક નો અર્થ થાય કે
A. 1000 મી થી વધુ ઊંડાણોમાં રહેતા પાણીના સજીવો
B. 2000 મી થી વધુ ઊંડાણોમાં રહેતા પાણીના સજીવો
C. 3000 મી થી વધુ ઊંડાણોમાં રહેતા પાણીના સજીવો
D. 4000 મી થી વધુ ઊંડાણોમાં પર રહેતા પાણીના સજીવો

19. __________ એ ઇકોલોજીકલ નિશ શબ્દ આપ્યો.
A. ગ્રિનલ B. મોહિત હુસેન
C. એલન D. પી. ઓડમ

20. સ્પોડોસોલ્સ શું છે ?
A. એસિડિક અને ઈન્ફર્ટાઈલ
B. બેઈઝીક અને ઈન્ફર્ટાઈલ
C. એસિડિક અને ફર્ટાઈલ
D. બેઈઝીક અને ફર્ટાઈલ

15. __________ is an organism used to gauge the quality of an ecosystem.
A. Decomposers
B. Predator
C. Bioindicator
D. Bio-remediator

16. Identify the correctly matched pair
A. Basal Convention - Biodiversity Conservation
B. Montreal Protocol - Global warming
C. Kyoto protocol - Climatic change
D. Ramsar Convention - Ground water pollution

17. What is the indicator of BOD in a water ecosystem
A. Traught B. Daphnia
C. Gambusia D. Lichen

18. Abyssopelagic include
A. Organisms living at water depths greater than 1000 m
B. Organisms living at water depths greater than 2000 m
C. Organisms living at water depths greater than 3000m
D. Organisms living at water depths greater than 4000 m

19. __________ gave the Ecological Niche term
A. Grinnel B. Mohit Husain
C. Allen D. P. Odum

20. Spodosols are
A. Acidic and infertile
B. Basic and infertile
C. Acidic and fertile
D. Basic and fertile

21. જંગલની જમીનમાં ન્યુટ્રિયન્ટ સાયક્લિંગમાં ખલેલ અને ત્યારબાદના ફેરફારોના વાતાવરણીય સ્તરો પર સંભવિતપણે મોટી અસર થઈ શકે છે.
A. N_2 B. CO_2
C. CH_4 D. SO_2

22. ખીણોમાં જંગલની જમીન કેવી હોય છે
A. લોમી-સિલ્ટી
B. બરછટ-દાણાદાર
C. લોમી-દાણાદાર
D. બરછટ-સિલ્ટી

23. સ્પેરોડલ જમીનની રચના કેવી હોય છે?
A. ગ્રેન્યુલેર અને ક્રમ્બ
B. સબ એન્ગ્યુલર બ્લોકી
C. પ્રેઇઝમેટિક અને કોલુમનાર
D. એન્ગ્યુલર અને બ્લોકી

24. જમીનને પકડી શકે તેવા વિનિમયક્ષમ કેટયન્સ ને શું કહેવાય છે?
A. ECEC B. EVEV
C. ECEV D. EVEV

25. __________ બેક્ટેરિયાની હાજરીને લીધે, જંગલની જમીન વાતાવરણના મીથેનનું સિંક કરે છે.
A. થર્મોફિલિક B. મેથેનોટ્રોફિક
C. સાયકોટ્રોફિક D. મેસોફિલિક

26. રૂટ રેસ્પિરેશનથી જંગલની જમીનનું તાપમાન
A. વધે છે B. ઘટાડો થાય છે
C. અસર કરતું નથી D. નિયમન કરે છે

27. સેસ્કવીયોક્ષાઇટ માંથી બહાર નીકળવાની પ્રક્રિયા કયા નામે ઓળખવામાં આવે છે?
A. લેટરલાઈઝેશન
B. ઇલ્યુએશન
C. ઇલ્યુએશન
D. પોડઝોલાઇઝેશન

21. Disturbance and subsequent changes in nutrient cycling in forest soils can potentially have large impacts on atmospheric levels of
A. N_2 B. CO_2
C. CH_4 D. SO_2

22. In valleys the forest soils are
A. Loamy-silty
B. Coarse-grained
C. Loamy-grained
D. Coarse-silty

23. Spheroidal soil structures are
A. granular and crumb
B. sub-angular blocky
C. prismatic and columnar
D. angular blocky

24. The total amount of exchangeable cations that a soil can hold is referred to as
A. ECEC B. EVEC
C. ECEV D. EVEV

25. Due to the presence of __________ bacteria, forest soils represent a sink of atmospheric methane.
A. Thermophilic B. Methanotrophic
C. Psychrotrophic D. Mesophilic

26. Root respiration __________ the temperature of forest soils
A. Increases B. Decreases
C. Does not affect D. Regulates

27. Process of moving out of Sesquioxide is known as
A. Lateralization
B. Eluviation
C. Illuviation
D. Podzolization

28. એગ્રિલિએશન એક પ્રક્રિયા છે જેમાં વિખરાયેલા માટીના કણો ક્યાંથી સ્થાનાંતરિત થાય છે?

A. નીચલાથી ઉપરના હોરાઈઝનમાં
B. મધ્યથી ઉપરની હોરાઈઝનમાં
C. ઉપરથી નીચલા હોરાઈઝનમાં
D. ઉપરથી મધ્ય હોરાઈઝનમાં

29. ગ્રેવીમેટ્રિક પાણી ક્યાં હોય છે ?

A. મેસો પોરમાં
B. મેક્રો પોરમાં
C. માઈક્રો પોરમાં
D. B અને C બન્ને

30. સમાન કદના કેટાયન બદલવાની પ્રક્રિયા કયા નામથી ઓળખવામાં આવે છે?

A. આઇસોમોર્ફસ સબસ્ટિટ્યુશન
B. આઇસોમોર્ફિક રિપ્લેસમેન્ટ
C. કેટાયનિક સબસ્ટિટ્યુશન
D. કેટાયનિક રિપ્લેસમેન્ટ

31. ક્યાં બે ઓર્ડર માટીના 12 મોટા ઓર્ડરમાંથી હોરાઈઝનની ઉત્ક્રાંતિ દ્વારા વર્ગીકૃત થયેલ છે?

A. મોલિસોલ્સ અને એરિડીસોલ્સ
B. અલ્ફિસોલ્સ અને મોલિસોલ્સ
C. સ્પોડોસોલ્સ અને અલ્ફિસોલ્સ
D. એન્ટિસોલ્સ અને ઇનસેપ્ટિસોલ

32. ________ એ ધુમ્મસ શબ્દ આપ્યો હતો?

A. નિકોલસ કોપરનીકસ
B. જો રોમ
C. હેનરી એન્ટોન
D. જેમ્સ હેન્સન

33. એસિડ વરસાદ કયા વાયુ પ્રદૂષકને કારણે થાય છે?

A. CO_2 B. CO
C. મિથેન D. NO અને SO_2

28. Agrilliation is a process in which the dispersed clay particles are migrated from

A. Lower to upper horizon
B. Middle to upper horizon
C. Upper to lower horizon
D. Upper to middle horizon

29. Gravimetric water is present in

A. Meso pores
B. Macro pores
C. Micro pores
D. Both B and C

30. The process of replacing one structural cation for another of similar size is referred to as

A. isomorphous substitution
B. isomorphic replacement
C. Cationic substitution
D. Cationic replacement

31. Of the 12 major soil orders, two are characterized by minor development of horizons, those are Mollisols and Aridisols

A. Alfisols and Mollisols
B. Spodosols and Alfisols
C. Entisols and Inceptisols

32. Who had coined the term SMOG

A. Nicolaus Copernicus
B. Joe Romm
C. Henry Antoine
D. James Hanson

33. Acid Rain is due to the air pollution of

A. CO_2 B. CO
C. Methane D. NO and SO_2

34. સૌથી વધુ મિથેન ઉત્પન્ન કરનારા પ્રાણીઓ કયા છે?

A. મરઘાં B. રુમીનન્ટ્સ
C. સમુદ્રના પ્રાણીઓ D. સરિસૃપ

35. કેટલાક સાયનોબેક્ટેરિયા, જળ સંસ્થાઓમાં ન્યુરોટોક્સિનનું ઉત્પાદન કરે છે, વર્ષ 2020 માં જેનાથી ક્યાં 300 થી વધુ હાથીઓને માર્યા ગયા?

A. બોત્સ્વાના B. ભારત
C. નમિબીઆ D. ઝિમ્બાબ્વે

36. ઓઝોન શેના દ્વારા રચાય છે?

A. ઓક્સિજન સાથે યુવી કિરણોત્સર્ગની ક્રિયાપ્રતિક્રિયા
B. ઓક્સિજન પર ડેલાઇટની ક્રિયા
C. ઓક્સિજન પરમાણુ અને મુક્ત ઓક્સિજન અણુના સંયોજન દ્વારા
D. ઉપરના બધા

37. WCEDનો શો અર્થ થાય છે?

A. ઊર્જા અને વિકાસ પર વિશ્વ પંચ
B. પર્યાવરણ અને વિકાસ પર વિશ્વ પંચ
C. વિશ્વ સ્વચ્છ પર્યાવરણ દિવસ
D. વિશ્વ સ્વચ્છ ઊર્જા દિવસ

38. માછલી કઈ ભારે ધાતુ ને વધુ માત્રામાં સંગ્રહ કરી શકે છે?

A. બિસ્મથ B. પારો
C. પેલેડિયમ D. ક્લોરિન

39. હાર્મફુલ એલ્ગલ બ્લૂમ્સ ________ સાથે પ્રતિક્રિયા કરે છે જે માછલીઓ અને માનવો માટે જોખમી છે.

A. પાણીમાં ઓક્સિજન
B. પાણીમાં હાનિકારક રસાયણો
C. પાણીમાં ઝેર
D. પાણીની ક્ષારતા

34. Highest methane producing animals are

A. Poultry B. Ruminants
C. Sea animals D. Reptiles

35. Some Cyanobacteria produce neurotoxins in water bodies, In the year 2020 that killed more than 300 elephants in

A. Botswana B. India
C. Namibia D. Zimbabwe

36. Ozone is formed by

A. interaction of UV radiation with oxygen
B. action of daylight on oxygen
C. By combination of oxygen molecule and free oxygen atom
D. All of the above

37. WCED stands for

A. World Commission on Energy and Development
B. World Commission on Environment and Development
C. World Clean Environment Day
D. World Clean Energy Day

38. Which heavy metal fish can stored in more quantity?

A. Bismuth B. Mercury
C. Palladium D. Chlorine

39. What does Harmful Algal Blooms (HABs) reacts to which is dangerous to fish and humans?

A. Deplete oxygen in water
B. Add harmful chemicals to water
C. Add toxicity to water
D. Reduce alkalinity of water

40. કયા બેક્ટેરિયમનો ઉપયોગ અણુ કચરામાંથી ટોલ્યુએન અને આયનીક પારાને ડિટોક્સિફાઇ કરવા માટે થાય છે?

A. ડીનોકોકસુરાદોડ્યુરાન્સ

B. આઇડિઓનેલાસ્કેઇનેન્સીસ

C. *મોરેક્સેલા* સ્પી.

D. ઉપર્યુક્ત તમામ

40. Which bacterium is used to detoxify toluene and ionic mercury from nuclear waste?

A. Deinococcusradiodurans

B. Ideonellasakainensis

C. *Moraxella* sp.

D. All of the above

41. ભારતમાં હાલમાં કેટલા CAAQMS સ્ટેશનો કાર્યરત છે?

A. 85 B. 110

C. 90 D. 120

41. How many CAAQMS stations are currently operating in India?

A. 85 B. 110

C. 90 D. 120

42. BOD અને COD એ શેના માપદંડ છે?

A. જળચર ઇકોસિસ્ટમ્સમાં પ્રદૂષણનું પૃથક્કરણ

B. લોહીમાં ઓક્સિજનનું સ્તર માપવા

C. વન ઇકોસિસ્ટમ્સમાં ઓક્સિજન સ્તરની ગણતરી

D. ઉંચાઇવાળા પ્રદેશોમાં ઓક્સિજનના સ્તરોનું મૂલ્યાંકન

42. BOD and COD are standard criterion for

A. Pollution assay in aquatic ecosystems

B. Measuring oxygen levels in blood

C. Computing oxygen levels in forest ecosystems

D. Assessing oxygen levels in high altitude regions

43. ઝાડમાંથી મળતો ઘાસચારો સામાન્ય રીતે ____________ માનવામાં આવે છે.

A. વરસાદી ઘાસચારો

B. શ્રેષ્ઠ ચારો

C. ઉનાળું પાવડર

D. અછતનો ઘાસચારો

43. Fodder from tree is usually considered as

A. Rainy fodder

B. Best fodder

C. Summer powder

D. Scarcity fodder

44. નીચેનામાંથી કયી પ્રજાતિને સાલના સૂચક છોડ તરીકે ગણવામાં આવે છે ?

A. *Clerodendron viscosum*

B. *Flamengia* sp

C. *Mogharia chappar*

D. ઉપર્યુક્ત તમામ

44. Which of the following is considered as Indicator plant of Sal?

A. *Clerodendron viscosum*

B. *Flamengia* sp

C. *Mogharia chappar*

D. All of the above

45. સેલ્ફ થિનિંગ નો સિદ્ધાંત કોણે આપ્યો હતો?

A. યોદા B. ક્લેમેન્ટ

C. ડાર્વિન D. ડોસન

45. The law of self-thinning was given by?

A. Yoda B. Clement

C. Darwin D. Dowson

46. ન્યુમેટોફોર્સ સામાન્ય રીતે કઈ પ્રજાતિ માં જોવા મળે છે?
A. સુકા પાનખર જંગલ
B. મેંગ્રોવ વન
C. ખુલ્લા વન
D. સબ-ઉષ્ણકટિબંધીય વન

47. સાઈટ ઈન્ડેક્સ શબ્દ એ શેના સંબંધને સમજાવવા માટે વપરાય છે?
A. ઊંચાઈ થી ઉંમર
B. ડાયામીટર થી ઉંમર
C. ડાયામીટર થી ઊંચાઈ
D. વોલ્યુમ થી ઉંમર

48. પ્રથમ વખત "ફોરેસ્ટ સકસેશન" શબ્દનો ઉપયોગ કોણે કર્યો?
A. ક્લેમેન્ટ
B. ડોસન
C. થોરો
D. ઉપર્યુક્તમાંથી એક પણ નહિ

49. નીચેનીમાંથી કઇ પ્રજાતિ ડ્રોટ સેન્સિટિવ નથી?
A. Mangifera indica
B. Pterocarpus marsupium
C. Hardwikia binate
D. Madhuca indica

50. લાઈન પ્લાન્ટિંગમાં છોડની સંખ્યા (n) ગણવામાટેનું સૂત્ર કયું છે?
A. N = 100 * 100 / X* Y
B. $N = 100 * 100 / (X)^2$
C. $N = 100 * 100 * 1.155 / (X)^2$
D. $N = 2 * 100 * 100 / (X)^2$

51. વુડલોટનો બેઝલ એરિયા શું દર્શાવે છે?
A. ઝાડનો વિસ્તાર ચોરસ ફૂટમાં
B. ક્રાઉનની ડેન્સીટી
C. સ્ટેન્ડ ડેન્સીટી
D. ટ્રંકનો ટેપર

46. Pneumatophores are commonly seen in
A. Dry deciduous forest
B. Mangrove forest
C. Open forest
D. Sub-tropical forest

47. Site index is a term used by foresters to explain the relationship of ——— in forest stands.
A. Height to Age
B. Diameter to Age
C. Diameter to Height
D. Volume to Age

48. By whom "Forest Succession" term is used first time?
A. Clement
B. Dawson
C. Thoreau
D. None of the above

49. Which of the following is not a Drought sensitive species.
A. *Acacia nilotica*
B. *Boswellia serrata*
C. *Hardwickia binata*
D. *Madhuca indica*

50. What is the formula to calculate number of plants (N) for Line planting pattern.
A. N = 100*100/ X*Y
B. $N = 100*100 / (X)^2$
C. $N = 100*100*1.155/ (X)^2$
D. $N = 2*100*100/ (X)^2$

51. Basal area of a woodlot represents
A. Area occupied by the tree in square feet
B. Crown density
C. Stand density
D. Taper of trunk

52. સ્થળાંતર ખેતી ______ તરીકે પણ ઓળખાય છે.

A. ઝુમ કલ્ટીવેશન

B. સ્લેશ અને બર્ન કલ્ટીવેશન

C. બંને A અને B

D. ટફ કલ્ટીવેશન

53. સેન્ટર ફોર સોશ્યલ ફોરેસ્ટ્રી એન્ડ ઇકો-રિહેબિલીટીશન ભારત માં ક્યાં સ્થિત છે ?

A. દહેરદૂન

B. ગાંધીનગર

C. નવી દિલ્હી

D. અલ્હાબાદ

54. ગુજરાતમાં, ઝાડનું આવરણ લગભગ 11 % ની મર્યાદામાં છે, જેમાં તેના ભૌગોલિક ક્ષેત્રના જંગલ (TOF) ની બહાર કેટલું વૃક્ષ આવરણ છે?

A. 1 % B. 3.5%

C. 8 % D. 10 %

55. વન પંડિત એવોર્ડ " ગુજરાત વન વિભાગ દ્વારા કોને આપવામાં આવે છે?

A. જંગલો વિશે જાણકાર વ્યક્તિને

B. જંગલો વિશે ઉચ્ચ શિક્ષણ પ્રાપ્ત કરેલ વ્યક્તિને

C. ખાનગી અને સામાન્ય જમીન પર મોટા વાવેતરમાં ફાળો આપનાર વ્યક્તિને

D. વન સંરક્ષણ માં ફાળો આપનાર વ્યક્તિને

56. ISFR 2019 મુજબ, કયા વૃક્ષની પ્રજાતિઓ ગુજરાતમાં જંગલની બહાર સૌથી વધુ સંખ્યા ધરાવે છે?

A. દેશી બાવળ

B. આંબો

C. લીમડો

D. નીલગિરિ

52. Shifting cultivation is also known as

A. Jhuming cultivation

B. Slash and burn cultivation

C. Both

D. Tough cultivation

53. Centre for Social Forestry and Eco-Rehabilitation is located at ______ in India.

A. Dehra Dun

B. Gandhinagar

C. New Delhi

D. Allahabad

54. In Gujarat, the tree cover is in the extent of around 11 %, in which how much tree cover outside forest (TOF) of its geographical area?

A. 1 % B. 3.5%

C. 8% D. 10 %

55. "Van Pandit Award" is being given by the Gujarat forest department to the person, who

E. having deep knowledge about the forests

F. having higher education about the forests

G. contributed in mass plantations on private and common lands

H. contributed in natural forest protection

56. As per last ISFR 2019, the dominant tree species is having the highest number outside the forest in Gujarat, is;

A. Desi Baval

B. Mango

C. Neem

D. Nilgiri

57. કયા સમુદાયો કુશળ વૃક્ષ કાપનારા છે?
A. આફ્રિકાના માસાઈ
B. ઓરિસ્સાના ગોન્ડ
C. જાવા ના કલન્ગ
D. છોટા નાગપુર ના મુંડા

58. “ઇંધણના લાકડાની ઘનતા” નું પ્રમાણ કઈ પદ્ધતિ દ્વારા નક્કી કરી શકાય છે?
A. સ્પેસિફિક ગ્રેવિટી પદ્ધતિ
B. ઝાયલોમેટ્રિક પદ્ધતિ
C. બંને A અને B
D. સ્ટેન્ડ પદ્ધતિ

59. કયા પ્રકારનાં લાકડાને રેપીડ સીઝનિંગની પ્રક્રિયા સામે ઓછા રક્ષણની જરૂર હોય છે?
A. પરવર્ત્ય લાકડું
B. સાધારણ પરવર્ત્ય લાકડું
C. બિન પરવર્ત્ય લાકડું
D. અત્યંત પરવર્ત્ય લાકડું

60. ઝાડની ડાળીના વ્યાસને સચોટ રીતે માપવા માટે કયા સાધનનો ઉપયોગ કરવામાં આવે છે?
A. વ્હીલર પેન્ટા પ્રિઝમ
B. સ્પીગેલ રેલાસ્કોપ
C. ડિજિટલ કેલિપર
D. ક્લીનોમીટર

61. ટેનિન અને કેટેચિન બાવળના લાકડાના કયા ભાગમાંથી કાઢવામાં આવે છે?
A. નવી ડાળીઓ B. છાલ
C. સેપવૂડ D. હાર્ટવૂડ

62. કોઇર પિથ મુખ્યત્વે કઈ વનસ્પતિમાંથી મેળવવામાં આવે છે?
A. *Bombax ceiba*
B. *Cocos nucifera*
C. *Ceiba patendra*
D. *Cochlospermum religiosum*

57. Which communities are of skilled forest cutters?
A. Maasais of Africa
B. Gonds of Orissa
C. Kalangs of Java
D. Mundas of Chotanagpur

58. The volume of solid firewood can determine by which method?
A. Specific gravity method
B. Xylometric method
C. Both A & B
D. Stand method

59. Which type of wood require little protection against rapid seasoning?
A. Refractory wood
B. Moderately refractory wood
C. Non-refractory wood
D. High refractory wood

60. Which instrument is used to measure stem diameter accurately?
A. Wheeler Penta Prism
B. Spiegel Relaskop
C. Digital calliper
D. Clinometer

61. Tannis and catechin are abstracted from which wood part of Acacia?
A. Early wood B. Bark
C. Sapwood D. Heartwood

62. Coir pith are mainly obtained from which plant species?
A. *Bombax ceiba*
B. *Cocos nucifera*
C. *Ceiba patendra*
D. *Cochlospermum religiosum*

63. ઝાડની કઈ પ્રજાતિમાંથી "ટ્રી ડામર" પ્રાપ્ત થાય છે?
A. *અગાથીસ લોરેન્થિફોલિયા*
B. *હોપ ઓડોરેટા*
C. *શોરિયા રોબસ્ટા*
D. *કેનેરિયમ સ્ટ્રિક્ટમ*

64. ઝાડ ને લાકડાના લોગમાં કાપવાને ______ કહેવાય છે?
A. બકિંગ B. સોવિંગ
C. ફેલિન્ગ D. ઉપરોક્ત તમામ

65. નીચેનામાંથી કઈ પ્રજાતિના મૂળમાંથી, મૂલ્યવાન લાલ રંગ મેળવવામાં આવે છે?
A. *Rubia corodifolia*
B. *Lawsonia inermis*
C. *Mallotus philippensis*
D. *Nyctanthes arbotristis*

66. બેટ્સ ગુઆનો એ શેનો એક સમૃદ્ધ સોત છે?
A. નાઇટ્રેટ B. ફોસ્ફેટ
C. ફેરસ D. બંને A અને B

67. નીચેનીમાંથી કઈ પ્રજાતિનો ઉપયોગ "ડીવી-ડીવી" ટેન મેળવવા માટે થાય છે?
A. *કેસિઆ ઓરિક્યુલાટા*
B. *સીઝાલ્પિનિયા કોરીરિયા*
C. *સેરીઓપ્સ રોક્સબર્ગિઆના*
D. *અલેઉરાઇટ્સ ફ્રોડીઆઈ*

68. કિતુલ ફાઇબર કઈ પ્રજાતિમાંથી મેળવવામાં આવે છે?
A. *એગાવ*
B. *કેરીઓટા યુરેનસ*
C. *કોર્ચરસ કેપ્સ્યુલરિસ*
D. ઉપર્યુક્તમાંથી કોઈ નહિ

69. નીચેનામાંથી કયા વિશ્વના શ્રેષ્ઠ લાકડા માનવામાં આવે છે?
A. બાલસા B. સાગ
C. સાલ D. દેવદાર

63. From which of the fooling species a true dammar is obtained?
A. *Agathis loranthifolia*
B. *Hopea odorata*
C. *Shorea robusta*
D. *Canarium strictum*

64. The cutting of trees into wooden logs is called as?
A. Bucking B. Sawing
C. Felling D. All of the above

65. From which of the following species roots, the valuable red dye is obtained?
A. *Rubia cordifolia*
B. *Lawsonia inermis*
C. *Mallotus philippensis*
D. *Nyctanthes arbotristis*

66. Bats guano is a rich source of 86
A. Nitrate B. Phosphate
C. Ferrous D. Both A & B

67. Which of the following species is used to obtain "Divi-Divi" tans?
A. *Cassia auriculata*
B. *Caesalpinia coriaria*
C. *Ceriops roxburghiana*
D. *Aleurites fordii*

68. Kittul fibre is obtained from which species?
A. *Agave* spp.
B. *Caryota urenus*
C. *Corchorus capsularis*
D. None of the above

69. Which of the following is considered as the world best timber?
A. Balsa B. Teak
C. Sal D. Deodar

70. ઇમારતી લાકડાનું ગ્રેડિંગ કયા સૂત્ર પર આધારિત છે?
A. ક્વાર્ટર ગર્થ સૂત્ર
B. પ્રિઝમોઇડલ સૂત્ર
C. ન્યુટ્નનું સૂત્ર
D. ક્વાર્ટર સૂત્ર

71. વૂડ્સને સૂકવવા માટે નીચેનામાંથી કયુ મહત્વપૂર્ણ છે?
A. FSP (Fibre Saturation Point)
B. AWP (Absorbed water point)
C. FWP (Free water point)
D. EMP (Equilibrium moisture point)

72. નીચેનામાંથી કઈ પ્રજાતિના બીજમાં ઝેરી "અબ્રિન" છે?
A. *Ocimum killimandscharicum*
B. *Abrus precatorius*
C. *Myrica esculanta*
D. *Ougeina oojeinensis*

73. લાકડાના વૈજ્ઞાનિક અભ્યાસને શું તરીકે કહેવામાં આવે છે
A. લાકડ રચના B. ઝાયલોટોમી
C. ડેંડ્રોલોજી D. હિસ્ટોલોજી

74. કોઈ વિશિષ્ટ સંસ્કૃતિ અને પ્રદેશના લોકો દેશી છોડનો ઉપયોગ કેવી રીતે કરે છે તેનો અભ્યાસ __________ ઓળખાય છે.
A. વર્ગીકરણ
B. પેથોલોજી
C. પ્લાન્ટ ઇકોલોજી
D. ઈથનોબોટની

75. ભારતમાં કેન અને વાંસ ટેકનોલોજી સેન્ટર ક્યાં આવેલું છે?
A. ગુવાહાટી, આસામ
B. રુદ્રપુર, ઉત્તરાખંડ
C. ઝાંસી, ઉત્તર પ્રદેશ
D. જબલપુર, મધ્યપ્રદેશ

70. Grading of the timber is based on which formula?
A. Quarter girth formula
B. Prismoidal formula
C. Newton's formula
D. Quarter formula

71. Which of the following is critical for drying of woods?
A. FSP (Fibre Saturation Point)
B. AWP (Absorbed water point)
C. FWP (Free water point)
D. EMP (Equilibrium moisture point)

72. Which of the following species contains poisonous "Abrin" in their seeds?
A. *Ocimum killimandscharicum*
B. *Abrus precatorius*
C. *Myrica esculanta*
D. *Ougeina oojeinensis*

73. Scientific study of wood is called as
A. Wood anatomy B. Xylotomy
C. Dendrology D. Histology

74. The study of how people of a particular culture and region make of use of indigenous plants is known as.
A. Taxonomy
B. Pathology
C. Plant Ecology
D. Ethnobotany

75. Where is the Cane and Bamboo Technology Centre (CBTC) is located in India?
A. Guwahati, Assam
B. Rudrapur, Uttarakhand
C. Jhansi, Uttar Pradesh
D. Jabalpur, Madhya Pradesh

76. કયો દેશ *એકેશિયાના* ગુંદર ના વિશ્વભરમાં અગ્રણી ઉત્પાદક છે?
A. જાપાન B. ભારત
C. સુદાન D. ઓસ્ટ્રેલિયા

77. અન્નાટો (પીળો રંગ) ________ ના બીજમાંથી મેળવવામાં આવે છે?
A. *Terminalia chebula*
B. *Bixa Orellana*
C. *Terminalia arjuna*
D. *Berberis aristate*

78. નીચેનામાંથી કયું ફળ ટેન આપનાર છોડ માનવામાં આવે છે?
A. *ટર્મિનલિયા ચેબ્યુલા*
B. *ટર્મિનલિયા બેલિરિકા*
C. *એમ્બ્લિકા ઓફિસિનાલિસ*
D. ઉપરોક્ત તમામ

79. આંતરરાષ્ટ્રીય ટ્રોપિકલ ટીમ્બર ઓર્ગેનાઇઝેશન ક્યાં સ્થિત છે?
A. કાનાગાવા, જાપાન
B. જિનીવા, સ્વિટ્ઝર્લેન્ડ
C. ન્યૂ યોર્ક, યુનાઇટેડ સ્ટેટ્સ
D. ગ્લેન્ડ, સ્વિટ્ઝર્લેન્ડ

80. ખેર હાર્ટવુડના મુખ્ય ઘટકો ________ (કાથા) છે.
A. કેટેચિન
B. એઝાડીરેક્ટીન
C. ક્વેર્સિટિન
D. ઉપરોક્ત તમામ

81. કયા લાકડાનો કોલસો વધુ કેલરીફિક મૂલ્ય ધરાવે છે?
A. *Prosopis juliflora*
B. *Bombex ceiba*
C. *Ailenthus excela*
D. *Butea monosperma*

76. Which Country is the leading producer of *Acacia* Gums worldwide?
A. Japan B. India
C. Sudan D. Australia

77. Annatto (Yellow Dye) is obtained from the seed of ________?
A. *Terminalia chebula*
B. *Bixa orellana*
C. *Terminalia chebula*
D. Berberis aristate

78. Which of the followings is considered to be a Fruit Tans yielding plants?
A. *Terminalia chebula*
B. *Terminalia bellirica*
C. *Emblica officinalis*
D. All of the above

79. International Tropical Timber Organization (ITTO) is located at ____
A. Kanagawa, Japan
B. Geneva, Switzerland
C. New York, United States
D. Gland, Switzerland

80. The principal constituents of Khair heartwood are ________ (Katha).
A. Catechin
B. Azadirachtin
C. Quercetin
D. All of the above

81. Which of the wooden coal is having higher calorific value?
A. *Prosopis juliflora*
B. *Bombex ceiba*
C. *Ailenthus excela*
D. *Butea monosperma*

82. વૃક્ષો, લેન્ડસ્કેપ્સ અને ગીચ ઝાડ જેવા હિમાયતી પરિબળોને લગતું કયું નિવાસસ્થાન છે જે વન્યપ્રાણી વસ્તીને અસર કરે છે?
A. ઇશ્ચ્યુરિન હેબીટાટ
B. માઇક્રોહેબીટાટ
C. ટેરેસ્ટ્રિયલ હેબીટાટ
D. મેક્રોહેબીટાટ

83. ભારતના પ્રખ્યાત હર્પેટોલોજિસ્ટ કોણ છે?
A. સલીમ અલી
B. વિવેક મેનન
C. રિષદ નૌરોજી
D. જે. સી. ડેનિયલ

84. ગુજરાત રાજ્યના પ્રોટેક્ટેડ એરિયા નેટવર્ક _____ સંરક્ષિત વિસ્તારોનો સમાવેશ કરે છે.
A. 28 B. 23
C. 30 D. 24

85. બક્સા ટાઇગર રિઝર્વમાં શેનુ ખાણકામ ગંભીર રીતે નુકસાન પોંહચાડે છે ?
A. કોલસો
B. કોપર
C. ડોલોમાઇટ
D. આયર્ન

86. વિકાસલક્ષી કામો માટે જંગલની જમીનના પરિવર્તન હેઠળ કયો એક્ટ લાગુ પડે છે?
A. ફોરેસ્ટ (કંઝર્વેશન) એક્ટ 1980
B. ઇન્ડિયન ફોરેસ્ટ એક્ટ 1927
C. ફોરેસ્ટ રાઇટ એક્ટ 2006
D. સૌરાષ્ટ્ર ફેલિંગ ઓફ ટ્રી એક્ટ 1951

87. સૌરાષ્ટ્ર ફેલિંગ એક્ટ 1951 મુજબ ગુજરાતમાં કયું વૃક્ષ રિઝર્વડ કેટેગરીમાં સામેલ નથી?
A. સાગ B. સીસમ
C. પીપળો D. ખેર

82. Which habitat pertains to onsite factors like trees, landscapes and thickets which affect wildlife populations?
A. Estuarine habitat
B. Microhabitat
C. Terrestrial habitat
D. Macrohabitat

83. Who is famous herpetologist of India?
A. Salin Ali
B. Vivek Menon
C. Rishad Naoroji
D. J. C. Denial

84. The Protected Area (PA) network of the State of Gujarat comprises of _____ Protected Areas.
A. 28 B. 23
C. 30 D. 24

85. Buxa Tiger Reserve is seriously threatened by the mining of
A. Coal
B. Copper
C. Dolomite
D. Iron

86. The diversion of forest land for developmental works has to processed under
A. Forest (Conservation) Act 1980
B. Indian Forest Act 1927
C. Forest Right Act 2006
D. Saurastra felling of tree Act 1951

87. As per Saurastra Felling of Trees Act 1951, which is not the Reserved Tree in Gujarat
A. Teak B. Shisham
C. Pipal D. Khair

88. 2008 માં કચ્છ બાયોસ્ફિયર રિઝર્વને કયા નિયમ હેઠળ જાહેર કરવામાં આવ્યું?
A. વન્યપ્રાણી (સંરક્ષણ) અધિનિયમ 1972
B. ભારતીય વન અધિનિયમ 1927
C. યુનેસ્કો, એમએબી (MAB) પ્રોગ્રામ
D. પર્યાવરણ (સંરક્ષણ) અધિનિયમ 1986

89. ગુજરાત બાયોડાયવર્સીટી બોર્ડ ની સ્થાપના કયા વર્ષમાં કરવામાં આવી હતી?
A. 2006 B. 2005
C. 2002 D. 2010

90. "ડાંગ દરબાર" માં સ્થાનિક લોકોને શેના માટે એવોર્ડ આપવામાં આવે છે?
A. વન અગ્નિ નિયંત્રિત કરવા અને મદદ કરવા માટે
B. વન્યપ્રાણી શિકારને નિયંત્રિત કરવા
C. સહભાગી વન વ્યવસ્થાપન
D. અપરાધીઓથી વનનું રક્ષણ

91. ગુજરાતમાં કરુણા અભિયાન શેના બચાવ માટે કરવામાં આવી રહ્યું છે?
A. શહેર વિસ્તારના પાશવી કૂતરા
B. શહેર વિસ્તારની ગાય
C. પક્ષીઓ
D. શહેરમાંના સાપ

92. ગુજરાતના જંગલોમાં આગની ઘટનાના મુખ્ય કારણો કયા છે?
A. વાંસના ઘર્ષણથી
B. જંગલોમાં ભૂગર્ભ ગરમીથી
C. માનવસર્જિત આગ
D. જંગલોમાં કેટલાક પ્રકાશિત કીટકોને કારણે

93. હર્પેટોલોજીમાં તેમના નોંધપાત્ર કામ બદલ વન્યપ્રાણી સંરક્ષણમાં પદ્મશ્રી કોને મળ્યો છે?
A. જે. સી. ડેનિયલ
B. અભિજિત દાસ
C. રોમ્યુલસ વ્હાઇટેકર
D. અજય દેસાઇ

88. Kutch Biosphere Reserve has been declared in 2008, under
A. Wildlife (Protection) Act 1972
B. Indian Forest Act 1927
C. UNESCO, MAB Programme
D. Environment (Protection) Act 1986

89. Gujarat Biodiversity Board (GBB) was established in the year
A. 2006 B. 2005
C. 2002 D. 2010

90. In "Dangs darbar" the local villagers are being awarded for
A. To control and help in forest fires
B. To control the wildlife poaching
C. Joint Forest Management
D. Protection of forest from offenders

91. Karuna Abhiyan in Gujarat is beinAg launched to save the
A. Feral dogs in city area.
B. Cows in city area
C. Birds
D. Snakes in city

92. In Gujarat, the fire occurrence in forests, are because of
A. Rubbing of bamboos
B. Ignition of underground heats in forests
C. Manmade fires
D. Because of some lighting insects in forests

93. Who has been awarded Padma Shri in wildlife conservation for his notworthy work in herpetology?
A. J. C. Daniel
B. Abhijeet Das
C. Romulus Whitaker
D. Ajay Desai

94. નવેમ્બર 2020 માં મહારાષ્ટ્રમાં કઈ રામસર સાઇટ ઉમેરાઈ?
A. લોનાર સરોવર
B. સુર સરોવર સરોવર
C. નંદુર માધમેશ્વર
D. સરસાઈ નવર સરોવ

95. "ભારતમાં વાઘની સ્થિતિ" પર MoEFCC નો તાજેતરનો અહેવાલ કયા વર્ષ માં બહાર પાડવામાં આવ્યો?
A. 2020 B. 2019
C. 2018 D. 2017

96. બિશ્નોઇ સમુદાયની સ્થાપના કોને કરી હતી?
A. ભગવાન બિશ્નેશ્વર
B. ભગવાન જાડેશ્વર
C. ભગવાન જંભેશ્વર
D. ભગવાન કરુણેશ્વર

97. MOSAIC એ ________ ના અભ્યાસ માટે મલ્ટિડિસિપ્લિનરી ડ્રિફ્ટિંગ વેધશાળા છે
A. એન્ટાર્કટિકાનો ક્લાઈમેટ
B. આર્કટિકનો ક્લાઈમેટ
C. હિમાલયનો ક્લાઈમેટ
D. શુષ્ક રણનો ક્લાઈમેટ

98. નીચેનામાંથી કયા સ્થાનને ભારતનું 'કોલ્ડ ડેઝર્ટ' કહેવામાં આવે છે ?
A. લદાખ B. લાહૌલ
C. લેહ D. મનાલી

99. 1986માં અમલમાં આવેલ પર્યાવરણ (સંરક્ષણ) અધિનિયમનો ઉદ્દેશ્ય શું હતો?
A. પર્યાવરણ રક્ષણ
B. પર્યાવરણમાં સુધારો
C. રક્ષણ અને પર્યાવરણ સુધારણા
D. ક્લાઈમેટ ચેન્જથી રક્ષણ

94. In November 2020, ________ was added Ramsar site in Maharashtra.
A. Lonar Lake
B. Sur Sarovar lake
C. Nandur Madhameshwar
D. Sarsai Nawar Jheel

95. The latest report of MoEFCC on "Status of Tigers in India" was released in
A. 2020 B. 2019
C. 2018 D. 2017

96. Bishnoi community was founded by
A. Lord Bishneshwar
B. Lord Jadeshwar
C. Lord Jambheshwar
D. Lord Karuneshwar

97. MOSAIC is Multidisciplinary drifting Observatory for the Study of
A. Antarctica Climate
B. Arctic Climate
C. Himalayan Climate
D. Arid Desert Climate

98. Which of the following places is termed as 'Cold Desert' of India?
A. Ladakh B. Lahaul
C. Leh D. Manali

99. The Environment (Protection) Act was enacted in 1986 with the objective of providing for the
A. protection of the environment
B. improvement of the environment
C. protection and improvement of the environment
D. protection for the climate change

100. પૃથ્વી દિવસ ક્યારે ઉજવાય છે ?

A. 21 ડિસેમ્બર

B. 12 સપ્ટેમ્બર

C. 22 એપ્રિલ

D. 31 ઓક્ટોબર

100. Which day is Earth Day?

A. 21 December

B. 12 September

C. 22 April

D. 31 October

Refer concern chapters for answers
જવાબો માટે સંબંધિત પ્રકરણો નો અભ્યાસ કરો

References

સંદર્ભો

General Referencs:

All the documents of Environmental Information System, Ministry of Environment, Forests and Climate Change, Government of India.

Annual Reports, Notifications, Official documents, Publications, Reports and Website of Ministry of Environment, Forests and Climate Change, Government of India.

Annual Reports, Notifications, Official documents, Publications, Reports and Website of Department of Environment and Forests, Government of Gujarat.

Annual Reports, Notifications, Official documents, Publications, Reports and Website of Department of Climate Change, Government of Gujarat.

Annual Reports, Notifications, Official documents, Publications, Reports and Website of Department of Agriculture, Farmers Welfare and Cooperation, Government of Gujarat.

Annual Reports, Notifications, Official documents, Publications, Reports and Website of Gujarat Forest Department, Gandhinagar.

Annual Reports, Notifications, Official documents, Publications, Reports and Website of Central Zoo Authority, New Delhi.

Annual Reports, Notifications, Official documents, Publications, Reports and Website of National Biodiversity Authority, Chennai.

Annual Reports, Notifications, Official documents, Publications, Reports and Website of National Green Tribunal, New Delhi.

Annual Reports, Notifications, Official documents, Publications, Reports and Website of National Tiger Conservation Authority, New Delhi.

Annual Reports, Notifications, Official documents, Publications, Reports and Website of Central Pollution Control Board, New Delhi.

Annual Reports, Notifications, Official documents, Publications, Reports and Website of Gujarat Ecological Education and Research Foundation, Gandhinagar.

Annual Reports, Notifications, Official documents, Publications, Reports and Website of Gujarat Ecology Commission, Gandhinagar.

Annual Reports, Notifications, Official documents, Publications, Reports and Website of Gujarat Forestry Research Foundation, Gandhinagar.

Annual Reports, Notifications, Official documents, Publications, Reports and Website of Indian Council of Forestry Research and Education, Dehradun.

Annual Reports, Notifications, Official documents, Publications, Reports and Website of Wildlife Institute of India, Dehradun.

Annual Reports, Notifications, Official documents, Publications, Reports and Website of FAO (Food and Agriculture Organisation of United Nations)

Annual Reports, Notifications, Official documents, Publications, Reports and Website of Global Forest Resource Assessment (FRA).

Annual Reports, Notifications, Official documents, Publications, Reports and Websites of Planning Commission of India, GOI, New Delhi.

References:

Champion HG and Seth SK 1968 A Revised Survey of the Forest Types of India, Government of India Press (Reprinted in 2005 by Natraj Publishers, Dehradun)

CITES (the Convention on International Trade in Endangered Species of Wild Fauna and Flora) 2020 https://cites.org/eng/contact.php. Accessed on 18 December 2020

CPCB (Central Pollution Control Board) 2016 Final Document on Classification of Industrial Sectors Under Red, Orange, Green and White Categories, Central Pollution Control Board, New Delhi.

Dhyani SK 2009 Challenges and scope of agroforestry research in India, In (eds) OP Chaturvedu, A Ventakesh et al., Agroforestry – Natural Resource sustainability, livelihood and climate moderation, 13-25 pages

Edward Mohr and Van Baren FA 1954 Tropical Soils: A critical study of Soil genesis as related to climate, rock and vegetation, New York: Interscience pub. Inc

FAO (Food and Agricutlure Organisation) 2019 India State Forest Report, Forest Survey of India, MoEF & CC, Dehradun

FAO (Food and Agricutlure Organisation) 2020 Global Forest Resources Assessment 2020-Key findings. Rome, Italy

GEER (Gujarat Ecological Education and Research Foundation) 2018 Important Wetland Destination of Gujarat, Gujarat Ecological Education and Research Foundation, Gandhinagar

GFRA (Global Forest Resource Assessment) 2020 Global Forest Resource Assessment, Food and Agriculture Organization, Rome, Italy

GoG (Government of Gujarat) 2019 Gujarat Forest Statistics:2018-19. Gandhinagar: Monitoring and Evaluation Wing, PCCF & HoFF, Gujarat.

GoI (Government of India) 2014 National Agroforestry Policy, Department of Agriculture, Cooperation & Farmers Welfare, New Delhi

Gopal R 1992 Fundamentals of Wildlife management. Natraj Publishers, Dehra Dun, India. 1295 pages

Hans Jenny 1941 Factors of Soil Formation, New York and London: Mc Graw-Hill Book Company, Inc

IUCN (International Union for Conservation of Nature) 2020 The IUCN Red List of Threatened Species. Version 2. https://www.iucnredlist.org. Accessed on 13 December 2020

Khanna LS 1984 Principles and Practices of Silviculture, Dehradun: Khanna Bhandu

Menon V 2014 Indian Mammals – A field guide. Hackett Book Publishing India Pvt Ltd, Gurgaon. 439 pages

Mohit Husain 2020 Universal Objective Forestry. Scientific publishers, Jodhpur, India.

NWAP (National Wildlife Action Pan) 2017 National Wildlife Action Plan 2017-2031, Ministry of Environment, Forests and Climate Change, New Delhi

NWLD (National Wildlife Database) 2020 Wildlife Institute of India, Dehradun

Pandit SJ, Uday Vora, Neeta Shah and Yogendra Shah 2020 9th Wild Ass Population Estimation-2020, Final Report, Forest Department, Gujarat

Pathak PS, Pateria HM and Solaki KR 2000 Agroforestry system in India – A diagnosis and Design Approach, NRCAF, ICAR, New Delhi, 184 page

Prakash R and Khanna L S 1983 Theory and Practices of Silviculture system. International Book Distributors, Dehradun

Raj AJ and Lal S 2018 Indian forestry - Principles and Applications, Scientific Publishers, Jodhpur

Raunkiaer C 1934 The Life forms of Plants and Statistical Plant Geography, Oxford: Clarendon Press. 632 pages

Rodgers WA, Panwar HS and Mathur VB 2000 Wildlife Protected Area Network in India: A review. Wildlife Institute of India, Dehradun

Saha GK and Mazumdar S 2017 Wildlife Biology- An Indian Perspective. PHI Learing, New Delhi. 313 pages

Sara J. Scherr and Satya Yadav 1996 Land Degradation on the Developing World: Implication for Food, Agriculture and Environment to 2020, IFPRI Food Agriculture and the Environment Discussion paper 14, Washington DC: International Food Policy Research Institute

Sharma, P D 2013 Ecology and Environment, Rastogi Publications, Merrut. 643 pages

Singh HS 2001 National Heritage of Gujarat (Forests and Wildlife), Published by Gujarat Ecological Education and Research (GEER) foundation, Gandhinagar. 211 pages

Singh SK 2005 Text book of Wildlife management (2nd Revised and Enlarged edition). International Book Distributing Co., Lucknow. 546 pages

Wallace A R. 1876 The geographical distribution of animals; with a study of the relations of living and extinct faunas as elucidating the past changes of the Earth's surface. New York: Harper & Brothers. Volume 1

Yadav RK, Sinha PR and Pandey A 2018 All-in-One Objective Forestry. Renu Publishers, New Delhi. 411 pages

E- Resources:

http://bsi.gov.in/

http://cza.nic.in/

http://envis.nic.in/

http://frienvis.nic.in/

http://moef.gov.in

http://nbaindia.org/

http://wiienvis.nic.in/

http://www.bsienvis.nic.in/

http://www.imdahm.gov.in/aau.htm

http://zsienvis.nic.in/
https://agri.gujarat.gov.in/
https://fed.gujarat.gov.in/
https://forests.gujarat.gov.in/
https://fsi.nic.in/
https://gujaratindia.gov.in/
https://unfccc.int
https://www.britannica.com/
https://www.un.org/